网络工程师的Python之路

网络运维自动化实战

王印　朱嘉盛●著

電子工業出版社
Publishing House of Electronics Industry
北京•BEIJING

内 容 简 介

本书第 1 版于 2020 年 11 月出版发行后大获成功，Python 及 NetDevOps 技术在国内计算机网络圈子里迅速走红，国内与 NetDevOps 相关的技术文章在网络上如雨后春笋般大量涌现，行业里也出现了很多要求掌握 NetDevOps 技术的网络工程师职位，并且需求越来越多，NetDevOps 是“下一代网络工程师”必须掌握的技能已经成为既定且不可逆的事实。

以 Python 为主的 NetDevOps 技术知识更迭很快，第 2 版将在第 1 版的基础上添加 TextFSM 详解、Netmiko 详解、Nornir 详解、NETCONF 详解及 RESTCONF 详解等 5 章内容，并去掉 pyping、pyntc、netdev 等一些不再流行或者作者已经公开宣布不再维护的模块内容。

另外，第 2 版将在思科设备的基础上添加华为设备的内容，由朱嘉盛老师执笔，相信能对主要使用国产设备的国内网络工程师更加有益。

图书在版编目（CIP）数据

网络工程师的 Python 之路：网络运维自动化实战 / 王印，朱嘉盛著. —2 版. —北京：电子工业出版社，2023.3
ISBN 978-7-121-45027-3

Ⅰ. ①网… Ⅱ. ①王… ②朱… Ⅲ. ①软件工具－程序设计 Ⅳ. ①TP311.561

中国国家版本馆 CIP 数据核字（2023）第 022520 号

责任编辑：李秀梅
印　　刷：三河市良远印务有限公司
装　　订：三河市良远印务有限公司
出版发行：电子工业出版社
　　　　　北京市海淀区万寿路 173 信箱　　邮编 100036
开　　本：787×980　1/16　印张：37　字数：828.8 千字
版　　次：2020 年 11 月第 1 版
　　　　　2023 年 3 月第 2 版
印　　次：2023 年 3 月第 1 次印刷
定　　价：138.00 元

凡所购买电子工业出版社图书有缺损问题，请向购买书店调换。若书店售缺，请与本社发行部联系，联系及邮购电话：（010）88254888，88258888。

质量投诉请发邮件至 zlts@phei.com.cn，盗版侵权举报请发邮件至 dbqq@phei.com.cn。

本书咨询联系方式：010-51260888-819，faq@phei.com.cn。

本书赞誉

（排名不分先后，以姓氏拼音为序）

身为一名在网络安全行业从业的工程师，经常面临客户和现有安全产品整合或大批量群编辑之类的需求，手动配置耗时费力还容易出错，该书由浅入深地让我这样一个 Python 新手对脚本自动化编排有了快速上手和实战中调用的经验，非常实用，极力推荐!

——刁文杰，新加坡派拓网络（Palo Alto）亚太区客户体验架构师

计算机网络的自动化和可编程化是未来的发展趋势。王印先生在网络和编程的融合上有着非常丰富的经验。相信本书可以很好地帮助网络工程师们开启网络自动化和可编程化的大门。

——董海宇，新加坡新传媒（Mediacorp）首席网络工程师

第一次认识王印兄弟是在基于智慧校园的合作项目中，我们就中国的数字化技术走向国际市场进行了探讨。王印不仅在网络领域的专业性毋庸置疑，还能孜孜不倦地分享知识，值得点赞。本书系统地讲述了 Python 编程知识，是普惠大众的好书，不论是在校大学生还是职场人士，都能够通过此书获取编程技能，强力推荐！

——樊宪政，阿里巴巴旗下浩鲸科技华南区原总经理、

腾讯云（南中国区）原战略合作总监

如今的 IT 从业者正处在一个最好的时代，一个技术相互连接的时代。传统的网络工程师可以通过学习编程来扩展传统网络技术，自动执行重复而烦琐的工作任务，简化人力的投入和规避误操作的风险。程序员也可以通过学习网络技术来更加深入地了解数据包的传递方式和协议之间的优劣，从而写出更加优质且适用于真实环境的代码。本书作者从网络工程师的角度出发，将程序员编程的思维带入了网络领域。不仅如此，大量真实案例的引用将引起你思维的共鸣与对额外设想可能性的思考。更让我惊讶的是，本书使用类似笔记的形式通过通俗易懂的文字详述了每一个实用案例的完整配置步骤，这将是非常适合网络工程师的 Python 入门教程。

——李小沛，思科认证讲师、亚马逊 AWS 认证讲师

新形势下，虚拟网络发展、SDN 兴起、云计算复杂环境及海量操作，对传统网络工程师来说可谓不可承受之痛。且在网络工程领域，受限于网络工程师的技术栈，如何科学有效地提高 DevOps 能力，仍存在不少挑战。本书是市面上为数不多的作者有深厚功底及丰富项目实操经验，且以网络工程师的视角和思路指出了一条网络自动化从 0 到 1 切实可行的转型道路的书。尤其对于没有任何编程能力的网络工程师满足企业网络可视化、自动化、智能化的运维的需求，本书将是不二之选。

——李耀，百度原网络工程师

这是一本以 Python 语言作为工具，从网络工程师的视角讲解如何利用 Python 来实现网络运维自动化的工程实践图书。全书以操作实践为基础，系统地梳理了 Python 语言的基础知识，并在此基础上将其运用于网络运维实战，是一本为网络工程师量身打造的网络运维自动化实践之书，值得一读。

——茹炳晟，腾讯 Tech Lead、中国计算机学会 CCF TF 研发效能 SIG 主席

印兄于我，亦师亦友，他为人诚恳谦和，技术扎实过硬，每次探讨难题，即便十分忙碌，他也依然第一时间抽身为我排忧解难。寒门出贵子，十年磨一剑，通过网络平台，印兄倾囊分享其细腻的技术细节、励志的人生经历、前沿的技术精华和不对一般人外传的职场秘诀，他是华人网络工程师领域的众人福音、闪耀明灯。

——王昊颖 Henry Wang，摩根大通（新加坡）ECN 网络经理

2006 年 Google 在搜索引擎大会上正式提出“云计算”后，接踵而至的就是把 DevOps、SysOps 推到了台前。借助越来越成熟的云平台技术，DevOps 和 SysOps 也越来越凸显出其强大的一面——自动化。而今 NetOps 作为其中的一个分支，也慢慢走到台前。本书作者是非常有先见之明的，从 2013 年就开始开发并实践自己的网络自动化脚本，而且从中获得了很大的成效。对于一个上千台网络设备的架构，按照传统人为操作更新，可能需要多人合作一个星期才能完成，而作者通过网络自动化进行多线程并行操作，大大缩减了对人力资源的需求。在此我还要替本书电子版的读者小田感谢作者。通过两个星期学习搭建环境和体验作者的书内实例，他获得了法国巴黎银行的青睐，加入了其日常网络运维团队。由此可见，NetOps 已经成为未来网络工程师的一个趋势。本书将会成为你从一个普通网络工程师转型为自动化网络工程师很好的启蒙教材。

——王渊浩，新加坡 Lazada IT 基础设施经理

本书作者是网络行业的“新人老兵”，也是我见过英文能力最突出的网络工程师。说新人是因为他年纪确实不大，说老兵是因为他的工作经历及参与项目的规模和质量是他这个年龄的人所无法企及的。本书是对他十多年海外工作经验的最好总结，具有很强的实操性，相信能够为下一代网络工程师提供不可估量的精神和业务食粮。

——温健，澳大利亚蒙纳士大学苏州校区首席运营官、总经理、副校长

对于很多跨专业的读者而言，一般 IT 编程语言类图书往往难以入手及持续阅读。很大一部分原因在于读者背景的广泛性造就了其对基本概念的了解参差不齐。在此影响下，阅读与

学习的效果往往大打折扣。而本书则有针对性地解决了这个问题。王兄从一个传统网络工程师的角度入手，指导大家学习 Python 编程语言，并利用其实现网络自动化任务。本书运用了网络工程师最熟悉的路由器、交换机来配合展示 Python 语言的特性，并适当结合生产环境中常常出现的自动化需求，让拥有基本网络背景的工程师们可以迅速掌握 Python 的精髓，理解网络自动化的优势，并且轻松展开模拟练习。在网络自动化的大趋势下，王兄的力作不仅启发、激励我为网络自动化进程添一份绵力，更值得所有有志于此的同行们阅读。未来，期待王兄的更多力作。

——吴茱萸 Conor Wu，爱尔兰爱彼迎（Airbnb）全球网络工程师

我和王印相识于新加坡，他是个网络红人，亲自参与了很多网络运维环境和方案的设计，合作方有国际知名投行，有政府部门，也有巨头企业。更可贵的是，他乐于分享网络实战经验和方案，并且热心给网络同行很多帮助和启发。随着互联网的高速发展，在维护和管理成百上千的网络设备时，传统网络工程师面临着很多挑战和局限，SDN 的兴起也要求网络工程师具有编程能力。传统网络工程师往往不懂编程，而编程人员又往往不了解网络运维和方案。在这个跨界的年代，懂网络又懂 Python 会让作为网络工程师的你变得更有竞争力，同时得到猎头公司的青睐。市面上很难找出一本专门针对网络工程师学习 Python 的书，而本书由浅入深，并在后期以实验的方式讲解 Python 在网络运维中的应用，可以让你更快地理解如何通过 Python 实现网络运维自动化，强烈推荐！

——熊希仁 Ryan Xiong，华为（新加坡）原网络方案经理、

梭子鱼网络（Barracuda Networks）售前工程师

我看过很多教 Python 编程的书，不是逻辑混乱就是干货不够，本书作者王印是资深网络老兵，从新加坡开始先后任职于美国 AT&T、苹果、苏格兰皇家银行、沙特 KAUST 等世界一流公司和大学，从行业经验来看，少有人能匹敌。整本书逻辑清晰，由浅入深，连配图的标记都很用心，是一本很难得的实用型编程工具书，有料有颜值，值得拥有。

——杨雯莉，深圳申鲲科技有限公司 CBO、中国移动互联网产品原品牌管理师

随着科技和商业的快速发展、第四次工业革命（4IR）的到来，网络技术的演变日新月异。

在今天的网络技术领域，如何能够满足大规模网络灵活高效、高质稳定运维的极致需求，是所有 IT 管理者都应该深入思考的问题，我相信网络基础设施建设和运维管理的标准化、自动化、数据可视化、智能化是大势所趋。本书作者结合丰富的实际案例，通过分享亲身经历及其转型经验，由浅入深地讲解了网络运维与 Python 的完美结合，并展示了其广阔的国际视野。本书干货满满，不仅能够带你走上新一代网络工程师快速转型之路，还能助你在未来国际化网络技术的舞台上蓬勃发展。

此外，作者本人不但是有着 10 多年经验的资深网络工程师，还为许多 IT 从业者带来了鼓舞和启示，此书是所有网络从业人员的“Go-To Book”，值得推荐。

——叶龙 Leon Ye，腾讯亚太区（新加坡）IT 总监

认识王印是在深圳城市数字化转型委员会的一次活动上。这位辗转了多个国家的 IT 精英，现在应该已成为沙特阿拉伯现代化转型的栋梁之材。我虽与他交流不多，但有幸看到他的书稿还是十分高兴。回想 20 世纪 80 年代中期我在国内大学教企业管理信息化，至今还对 Fortran、ALGOL、COBOL、BASIC 等应用程序记忆犹新。我虽未接触过 Python，但看王印的书，感觉思维逻辑和表述层次非常清晰，相信本书会帮到以中文为母语的编程人士。

——易全，深圳市信息化与工业化融合研究院执行院长、深圳市科技金融促进会创会会长、城市数字化转型委员会高级顾问

关于作者

王印，知乎专栏“网路行者”作家，知乎 ID：弈心，沙特阿卜杜拉国王科技大学（King Abdullah University of Science and Technology，KAUST）高级网络工程师，14 年海外从业经验，CCIE#40245。2009 年起在新加坡先后任职于美国电信运营商 AT&T、美国数据中心公司 Equinix、新加坡陆路交通管理局（LTA）、新加坡石林 IT 咨询公司（SFIT）、美国苹果公司、苏格兰皇家银行（RBS），对大型园区、金融行业的网络设计与运维具有丰富的实战经验。

朱嘉盛，知乎专栏“网工手艺”作者，2010 年毕业于华南理工大学，网络工程师 13 载，长期扎根网络运维第一线。通信网从 2G 到 5G，互联网从几乎空白建设运维到百万级用户规模，在这些领域均具备丰富的大网络实战经验。工作之余，热爱阅读与写作，善于梳理知识，乐于分享和帮助新手。近年来深耕 Python 网络自动化领域。在本书第 1 版发行后，将其中的思科设备实验适配成华为设备实验，并加入大量自己的理解与思考，同时在相关读者群中坚持答疑和提供支撑服务，引领众多读者成功将技术应用到其他国产设备。

再版前言

光阴似箭，岁月如梭。转眼之间，距离本书最早的电子书出版已经过去了三年之久。承蒙广大读者的厚爱，电子书和第 1 版纸质书的发行量远远地超出了我的预期。这也使我感到有必要更新内容，再版此书，以飨读者。

IT 技术日新月异，Python 和 NetDevOps 领域也不例外。第 2 版基于截稿前最稳定的 Python 3.10.6，对基于 Python 3.8.6 的第 1 版里的每一章内容都作了修改、更新、补充或删减，去掉了一些已经不再流行或作者已经公开宣布不再维护的模块，如 pyping、pyntc、netdev 等，在第 1 版总共 6 章内容的基础上新加入了 TextFSM 详解、Netmiko 详解、Nornir 详解、NETCONF 详解及 RESTCONF 详解 5 章内容，让第 2 版的内容及书的厚度相较于第 1 版完全翻倍。

本书第 1 版发行后，Python 及 NetDevOps 技术在国内计算机网络圈子里迅速走红，国内与 NetDevOps 相关的技术文章在网络上如雨后春笋般大量涌现。鉴于笔者工作环境的限制，第 1 版所有实验部分的内容是完全基于思科的设备创作的，很多第 1 版的读者针对本书的内容以华为、H3C 等国产设备做了相应的补充，其中最引人注目的是读者朱嘉盛老师，在本书第 1 版发行两年来，笔耕不辍的他在知乎个人专栏里将笔者第 1 版几乎所有的内容都用华为设备“重写”了一遍，加入了很多他个人的理解和心得，并在读者群里耐心解答新人们的各类提问。有鉴于此，第 2 版特意邀请了朱嘉盛老师作为联合作者，朱嘉盛老师将在本书实验篇章里加入华为设备的使用说明，第 7 章 TextFSM 详解也完全由他执笔。另外，也非常感谢电子工业出版社李秀梅老师对本书出版的热情支持与多方面的协助，和这样能力卓著、认真负责的编辑共事是一个作者的幸运。

人类社会在发展，科学技术在进步，虽然国内 NetDevOps 技术起步较晚，总体落后国外先进国家，但是中国地大物博、人才济济，本书第 1 版发行两年来，这一领域已经涌现出一大批深入钻研、乐于分享的国内读者。中国已经逐步从 2019 年之前的 NetDevOps 学“荒漠”成长为 NetDevOps 学“大国”。然而要从一个 NetDevOps 学“大国”成长为一个 NetDevOps 学“强国”，尚需国内同仁的勇于探索、广泛交流、求真务实，杜绝急功近利，摒弃浮躁之风。若能如此，我们的目标不但能够实现，且将为期不远。希望本书能为我们实现这一目标起到抛砖引玉、添砖加瓦的作用，为国内广大年轻的网络工程专业学子和职业人士提供一个比较准确、系统并富有理念的 NetDevOps 学读本，不求包罗万象，旨在言简意赅、提纲挈领。这是第 1 版的宗旨，也是第 2 版的指南。

王　印

2022 年 8 月 23 日于沙特阿拉伯

读者服务

微信扫码回复：45027

- 获取本书配套视频资源
- 加入本书读者交流群，与本书作者互动
- 获取【百场业界大咖直播合集】（持续更新），仅需 1 元

致谢（王印）

首先感谢我的亲人，感谢你们对我学业、事业的支持，一步一步陪伴我走到今天。感谢你们长期以来默默地在生活及其他方面对我的照顾和关怀，让我能够专心完成此书。感谢刚满六岁的儿子弈仁，你的笑容化解了我生活中的一切烦恼。

感谢在新加坡 12 年的学习、生活、工作中给予我无私帮助的同学、朋友、同事们。感谢王渊浩和 Lawrence Lee，感谢你们在大学四年同窗及毕业后的生活中给予我方方面面的鼓励和帮助，让独自远离家乡出国留学的我并不感到孤独。感谢 Newmedia Express 的老板马来西亚人 Alan Woo 和 Shirley Lee，毕业后的半年时间里求职四处碰壁的那段日子是我人生中最灰暗的一段回忆，感恩你们在我人生的最低谷给了我一个证明自己的工作机会，它改变了我的人生轨迹。感谢 Wired-Media 公司的新加坡前辈杨绍鹏（Kenneth Yeo），感谢你当年每晚不辞辛劳地绕路驾车送上晚班的我回家，给了我这个刚刚踏入社会、远离家乡在狮城打拼的小职员很多温暖。感谢来自马来西亚的挚友卢忠声，相识十三年来，与你在工作和生活中一起努力、相互勉励、相持而笑的日子是我一生中最珍贵的回忆。感谢新加坡同事 Darry Tan，感谢你在我任职于苹果公司的那段时间里对我在工作和技术上无私的指导和帮助，感谢你毫无保留地同我分享了你备考 JNCIE、CCDE 的所有笔记和资料，以前辈的身份同我分享了许多宝贵的人生经验，感恩自己能有幸遇到像你这样真正以德服人、充满正能量的导师和贵人。

感谢我在 KAUST 的领导 Gary Corbett、Khalid Mustafa 及 Kevin Sale，没有你们在工作中及工作外给予我充分的支持、关心和信任，我将无法从一名传统网络工程师转型成为 NetDevOps 工程师，自然也就没有本书的诞生。

I would like to express my sincere gratitude by dedicating this book to Khalid Mustafa，Kevin Sale, Gary Corbett and whoever works with me at King Abdullah University of Science and Technology, without your selfless and continuous support, guidance and encouragement, this book won't be born.

最后感谢所有致力于在计算机网络这一行默默传授知识和分享经验的每一个人，你们改变了整个世界！

致谢（朱嘉盛）

感谢互联网时代，如今身处家中即可对接世界；感谢家乡汕头，海纳百川、自强不息，探寻对接世界早已成为传统。

感谢各亲友在我成长过程中给予支持与帮助。感谢我读过的学校，给予我启发的老师们和同学们。感谢霍霞教授早年到美国得克萨斯州大学做访问学者期间与后来到南极科考期间均与我保持联系，打开了我放眼看世界的大门。感谢“继往开来”成员团在各自领域中脚踏实地奋勇拼搏、相互打气；感谢许超、黄洁菡、陈斯、谢旭楠、陈宇等好友在我写作期间提供了很多帮助与指导。

感谢我的供职单位（各级），它不仅给了我一个营生手段，也给了我一个不断学习和实践网络技能的平台。感谢张雄波早年间与我分享网络自动化思想，让我得以启蒙。感谢张振渠在购买本书第 1 版时，还不忘多买一本送给我阅读，让我钻研入戏。感谢长期与我并肩作战的网络运维同伴们。感谢长期与我有网络业务对接的上下游伙伴们。

感谢王印老师在 NetDevOps 领域不断引领与布道。他持续支持我创作，并直接帮我引流，帮我推广，还让我从读者变成了联合作者。感谢编辑李秀梅老师，她最早向王印老师建议把我写的内容纳入本书，并在文稿拟写过程中对我做了细致指导。

感谢愿意与我一起自学自驱、同向而行、相互勉励的读者们。他们有何平、林绍岳、丘兆杰、罗晨曦、李波、李习磊、沈卡、冉茂林、岳国宾、谭振波、曹文欣、杨林森、高磊、裴飞、马天宇、何丹、袁丹鹏、袁泽海、罗升华、瞿祖强、孟令沛、梁勇、宫琦、农雨衡、刘旭光、胡谦、刘学坤、李兵、刘金丰、胡元新、谭永明、刘元成、高华伟、胡玉少、杜鹏飞、岳飞宇、刘崇杰、万质鑫、唐志强、杨华……（排名不分先后）。我们身处不同地域，分属不同行业，从事不同工作，但都围绕着 NetDevOps 努力实践，以文会友，互为读者，交流切磋，相互鼓励，共同为这一领域做出自己的一点贡献。他们同样是本书的联合作者，我仅因幸运，成为代表！

最后，一起感谢我们对接世界的开放心态、分享知识的开源行动！我们在努力缩小国内国外在本领域的差距，并在试图追赶超越的过程中，同时成就了自己！

目录

第1章

Python的安装和使用

工欲善其事，必先利其器，鉴于很多网络工程师读者都是第一次接触 Python，本书开篇将配图详细介绍 Python 在不同操作系统下的安装和使用方法。Python 在 Windows、Linux 及 macOS 下都可以使用，目前最新的 macOS 本身已经内置了 Python，打开命令行终端输入命令 python 即可使用。本章主要介绍 Python 在 Windows 和 Linux（CentOS）下的安装和使用方法。

Python 的运行模式大致分为两种：一种是使用解释器（Interpreter）的交互模式（Interactive Mode），另一种是运行脚本的脚本模式（Script Mode）。使用解释器和脚本来运行 Python 的最大区别是，前者能在你执行一行或一段代码后提供"即时反馈"，让你看到是否得到了想要的结果，或者告诉你代码是否有误，而后者则是将整段代码写入一个扩展名为.py 的文本文件中"打包执行"。脚本模式在实际的网络运维工作中很常见，但是从学习的角度来讲，肯定是能提供"即时反馈"的解释器更利于初学者，因此本章大部分内容将基于解释器的交互模式来讲解，当然，也有部分代码案例必须用脚本模式来进行演示。

1.1 安装 Python

本书所有内容以 Windows 10.0（64 位）和 CentOS 8 分别作为 Windows 和 Linux 两大操作系统的演示平台。

1.1.1 在 Windows 下安装 Python 3.10.6

首先在 Python 官网下载 Windows 版的 Python 3（注意，从 Python 3.5 开始，Python 3 已经不再支持 Windows XP 及更早版本的 Windows 系统）。截至 2020 年 4 月，最新的版本为 3.10.6，读者可根据自身情况选择 32 位和 64 位版本，安装文件有.zip、.exe 和.web-based 3 种格式可选，这里推荐选择.exe 格式，如下图所示。

Files

Version	Operating System	Description	MD5 Sum	File Size	GPG
Gzipped source tarball	Source release		d76638ca8bf57e44ef0841d2cde557a0	25986768	SIG
XZ compressed source tarball	Source release		afc7e14f7118d10d1ba95ae8e2134bf0	19600672	SIG
macOS 64-bit universal2 installer	macOS	for macOS 10.9 and later	2ce68dc6cb870ed3beea8a20b0de71fc	40826114	SIG
Windows embeddable package (32-bit)	Windows		a62cca7ea561a037e54b4c0d120c2b0a	7608928	SIG
Windows embeddable package (64-bit)	Windows		37303f03e19563fa87722d9df11d0fa0	8585728	SIG
Windows help file	Windows		0aee63c8fb87dc71bf2bcc1f62231389	9329034	SIG
Windows installer (32-bit)	Windows		c4aa2cd7d62304c804e45a51696f2a88	27750096	SIG
Windows installer (64-bit)	Windows	Recommended	8f46453e68ef38e5544a76d84df3994c	28916488	SIG

安装过程中有一个很重要的步骤，如下图中的“Add Python 3.10 to PATH”，这里默认是没有勾选的，请务必勾选，它会自动帮你设置好环境变量，也就是说将来在你打开命令行运行 Python 脚本时，你可以在任意盘符和文件夹下直接输入命令 python xxx.py 来运行脚本，而无须输入 Python 执行程序所在的完整路径，例如 C:\Python38\python xxx.py。不要小看这一选项提供的自动环境变量设置功能，它能帮助 Python 初学者节省很多很多时间！

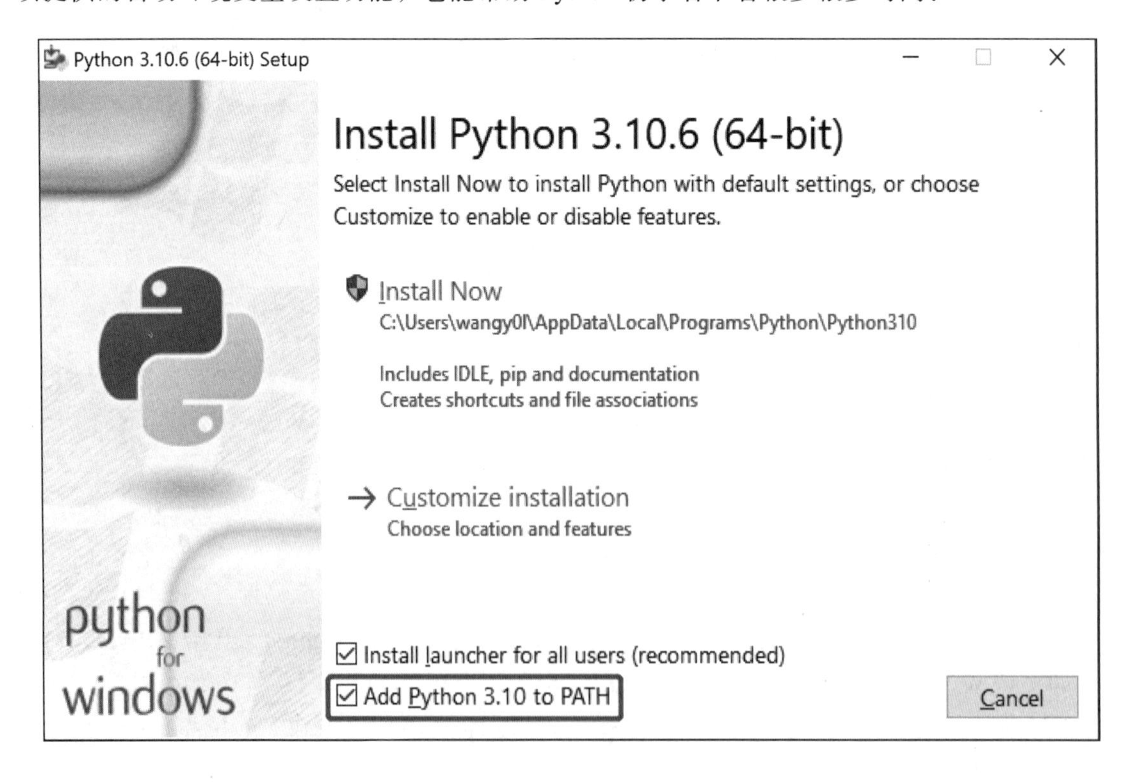

之后选择“Customize installation”进入自定义安装，如下图所示。

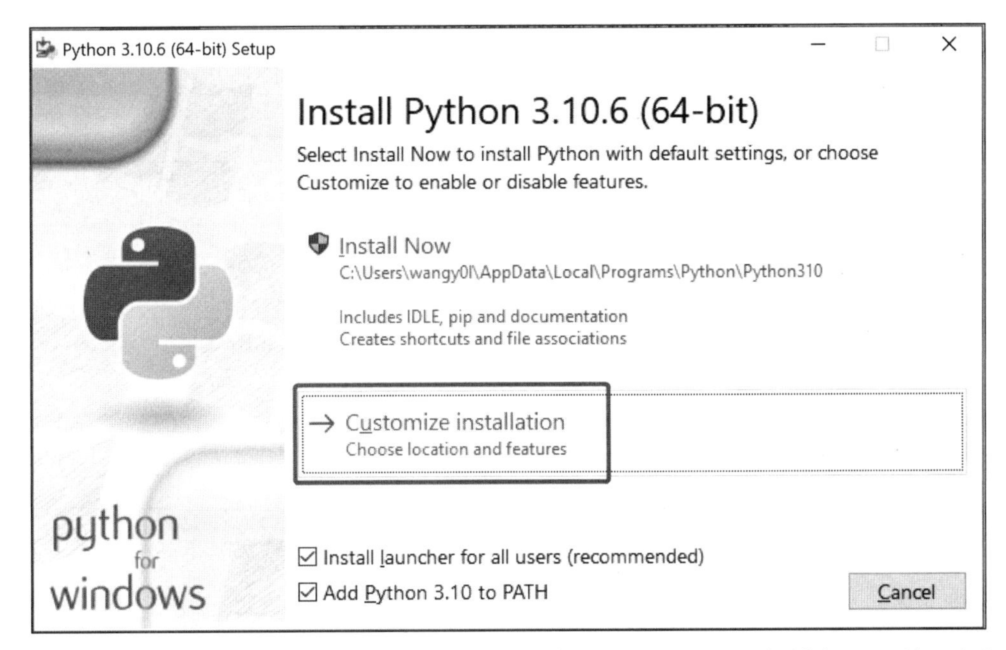

在 Optional Features 的选项中确保“pip”和“tcl/tk and IDLE”都被勾选，关于它们的作用后面会提到，其他选项使用默认配置即可，然后单击“Next”按钮，如下图所示。

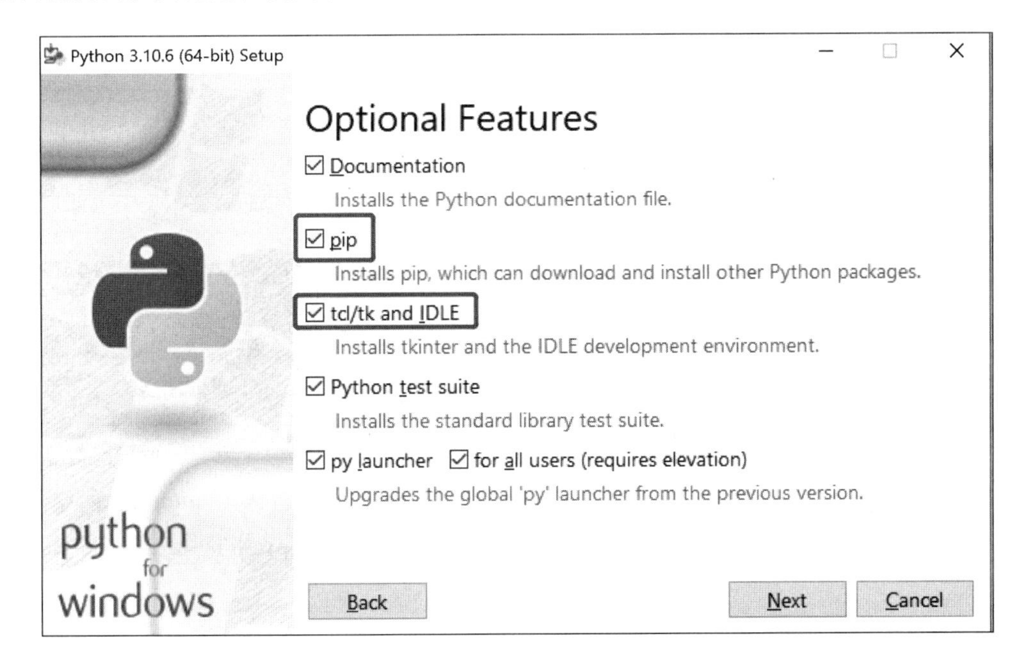

如下图所示，在 Advanced Options 中，推荐将“Install for all users”勾选上，它会将 Python 的安装路径从 C:\Users\admin\AppData\Local\Programs\Python\Python310 切换成 C:\Program Files\Python310，方便将来查找和访问。当然读者也可以自定义安装路径，以及根据自身情况决定是否给所有用户都安装 Python 3。

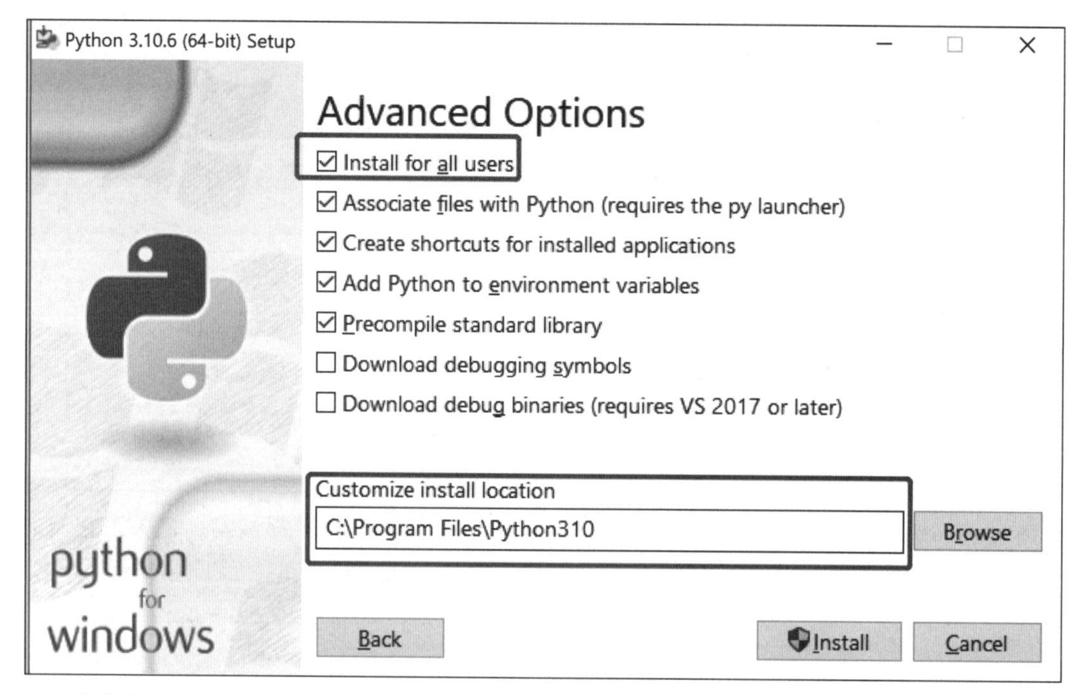

在安装好 Python 3 后，打开命令行，输入 py 或者 python，如果可以进入 Python 3.10.6 的解释器，则说明 Python 3 安装成功，如下图所示。

```
Command Prompt - py
Microsoft Windows [Version 10.0.18363.2274]
(c) 2019 Microsoft Corporation. All rights reserved.

P:\>py
Python 3.10.6 (tags/v3.10.6:9c7b4bd, Aug  1 2022, 21:53:49) [MSC v.1932 64 bit (AMD64)] on win32
Type "help", "copyright", "credits" or "license" for more information.
>>> _
```

注：如果之前你已经安装过 Python 2，则输入命令 python 会进入 Python 2，两个版本之间的使用互不影响。如果只安装了 Python 3，则只能通过命令 py 来进入 Python 3，如下图所示。

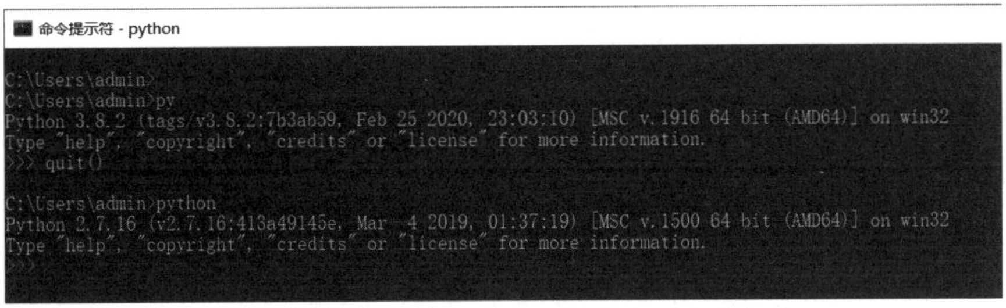

```
命令提示符 - python

C:\Users\admin>
C:\Users\admin>py
Python 3.8.2 (tags/v3.8.2:7b3ab59, Feb 25 2020, 23:03:10) [MSC v.1916 64 bit (AMD64)] on win32
Type "help", "copyright", "credits" or "license" for more information.
>>> quit()

C:\Users\admin>python
Python 2.7.16 (v2.7.16:413a49145e, Mar  4 2019, 01:37:19) [MSC v.1500 64 bit (AMD64)] on win32
Type "help", "copyright", "credits" or "license" for more information.
>>>
```

1.1.2　在 Linux 下安装 Python 3.10.6

本书将使用 CentOS 8 作为 Linux 版本的演示平台（在 Windows 上运行的 VMware 虚拟机），这里只介绍在 CentOS 命令行终端里使用 Python 的方法，在 GNOME 桌面环境下使用 Python 的方法不在本书的讨论范围内。因为是实验环境，所以直接使用 Root 用户，免去了 sudo 命令，读者请根据自身情况决定是否使用 sudo。

和 macOS 一样，最新的 CentOS 8 已经内置了 Python 2 和 Python 3，输入 python2 和 python3 两个命令可以分别进入 Python 2 和 Python 3，如下图所示。

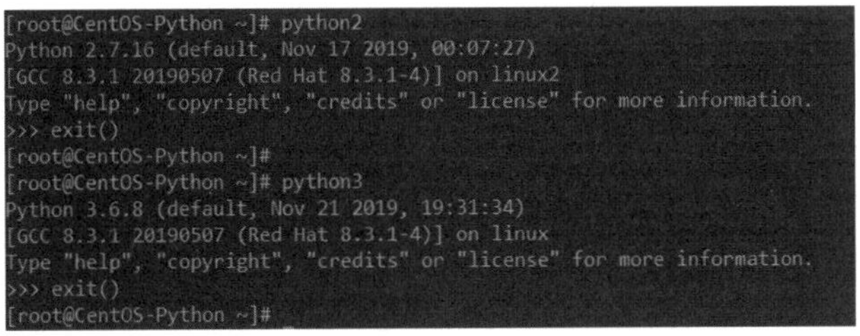

```
[root@CentOS-Python ~]# python2
Python 2.7.16 (default, Nov 17 2019, 00:07:27)
[GCC 8.3.1 20190507 (Red Hat 8.3.1-4)] on linux2
Type "help", "copyright", "credits" or "license" for more information.
>>> exit()
[root@CentOS-Python ~]#
[root@CentOS-Python ~]# python3
Python 3.6.8 (default, Nov 21 2019, 19:31:34)
[GCC 8.3.1 20190507 (Red Hat 8.3.1-4)] on linux
Type "help", "copyright", "credits" or "license" for more information.
>>> exit()
[root@CentOS-Python ~]#
```

CentOS 8 内置的 Python 3 的版本为 3.6.8，我们需要将它升级到 3.10.6，方法如下。

在 CentOS 中安装 Python 需要确保系统安装了 GCC 编译器，首先通过下列命令来安装系统依赖包。

```
dnf install wget yum-utils make gcc openssl-devel bzip2-devel libffi-devel zlib-devel
```

运行该命令后如下图所示。

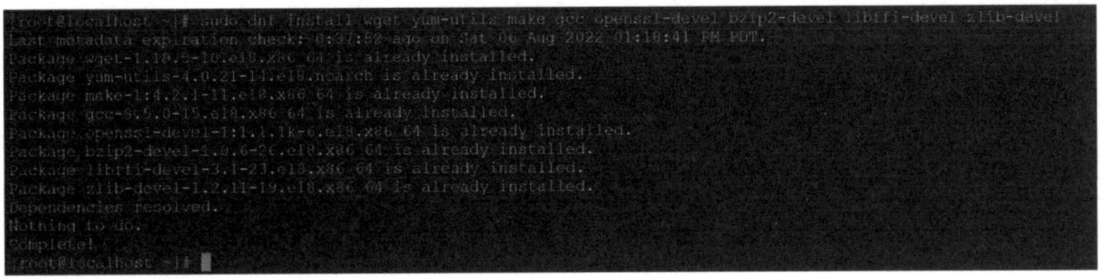

```
[root@localhost ~]# sudo dnf install wget yum-utils make gcc openssl-devel bzip2-devel libffi-devel zlib-devel
Last metadata expiration check: 0:37:52 ago on Sat 06 Aug 2022 01:18:41 PM PDT.
Package wget-1.19.5-10.el8.x86_64 is already installed.
Package yum-utils-4.0.21-14.el8.noarch is already installed.
Package make-1:4.2.1-11.el8.x86_64 is already installed.
Package gcc-8.5.0-15.el8.x86_64 is already installed.
Package openssl-devel-1:1.1.1k-6.el8.x86_64 is already installed.
Package bzip2-devel-1.0.6-26.el8.x86_64 is already installed.
Package libffi-devel-3.1-23.el8.x86_64 is already installed.
Package zlib-devel-1.2.11-19.el8.x86_64 is already installed.
Dependencies resolved.
Nothing to do.
Complete!
[root@localhost ~]#
```

然后通过下列命令来下载 Python 3.10.6 的安装包。

```
wget https://www.python.org/ftp/python/3.10.6/Python-3.10.6.tgz
```

运行该命令后如下图所示。

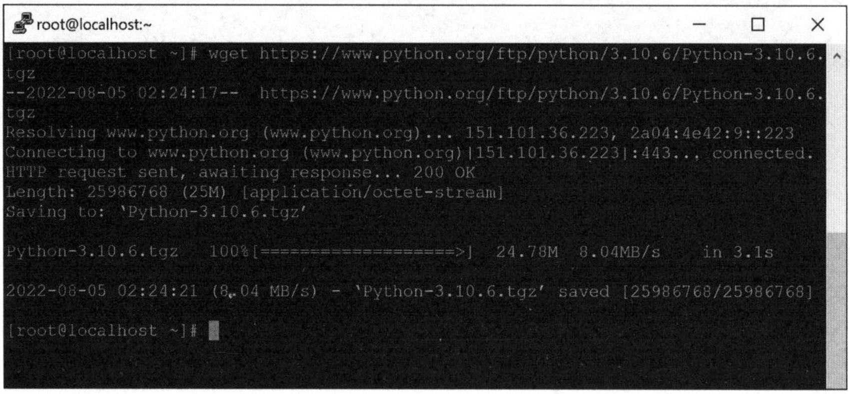

```
root@localhost:~
[root@localhost ~]# wget https://www.python.org/ftp/python/3.10.6/Python-3.10.6.tgz
--2022-08-05 02:24:17--  https://www.python.org/ftp/python/3.10.6/Python-3.10.6.tgz
Resolving www.python.org (www.python.org)... 151.101.36.223, 2a04:4e42:9::223
Connecting to www.python.org (www.python.org)|151.101.36.223|:443... connected.
HTTP request sent, awaiting response... 200 OK
Length: 25986768 (25M) [application/octet-stream]
Saving to: 'Python-3.10.6.tgz'

Python-3.10.6.tgz   100%[===================>]  24.78M  8.04MB/s    in 3.1s

2022-08-05 02:24:21 (8.04 MB/s) - 'Python-3.10.6.tgz' saved [25986768/25986768]

[root@localhost ~]#
```

接下来用 tar 命令对刚才下载的 Python-3.10.6.tgz 包解压缩，解压缩完成后，当前盘符下会多出一个 Python 3.10.6 的文件夹，执行 cd 命令进入该文件夹，如下图所示。

```
[root@localhost ~]# tar zxf Python-3.10.6.tgz
[root@localhost ~]# ls
anaconda-ks.cfg          Documents  original-ks.cfg  Python-3.10.6      test.py
custome_ansible_modules  Downloads  Pictures         Python-3.10.6.tgz  Videos
Desktop                  Music      Public           Templates
[root@localhost ~]# cd Python-3.10.6/
[root@localhost Python-3.10.6]#
```

然后依次输入下列命令来完成 Python 3.10.6 的安装。

```
./configure --enable-optimizations
make -j ${nproc}
make altinstall
```

运行命令后如下图所示。

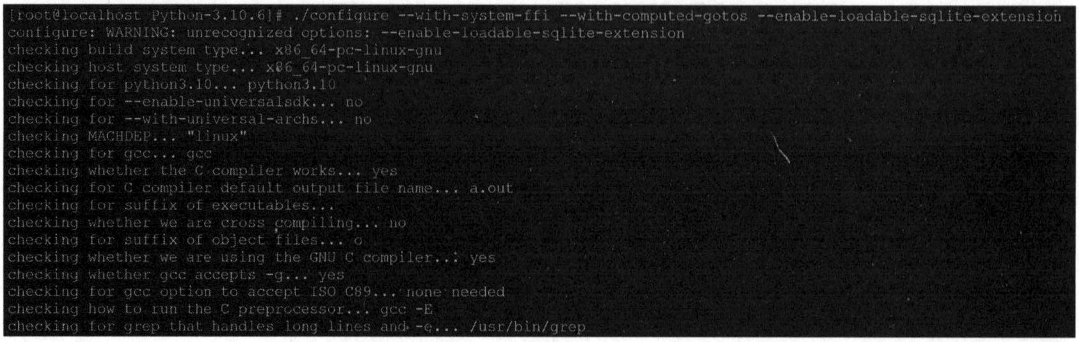

```
[root@localhost Python-3.10.6]# ./configure --with-system-ffi --with-computed-gotos --enable-loadable-sqlite-extension
configure: WARNING: unrecognized options: --enable-loadable-sqlite-extension
checking build system type... x86_64-pc-linux-gnu
checking host system type... x86_64-pc-linux-gnu
checking for python3.10... python3.10
checking for --enable-universalsdk... no
checking for --with-universal-archs... no
checking MACHDEP... "linux"
checking for gcc... gcc
checking whether the C compiler works... yes
checking for C compiler default output file name... a.out
checking for suffix of executables...
checking whether we are cross compiling... no
checking for suffix of object files... o
checking whether we are using the GNU C compiler... yes
checking whether gcc accepts -g... yes
checking for gcc option to accept ISO C89... none needed
checking how to run the C preprocessor... gcc -E
checking for grep that handles long lines and -e... /usr/bin/grep
```

```
[root@localhost ~]# cd Python-3.10.6/
[root@localhost Python-3.10.6]# make -j $(nproc)
 CC='gcc -pthread' LDSHARED='gcc -pthread -shared    ' OPT='-DNDEBUG -g -fwrapv -O3 -Wall'      _TCLTK_INCLUDES='' _TCLTK_LIBS=''       ./python -E ./setup.py  build
running build
running build_ext
INFO: Can't locate Tcl/Tk libs and/or headers

The necessary bits to build these optional modules were not found:
_curses               _curses_panel         _dbm
_gdbm                 _lzma                 _sqlite3
_tkinter              _uuid                 nis
readline
To find the necessary bits, look in setup.py in detect_modules() for the module's name.


The following modules found by detect_modules() in setup.py, have been
built by the Makefile instead, as configured by the Setup files:
_abc                  pwd                   time

running build_scripts
copying and adjusting /root/Python-3.10.6/Tools/scripts/pydoc3 -> build/scripts-3.10
copying and adjusting /root/Python-3.10.6/Tools/scripts/idle3 -> build/scripts-3.10
copying and adjusting /root/Python-3.10.6/Tools/scripts/2to3 -> build/scripts-3.10
changing mode of build/scripts-3.10/pydoc3 from 644 to 755
changing mode of build/scripts-3.10/idle3 from 644 to 755
changing mode of build/scripts-3.10/2to3 from 644 to 755
renaming build/scripts-3.10/pydoc3 to build/scripts-3.10/pydoc3.10
renaming build/scripts-3.10/idle3 to build/scripts-3.10/idle3.10
renaming build/scripts-3.10/2to3 to build/scripts-3.10/2to3-3.10
[root@localhost Python-3.10.6]# 
```

安装完毕后，输入命令 python3.10，如果可以进入 Python 的解释器，则说明 Python 3.10.6 安装成功，如下图所示。

```
[root@localhost Python-3.10.6]# make altinstall
if test "no-framework" = "no-framework" ; then \
        /usr/bin/install -c python /usr/local/bin/python3.10; \
else \
        /usr/bin/install -c -s Mac/pythonw /usr/local/bin/python3.10; \
fi
if test "3.10" != "3.10"; then \
        if test -f /usr/local/bin/python3.10 -o -h /usr/local/bin/python3.10; \
        then rm -f /usr/local/bin/python3.10; \
        fi; \
        (cd /usr/local/bin; ln python3.10 python3.10); \
fi
if test "x" != "x" ; then \
        rm -f /usr/local/binpython3.10-32; \
        lipo  \
                -output /usr/local/bin/python3.10-32 \
                /usr/local/bin/python3.10; \
fi
if test "x" != "x" ; then \
        rm -f /usr/local/binpython3.10-intel64; \
        lipo  \
                -output /usr/local/bin/python3.10-intel64 \
                /usr/local/bin/python3.10; \
fi
./python -E -S -m sysconfig --generate-posix-vars ;\
if test $? -ne 0 ; then \
        echo "generate-posix-vars failed" ; \
        rm -f ./pybuilddir.txt ; \
        exit 1 ; \
fi
```

注：安装 Python 3.10.6 并不会覆盖 CentOS 内置的 Python 3.6.8，使用命令 python3 仍然可以进入 3.6.8 版本，必须使用命令 python3.10 才能进入 3.10.6 版本，如下图所示。

```
[root@localhost Python-3.10.6]# python3
Python 3.6.8 (default, Aug 24 2020, 17:57:11)
[GCC 8.3.1 20191121 (Red Hat 8.3.1-5)] on linux
Type "help", "copyright", "credits" or "license" for more information.
>>> exit()
[root@localhost Python-3.10.6]# python3.10
Python 3.10.6 (main, Aug  5 2022, 02:35:51) [GCC 8.3.1 20191121 (Red Hat 8.3.1-5)] on linux
Type "help", "copyright", "credits" or "license" for more information.
>>> exit()
[root@localhost Python-3.10.6]#
```

1.2　在 Windows 下使用 Python 3.10.6

前面提到 Python 运行模式分为使用解释器的交互模式和运行脚本的脚本模式，下面分别举例介绍这两种运行模式在 Windows 中的使用方法。

1.2.1　交互模式

在 Windows 下，有两种进入 Python 解释器使用交互模式的方法：一种是通过命令行输入命令 py 或者 python；另一种是打开 Python 软件包自带的集成开发环境（IDE），也就是 IDLE。两种方法进入的解释器的界面稍有不同，但是功能完全一样。

1. 使用命令行进入 Python 解释器

首先来看第一种方法，打开 Windows 的命令行（CMD），输入命令 py 或者 python 即可进入 Python 解释器，如下图所示。

```
Command Prompt - py
Microsoft Windows [Version 10.0.18363.2274]
(c) 2019 Microsoft Corporation. All rights reserved.

P:\>py
Python 3.10.6 (tags/v3.10.6:9c7b4bd, Aug  1 2022, 21:53:49) [MSC v.1932 64 bit (AMD64)] on win32
Type "help", "copyright", "credits" or "license" for more information.
>>> print ("hello,world!")
hello,world!
>>>
```

我们在 Python 解释器中输入第一段代码 print("hello，world!")，解释器随即打印出“hello，world!”。这种“即时反馈”的特性是交互模式下特有的，脚本模式下不具备。

注：在 Python 2 中，print ("hello，world! ")也可以省去括号写成 print "hello，world! "，但是在 Python 3 中，print 后面的内容必须加上括号，否则 Python 会报错，提醒你加上括号，如下图所示。

```
P:\>python
Python 3.10.6 (tags/v3.10.6:9c7b4bd, Aug  1 2022, 21:53:49) [MSC v.1932 64 bit (AMD64)] on win32
Type "help", "copyright", "credits" or "license" for more information.
>>> print "hello,world!"
  File "<stdin>", line 1
    print "hello,world!"
    ^^^^^^^^^^^^^^^^^^^^
SyntaxError: Missing parentheses in call to 'print'. Did you mean print(...)?
>>>
```

2. 使用 IDLE 进入 Python 解释器

现在介绍使用 IDLE 进入解释器的方法。以 Windows 10 为例，单击计算机屏幕左下角的“开始”按钮后搜索“idle”即可找到 IDLE（Python 3.10 64-bit）这个桌面应用程序，如下图所示。

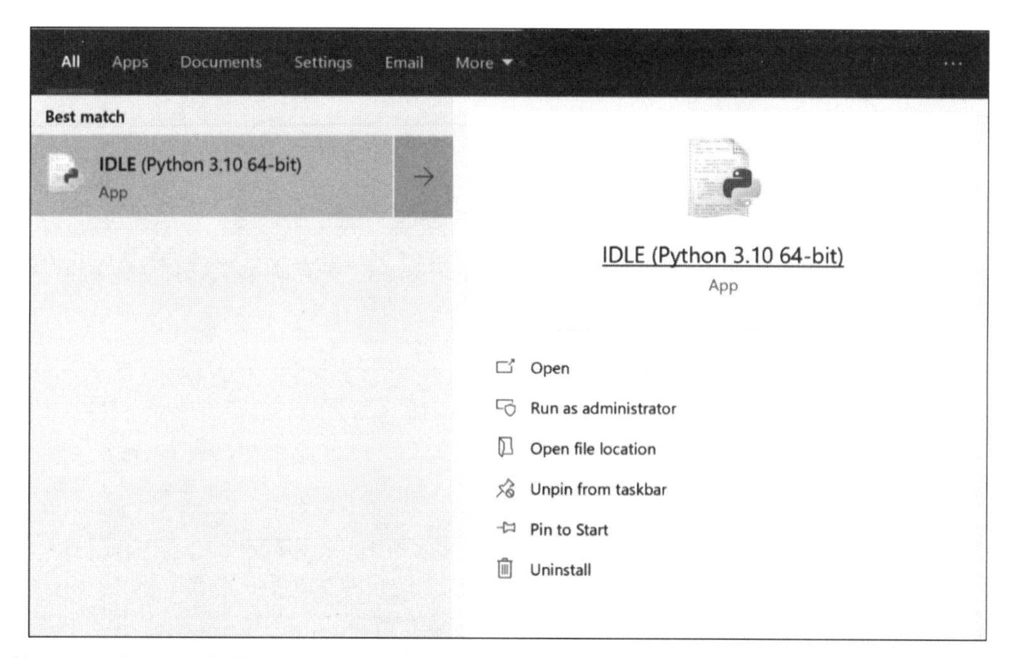

将 IDLE 打开后会弹出如下图所示的窗口。再次输入代码 print ('hello，world!')，可以看到解释器同样立即打印出“hello，world!”，并且默认支持语法和代码高亮。

```
IDLE Shell 3.10.6
File Edit Shell Debug Options Window Help
    Python 3.10.6 (tags/v3.10.6:9c7b4bd, Aug  1 2022, 21:53:49) [MSC v.1932 64 bit (
    AMD64)] on win32
    Type "help", "copyright", "credits" or "license()" for more information.
>>> print ('hello, world!')
    hello, world!
>>>
```

1.2.2　脚本模式

在 Windows 里，有两种方法可用于创建 Python 脚本，一种是将代码写进 Windows 记事本里，另一种是借助第三方编辑器。两种方法分别介绍如下。

1. 使用记事本创建 Python 脚本

在桌面上新建一个记事本文件，将代码 print ('hello,world!')写入，如下图所示。

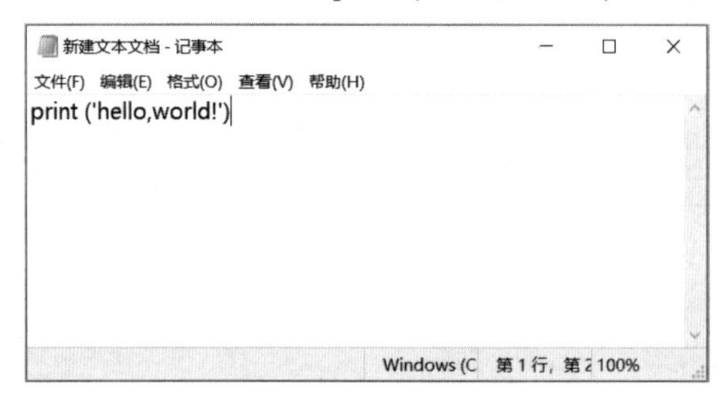

然后将其另存为.py 格式，保存在桌面上。这里需要将“保存类型”选择为“所有文件”，否则该文件的类型依然为.txt，如下图所示。

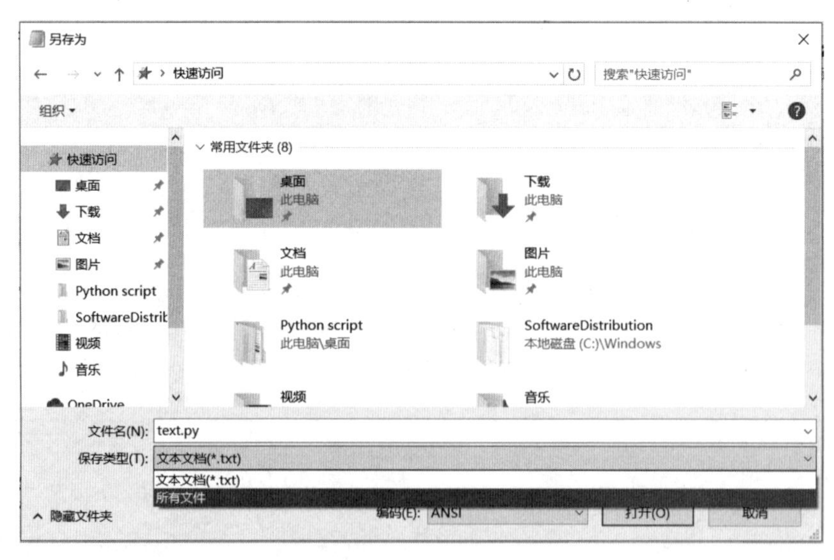

回到桌面，可以发现第一个 Python 脚本已经创建成功，如下图所示。

2. 使用第三方编辑器创建 Python 脚本

支持 Python 的第三方编辑器有很多，Pycharm、Sublime Text 2/3、Notepad ++、vim（Linux 系统）和 Python 自带的 IDLE 等都是很优秀也很常用的编辑器。这里以 Sublime Text 3 为例简单介绍使用第三方编辑器创建 Python 脚本的方法。

首先在 Sublime Text 官网下载 Sublime Text 3。Sublime Text 为付费软件，但是也可以免费使用，免费版本每使用几次后会弹出一个窗口问你是否愿意购买付费版本，如果你不愿意付费，将窗口关闭即可，基本不会影响使用体验。

Sublime Text 支持近 50 种编程语言，默认句法（Syntax）是 Plain Text。在 Plain Text 下写出来的 Python 代码的效果和记事本没有区别，依然只有黑白两色，而且保存文件的时候依然需要手动将文件另存为.py 格式，如下图所示。

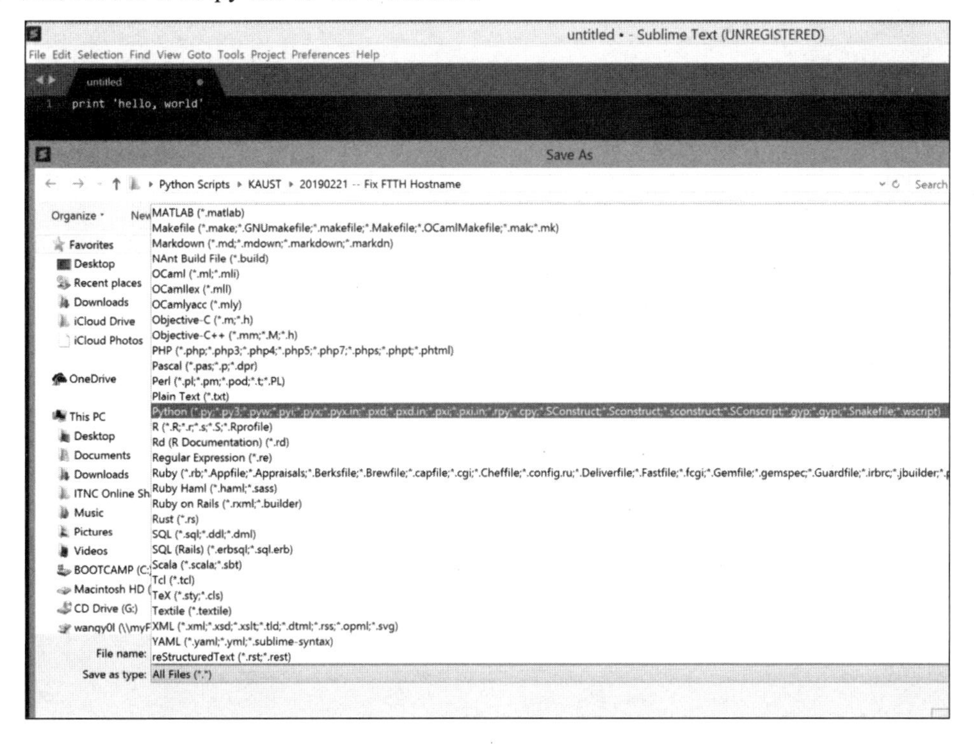

因此，在进入 Sublime Text 后需要做的第一件事是选择 View→Syntax→Python 将句法改为 Python，如下图所示。这样才能获得对 Python 最好的支持，包括代码高亮、语法提示、代码自动补完、默认将脚本保存为.py 格式等诸多实用功能。

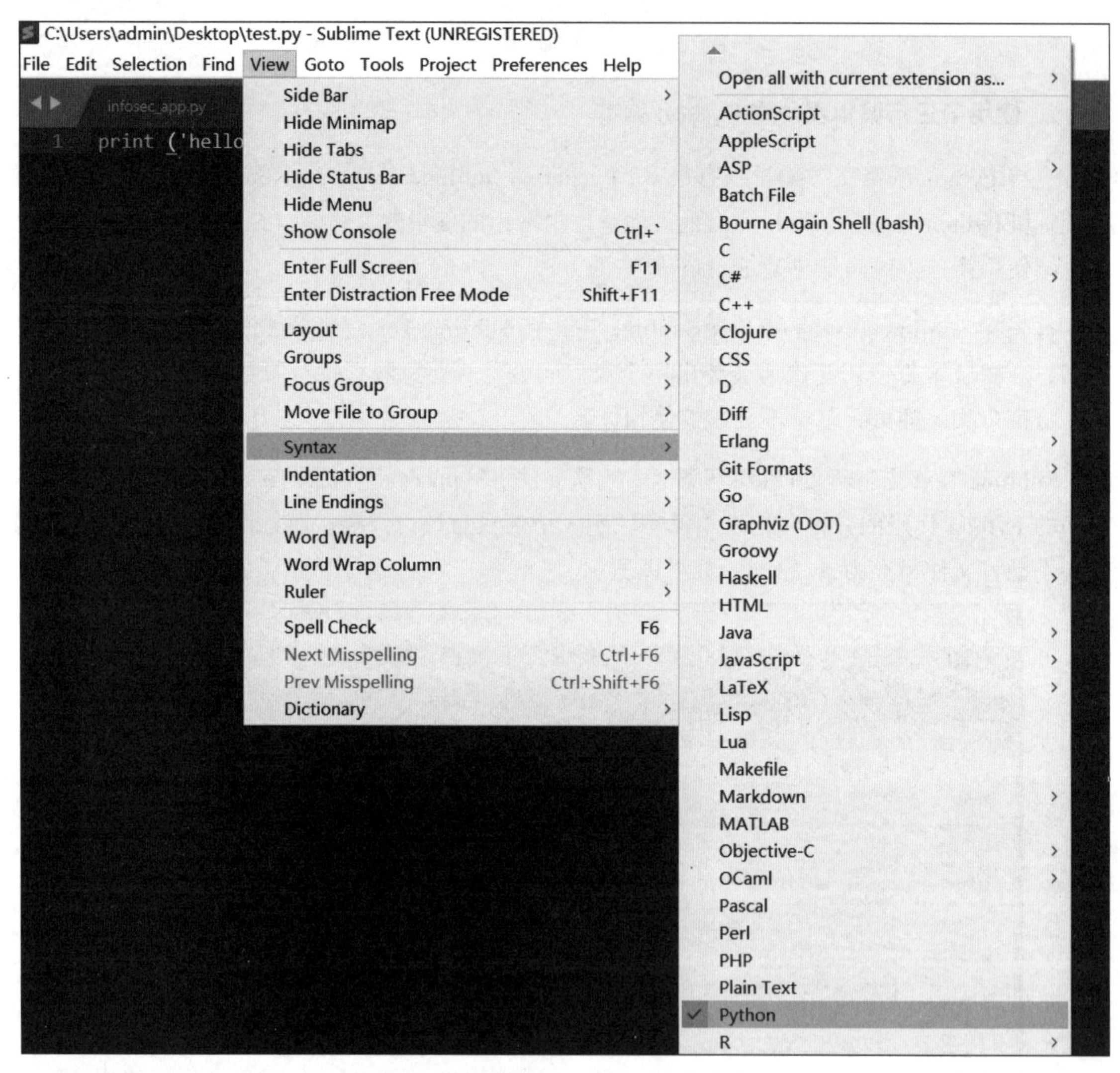

将句法改为 Python 后，代码立刻变为高亮，并且保存代码的时候文件类型已经自动默认为.py 格式，如下图所示。

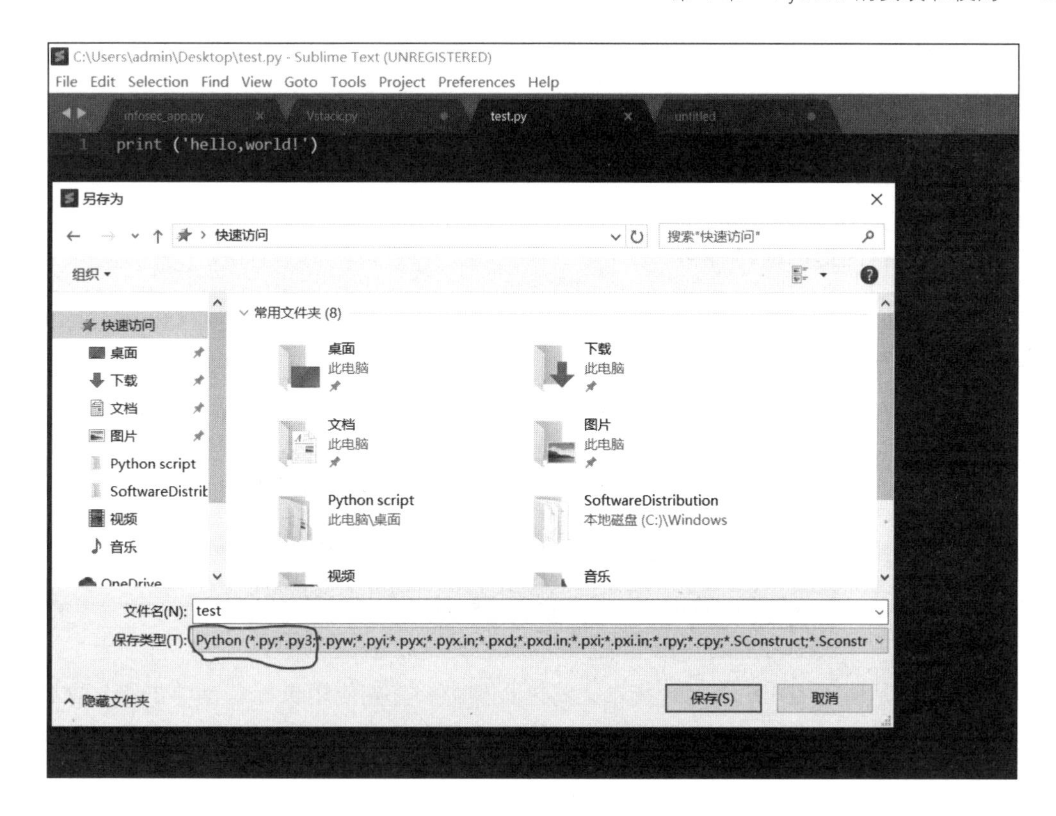

1.2.3　运行 Python 脚本

在 Windows 系统里，有四种运行脚本的方法。

第一种方法是双击.py 文件，这种方法的缺点是在双击运行脚本后，你会看到一个“闪退”的命令行窗口，窗口闪退速度很快，从弹出到消失只有 0.1 ~ 0.2s，肉眼刚刚能看到窗口的轮廓，但是无法看清窗口中的内容。这是因为 Python 脚本程序执行完后自动退出了，要想让窗口停留，需要在代码最后写上一个 input()，如下图所示。

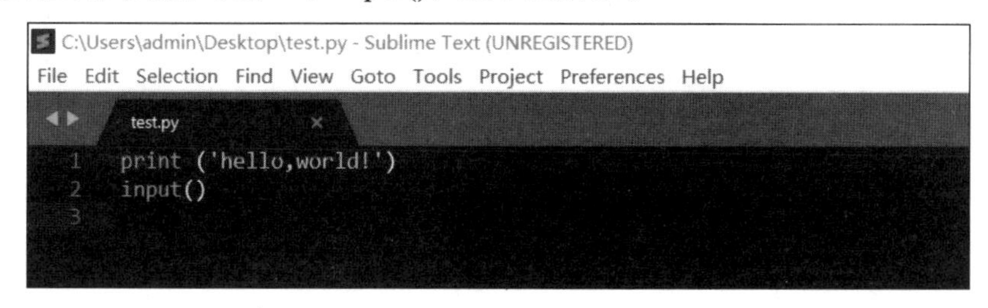

然后用同样的方法将该脚本另存为.py 文件，再次双击可运行该脚本，效果如下图所示。

关于 input()会在 2.4.1 节中详细解释，这里只需要知道可以用它来解决通过双击运行 Python 脚本时窗口闪退的问题即可。

第二种方法是在命令行里移动到脚本文件所在的文件夹下，输入 py xxx.py 或 python xxx.py 命令来运行脚本，结果如下图所示。

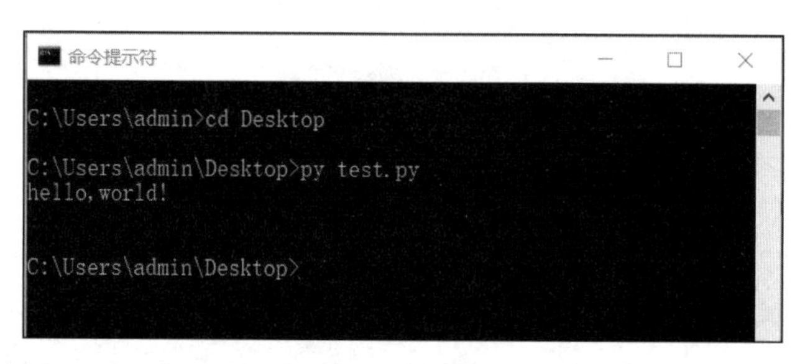

第三种方法是使用 IDLE 来运行脚本，具体步骤为：首先使用鼠标右键单击脚本文件，选择"Edit with IDLE"，进入 IDLE，如下图所示。

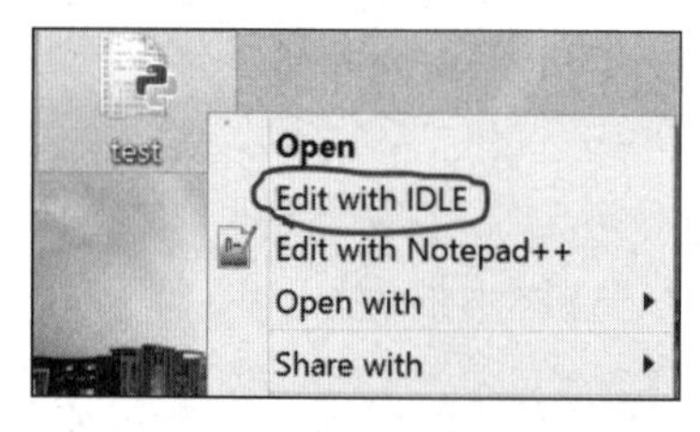

然后在 IDLE 里依次选择 Run → Run Module 来运行脚本，如下图所示。

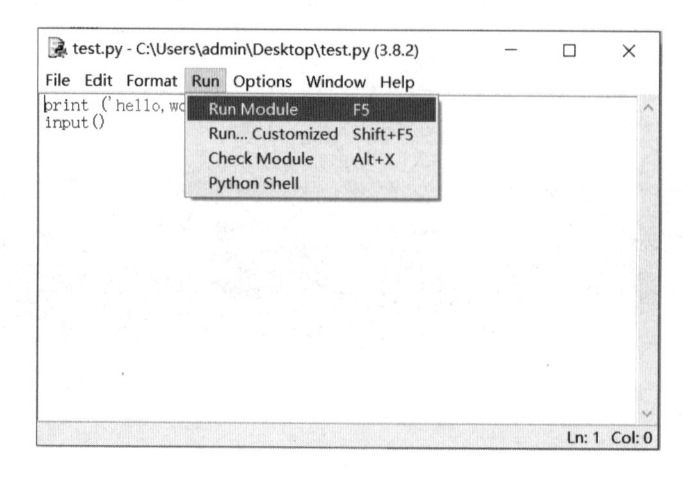

得到的效果如下图所示。

```
*IDLE Shell 3.10.6*
File Edit Shell Debug Options Window Help
Python 3.10.6 (tags/v3.10.6:9c7b4bd, Aug  1 2022, 21:53:49) [MSC v.1932 64 bit (
AMD64)] on win32
Type "help", "copyright", "credits" or "license()" for more information.
>>>
================== RESTART: C:\Users\wangy01\Desktop\test.py =================
hello,world!
```

可以发现，在 IDLE 里即使不使用 input()，运行脚本时也不会出现窗口闪退的问题，因此，通常建议使用 IDLE 来运行脚本。

第四种方法是在第三方编辑器里运行脚本。依然以 Sublime Text 3 为例，方法很简单，首先进入 Sublime Text 3，如下图所示，依次选择 Tools → Build System → Python。

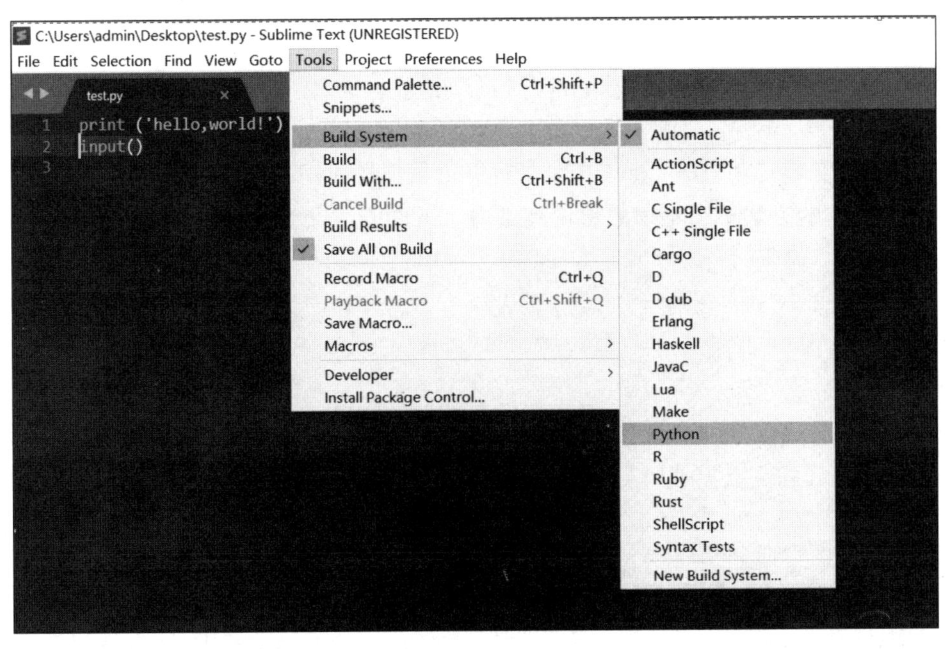

写好代码并保存后，打开 Tools → Build 或者使用快捷键 Ctrl + B 就可以在窗口底部看到运行脚本的结果了，如下图所示。

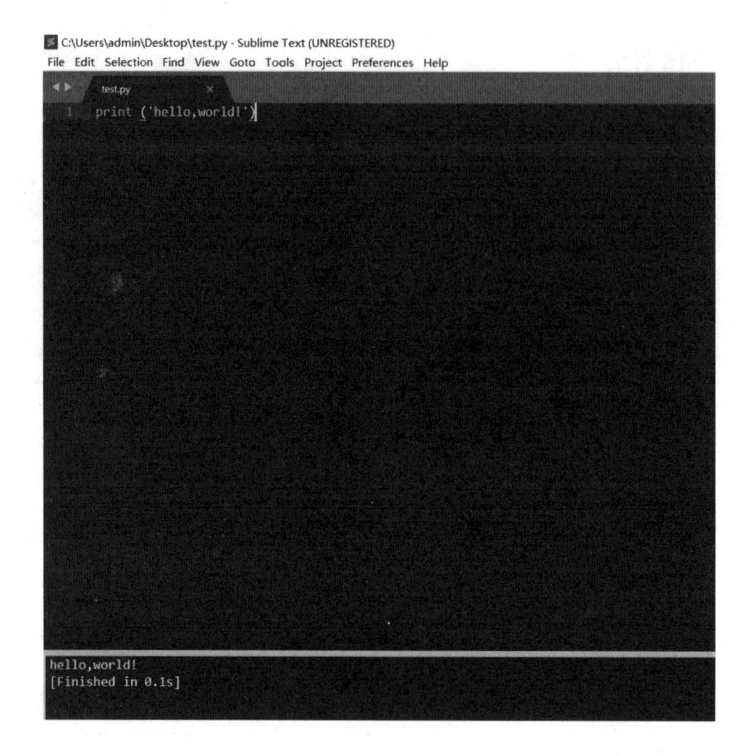

1.3　在 Linux 下使用 Python 3.10.6

前面提到，本书只介绍在 CentOS 命令行终端里使用 Python 的方法，在 GNOME 桌面环境下使用 Python 的方法不在本书的讨论范围内。下面介绍交互模式和脚本模式在 CentOS 中的使用方法。

1.3.1　交互模式

我们知道，在 CentOS 的命令行终端里输入命令 python3.10 即可进入 Python 3.10.6 的解释器，即进入 Python 的交互模式，如下图所示。

```
root@localhost:~
[root@localhost ~]# python3.10
Python 3.10.6 (main, Aug  5 2022, 02:35:51) [GCC 8.3.1 20191121 (Red Hat 8.3.1-5
)] on linux
Type "help", "copyright", "credits" or "license" for more information.
>>>
```

在 Python 解释器里输入第一段代码 print ('hello, world!')，解释器随即打印出了“hello, world!”的内容。这种“即时反馈”的特性是交互模式下特有的，脚本模式下不具备，如下图所示。

```
root@localhost:~
[root@localhost ~]# python3.10
Python 3.10.6 (main, Aug  5 2022, 02:35:51) [GCC 8.3.1 20191121 (Red Hat 8.3.1-5
)] on linux
Type "help", "copyright", "credits" or "license" for more information.
>>> print ('hello,world!')
hello,world!
>>>
```

1.3.2 脚本模式

在 CentOS 的命令行终端里，我们可以使用文本编辑器来创建脚本，CentOS 支持几种常见的文本编辑器，如 emacs、nano、vi 等。这里介绍使用 vi 创建 Python 脚本的方法。

关于 vi 的用法本书将只做简单介绍。另外，vi 有一个加强版本叫作 vim，两者的具体区别不在本书的讨论范围内。读者只需要知道在创建 Python 脚本时，vim 支持语法高亮，而 vi 不支持。

vi 不支持语法高亮，仅显示 PuTTY 默认字体的颜色，如下图所示。

vim 支持语法高亮，显示彩色字体，如下图所示。

除此之外，两者对 Python 的支持并无本质区别，是否喜欢语法高亮全凭个人。只需要注意一点：vi 是 CentOS 安装时自带的文本编辑器，vim 则需要通过输入命令 yum install vim 安装后才能使用。

下面用实例介绍使用 vi 创建 Python 脚本的方法。

首先输入命令 vi test.py，创建一个名为 test.py 的 Python 脚本，如下图所示。

进入 vi 后，按“i”键进入输入模式（屏幕左下角会显示“- - INSERT - -”），输入第一段代码 print ('hello, world!')，如下图所示。

然后按“ESC”键，在屏幕左下角的“- - INSERT - -”消失后，接着输入:wq，按回车键后即可保存文件并退出 vi，如下图所示。

之后回到命令行终端，输入 ls 即可看到刚刚创建成功的 Python 脚本 test.py，如下图所示。

```
root@localhost:~
[root@localhost ~]# vi test.py
[root@localhost ~]# ls
anaconda-ks.cfg          Documents  original-ks.cfg  Python-3.10.6      test.py
custome_ansible_modules  Downloads  Pictures         Python-3.10.6.tgz  Videos
Desktop                  Music      Public           Templates
[root@localhost ~]#
```

1.3.3　运行 Python 脚本

与 Windows 命令行终端一样，在 CentOS 命令行终端也是通过输入 python xxx.py 来运行 Python 脚本的，不同的是，因为本书以 Python 3.10.6 为例，因此，这里需要把 python 换成 python3.10，写成 python3.10 xxx.py 的形式来运行脚本，如下图所示。

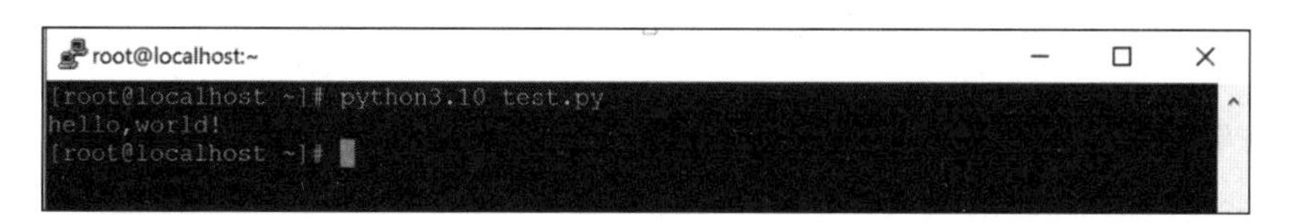

1.3.4　Shebang 符号

在 Linux 和 UNIX 里，符号#!叫作 Shebang，通常可以在 Linux/UNIX 系统脚本中第一行的开头处看到它。它的作用是指明执行脚本文件的解释程序。写在 Shebang 后面的解释程序如果是一个可执行文件，则当执行脚本时，Shebang 会把文件名作为参数传递给解释程序去执行。比如 python3.10 test.py 中的 python3.10 是解释程序，test.py 是文件名，使用 Shebang 后，可以省去解释程序，把 python3.10 test.py 写成./test.py 就可以运行 Python 脚本。另外，Shebang

指定的解释程序必须为可执行程序，否则系统会报错“Permission denied.”。

因此，如果你觉得每次都需要输入命令 python2、python3 或者 python3.10 来运行脚本比较麻烦，则可以在脚本的开头部分使用 Shebang 符号，然后在其后面加上/usr/bin/env python3 来指定 python3 为解释程序（同理，如果你想使用 python2 作为解释程序，则可以写成#!/usr/bin/env python2），如下图所示。

将脚本保存并退出后，用 chmod 命令将 test.py 改为可执行，如下图所示。

```
root@localhost:~
[root@localhost ~]# chmod 777 test.py
[root@localhost ~]# ls
anaconda-ks.cfg          Documents  original-ks.cfg  Python-3.10.6      test.py
custome_ansible_modules  Downloads  Pictures         Python-3.10.6.tgz  Videos
Desktop                  Music      Public           Templates
[root@localhost ~]# ./test.py
hello,world!
[root@localhost ~]#
```

然后就可以用./test.py 来运行脚本，省去了每次都必须输入命令 python2、python3 或者 python3.10 的麻烦，如下图所示。

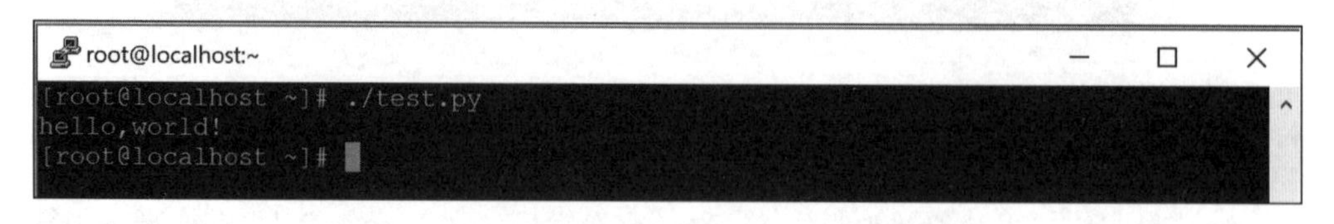

第2章

Python基本语法

本章主要介绍 Python 的基本语法知识，举例演示的所有代码都将在 CentOS 中执行，如果代码前面带有>>>符号，则表示在解释器下运行，不带>>>符号即表示在脚本模式下运行。

2.1 变量

所谓变量（Variable），顾名思义，是指在程序运行过程中，值会发生变化的量。与变量相对应的是常量，也就是在程序运行过程中值不会发生变化的量，不同于 C/C++等语言，Python 并没有严格定义常量这个概念，在 Python 中约定俗成的方法是使用全大写字母的命名方式来指定常量，如圆周率 Pi=3.1415926。

变量是存储在内存中的一个值，创建一个变量，也就意味着在内存中预留了一部分空间给它。变量用来指向存储在内存中的一个对象，每个对象根据自身情况又可以代表不同的数据类型（Data Type）。我们可以通过**变量赋值**这个操作将变量指向一个对象，比如下面的 a = 10 即一个最简单的变量赋值的示例。

```
>>> a = 10
```

在 Python 中，我们使用等号=来连接变量名和值，进而完成变量赋值的操作。这里将 10 这个整数（也就是内存中的对象）赋给变量 a，因为 10 本身是"**整数**"（**Integer**），所以变量 a 此时就代表了"整数"这个数据类型的值。我们可以使用 type()函数来确认 a 的数据类型，发现变量 a 的数据类型此时为 int，也就是 integer 的缩写，代码如下。

```
>>> type(a)
<class 'int'>
>>>
```

Python 是一门动态类型语言，和 C、Java 等不同，我们无须手动指明变量的数据类型，根据赋值的不同，Python 可以随意更改一个变量的数据类型。举例来说，刚才我们把“整数”这个数据类型的值赋给变量 a，现在再次将一个内容为 test 的**字符串（String）**数据类型的值赋给变量 a，然后用 type()函数确认，这时 a 的数据类型已经从 int 变为了 str，即字符串，代码如下。

```
>>> a = 'test'
>>> type(a)
<class 'str'>
>>>
```

变量名可以用大小写英文字母、下画线、数字来表示，但是不能包含标点符号、空格及各类其他特殊符号，如括号、货币符号等。

变量名可以以字母和下画线开头，**但是不能以数字开头**，举例如下。

```
>>> test = 'test'
>>> _a_ = 1
>>> 123c = 10
  File "<stdin>",  line 1
    123c = 10
       ^
SyntaxError: invalid syntax
>>>
```

这里 Python 解释器返回了“SyntaxError: invalid syntax”这个无效语法的错误提示，告诉我们 123c 为无效的变量名。这也是使用解释器来学习 Python 的优势，无论代码里出现什么问题，都能得到“即时反馈”。

变量名区分大小写，举例如下。

```
>>> a = 10
>>> print (A)
Traceback (most recent call last):
  File "<stdin>", line 1, in <module>
NameError: name 'A' is not defined
>>> print (a)
10
>>>
```

如果变量名中间出现两个或两个以上的单词，则只能用下画线将它们连接，不可以使用空格将它们隔开，举例如下。

```
>>> ip_address = '192.168.1.1'
>>> ip address = '192.168.1.1'
  File "<stdin>", line 1
    ip address = '192.168.1.1'
         ^
SyntaxError: invalid syntax
>>>
```

最后，不是所有的英文单词都能用作变量名，Python 中有**保留字**（**Reserved Word**）的概念。保留字通常是 Python 中常用的关键词，比如用于创建函数的“def”，用于 while 循环和 for 循环的“while”和“for”，等等。可以用下面的方法来查询当前的 Python 版本中有哪些保留字。

```
>>> import keyword
>>> print (keyword.kwlist)
['False', 'None', 'True', 'and', 'as', 'assert', 'async', 'await', 'break', 'class',
'continue', 'def', 'del', 'elif', 'else', 'except', 'finally', 'for', 'from', 'global',
'if', 'import', 'in', 'is', 'lambda', 'nonlocal', 'not', 'or', 'pass', 'raise', 'return',
'try', 'while', 'with', 'yield']
>>>
```

看不懂上面的代码没关系，本书后面会讲到这些保留字的用法。这里只需注意输入 print (keyword.kwlist)后的返回值为一个列表，该列表中的元素（列表和元素的概念后文会讲到）即当前 Python 版本中的保留字，这些保留字均不能用作变量名，举例如下。

```
>>> and = 1
  File "<stdin>", line 1
    and = 1
      ^
SyntaxError: invalid syntax
>>> as = 1
  File "<stdin>", line 1
    as = 1
     ^
SyntaxError: invalid syntax
>>> else = 1
  File "<stdin>", line 1
    else = 1
```

```
      ^
SyntaxError: invalid syntax
>>>
```

2.2 注释

在编程时，尤其是当编写代码内容较多的大型程序时，**注释（Comments）**可以起到备注的作用，在回顾代码时帮助回忆和理解。除此之外，在团队合作的时候，在代码中使用注释也是极其重要的一项要求，因为你写的代码可能会被他人调用、维护，为了让他人更容易理解你写的代码的目的和用途，在代码中使用注释是非常必要的。

在 Python 中，我们使用#来做注释符号。和#写在同一排，并且写在#后面的代码将只做注释使用，不会被当作代码的一部分，也就不会被执行。大多数情况下我们会在脚本模式中用到注释，在交互模式中使用注释的情况很少，举例如下。

```
#coding=utf-8
#生成一个整数列表，该列表为整数 1～10 的平方数的集合
test_list = [i ** 2 for i in range(1,11)]
print (test_list)
```

这段代码也许你看不懂，但是通过注释可以知道它的作用是"生成一个整数列表，该列表为整数 1～10 的平方数的集合"，也就是 1，4，9，16，…，100。我们可以运行该脚本查看输出内容。

需要注意的是，在 Python 2 中，如果使用脚本模式运行 Python，并且代码中出现了中文，则必须在代码的开头加上"#coding=utf-8"，因为 Python 默认的编码格式是 ASCII，如果不修改编码格式，则 Python 将无法正确显示中文，但是在 Python 3 中，字符串的默认编码格式已经更改为 Unicode，因此无须在脚本开头键入#coding=utf-8。

运行之前的脚本，效果如下图所示。

因为写在#后面的代码只做注释使用，并不会被当作代码的一部分执行，因此有时我们还可以巧用#来"遮盖"我们不想执行的代码。比如可以选择性地在脚本的某些 print()函数前加上#，不看其输出内容，想看其输出内容时再把#删除，而不至于每次都要在脚本里反复删除、重写该段 print()函数。

```
root@CentOS-Python:~
[root@CentOS-Python ~]# cat demo.py
# coding=utf-8
#生成一个整数列表，该列表为整数1-10的平方数的集合
test_list = [i ** 2 for i in range(1,11)]
print (test_list)
[root@CentOS-Python ~]#
[root@CentOS-Python ~]# python3 demo.py
[1, 4, 9, 16, 25, 36, 49, 64, 81, 100]
[root@CentOS-Python ~]#
```

2.3　方法和函数

在 Python 中，方法（Method）和函数（Function）大体来说是可以互换的两个词，它们之间有一个细微的区别：函数是独立的功能，无须与对象关联；方法则与对象有关，不需要传递数据或参数就可以使用。举个例子，前面讲到的 type()就是一个函数，代码如下。

```
>>> a = 123
>>> type(a)
<class 'int'>
>>>
>>> type('xyz')
<class 'str'>
>>>
```

方法则需要与一个对象（变量或数据）关联，比如 upper()是一个方法，它的作用是将字符串里的小写英文字母转换为大写英文字母，代码如下。

```
>>> vendor = 'Cisco'
>>> vendor.upper()
'CISCO'
>>>
```

这里我们创建了一个名为 vendor 的变量，并将字符串内容“Cisco”赋值给它，随后对该变量调用 upper()方法，返回值即所有字母都变为大写的“CISCO”。

在 Python 中，每种数据类型都有自己默认自带的函数、方法和变量，要查看某一数据类型本身具有的函数、方法和变量，可以使用 dir()函数，这里以字符串和整数为例，代码如下。

```
>>> dir(str)
['__add__', '__class__', '__contains__', '__delattr__', '__doc__', '__eq__',
'__format__', '__ge__', '__getattribute__', '__getitem__', '__getnewargs__',
```

```
'__getslice__', '__gt__', '__hash__', '__init__', '__le__', '__len__', '__lt__',
'__mod__', '__mul__', '__ne__', '__new__', '__reduce__', '__reduce_ex__', '__repr__',
'__rmod__', '__rmul__', '__setattr__', '__sizeof__', '__str__', '__subclasshook__',
'_formatter_field_name_split', '_formatter_parser', 'capitalize', 'center', 'count',
'decode', 'encode', 'endswith', 'expandtabs', 'find', 'format', 'index', 'isalnum',
'isalpha', 'isdigit', 'islower', 'isspace', 'istitle', 'isupper', 'join', 'ljust',
'lower', 'lstrip', 'partition', 'replace', 'rfind', 'rindex', 'rjust', 'rpartition',
'rsplit', 'rstrip', 'split', 'splitlines', 'startswith', 'strip', 'swapcase', 'title',
'translate', 'upper', 'zfill']

>>> dir(int)
['__abs__', '__add__', '__and__', '__class__', '__cmp__', '__coerce__', '__delattr__',
'__div__', '__divmod__', '__doc__', '__float__', '__floordiv__', '__format__',
'__getattribute__', '__getnewargs__', '__hash__', '__hex__', '__index__', '__init__',
'__int__', '__invert__', '__long__', '__lshift__', '__mod__', '__mul__', '__neg__',
'__new__', '__nonzero__', '__oct__', '__or__', '__pos__', '__pow__', '__radd__',
'__rand__', '__rdiv__', '__rdivmod__', '__reduce__', '__reduce_ex__', '__repr__',
'__rfloordiv__', '__rlshift__', '__rmod__', '__rmul__', '__ror__', '__rpow__',
'__rrshift__', '__rshift__', '__rsub__', '__rtruediv__', '__rxor__', '__setattr__',
'__sizeof__', '__str__', '__sub__', '__subclasshook__', '__truediv__', '__trunc__',
'__xor__', 'bit_length', 'conjugate', 'denominator', 'imag', 'numerator', 'real']
>>>
```

以上即使用 dir()函数列出的字符串和整数所自带的函数、方法和变量。注意，其中前后都带单下画线或双下画线的变量不会在本书中介绍，比如“_formatter_parser”和“__contains__”，初学 Python 的网络工程师只需要知道它们在 Python 中分别表示私有变量与内置变量，学有余力的网络工程师可以自行阅读其他 Python 图书深入学习，其他不带下画线的函数与方法并不是每一个都会在网络运维中用到，笔者将在下一章中选取部分进行讲解。

2.4　数据类型

前面讲到，我们可以使用变量来指定不同的数据类型，对网络工程师来说，常用的数据类型有字符串（String）、整数（Integer）、列表（List）、字典（Dictionary）、浮点数（Float）、布尔类型（Boolean）。另外，不是很常用但需要了解的数据类型包括集合（Set）、元组（Tuple）及空值（None）。下面一一举例讲解。

2.4.1　字符串

字符串即文本，可以用单引号''、双引号""和三引号''' '''表示，下面分别介绍三者的区别和用法。

1. 单引号和双引号

当表示内容较短的字符串时，单引号和双引号比较常用且两者用法相同，比如'cisco'和"juniper"，需要注意的是，**单引号和双引号不可以混用。**

```
>>> vendor1 = 'Cisco'
>>> vendor2 = "Juniper"
>>> vendor3 = 'Arista"
  File "<stdin>", line 1
    vendor3 = 'Arista"
                     ^
SyntaxError: EOL while scanning string literal
>>> vendor3 = 'Arista'
```

这里创建了 3 个变量：vendor1、vendor2 和 vendor3，分别将字符串 “Cisco”、“Juniper” 和 “Arista” 赋值给它们，因为字符串 Arista 混用了单引号和双引号，导致解释器报错，重新给 vendor3 赋值并且只使用单引号后解决了这个问题。这时我们可以用 print 语句(Statements) 将 3 个变量的内容打印出来，如下所示。

```
>>> print (vendor1)
Cisco
>>> print (vendor2)
Juniper
>>> print (vendor3)
Arista
>>>
```

除了使用 print()函数，我们还可以在解释器里直接输入变量名来获取它的值，这是编辑器交互模式下特有的功能，脚本模式做不到，举例如下。

```
>>> vendor1
'Cisco'
>>> vendor2
'Juniper'
>>> vendor3
```

```
'Arista'
>>>
```

需要指出的是，如果变量中存在换行符\n，则 print 会执行换行动作，但如果在解释器里直接输入变量名，则解释器会把换行符\n 当作字符串内容的一部分一起返回，举例如下。

```
>>> banner = "\n\n  Warning: Access restricted to Authorised users only. \n\n"
>>> print banner


  Warning: Access restricted to Authorised users only.


>>> banner
'\n\n  Warning: Access restricted to Authorised users only. \n\n'
>>>
```

看出区别了吗？

在 Python 中，我们还可以通过加号+来拼接（Concatenate）字符串，举例如下。

```
>>> ip = '192.168.1.100'
>>> statement = '交换机的 IP 地址为'
>>>
>>> print (statement + ip)
交换机的 IP 地址为 192.168.1.100
```

注意，在使用加号+将变量拼接合并时，如果其中一个变量为字符串，那么其他所有要与之拼接的变量也都必须为字符串，否则 Python 会报错，举例如下。

```
>>> statement1='网段 192.168.1.0/24 下有'
>>> quantity = 60
>>> statement2='名用户'
>>>
>>> print (statement1 + quantity + statement2)
Traceback (most recent call last):
  File "<stdin>", line 1, in <module>
TypeError: cannot concatenate 'str' and 'int' objects
>>>
```

这里 statement1 和 statement2 两个变量都为字符串，但是 quantity 这个变量为整数，因此 print statement1 + quantity + statement2 会报错"TypeError: cannot concatenate 'str' and 'int' objects"，提示不能将字符串和整数拼接合并。解决办法是使用 str()函数将 quantity 从整数转化为字符串，代码如下。

```
>>> print (statement1 + str(quantity) + statement2)
网段 192.168.1.0/24 下有 60 名用户
>>>
```

2. 三引号

三引号形式的字符串通常用来表示内容较长的文本文字，它最大的好处是如果遇到需要换行的文本，文本内容里将不再需要换行符\n。比如路由器和交换机中用来警告非授权用户非法访问设备后果的 MOTD（Message of The Day）之类的旗标（Banner）配置，此类文本内容通常比较长且需要换行，这时用三引号来表示该文本内容是最好的选择，举例如下。

```
>>> motd = ''' ---------------------------------------------------------------
...
... Warning: You are connected to the Cisco systems,  incorporated network.
... Unauthorized access and use of this network will be vigorously prosecuted.
...
... ---------------------------------------------------------------'''
>>>
>>> print (motd)
----------------------------------------------------------------------------
Warning: You are connected to the Cisco systems,  incorporated network.
Unauthorized access and use of this network will be vigorously prosecuted.
----------------------------------------------------------------------------
>>>
```

3. 与字符串相关的方法与函数

◎　upper()

前面提到过 upper()方法，它的作用是将字符串里的小写英文字母转换为大写英文字母。**upper()的返回值是字符串**，举例如下。

```
>>> vendor = 'Cisco'
>>> vendor.upper()
```

```
'CISCO'
>>>
```

◎ lower()

顾名思义，与 upper()相反，lower()方法的作用是将字符串里的大写英文字母转换为小写英文字母。**lower()的返回值是字符串**，举例如下。

```
>>> vendor = 'CISCO'
>>> vendor.lower()
'cisco'
>>>
```

◎ strip()

strip()用来在字符串的开头和结尾移除指定的字符（如字母、数字、空格、换行符\n、标点符号等）。如果没有指定任何参数，则默认移除字符串开头和结尾处的所有空格和换行符\n。有时字符串的开头和结尾处会夹杂一些空格，如“ 192.168.100.1 ”，要去掉这些多余的空格，可以使用 strip()。**strip()的返回值是字符串**，举例如下。

```
>>> ip='   192.168.100.1   '
>>> ip.strip()
'192.168.100.1'
>>>
```

有时字符串末尾会有换行符\n（比如使用 open()函数的 readlines()方法来读取文本文件里的内容后所返回的列表里的元素，后面会讲到），我们也可以使用 strip()来移除这些换行符，举例如下。

```
>>> ip='192.168.100.1\n'
>>> ip.strip()
'192.168.100.1'
>>>
```

◎ count()

count()用来判断一个字符串里给定的字母或数字具体有多少个，比如要找出“39419591034989320”这个字符串里有多少个数字 9，就可以用 count()。**count()的返回值是整数**，举例如下。

```
>>> '39419591034989320'.count('9')
```

```
5
>>>
```

◎　len()

len()用来判断字符串的长度，比如要回答上面提到的“39419591034989320”总共有多少位，就可以用 len()。**len()的返回值是整数，**举例如下。

```
>>> a='39419591034989320'
>>> len(a)
17
>>>
```

◎　split()和 join()

之所以把这两个方法放在一起讲，是因为它们俩的关系比较接近，在字符串、列表的转换中互成对应的关系，split()将字符串转换成列表，join()将列表转换成字符串。

到目前为止，我们还没有讲到列表（List），这里简单讲解一下：在 Python 中，列表是一种有序的集合，用中括号[]表示，该集合里的数据又被叫作元素，比如[1,3,5,7,9]就是一个最简单的列表，其中的整数 1、3、5、7、9 都属于该列表中的元素。下面我们把该列表赋值给变量 list1，用 type()来确认该变量的数据类型，可以发现它的数据类型为 list。

```
>>> list1 = [1,3,5,7,9]
>>> type(list1)
<class 'list'>
```

我们可以使用索引来访问和指定列表中的每个元素，索引的顺序是从数字 0 开始的。列表索引的用法举例如下。

```
>>> list1[0]
1
>>> list1[1]
3
>>> list1[2]
5
>>> list1[3]
7
>>> list1[4]
9
>>>
```

讲完列表后，为了配合下面的案例，需要讲一下 input()函数。

- **input()的返回值是字符串。**它的作用是提示用户输入数据与 Python 程序互动，比如你想询问用户的年龄，让用户自己输入年龄，可以写一段这样的脚本代码。

```
[root@CentOS-Python ~]#cat demo.py

age = input('How old are you? ')
print ('Your age is: ' + age)

[root@CentOS-Python ~]#
```

然后执行该脚本代码。

```
[root@CentOS-Python ~]#python demo.py

How old are you? 32
Your age is: 32

[root@CentOS-Python ~]#
```

注意这里的 32 是用户自己输入的,虽然它看着像整数,但是它实际的数据类型是字符串。

- 在 Python 2 中，上面提到的 Python 3 中的 input()函数实际对应的是 Python 2 的 raw_input()，而 Python 2 中的 input()函数只能用来接收数字（整数或浮点数），并且相对应的返回值也为整数或浮点数。实际上 Python 3 将 Python 2 中的 input()和 raw_input()整合成了 input()（在 Python 3 中已经没有 raw_input()，这一点需要注意）。

在了解了列表和 input()函数的大致原理和用法后，再来看网络工程师如何在网络运维中使用 split()和 join()。举例来说，在大中型公司里，IP 地址的划分一般是有规律可循的，比如某公司有一栋 10 层楼的建筑，一楼的 IP 子网为 192.168.1.0/24，二楼的为 192.168.2.0/24，三楼的为 192.168.3.0/24，依此类推。现在你需要做一个小程序，让用户输入任意一个属于公司内网的 IP 地址，然后让 Python 告诉用户这个 IP 地址属于哪一层楼。思路如下。

因为该公司内网 IP 地址的第一段都为 192，第二段都为 168，第四段不管用户输入任何 IP 地址都不影响我们对楼层的判断。换句话说，我们只能从该 IP 地址的第三段来判断属于哪一层楼，但是我们要怎样告诉 Python 哪一个数字属于 IP 地址的第三段呢？这时就可以用 split() 将用户输入的 IP 地址（字符串）转化成列表，然后通过列表的索引来指向 IP 地址的第三段，代码如下。

```
>>> floor1='192.168.1.0'
>>> floor1_list = floor1.split('.')
>>>
>>> print (floor1_list)
['192', '168', '1', '0']
>>>
>>> floor1_list[2]
'1'
>>>
```

我们先将 192.168.1.0 赋值给 floor1 这个变量，再对该变量调用 split()方法，然后将返回值赋给另一个变量 floor1_list。注意，split()括号里的“.”表示分隔符，该分隔符用来对字符串进行切片。因为 IP 地址的写法都是 4 段数字用 3 个“.”分开，所以这里分隔符用的是“.”。**因为 split()的返回值是列表，所以我们执行 print (floor1_list)后可以看到，IP 地址的 4 段数字已经被切片独立开来，分别成为组成 floor1_list 列表的 4 个元素的其中之一**，之后我们就可以通过 floor1_list[2]这种索引的方式来查询该列表的第三个元素，从而得到 IP 地址的第三段数字了，也就是这里的数字 1。

在知道怎么通过 split()来获取 IP 地址的第三段数字后，回到前面的需求：让用户输入任意一个属于公司内网的 IP 地址，然后让 Python 告诉用户这个 IP 地址属于哪一层楼。脚本代码如下。

```
[root@CentOS-Python ~]#cat demo.py

#coding=utf-8
ip = input('请输入要查询的 IP 地址：')
ip_list = ip.split('.')
print ('该 IP 地址属于' + ip_list[2] + '楼.')

[root@CentOS-Python ~]#
```

我们使用 input()函数提示用户输入想要查询的 IP 地址，然后将得到的值（字符串）赋给变量 ip。随后对其调用 split()函数，并将返回值（列表）赋给另一个变量 ip_list。接着通过 ip_list[2]做索引，得到该列表的第三个元素，也就是用户输入的 IP 地址的第三段。最后用 print 将查询的结果返回告知用户。

执行以下代码来看效果。

```
[root@CentOS-Python ~]#python demo.py

请输入要查询的 IP 地址: 192.168.3.100
该 IP 地址属于 3 楼

[root@CentOS-Python ~]#
```

讲完 split()后，再来看 join()怎么用。首先来看下面这个列表，它包含了开启思科交换机端口的几条最基本的命令。

```
>>> commands = ['configure terminal', 'interface Fa0/1', 'no shutdown']
```

这几条命令缺少了关键的一点：换行符\n（也就是回车键），这时我们可以使用 join()将换行符\n 加在每条命令的末尾，注意 **join()的返回值是字符串。**

```
>>> '\n'.join(commands)
'configure terminal\ninterface Fa0/1\nno shutdown\n'
>>>
```

再举个例子，如果我们要把之前的列表['192', '168', '1', '0']转换回字符串“192.168.1.0”，可以采用如下方法。

```
>>> '.'.join(['192', '168', '1', '0'])
'192.168.1.0'
>>>
```

如果不加这个“.”会怎样？试试看。

```
>>> ''.join(['192', '168', '1', '0'])
'19216810'
>>>
```

◎ startswith()，endswith()，isdigit()，isalpha()

之所以把上述 4 个字符串函数和方法放在一起讲，是因为**它们的返回值都是布尔值（Boolean）**。布尔值只有两个：True 和 False，且首字母必须大写，true 和 false 都不是有效的布尔值。布尔值通常用来判断条件是否成立，如果成立，则返回 True；如果不成立，则返回 False。

首先来看 startswith()，startswith()用来判断字符串是否以给定的字符串开头，举例如下。

```
>>> ip = '172.16.5.12'
>>> ip.startswith('17')
True
>>> ip.startswith('172.')
True
>>> ip.startswith('192')
False
>>>
```

endswith()与 startswith()恰好相反，用来判断字符串是否以给定的字符串结尾，举例如下。

```
>>> ip = '192.168.100.11'
>>> ip.endswith('1')
True
>>> ip.endswith('11')
True
>>> ip.endswith('2')
False
>>>
```

字符串的内容包罗万象，可以为空，可以为中文汉字或英文字母，可以为整数或小数，可以为任何标点符号，也可以为上述任意形式的组合。而 isdigit()就是用来判断字符串是否为整数的，举例如下。

```
>>> year='2019'
>>> year.isdigit()
True
>>> vendor='F5'
>>> vendor.isdigit()
False
>>> PI='3.1415926'
>>> PI.isdigit()
False
>>> IP='1.1.1.1'
>>> IP.isdigit()
False
>>>
```

isalpha()用来判断字符串是否为文字（包括汉字和英文字母），举例如下。

```
>>> chinese = '中文'
```

```
>>> chinese.isalpha()
True
>>> english = 'English'
>>> english.isalpha()
True
>>> family_name = 'Wang'
>>> family_name.isalpha()
True
>>> full_name = 'Parry Wang'
>>> full_name.isalpha()
False
>>> age = '33'
>>> age.isalpha()
False
>>>
```

注意，isalpha()很严格，只要字符串中出现了哪怕一个非英文字母，isalpha()就会返回 False，如'Parry Wang'（包含了空格）、'33'（包含了数字）等。

2.4.2　整数和浮点数

在 Python 中，有 5 种数值类型（Numeric Type），分别为整数（Integer）、浮点数（Float）、布尔类型（Boolean）、长整数（Long）和复数（Complex）。对网络工程师来说，掌握前面三种就够了，后面两种不是我们需要关心的。

所谓整数，即通常理解的不带小数点的正数或负数，浮点数则是带小数点的正数或负数。可以通过 type()函数来验证，如下面的 1 为整数，1.0 则为浮点数。

```
>>> type(1)
<class 'int'>
>>> type(1.0)
<class 'float'>
>>>
```

我们可以把 Python 当成一个计算器，使用+、-、*、//、**等**算术运算符**做加、减、乘、除、求幂等常见的数学运算，举例如下。

```
>>> 256 + 256
512
```

```
>>> 1.2 + 3.5
4.7
>>> 1024 - 1000
24
>>> 16 * 16
256
>>> 100/10
10
>>> 12 // 10
1
>>> 12 % 10
2
>>> 8**2
64
>>> 3**3
27
>>>
```

在 Python 中，可以通过运算符**做幂运算，比如 8 的 2 次方可以表示为 8**2，3 的 3 次方可以表示为 3**3。

在做除法运算时，可以看到示例中分别使用了/、//和% 三个运算符，它们的区别如下。

/ 表示正常的除法运算，注意在 Python 2 中，如果碰到整数除以整数，结果出现小数部分的时候，比如 12 / 10，Python 2 只会返回整数部分，即 1，要想得到小数点后面的部分，必须将除数或被除数通过 float()函数换成浮点数来运算，举例如下。

```
[root@CentOS-Python ~]#python2
Python 2.7.16 (default, Nov 17 2019, 00:07:27)
[GCC 8.3.1 20190507 (Red Hat 8.3.1-4)] on linux2
Type "help", "copyright", "credits" or "license" for more information.
>>> 12/10
1
>>> 12/float(10)
1.2
>>> float(12)/10
1.2
>>>
```

而在 Python 3 中，12/10 则直接返回浮点数 1.2。

```
[root@CentOS-Python ~]#python3.10
Python 3.10.6 (default, Apr 27 2020, 23:06:10)
[GCC 8.3.1 20190507 (Red Hat 8.3.1-4)] on linux
Type "help", "copyright", "credits" or "license" for more information.
>>> 12/10
1.2
>>>
```

// 表示向下取整，求商数。

```
>>> 12//10
1
>>>
```

% 则表示求余数。

```
>>> 12%10
2
>>>
```

整数也不单单用来做数学运算，通过加号+或乘号*两种运算符，**还可以与字符串互动**，适合用来画分割线，举例如下。

```
>>> print ('CCIE ' * 8)
CCIE CCIE CCIE CCIE CCIE CCIE CCIE CCIE
>>>
>>> print ('CCIE ' + 'CCIE')
CCIE CCIE
>>>
>>> print ('*' * 50)
**************************************************
>>>
```

在网络运维中，有时会遇到需要用计数器做统计的时候，比如某公司有 100 台思科 2960 交换机，由于长期缺乏系统性的运维管理，交换机的 IOS 版本并不统一。为了统计其中有多少台交换机的 IOS 版本是最新的，需要登录所有的交换机，每发现一台 IOS 版本为最新的交换机就通过计数器加 1，直到结束。由于要完成这个脚本涉及 Paramiko、if、for 循环、正则表达式等进阶性的 Python 知识点，所以这里仅演示计数器的用法。

```
>>> counter = 0
>>> counter = counter + 1
```

```
>>> counter
1
>>> counter = counter + 1
>>> counter
2
>>> counter += 1
>>> counter
3
>>> counter += 1
>>> counter
4
>>>
```

首先我们创建一个变量 counter，将 0 赋给它，该变量就是我们最初始的计数器。之后如果每次发现有交换机的 IOS 版本为最新，就在该计数器上加 1，注意 **counter = counter + 1** 可以简写为 **counter + = 1**。

2.4.3　列表

列表（List）是一种有序的集合，用中括号[]表示，列表中的数据称为元素（Element），每个元素之间都用逗号隔开。列表中元素的数据类型可以不固定，举例如下。

```
>>> list1 = [2020, 1.23, 'Cisco', True, None, [1,2,3]]
>>>
>>> type(list1[0])
<class 'int'>
>>> type(list1[1])
<class 'float'>
>>> type(list1[2])
<class 'str'>
>>> type(list1[3])
<class 'bool'>
>>> type(list1[4])
<class 'NoneType'>
>>> type(list1[5])
<class 'list'>
>>>
```

由上例可知，我们创建了一个名为 list1 的变量，并将一个含有 6 个元素的列表赋值给它。

可以看到这 6 个元素的数据类型都不一样，我们使用 type()函数配合列表的索引来验证每个元素的数据类型，**列表的索引号从 0 开始，对应列表里的第 1 个元素**。可以发现从第 1 个到第 6 个元素的数据类型分别为整数、浮点数、字符串、布尔值、空值，以及列表。

注：一个列表本身也可以以元素的形式存在于另一个列表中，举例来说，上面的列表 list1 的第 6 个元素为列表[1,2,3]，我们可以通过使用两次索引的方法来单独调取列表[1,2,3]中的元素，也就是整数 1、2、3。

```
>>> list1 = [2020, 1.23, 'Cisco', True, None, [1,2,3]]
>>> list1[5][0]
1
>>> list1[5][1]
2
>>> list1[5][2]
3
>>>
```

下面介绍与列表相关的方法和函数。

◎　range()

range()函数在 Python 2 和 Python 3 中有较大区别。在 Python 2 中，range()函数用来创建一个整数列表，返回值为列表，举例如下。

```
[root@CentOS-Python ~]#python2
Python 2.7.16 (default, Nov 17 2019, 00:07:27)
[GCC 8.3.1 20190507 (Red Hat 8.3.1-4)] on linux2
Type "help", "copyright", "credits" or "license" for more information.
>>>
>>>a = range(10)
>>>type(a)
<type 'list'>  #Python 2 中的 range()函数返回值的类型为列表
>>> print a
[0, 1, 2, 3, 4, 5, 6, 7, 8, 9]
>>> range(1, 15)
[1, 2, 3, 4, 5, 6, 7, 8, 9, 10, 11, 12, 13, 14]
>>> range(1, 20, 2)
[1, 3, 5, 7, 9, 11, 13, 15, 17, 19]
>>>
```

- Python 2 中的 range()创建的整数列表从 0 开始，因此 range(10)返回的是一个包含整数 0 ~ 9 的列表，**并不包含 10**。

- 也可以在 range()中指定起始数和结尾数，返回的整数列表的最后一个元素为指定的结尾数减 1，如 range(1, 15)将返回一个包含整数 1 ~ 14 的列表，14 由结尾数 15 减 1 得来。
- range()还可以通过指定步长来得到我们想要的整数，比如我们只想选取 1 ~ 19 中所有的奇数，那么就可以使用 range (1, 20, 2)来实现（这里的 2 即步长）。

这种返回列表的 range()函数有一个缺点，即占用内存，列表所含元素数量不多时对主机的性能影响不大，但是当使用 range(10000000000000)来建立诸如这样巨大的列表时所占用的内存就非常“恐怖”了。因此在 Python 3 中，range()函数的返回值被改成了 range，这是一种可以被迭代的对象，这样改的目的就是节省内存。

```
[root@CentOS-Python ~]#python3.10
Python 3.10.6 (default, Apr 27 2020, 23:06:10)
[GCC 8.3.1 20190507 (Red Hat 8.3.1-4)] on linux
Type "help", "copyright", "credits" or "license" for more information.
>>> a = range(10)
>>> type(a)
<class 'range'> #Python 3 中的 range()函数的返回值不再是列表，而是 range 迭代值
>>> print (a)
range(0, 10)  #不再像 Python 2 那样返回整数列表[0,1,2,3,4,5,6,7,8,9]
>>> range (1,15)
range(1, 15) #同上
>>> range (1,20,2)
range(1, 20, 2) #同上
>>>
```

如果要使 Python 3 的 range()也返回列表，则需要对其使用 list()函数，举例如下。

```
>>> a = list(range(10))
>>> print (a)
[0, 1, 2, 3, 4, 5, 6, 7, 8, 9]
>>> list(range(1,15))
[1, 2, 3, 4, 5, 6, 7, 8, 9, 10, 11, 12, 13, 14]
>>> list(range(1,20,2))
[1, 3, 5, 7, 9, 11, 13, 15, 17, 19]
>>>
```

◎ append()

append()用来向列表中添加元素，举例如下。

```
>>> interfaces = []
>>> interfaces.append('Gi1/1')
>>> print (interfaces)
['Gi1/1']
>>> interfaces.append('Gi1/2')
>>> print (interfaces)
['Gi1/1', 'Gi1/2']
>>>
```

首先我们建立一个空列表（以[]表示），并把它赋值给 interfaces 变量，然后使用 append()方法将端口 Gi1/1 加入该列表，随后调用 append()方法将 Gi1/2 加入该列表，现在列表 interfaces 中就有 Gi1/1 和 Gi1/2 两个元素了。

◎ len()

列表的 len()方法和字符串的 len()方法大同小异，前者用来统计列表中有多少个元素，后者用来统计字符串的长度，其返回值也依然为整数，举例如下。

```
>>> len(interfaces)
2
>>>
>>> cisco_switch_models = ['2960', '3560', '3750', '3850', '4500', '6500', '7600', '9300']
>>> len(cisco_switch_models)
8
>>>
```

◎ count()

与字符串一样，列表也有 count()方法，列表的 count()方法用来找出指定的元素在列表中有多少个，返回值为整数，举例如下。

```
>>> vendors = ['Cisco', 'Juniper', 'HPE', 'Aruba', 'Arista', 'Huawei', 'Cisco',
'Palo Alto', 'CheckPoint', 'Cisco', 'H3C', 'Fortinet']
>>> vendors.count('Cisco')
3
>>>
```

◎　insert()

列表是有序的集合，前面讲到的 append()方法的作用是将新的元素添加到列表的最后，如果我们想自己控制新元素在列表中的位置，则要用 insert()方法。举例如下。

```
>>> ospf_configuration = ['router ospf 100\n', 'network 0.0.0.0 255.255.255.255 area
0\n']
>>> ospf_configuration.insert(0, 'configure terminal\n')
>>> print (ospf_configuration)
['configure terminal\n', 'router ospf 100\n', 'network 0.0.0.0 255.255.255.255 area
0\n']
>>>
```

首先创建一个名为 ospf_configuration 的变量，将配置 OSPF 的命令写在一个列表中赋值给该变量。随后发现遗漏了 configure terminal 命令，该命令要写在列表的最前面，这时我们可以用 insert(0, 'configure terminal\n')将该命令加在列表的最前面（记住列表的索引号是从 0 开始的）。

如果这时我们还想给该 OSPF 路由器配置一个 router-id，如把 router-id 这条命令写在 router ospf 100 的后面，可以再次使用 insert()，举例如下。

```
>>> ospf_configuration.insert(2, 'router-id 1.1.1.1\n')
>>> print (ospf_configuration)
['configure terminal\n', 'router ospf 100\n', 'router-id 1.1.1.1\n', 'network 0.0.0.0
255.255.255.255 area 0\n']
>>>
```

◎　pop()

pop()用来移除列表中的元素，如果不指定索引号，则 pop()默认将去掉排在列表末尾的元素；如果指定了索引号，则可以精确移除想要移除的元素，举例如下。

```
>>> cisco_switch_models = ['2960', '3560', '3750', '3850', '4500', '6500', '7600',
'9300']
>>> cisco_switch_models.pop()
'9300'  #排在末尾的 9300 被去掉
>>> print cisco_switch_models
['2960', '3560', '3750', '3850', '4500', '6500', '7600']
>>> cisco_switch_models.pop(1)
'3560' #使用索引号 1 将排在第 2 位的 3560 去掉
```

```
>>> print (cisco_switch_models)
['2960', '3750', '3850', '4500', '6500', '7600']
>>>
```

◎ index()

看了 pop()的用法后，你也许会问：在拥有很多元素的列表中，怎么知道想要移除的元素的索引号是多少呢？这时就需要用 index()，如想从 cisco_switch_models 列表中移除元素 4500，可以如下操作。

```
>>> cisco_switch_models = ['2960', '3560', '3750', '3850', '4500', '6500', '7600', 
'9300']
>>> cisco_switch_models.index('4500')
4
>>> cisco_switch_models.pop(4)
'4500'
>>> print (cisco_switch_models)
['2960', '3560', '3750', '3850', '6500', '7600', '9300']
>>>
```

先通过 index()找出 4500 的索引号为 4，然后配合 pop(4)将它从列表中移除。

2.4.4 字典

在 Python 里，字典（Dictionary）是若干键值对（Key-Value pair）的集合，用大括号{}表示，每一组键值对都用逗号隔开，举例如下。

```
>>> dict = {'Vendor':'Cisco', 'Model':'WS-C3750E-48PD-S', 'Ports':48, 
'IOS':'12.2(55)SE12', 'CPU':36.3}
```

这里我们创建了一个名为 dict 的字典，该字典中有 5 组键值对，分别如下。

```
'Vendor':'Cisco'
'Model':'WS-C3750E-48PD-S'
'Ports':48
'IOS':'12.2(55)SE12'
'CPU':36.3
```

- 键值对里的键（Key）和值（Value）用冒号:隔开，冒号的左边为键，右边为值。
- 键的数据类型可为字符串、整数、浮点数或者元组，对网络工程师来说，最常用的肯定是字符串，如 “Vendor” “Model” 等。

- 值可为任意的数据类型，比如这里的 “Cisco” 为字符串，48 为整数，36.3 为浮点数。

与列表不同，Python3.7 版本之前的字典是无序的，举例如下。

```
>>> a = [1, 2, 3, 'a', 'b', 'c']
>>> print (a)
[1,  2,  3,  'a',  'b',  'c']
>>>
>>> dict = {'Vendor':'Cisco',  'Model':'WS-C3750E-48PD-S',  'Ports':48,
'IOS':'12.2(55)SE12',  'CPU':36.3}
>>> print (dict)
{'IOS': '12.2(55)SE12',  'CPU':36.3,  'Model': 'WS-C3750E-48PD-S',  'Vendor': 'Cisco',
'Ports': 48}
>>>
```

这里我们创建一个内容为[1，2，3，'a'，'b'，'c']的列表 a，将它打印出来后，列表中元素的位置没有发生任何变化，因为列表是有序的。但是如果我们将刚才的字典 dict 打印出来，会发现字典里键值对的顺序已经彻底被打乱了，没有规律可循，正因为字典是无序的，我们自然也不能像列表那样使用索引来查找字典中某个键对应的值。

而在 Python 3.7 及之后的版本里，字典变为了有序的，举例如下。

```
>>> dict = {'Vendor':'Cisco',  'Model':'WS-C3750E-48PD-S',  'Ports':48,
'IOS':'12.2(55)SE12',  'CPU':36.3}
>>> print (dict)
{'Vendor': 'Cisco', 'Model': 'WS-C3750E-48PD-S', 'Ports': 48, 'IOS': '12.2(55)SE12',
'CPU': 36.3}
>>>
```

在字典里，查找某个值的格式为'字典名[键名]'，举例如下。

```
>>> dict = {'Vendor':'Cisco',  'Model':'WS-C3750E-48PD-S',  'Ports':48,
'IOS':'12.2(55)SE12',  'CPU':36.3}
>>> print (dict['Vendor'])
Cisco
>>> print (dict['CPU'])
36.3
>>> print (dict['Ports'])
48
>>>
```

如果要在字典里新添加一组键值对，则格式为'字典名[新键名]' = '新值'，举例如下。

```
>>> dict['Number of devices']=100
>>> print (dict)
{'Vendor': 'Cisco',  'Number of devices': 100,  'IOS': '12.2(55)SE12',  'CPU': 36.3,
'Model': 'WS-C3750E-48PD-S',  'Ports': 48}
>>>
```

如果要更改字典里某个已有键对应的值，则格式为'字典名[键名]' = '新值'，举例如下。

```
>>> dict['Model'] = 'WS-C2960X-24PS-L'
>>> dict['Ports'] = '24'
>>> print (dict)
{'IOS': '12.2(55)SE12',  'Model': 'WS-C2960X-24PS-L',  'Vendor': 'Cisco',  'Ports':
'24',  'CPU': 36.3}
>>>
```

如果要删除字典里的某组键值对，则格式为 del '字典名[键名]'，举例如下。

```
>>> del dict['Number of devices']
>>> print (dict)
{'Vendor':'Cisco',  'IOS':'12.2(55)SE12',  'CPU':36.3,  'Model':'WS-C3750E-48PD-S',
'Ports': 48}
>>>
```

下面介绍与字典相关的函数和方法。

◎　len()

len()用来统计字典里有多少组键值对。**len()的返回值是整数**，举例如下。

```
>>> print (dict)
{'Vendor':'Cisco',  'IOS':'12.2(55)SE12',  'CPU':36.3,  'Model':'WS-C3750E-48PD-S',
'Ports': 48}
>>> len(dict)
5
>>>
```

◎　keys()

keys()用来返回一个字典里所有的键。**keys()在 Python 2 中的返回值为列表；**在 Python 3 中的返回值是可迭代的对象，需要使用 list()将它转换为列表，了解即可。举例如下。

```
>>> print (dict)
{'Vendor': 'Cisco', 'IOS': '12.2(55)SE12', 'CPU': 36.3, 'Model': 'WS-C3750E-48PD-S',
'Ports': 48}
>>> print (dict.keys())
['Vendor', 'IOS', 'CPU', 'Model', 'Ports']
>>>
```

◎ values()

values()用来返回一个字典里所有的值。**注意，values()在 Python 2 中的返回值为列表,在 Python 3 中的返回值是可迭代的对象**，在有必要的情况下需要使用 list()将它转换为列表，举例如下。

```
[root@CentOS-Python ~]#python2
Python 2.7.16 (default, Nov 17 2019, 00:07:27)
[GCC 8.3.1 20190507 (Red Hat 8.3.1-4)] on linux2
Type "help", "copyright", "credits" or "license" for more information.
>>> dict = {'Vendor':'Cisco', 'Model':'WS-C3750E-48PD-S', 'Ports':48,
'IQS':'12.2(55)SE12', 'CPU':36.3}
>>> print (dict.values())
['Cisco', '12.2(55)SE12', 36.3, 'WS-C3750E-48PD-S', 48] #Python 2 里返回列表
>>>

[root@CentOS-Python ~]#python3.10
Python 3.10.6 (default, Apr 27 2020, 23:06:10)
[GCC 8.3.1 20190507 (Red Hat 8.3.1-4)] on linux
Type "help", "copyright", "credits" or "license" for more information.
>>> dict = {'Vendor':'Cisco', 'Model':'WS-C3750E-48PD-S', 'Ports':48,
'IOS':'12.2(55)SE12', 'CPU':36.3}
>>> print (dict.values())
dict_values(['Cisco', 'WS-C3750E-48PD-S', 48, '12.2(55)SE12', 36.3]) #Python 3 里返回
可迭代对象
>>>
```

◎ pop()

前面讲到，要删除字典中某组键值对可以用命令 del '字典名[键名]'。我们也可以使用 pop()来达到同样的目的。与列表的 pop()不同，字典的 pop()不能导入索引号，需要导入的是键名，

而且字典的 pop()的返回值不是列表，而是键名对应的值（比如下面的 48），举例如下。

```
>>> print (dict)
{'Vendor': 'Cisco', 'IOS': '12.2(55)SE12', 'CPU': 36.3, 'Model': 'WS-C3750E- 48PD-S',
'Ports': 48}
>>> dict.pop('Ports')
48
>>> print (dict)
{'Vendor': 'Cisco', 'IOS': '12.2(55)SE12', 'CPU': 36.3, 'Model': 'WS-C3750E-48PD-S'}
>>>
```

◎　get()

前面讲到，我们可以使用 values()方法返回一个字典里所有的值。除此之外，还可以使用 get()来返回字典里具体键名对应的值，**get()的返回值是所导入的键名对应的值，**举例如下。

```
>>> print (dict)
{'Vendor': 'Cisco', 'IOS': '12.2(55)SE12', 'CPU': 36.3, 'Model': 'WS-C3750E-48PD-S'}
>>> dict.get('Vendor')
'Cisco'
>>> dict.get('CPU')
36.3
>>>
```

2.4.5　布尔类型

布尔类型（Boolean）用来判断条件是否成立，布尔值只有两种：True 和 False，如果条件成立，则返回 True；如果条件不成立，则返回 False。**两种布尔值的首字母（T 和 F）必须大写，true 和 false 都不是有效的布尔值。**布尔类型在判断语句中常用，Python 的判断语句将在进阶语法中详细讲解。

1. 比较运算符

既然布尔类型用来判断条件是否成立，那就不得不提一下 Python 中的**比较运算符（Comparison Operators）**。比较运算符包括等于号==、不等于号!=、大于号>、小于号<、大于等于号>=、小于等于号<=。**比较运算符**和+、-、*、//、**这些**算术运算符**（在 2.3.2 节中提到过）**最大的区别是：前者用来判断符号左右两边的变量和数据是否满足运算符本身的**

条件，并且返回值是布尔值，后者则单纯用于做加减乘除等运算，返回值是整数或浮点数。

在编辑器模式下，使用比较运算符后可以马上看到返回的布尔值 True 或者 False。如果是脚本模式，则需要配合 print 命令才能看到，举例如下。

```
>>> a = 100
>>> a == 100
True
>>> a == 1000
False
>>> a != 1000
True
>>> a != 100
False
>>> a > 99
True
>>> a > 101
False
>>> a < 101
True
>>> a < 99
False
>>> a >= 99
True
>>> a >= 101
False
>>> a <= 100
True
>>> a <= 99
False
>>>
```

2. 逻辑运算符

除了比较运算符，使用**逻辑运算符（Logical Operators）**也能返回布尔值。逻辑运算符有 3 种：与（and）、或（or）、非（not）。学过离散数学的读者一定不会对与、或、非的逻辑运算感到陌生，在逻辑运算中使用的真值表（Truth Table）如下。

真值表（Truth Table）				
P	Q	P 与 Q（and）	P 或 Q（or）	非 P（not）
True	True	True	True	False
True	False	False	True	False
False	True	False	True	True
False	False	False	False	True

逻辑运算符在 Python 中的使用举例如下。

```
>>> A = True
>>> B = True
>>> A and B
True
>>> A or B
True
>>> not A
False
>>>
>>>
>>> A = False
>>> B = True
>>> A and B
False
>>> A or B
True
>>> not A
True
>>>
>>> A = False
>>> B = False
>>> A and B
False
>>> A or B
False
>>> not A
True
>>>
```

2.4.6　集合、元组、空值

作为同样需要网络工程师掌握的 Python 数据类型，**集合**(Set)、**元组**(Tuple)、**空值**(None) 相对来说使用频率不如字符串、整数、浮点数、列表、字典及布尔类型那么高，这里进行简单介绍。

1. 集合

- 集合是一种特殊的列表，里面**没有重复的元素**，因为每个元素在集合中都只有一个，所以集合没有 count()方法。
- 集合可以通过大括号{}（与字典一样，但是集合中没有键值对）或者 set()函数创建。

```
>>> interfaces = {'Fa0/0', 'Fa0/1', 'Fa0/2'}
>>> type(interfaces)
<class 'set'>
>>> vendors = set(['Cisco', 'Juniper', 'Arista', 'Cisco'])
>>> type(vendors)
<class 'set'>
>>> print (vendors)
{'Cisco', 'Arista', 'Juniper'}
>>>
```

vendors 列表中有两个重复的元素，即“Cisco”，在用 set()函数将它转换成集合后，多余的“Cisco”将被去掉，只保留一个。

- 集合是无序的，不能像列表那样使用索引号，也不具备 index()函数。

```
>>> vendors[2]
Traceback (most recent call last):
  File "<stdin>", line 1, in <module>
TypeError: 'set' object is not subscriptable
>>>
>>> vendors.index('Cisco')
Traceback (most recent call last):
  File "<stdin>", line 1, in <module>
AttributeError: 'set' object has no attribute 'index'
>>>
```

下面介绍与集合有关的方法和函数。

◎ add()

add()用来向一组集合中添加新元素，**其返回值依然是集合**，举例如下。

```
>>> vendors.add('Huawei')
>>> vendors
{'Huawei', 'Cisco', 'Arista', 'Juniper'}
>>>
```

◎ remove()

remove()用来删除一组集合中已有的元素，**其返回值依然是集合**，举例如下。

```
>>> vendors.remove('Arista')
>>> vendors
{'Huawei', 'Cisco', 'Juniper'}
>>>
```

2. 元组

- 与集合一样，元组也是一种特殊的列表。它与列表最大的区别是：可以任意对列表中的元素进行增加、删除、修改，而元组不可以。一旦创建元组，将无法对其做任何形式的更改，所以元组没有 append()、insert()、pop()、add()和 remove()，只保留了 index()和 count()两种方法。
- 元组可以通过小括号()创建，也可以使用 tuple()函数创建。
- 与列表一样，元组是有序的，可以对元素进行索引。

```
>>> vendors = ('Cisco', 'Juniper', 'Arista')
>>> print (vendors)
('Cisco', 'Juniper', 'Arista')
>>> print (vendors[1])
Juniper
>>> vendors[2] = 'Huawei'
Traceback (most recent call last):
 File "<stdin>", line 1, in <module>
TypeError: 'tuple' object does not support item assignment  #不能对元组做任何形式的修改
>>>
```

下面介绍与元组有关的方法和函数。

◎　index()

元组的 index()与列表用法相同，都用来查询指定元素的索引号。**index()的返回值为整数**，举例如下。

```
>>> vendors
('Cisco', 'Juniper', 'Arista')
>>> vendors.index('Cisco')
0
>>>
```

◎　count()

元组的 count()与列表用法相同，都用来查询指定元素在元组中的数量。**count()的返回值为整数**，举例如下。

```
>>> vendors = ('Cisco', 'Juniper', 'Arista', 'Cisco')
>>> vendors.count('Cisco')
2
>>> vendors.count('Juniper')
1
>>>
```

3. 空值

空值是比较特殊的数据类型，它没有自带的函数和方法，也无法做任何算术和逻辑运算，但是可以被赋值给一个变量，举例如下。

```
>>> type(None)
<type 'NoneType'>
>>> None == 100
False
>>> a = None
>>> print (a)
None
```

空值（None）较常用在判断语句和正则表达式中。对于网络工程师来说，日常工作中需要经常使用显示命令（show 或者 display）来对网络设备进行排错或者查询网络信息，通常这类显示命令都会给出很多回显内容，而大多数时候我们只需要关注其中的一两项参数即可。如果用 Python 来实现网络运维自动化，则需要使用正则表达式来告诉 Python 应该抓取哪一个“关键词”（即我们想要的参数）；而空值则可以用来判断“关键词”是否抓取成功。关于判断语句和正则表达式的用法将会在第 3 章中详细介绍。

第 3 章

Python 进阶语法

前两章分别介绍了 Python 在 Windows 和 Linux 里的安装和使用方法，详细讲解了网络工程师需要掌握的 Python 数据类型及每种数据类型自带的函数和方法的用法。本章将讲解 Python 中的条件（判断）语句、循环语句、文本文件的读写、自定义函数、模块、正则表达式及异常处理等网络工程师需要掌握的 Python 进阶语法知识。

3.1 条件（判断）语句

在 Python 中，条件语句（Conditional Statements）又称为判断语句，判断语句由 if、 elif 和 else 3 种语句组成，其中 if 为强制语句，可以独立使用，elif 和 else 为可选语句，并且不能独立使用。判断语句配合布尔值，通过判断一条或多条语句的条件是否成立（True 或者 False），从而决定下一步的动作，如果判断条件成立（True），则执行 if 或 elif 语句下的代码；如果判断条件不成立（False），则执行 else 语句下的代码；如果没有 else 语句，则不做任何事情。

布尔值是判断语句不可或缺的部分，在基本语法中讲到的比较运算符、逻辑运算符，以及字符串自带的 startswith()、endswith()、isdigit()、isalpha()等方法，还有下面将会讲到的成员运算符等都会返回布尔值。下面就举例讲解它们各自在 Python 判断语句中的应用场景。

3.1.1 通过比较运算符作判断

在讲布尔类型时，我们已经提到与布尔值 True 和 False 息息相关的各种比较运算符，包括等于号==、不等于号!=、大于号>、小于号<、大于等于号>=和小于等于号<=，因为使用比较运算符后会直接返回布尔值，所以比较运算符在判断语句中会经常被用到，举例如下。

首先我们用脚本模式写一段代码。

```
[root@CentOS-Python ~]#cat lab.py

#coding=utf-8
final_score = input('请输入你的 CCNA 考试分数:')
if int(final_score) > 811:
    print ('恭喜你通过考试。')
elif int(final_score) == 811:
    print ('恭喜你压线通过考试。')
else:
    print ('成绩不及格。')
```

这段代码用来让用户输入自己的 CCNA 考试成绩并作判断，假设及格线为 811 分，如果用户所得分数大于 811 分，则打印“恭喜你通过考试。”；如果分数刚好等于 811 分，则打印“恭喜你压线通过考试。”；如果低于 811 分，则打印“成绩不及格。”。这段代码需要注意以下几点。

- 写在 if、elif 和 else 下的代码都做了**代码缩进（Indentation）**，也就是 print()函数的前面保留了 4 个空格。不同于 C、C++、Java 等语言，**Python 要求严格的代码缩进**，目的是让代码工整并且具有可读性，方便阅读和修改。缩进不一定必须是 4 个空格，两个空格或者 8 个空格都是允许的，目前最常见的是 4 个空格的缩进。
- if、elif 和 else 语句的结尾必须接冒号，这点需要注意。
- 使用 input()函数让用户输入自己的分数，并把它赋值给 final_score 变量。在第 2 章里讲过，input()函数的返回值是字符串，因为要与 811 这个整数做比较，所以需要通过 int()函数先将 final_score 从字符串转换为整数。
- 与 if 和 elif 语句不同，else 后面不需要再给任何判断条件。

运行这段脚本看效果。

```
[root@CentOS-Python ~]#python3.10 lab.py
请输入你的 CCNA 考试分数:1000
恭喜你通过考试。
[root@CentOS-Python ~]#python3.10 lab.py
请输入你的 CCNA 考试分数:811
恭喜你压线通过考试。
[root@CentOS-Python ~]#python3.10 lab.py
请输入你的 CCNA 考试分数:700
```

```
成绩不及格。
[root@CentOS-Python ~]#
```

3.1.2　通过字符串方法+逻辑运算符作判断

当使用 input()函数让用户输入内容时，你无法保证用户输入的内容合乎规范。比如你给用户 6 个选项，每个选项分别对应一个动态路由协议的名称（选项 1：RIP；选项 2：IGRP；选项 3：EIGRP；选项 4：OSPF；选项 5：ISIS；选项 6：BGP），提示用户输入路由协议的选项号码来查询该路由协议的类型，然后让 Python 根据用户输入的选项号码告诉用户该路由协议属于链路状态路由协议、距离矢量路由协议，还是路径适量路由协议。

这里你无法保证用户输入的肯定是整数，即使用户输入的是整数，也无法保证输入的是 1～6 的数字。因为 input()函数的返回值是字符串，所以可以首先使用字符串的 isdigit()函数来判断用户输入的内容是否为整数，这是判断条件之一。然后通过 int()将该字符串数字转换成整数，继续判断该整数是否介于 1 和 6 之间（包含 1 和 6），这是判断条件之二。再将这两个判断条件通过逻辑运算符 and 来判断它俩是否同时成立。如果成立，则返回相应的答案；如果不成立，则提示用户输入的内容不符合规范并终止程序。代码如下。

```
[root@CentOS-Python ~]#cat lab.py

#coding=utf-8

print ('''请根据对应的号码选择一个路由协议：
1. RIP
2. IGRP
3. EIGRP
4. OSPF
5. ISIS
6. BGP ''')

option = input('请输入你的选项(数字 1-6)：')
if option.isdigit() and 1 <= int(option) <= 6:
    if option == '1' or option == '2' or option == '3':
        print ('该路由协议属于距离矢量路由协议。')
    elif option == '4' or option == '5':
        print ('该路由协议属于链路状态路由协议。')
    else:
```

```
        print ('该路由协议属于路径矢量路由协议。')
else:
    print ('选项无效，程序终止。')
```

这里我们用到了**嵌套 if 语句**。在某一个条件成立（判定为 True）后，如果还需要检查其他子条件，就可以用嵌套 if 语句来完成。在嵌套 if 语句中，一组 if、elif 和 else 可以构造在另一组 if、elif 和 else 中，不过需要注意缩进。

运行代码，测试效果如下。

```
[root@CentOS-Python ~]#python3.10 lab.py

请根据对应的号码选择一个路由协议:
1. RIP
2. IGRP
3. EIGRP
4. OSPF
5. ISIS
6. BGP
请输入你的选项(数字 1-6): 1
该路由协议属于距离矢量路由协议。

[root@CentOS-Python ~]#python lab.py
请根据对应的号码选择一个路由协议:
1. RIP
2. IGRP
3. EIGRP
4. OSPF
5. ISIS
6. BGP
请输入你的选项(数字 1-6): 4
该路由协议属于链路状态路由协议。

[root@CentOS-Python ~]#python lab.py
请根据对应的号码选择一个路由协议:
1. RIP
2. IGRP
3. EIGRP
4. OSPF
```

```
5. ISIS
6. BGP
请输入你的选项(数字 1-6)：6
该路由协议属于路径矢量路由协议。

[root@CentOS-Python ~]#python lab.py
请根据对应的号码选择一个路由协议：
1. RIP
2. IGRP
3. EIGRP
4. OSPF
5. ISIS
6. BGP
请输入你的选项(数字 1-6)：abc
选项无效，程序终止。

[root@CentOS-Python ~]#python lab.py
请根据对应的号码选择一个路由协议：
1. RIP
2. IGRP
3. EIGRP
4. OSPF
5. ISIS
6. BGP
请输入你的选项(数字 1-6)：8
选项无效，程序终止。
[root@CentOS-Python ~]#
```

3.1.3　通过成员运算符作判断

成员运算符用于判断是否可以在给定的一组字符串、列表、字典和元组中找到一个给定的值或变量，如果能找到，则返回布尔值 True；如果找不到，则返回布尔值 False。**成员运算符有两种：in 和 not in**，举例如下。

```
>>> netdevops = '网络工程师需要学习 Python 吗？'
>>> 'Python' in netdevops
True
>>> 'Java' in netdevops
```

```
False
>>> 'C++' not in netdevops
True
>>>
>>> interfaces = ['Gi1/1', 'Gi1/2', 'Gi1/3', 'Gi1/4', 'Gi1/5']
>>> 'Gi1/1' in interfaces
True
>>> 'Gi1/10' in interfaces
False
>>> 'Gi1/3' not in interfaces
False
>>>
```

依靠成员运算符，我们还可以将类似上一节的脚本简化。上一节的脚本给出的选项只有 6 种，我们尚且能够使用 if option == '1' or option == '2' or option == '3':此类的方法来列举所需的选项，但是如果所需选项超过 20 个甚至上百个，那么再使用这种方法岂不是太笨了？这时可以使用第 2 章讲的 range()函数配合 list()函数创造一个整数列表，然后配合成员运算符来判断用户输入的选项号码是否存在于该整数列表中。

按照这种思路，我们将上一节的脚本做如下简化。

```
[root@CentOS-Python ~]#cat lab.py
#coding=utf-8

print ('''请根据对应的号码选择一个路由协议：
1. RIP
2. IGRP
3. EIGRP
4. OSPF
5. ISIS
6. BGP ''')

option = input('请输入你的选项(数字 1-6)： ')
if option.isdigit() and int(option) in list(range(1, 7)):
   if int(option) in list(range(1, 4)):
       print ('该路由协议属于距离矢量路由协议。')
   elif int(option) in list(range(4, 6)):
```

```
        print ('该路由协议属于链路状态路由协议。')
    else:
        print ('该路由协议属于路径矢量路由协议。')
else:
    print ('选项无效，程序终止。')

[root@CentOS-Python ~]#
```

之前依靠比较运算符+逻辑运算符作判断的 if option == '1' or option == '2' or option == '3': 方法已经被成员运算符+range()函数简化为 if int(option) in list(range(1, 4)):了。同理，判断选项在整数 1 ~ 6 之间也已经通过成员运算符配合 range()和 list()函数写成 int(option) inrange(1, 7)了。需要注意的是，由于 input()函数返回的是字符串，因此需要把变量 option 先通过 int()函数转换为整数才能使用成员运算符 in 来判断它是否存在于 range()和 list()函数所创建的整数列表中。

3.2　循环语句

Python 中最常用的循环语句（Looping Statements）有两种：while 和 for。除此之外，还有文件迭代器（File Iterator）、列表解析式（List Comprehension）等循环工具，不过对于网络工程师来说，用得最多的还是 while 和 for，因此本节将只讲解这两种循环语句。

3.2.1　while 语句

在 Python 中，while 语句用于循环执行一段程序，它和 if 语句一样，两者都离不开判断语句和缩进。每当写在 while 语句下的程序被执行一次，程序就会自动回到“顶上”（也就是 while 语句的开头部分），根据 while 后的判断语句的返回值来决定是否要再次执行该程序，如果判断语句的返回值为 True，则继续执行该程序，一旦判断语句的返回值为 False，则该 while 循环随即终止，如此反复。如果需要中途强行中止 while 循环，则需要使用 break 语句。下面通过 3 个例子帮助大家理解。

◎　**例 1**

```
>>> a = 1
>>> b = 10
>>> while a < b:
...     print (a)
```

```
...     a += 1
...
1
2
3
4
5
6
7
8
9
>>>
```

在上面的代码中，我们用 while 循环来判断变量 a 是否小于 b，如果判断结果为 True，则打印出变量 a 的值，并且每次都让 a 的值加 1，如此反复循环，直到第 10 次执行该 while 循环，a = 10，因为 a < b 不再成立（10<10 不成立），程序随即终止。

◎ **例 2**

```
>>> vendors = ['Cisco', 'Huawei', 'Juniper', 'Arista', 'HPE', 'Extreme']
>>> while len(vendors) > 0:
...     vendors.pop()
...     print (vendors)
...
'Extreme'
['Cisco', 'Huawei', 'Juniper', 'Arista', 'HPE']
'HPE'
['Cisco', 'Huawei', 'Juniper', 'Arista']
'Arista'
['Cisco', 'Huawei', 'Juniper']
'Juniper'
['Cisco', 'Huawei']
'Huawei'
['Cisco']
'Cisco'
[]
>>>
```

在上面的代码中，我们用 while 循环配合 len()函数来判断列表 vendors 的长度是否大于 0，如果判断结果为 True，则用 pop()方法从列表中删掉一个元素，并且随即打印列表里剩余的元素。最终当列表中所有的元素都被移除时，列表的长度为 0，while len(vendors)>0:的返回值为 False，该 while 循环也就随即终止。

◎　**例 3**

在 3.1.2 节和 3.1.3 节的案例代码中，一旦用户输入的选项不符合规范，程序就会立即中止，用户必须再次手动运行一次脚本重新输入选项，这样显得很笨拙。借助 while 循环，可以不断地重复执行 input()函数，直到用户输入正确的选项号码。优化后的脚本代码如下。

```
#coding=utf-8

print ('''请根据对应的号码选择一个路由协议：
1. RIP
2. IGRP
3. EIGRP
4. OSPF
5. ISIS
6. BGP ''')

while True:
    option = input('请输入你的选项(数字 1-6): ')
    if option.isdigit() and int(option) in list(range(1, 7)):
        if int(option) in list(range(1, 4)):
            print ('该路由协议属于距离矢量路由协议。')
        elif int(option) in list(range(5, 7)):
            print ('该路由协议属于链路状态路由协议。')
        else:
            print ('该路由协议属于路径矢量路由协议。')
        break
    else:
        print ('选项无效，请再次输入。')

[root@CentOS-Python ~]#
```

这里我们使用了 **while True。while True 是一种很常见的 while 循环的用法，因为这里的判定条件的结果已经手动指定了 True，意味着判定条件将永久成立，也就意味着 while 下面的程序将会被无数次重复执行**，从而引起“无限循环”（Indefinite Loop）的问题。为了避免无限循环，我们必须在程序代码中使用 break 语句来终止 while 循环，注意 break 在上面的代码里的位置，带着这个问题去思考，你会更加明白缩进在 Python 中的重要性。

执行代码看效果。

```
[root@CentOS-Python ~]#python lab.py
请根据对应的号码选择一个路由协议：
1. RIP
2. IGRP
3. EIGRP
4. OSPF
5. ISIS
6. BGP
请输入你的选项(数字 1-6): 7
选项无效，请再次输入。
请输入你的选项(数字 1-6): a
选项无效，请再次输入。
请输入你的选项(数字 1-6): ..!!!
选项无效，请再次输入。
请输入你的选项(数字 1-6): 134ui134lkadjl
选项无效，请再次输入。
请输入你的选项(数字 1-6): 1
该路由协议属于距离矢量路由协议。

[root@CentOS-Python ~]#
```

3.2.2　for 语句

同为循环语句，for 语句的循环机制和 while 语句完全不同：while 语句需要配合判断语句来决定什么时候开始循环和中止循环，而 for 语句则用来遍历一组可迭代的序列，可迭代的序列包括字符串、列表、元组等。在将这些序列中的元素遍历完后，for 语句的循环也随即终止。for 语句的基本语法格式如下。

```
for item in sequence:
    statements
```

这里的 sequence 为可迭代的序列（如字符串、列表、元组），而 item 可以理解为该序列里的每个元素（item 名称可以任意选取），statements 则是循环体（将要循环的程序部分）。

举例如下。

◎　**例 1**

```
>>> for letter in 'Python':
...  print (letter)
```

```
...
P
y
t
h
o
n
>>>
```

我们用 letter 作为 for 语句中的 item 来遍历字符串 “Python”，并将该字符串中的元素依次全部打印出来，得到 P、y、t、h、o、n。

◎ **例 2**

```
>>> sum = 0
>>> for number in range(1, 6):
...     sum = sum + number
...     print (sum)
...
1
3
6
10
15
>>>
```

我们将 0 赋值给变量 sum，然后用 number 作为 for 语句中的 item 来遍历 range(1,6)，返回 1～5 的 5 个整数，将这 5 个整数依次与 sum 累加，每累加一次用 print()函数打印出结果，最后得到 1、3、6、10、15。

◎ **例 3**

```
>>> routing_protocols = ['RIP', 'IGRP', 'EIGRP', 'OSPF', 'ISIS', 'BGP']
>>> link_state_protocols = ['OSPF', 'ISIS']
>>> for protocols in routing_protocols:
...     if protocols not in link_state_protocols:
...         print (protocols + '不属于链路状态路由协议：' )
RIP 不属于链路状态路由协议：
IGRP 不属于链路状态路由协议：
EIGRP 不属于链路状态路由协议：
```

```
BGP 不属于链路状态路由协议:
>>>
```

我们分别创建两个列表：routing_protocols 和 link_state_protocols，列表中的元素是对应的路由协议和链路状态路由协议。首先用 protocols 作为 for 语句中的 item 来遍历第一个列表 routing_protocols，然后使用 if 语句来判断哪些 protocols 不属于第二个列表 link_state_protocols 中的元素，并将它们打印出来。

正如前面讲到的，上述 3 个例子中写在 for 后面的 letter、number 和 protocols 代表将要遍历的可迭代序列里的每一个元素（即 item 名称），**它们的名称可以由用户随意制定**，比如在例 1 中，我们把 letter 换成 a 也没问题。

```
>>> for a in 'Python':
...  print (a)
...
P
y
t
h
o
n
```

通常建议取便于理解的 item 名称，像 a 这类的 item 名称在做实验或者练习时可以偷懒使用，在实际的工作代码中用这种毫无意义的 item 名称肯定是会被人诟病的。

3.3　文本文件的读/写

在日常网络运维中，网络工程师免不了要和大量的文本文件打交道，比如用来批量配置网络设备的命令模板文件，存放所有网络设备 IP 地址的文件，以及备份网络设备 show run 输出结果之类的配置备份文件。正因如此，知道如何使用 Python 来访问和管理文本文件是学习网络运维自动化技术的网络工程师必须掌握的一项 Python 知识点。

3.3.1　open()函数及其模式

在 Python 中，我们可以通过 open()函数来访问和管理文本文件，open()函数用来打开一个文本文件并创建一个文件对象（File Object），通过文件对象自带的多种函数和方法，可以对文本文件执行一系列访问和管理操作。在讲解这些函数和方法之前，首先创建一个名为 test.txt 的测试文本文件，该文件包含 5 个网络设备厂商的名字，内容如下。

```
[root@CentOS-Python ~]#cat test.txt
Cisco
Juniper
Arista
H3C
Huawei
[root@CentOS-Python ~]#
```

然后用 open()函数访问该文件。

```
>>> file = open('test.txt', 'r')
```

我们通过 open()函数的 r 模式（只读）访问 test.txt 文件，并返回一个文件对象，再将该文件对象赋值给 file 变量。**r（reading）是默认访问模式**，除此之外，open()函数还有很多其他文件访问模式。这里只介绍网络工程师最常用的几种模式，如下表所示。

模　式	作　用
r	以只读方式打开文件，r 模式只能打开已存在的文件。如果文件不存在，则会报错
w	打开文件并只用于写入。如果文件已经存在，则原有内容将被删除覆盖；如果文件不存在，则创建新文件
a	以追加方式打开文件。如果文件已经存在，则原有内容不会被删除覆盖，新内容将添加在原有内容后面；如果文件不存在，则创建新文件
r+	以读写方式打开文件，r+模式只能打开已存在的文件。如果文件不存在，则会报错
w+	以读写方式打开文件。如果文件已经存在，则原有内容将被删除覆盖；如果文件不存在，则创建新文件
a+	以读写方式打开文件。如果文件已经存在，则原有内容不会被删除覆盖，新内容将添加在原有内容后面；如果文件不存在，则创建新文件

网络工程师必须熟练掌握 open()函数的上述 6 种模式，关于它们的具体使用将在下一节中举例讲解。

3.3.2　文件读取

在使用 open()函数创建文件对象之后，我们并不能马上读取文件里的内容。如下所示，在创建了文件对象并将它赋值给 file 变量后，如果用 print()函数将 file 变量打印出来，则只会得到文件名称、open()函数的访问模式及该文件对象在内存中的位置（0x7fa194215660）等信息。

```
>>> file = open('test.txt', 'r')
>>> print (file)
<open file 'test.txt', mode 'r' at 0x7fa194215660>
>>>
```

要想读取文件里的具体内容，我们还需要用 read()、readline()或者 readlines() 3 种方法中的一种。因为这 3 种方法都和读取有关，因此 open()函数中只允许写入的 w 模式和只允许追加的 a 模式不支持它们，而其他 4 种模式则都没有问题，举例如下。

```
#w 模式不支持 read(), readline(), readlines()
>>> file = open('test.txt', 'w')
>>> print (file.read())
Traceback (most recent call last):
  File "<stdin>", line 1, in <module>
io.UnsupportedOperation: not readable
>>>

#a 模式也不支持 read(), readline(), readlines()
>>> file = open('test.txt', 'a')
>>> print (file.readline())
Traceback (most recent call last):
  File "<stdin>", line 1, in <module>
io.UnsupportedOperation: not readable
>>>
```

read()、readline()和 readlines()是学习 open()函数的重点内容，三者的用法和差异很大，其中 readlines()更是重中之重（原因后面会讲到），网络工程师必须熟练掌握。下面对这 3 种函数一一进行讲解。

1. read()

read()方法读取文本文件里的全部内容，**返回值为字符串**。

```
>>> file = open('test.txt')
>>> print (file.read())
Cisco
Juniper
Arista
```

```
H3C
Huawei

>>> print (file.read())
>>>
```

我们尝试连续两次打印 test.txt 文件的内容，第一次打印出的内容没有任何问题，**为什么第二次打印的时候内容为空了呢**？这是因为在使用 read()方法后，**文件指针的位置从文件的开头移动到了末尾**，要想让文件指针回到开头，必须使用 seek()函数，方法如下。

```
>>> file.seek(0)
0
>>> file.tell()
0
>>> print (file.read())
Cisco
Juniper
Arista
H3C
Huawei

>>> file.tell()
32
>>>
```

我们用 seek(0)将文件指针从末尾移回开头，并且用 tell()方法确认文件指针的位置（文件开头的位置为 0），随后使用 read()方法打印文件内容并成功，之后再次使用 tell()方法确认文件指针的位置，可以发现指针现在已经来到文件末尾处（32）。这个 32 是怎么得来的？下面我们去掉 print()函数，再次通过 read()方法来读取一次文件内容。

```
>>> file.seek(0)
0
>>> file.read()
'Cisco\nJuniper\nArista\nH3C\nHuawei\n'
>>>
```

去掉 print()函数的目的是能清楚地看到换行符\n，如果这时从左往右数，则会发现 Cisco(5) + \n(1) + Juniper(7) + \n(1) + Arista(6) + \n(1) + H3C(3) + \n(1) + Huawei(6) + \n(1) = 32，这就解

释了为什么在文件指针移动到文件末尾后，tell()方法返回的文件指针的位置是 32。文件指针的位置及 seek()和 tell()方法的用法是文本文件访问和管理中很重要但又容易被忽略的知识点，网络工程师务必熟练掌握。

2. readline()

readline()与 read()的区别是它不会像 read()那样把文本文件的所有内容一次性都读完，而是会一排一排地去读。**readline()的返回值也是字符串**。举例如下。

```
>>> file = open('test.txt')
>>> print (file.readline())
Cisco

>>> print (file.readline())
Juniper

>>> print (file.readline())
Arista

>>> print (file.readline())
H3C

>>> print (file.readline())
Huawei

>>> print (file.readline())

>>>
```

readline()方法每次返回文件的一排内容，顺序由上至下（这里出现的空排部分是因为换行符的缘故）。另外，文件指针会跟随移动直到文件末尾，因此最后一个 print (file.readline())的返回值为空。

3. readlines()

readlines()与前两者最大的区别是它的返回值不再是字符串，而是列表。可以说 readlines()是 read()和 readline()的结合体，首先它同 read()一样把文本文件的所有内容都读完。另外，它会像 readline()那样一排一排地去读，并将每排的内容以列表元素的形式返回，举例如下。

```
>>> file = open('test.txt')
>>> print (file.readlines())
['Cisco\n', 'Juniper\n', 'Arista\n', 'H3C\n', 'Huawei\n']
>>> file.seek(0)
>>> devices = file.readlines()
>>> print (devices[0])
Cisco

>>> print (devices[1])
Juniper

>>> print (devices[2])
Arista

>>> print (devices[3])
H3C

>>> print (devices[4])
Huawei

>>>
```

同 read()和 readline()一样，使用一次 readlines()后，文件指针会移动到文件的末尾。为了避免每次重复使用 seek(0)的麻烦，可以将 readlines()返回的列表赋值给一个变量，即 devices，之后便可以使用索引号来一个一个地验证列表里的元素。**注意 readlines()返回的列表里的元素都带换行符\n，这与我们手动创建的普通列表是有区别的。**

同 read()和 readline()相比，笔者认为 readlines()应该是网络运维中使用频率最高的一种读取文本文件内容的方法，因为它的返回值是列表，通过列表我们可以做很多事情。举个例子，现在有一个名为 ip.txt 的文本文件，该文件保存了一大堆没有规律可循的交换机的管理 IP 地址，具体如下。

```
[root@CentOS-Python ~]#cat ip.txt
172.16.100.1
172.16.30.1
172.16.41.1
172.16.10.1
172.16.8.1
172.16.112.1
172.16.39.1
172.16.121.1
```

```
172.16.92.1
172.16.73.1
192.168.54.1
192.168.32.1
192.168.2.1
10.3.2.1
10.58.23.3
192.168.230.29
10.235.21.42
192.168.32.32
10.4.3.3
172.16.30.2
172.16.22.30
172.16.111.33
[root@CentOS-Python ~]#
```

现在需要回答 3 个问题：

（1）怎么使用 Python 来确定该文件有多少个 IP 地址？

（2）怎么使用 Python 来找出该文件中的 B 类 IP 地址（172.16 开头的 IP 地址），并将它们打印出来？

（3）怎么使用 Paramiko 来批量 SSH 登录这些交换机修改或查看配置（后面会讲到 Paramiko）？

答案（1）：最好的方法是使用 open()函数的 readlines()来读取该文件的内容，因为 readlines()的返回值是列表，可以方便我们使用 len()函数来判断该列表的长度，从而得到该文件中所包含 IP 地址的数量，举例如下。

```
>>> f = open('ip.txt')
>>> print (len(f.readlines()))
22
>>>
```

仅仅通过两行代码，就得到了结果：22 个 IP 地址，readlines()返回的列表是不是很方便？

答案（2）：最好的方法依然是使用 readlines()来完成，因为 readlines()返回的列表中的元素的数据类型是字符串，可以使用 for 循环来遍历所有的字符串元素，然后配合 if 语句，通过字符串的 startswith()函数判断这些 IP 地址是否以 172.16 开头。如果是，则将它们打印出来，举例如下。

```
>>> f = open('ip.txt')
>>> for ip in f.readlines():
...     if ip.startswith('172.16'):
...             print (ip)
...
172.16.100.1

172.16.200.1

172.16.130.1

172.16.10.1

172.16.8.1

172.16.112.1

172.16.39.1

172.16.121.1

172.16.92.1

172.16.73.1

172.16.30.2

172.16.22.30

172.16.111.33
>>>
```

需要注意的是，因为 readlines()返回的列表中的元素是带换行符\n 的，所以打印出来的每个 B 类 IP 地址之间都空了一排，影响阅读和美观，解决方法也很简单，只需要使用第 2 章讲过的 strip()函数去掉换行符即可，举例如下。

```
>>> f = open('ip.txt')
>>> for ip in f.readlines():
...     if ip.startswith('172.16'):
```

```
...             print (ip.strip())
...
172.16.100.1
172.16.200.1
172.16.130.1
172.16.10.1
172.16.8.1
172.16.112.1
172.16.39.1
172.16.121.1
172.16.92.1
172.16.73.1
172.16.30.2
172.16.22.30
172.16.111.33
>>>
```

答案（3）：同答案（2）一样，因为 readlines()返回的是列表，我们可以通过 for 循环来一一访问该列表中的每个元素，也就是交换机的每个 IP 地址，进而达到使用 Paramiko 来一一登录每个交换机做配置的目的，该脚本的代码如下。

```
import Paramiko

username = input('Username: ')
password = input('Password: ')
f = open('ip.txt')

for ip in f.readlines():
    ssh_client = Paramiko.SSHClient()
    ssh_client.set_missing_host_key_policy(Paramiko.AutoAddPolicy())
    ssh_client.connect(hostname=ip, username=username, password=password)
    print ("Successfully connect to ", ip)
```

- 关于 Paramiko 的用法在第 4 章会讲到，这里看不懂没关系。
- 这里通过使用 for 循环配合 readlines()返回列表的方式来访问 ip.txt 文件中的交换机管理 IP 地址。

- 该 for 循环会尝试一一登录 ip.txt 的所有交换机 IP 地址，每成功登录一个交换机都随即打印信息“"Successfully connect to ", ip”来提醒用户登录成功。

open()函数的 readlines()在网络运维中的用处远不止这 3 点，这里只是给出了几个比较典型的实用例子。再次强调，每一位学习 Python 的网络工程师都必须熟练掌握它。

3.3.3　文件写入

在使用 open()函数创建文件对象后，我们可以使用 **write()函数**来对文件写入数据。顾名思义，既然 write()函数与文件写入相关，**那么只允许只读的 r 模式并不支持它**，而其他 5 种模式则不受限制（包括 r+模式），举例如下。

```
>>> f = open('test.txt', 'r')
>>> f.write()
Traceback (most recent call last):
  File "<stdin>", line 1, in <module>
TypeError: write() takes exactly one argument (0 given)
>>>
```

write()函数在 r+、w/w+、a/a+这 5 种模式中的应用讲解如下。

1. r+

在 r+模式下使用 write()函数，新内容会添加在文件的开头部分，而且会覆盖开头部分原来已有的内容，举例如下。

```
#文本修改前的内容
[root@CentOS-Python ~]#cat test.txt
Cisco
Juniper
Arista
H3C
Huawei

[root@CentOS-Python ~]#

#在 r+模式下使用 write()函数修改文本内容
>>> f = open('test.txt', 'r+')
>>> f.write('Avaya')
```

```
>>> f.close()

#文本修改后的内容
[root@CentOS-Python ~]#cat test.txt
Avaya
Juniper
Arista
H3C
Huawei

[root@CentOS-Python ~]#
```

可以看到，文本开头的 Cisco 已经被 Avaya 覆盖了。这里注意使用 write()函数对文本写入新内容后，必须再用 close()方法将文本关闭，这样新写入的内容才能被保存。

2. w/w+

在 w/w+模式下使用 write()函数，新内容会添加在文件的开头部分，已存在的文件的内容将会完全被清空，举例如下。

```
#文本修改前的内容
[root@CentOS-Python ~]#cat test.txt
Avaya
Juniper
Arista
H3C
Huawei

[root@CentOS-Python ~]#

#在 w 模式下使用 write()函数修改文本内容
>>> f = open('test.txt', 'w')
>>> f.write('test')
>>> f.close()

#文本修改后的内容
[root@CentOS-Python ~]#cat test.txt
test
```

```
[root@CentOS-Python ~]#

#在 w+模式下使用 write()函数修改文本内容
>>> f = open('test.txt', 'w+')
>>> f.write('''Cisco
... Juniper
... Arista
... H3C
... Huawei\n''')
>>> f.close()

#文本修改后的内容
[root@CentOS-Python ~]#cat test.txt
Cisco
Juniper
Arista
H3C
Huawei

[root@CentOS-Python ~]#
```

3. a/a+

在 a/a+模式下使用 write()函数，新内容会添加在文件的末尾部分，已存在的文件的内容将不会被清空，举例如下。

```
#文本修改前的内容
[root@CentOS-Python ~]#cat test.txt
Cisco
Juniper
Arista
H3C
Huawei

[root@CentOS-Python ~]#

#在 a 模式下使用 write()函数修改文本内容
>>> f = open('test.txt', 'a')
>>> f.write('Avaya')
```

```
>>> f.close()

#文本修改后的内容
[root@CentOS-Python ~]#cat test.txt
Cisco
Juniper
Arista
H3C
Huawei
Avaya

[root@CentOS-Python ~]#

#在 a+模式下使用 write()函数修改文本内容
>>> f = open('test.txt', 'a+')
>>> f.write('Aruba')
>>> f.close()

#文本修改后的内容
[root@CentOS-Python ~]#cat test.txt
Cisco
Juniper
Arista
H3C
Huawei
Avaya
Aruba

[root@CentOS-Python ~]#
```

3.3.4　with 语句

每次用 open()函数打开一个文件，该文件都将一直处于打开状态。这一点我们可以用 closed 方法来验证：如果文件处于打开状态，则 closed 方法返回 False；如果文件已被关闭，则 closed 方法返回 True。

```
>>> f = open('test.txt')
>>> f.closed
```

```
False
>>> f.close()
>>> f.closed
True
>>>
```

这种每次都要手动关闭文件的做法略显麻烦，可以使用 with 语句来管理文件对象。用 with 语句打开的文件将被自动关闭，举例如下。

```
>>> with open('test.txt') as f:
...     print f.read()
...
Cisco
Juniper
Arista
H3C
Huawei

>>> f.closed
True
>>>
```

这里没有使用 close()来关闭文件，因为 with 语句已经自动将文件关闭（closed 方法的返回值为 True）。

3.4　自定义函数

函数是已经组织好的可以被重复使用的一组代码块，它的作用是用来提高代码的重复使用率。在 Python 中，有很多内建函数（Built-in Function），比如前面已经讲到的 type()、dir()、print()、int()、str()、list()、open()等，在安装好 Python 后就能立即使用。除了上述内建函数，我们也可以通过创建**自定义函数**（**User-Defined Function**）来完成一些需要重复使用的代码块，提高工作效率。

3.4.1　函数的创建和调用

在 Python 中，我们使用 def 语句来自定义函数。def 语句后面接函数名和括号()，括号里根据情况可带参数也可不带参数。在自定义函数创建好后，要将该函数调用才能得到函数的输出结果（即使该函数不带参数），举例如下。

```
#带参数的自定义函数
>>> def add(x, y):
...     result = x + y
...     print (result)
...
>>> add(1, 2)
3
>>>

#不带参数的自定义函数
>>> def name():
...     print ('Parry')
...
>>>
>>> name()
Parry
>>>
```

不管自定义函数是否带参数，**函数都不能在创建前就被调用**，比如下面这段用来求一个数的二次方的脚本。

```
[root@CentOS-Python ~]#cat test.txt
square(10)

def square(x):
 squared = x ** 2
 print (squared)

[root@CentOS-Python ~]#python3.10 test.txt
Traceback (most recent call last):
  File "test.txt", line 1, in <module>
    square(10)
NameError: name 'square' is not defined
[root@CentOS-Python ~]#
```

运行该脚本后报错，原因就是在还没有创建函数的情况下提前调用了该函数，正确写法如下。

```
[root@CentOS-Python ~]#cat test.txt

def square(x):
 squared = x ** 2
 print squared

square(10)

[root@CentOS-Python ~]#python test.txt
100
[root@CentOS-Python ~]#
```

3.4.2　函数值的返回

任何函数都需要返回一个值才有意义。自定义函数可以用 print 和 return 两种语句来返回一个值，**如果函数中没有使用 print 或 return，则该函数只会返回一个空值（None）**，举例如下。

```
>>> def add_1(x):
...   x = x+1
>>> print add_1(1)
None
>>>
```

◎　print 和 return 的区别

print 用来将返回值打印输出在控制端上，以便让用户看到，但是该返回值不会被保存下来。也就是说，如果将来把该函数赋值给一个变量，该变量的值将仍然为空值（None），举例如下。

```
[root@CentOS-Python ~]#cat test.txt
def name():
   print ('Parry')

name()
a = name()
print (a)

[root@CentOS-Python ~]#python3.10 test.txt
```

```
Parry
Parry
None
[root@CentOS-Python ~]#
```

注意这里返回了两个 Parry,是因为除了 name()直接调用函数,将函数赋值给一个变量时,比如这里的 a = name(), 也会触发调用函数的效果。

而 return 则恰恰相反，在调用函数后，return 的返回值不会被打印输出在控制端上，如果想看输出的返回值，则要在调用函数时在函数前加上 print。**但是返回值会被保存下来，如果把该函数赋值给一个变量，则该变量的值即该函数的返回值，**举例如下。

```
[root@CentOS-Python ~]#cat test.txt
def name():
    return 'Parry'

name()
a = name()
print (a)

[root@CentOS-Python ~]#python test.txt
Parry
[root@CentOS-Python ~]#
```

3.4.3　嵌套函数

函数支持嵌套，也就是一个函数可以在另一个函数中被调用，举例如下。

```
[root@CentOS-Python ~]#cat test.txt

def square(x):
    result = x ** 2
    return result

def cube(x):
    result = square(x) * x
      return result

cube(3)
```

```
print (cube(3))

[root@CentOS-Python ~]#python test.txt
27
```

我们首先创建一个求 2 次方的函数 square(x)（注意这里用的是 return，不是 print，否则返回值将会是 None，不能被其他函数套用）。然后创建一个求 3 次方的函数 cube(x)，在 cube(x)中我们套用了 square(x)，并且也使用了 return 来返回函数的结果。最后分别使用了 cube(3)和 print(cube(3))来展示和证明 3.4.2 节中提到的"在调用函数后，return 的返回值不会被打印输出在控制端上，如果想看输出的返回值，则要在调用函数时在函数前面加上 print()函数"，因为这里只能看到一个输出结果 27。

3.5　模块

在第 1 章已经讲过，Python 的运行模式大致分为两种：一种是使用解释器的交互模式，另一种是运行脚本的脚本模式。在交互模式下，一旦退出解释器，那么之前定义的变量、函数及其他所有代码都会被清空，一切都需要从头开始。因此，如果想写一段较长、需要重复使用的代码，则最好使用编辑器将代码写进脚本，以便将来重复使用。

不过在网络工程师的日常工作中，随着网络规模越来越大，需求越来越多，相应的代码也将越写越多。为了方便维护，我们可以把其中一些常用的自定义函数分出来写在一个独立的脚本文件中，然后在交互模式或脚本模式下将该脚本文件导入（import）以便重复使用，这种在交互模式下或其他脚本中被导入的脚本被称为**模块（Module）**。在 Python 中，我们使用 **import** 语句来导入模块，**脚本的文件名（不包含扩展名.py）即模块名。**

被用作模块的脚本中可能带自定义函数，也可能不带，下面将分别举例说明。

3.5.1　不带自定义函数的模块

首先创建一个脚本，将其命名为 script1.py，该脚本的代码只有一行，即打印内容"这是脚本 1."。

```
[root@CentOS-Python ~]#cat script1.py
#coding=utf-8
print ("这是脚本 1.")
[root@CentOS-Python ~]#
```

然后创建第二个脚本，将其命名为script2.py。在脚本2里，我们将使用import语句导入脚本1（import script1），打印内容“这是脚本2.”，并运行脚本2。

```
[root@CentOS-Python ~]#cat script2.py
#coding=utf-8
import script1
print ("这是脚本2.")

#运行script2
[root@CentOS-Python ~]#python3.10 script2.py
这是脚本1.
这是脚本2.
[root@CentOS-Python ~]#
```

可以看到，在运行脚本2后，我们同时得到了“这是脚本1.”和“这是脚本2.”的打印输出内容，其中，“这是脚本1.”正是脚本2通过import script1导入脚本1后得到的。

3.5.2　带自定义函数的模块

首先修改脚本1的代码，创建一个test()函数，该函数的代码只有一行，即打印内容“这是带函数的脚本1.”。

```
[root@CentOS-Python ~]#cat script1.py
#coding=utf-8
def test():
   print ("这是带函数的脚本1.")
[root@CentOS-Python ~]#
```

然后修改脚本2的代码，调用脚本1的test()函数，因为是脚本1的函数，所以需要在函数名前加入模块名，即script1.test()。

```
[root@CentOS-Python ~]#cat script2.py
#coding=utf-8
import script1
print ("这是脚本2.")
script1.test()

#运行script2
[root@CentOS-Python ~]#python script2.py
```

```
这是脚本 2.
这是带函数的脚本 1.
```

3.5.3　Python 内建模块和第三方模块

除了上述两种用户自己创建的模块，Python 还有内建模块及需要通过 pip 下载安装的第三方模块（也叫作第三方库），下面分别讲解。

1. Python 内建模块

Python 有大量的内建模块直接通过 import 就可以使用，后面实验部分的案例代码将会重点讲解这些内建模块的使用。这里仅举一例，我们可以使用 **os** 这个内建函数来发送 Ping 包，判断网络目标是否可达，脚本代码及解释如下。

```
import os
hostname = 'www.cisco.com'
response = os.system("ping -c 1 " + hostname)
if response == 0:
   print (hostname + ' is reachable.')
else:
   print (hostname + ' is not reachable.')
```

- os 是很常用的 Python 内建模块，os 是 operating system 的简称。顾名思义，它是用来与运行代码的主机操作系统互动的，后面实验部分的案例代码会介绍关于 os 在网络运维中的许多用法。
- 这里我们使用 os 模块的 system()函数来直接在主机里执行 ping -c 1 http://www.cisco.com 这条命令，也就是向 http://www.cisco.com 发送 1 个 Ping 包。注意，我们做代码演示的主机是 CentOS，CentOS 中指定 Ping 包个数的参数为-c。如果在 Windows 中运行这段脚本，因为 Windows 中指定 Ping 包个数的参数为-n，则需要将代码里的参数-c 换成-n，也就是"ping -n 1 " + hostname！
- **os.system()的返回值为一个整数。**如果返回值为 0，则表示目标可达；如果为非 0，则表示不可达。我们将 os.system()的返回值赋给变量 response，然后写一个简单的 if...else 判断语句来作判断。如果 response 的值等于 0，则打印目标可达的信息，反之则打印目标不可达的信息。

执行代码看效果。

```
[root@CentOS-Python ~]#python3.10 test.py
PING e2867.dsca.akamaiedge.net (104.86.224.155) 56(84) bytes of data.
64 bytes from a104-86-224-155.deploy.static.akamaitechnologies.com (104.86.224.155):
icmp_seq=1 ttl=128 time=115 ms

--- e2867.dsca.akamaiedge.net ping statistics ---
1 packets transmitted, 1 received, 0% packet loss, time 0ms
rtt min/avg/max/mdev = 115.206/115.206/115.206/0.000 ms
www.cisco.com is reachable.
[root@CentOS-Python ~]#
```

除了 os，能实现 ping 命令功能的 Python 内建模块还有很多，如 subprocess。当在 Python 中使用 os 和 subprocess 模块来执行 ping 命令时，两者都有个"小缺陷"，就是它们都会显示操作系统执行 ping 命令后的回显内容，如果想让 Python"静悄悄"地 ping，则可以使用第三方模块 pythonping，关于 pythonping 模块的用法在实验部分中将会提到。

2. Python 第三方模块

Python 第三方模块需要从 pip 下载安装，pip 随 Python 版本的不同有对应的 pip2、pip3 和 pip3.10，在 CentOS 8 里已经内置了 pip2 和 pip3，我们在第 1 章介绍 Python 3.10 在 Windows 和 Linux 的安装时已经安装了 pip3.10，可以在 Windows 命令行及 CentOS 8 下输入 pip3.10 来验证，如下图所示。

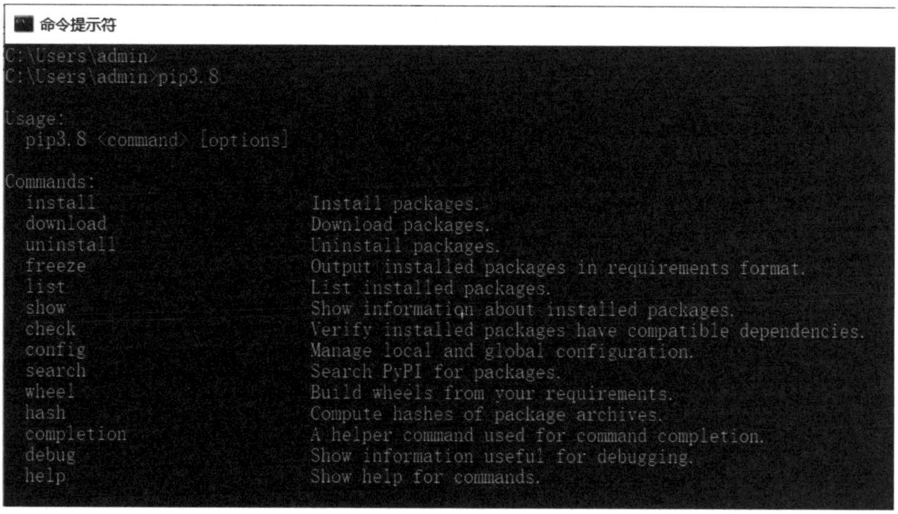

```
命令提示符
C:\Users\admin>
C:\Users\admin>pip3.8

Usage:
  pip3.8 <command> [options]

Commands:
  install                     Install packages.
  download                    Download packages.
  uninstall                   Uninstall packages.
  freeze                      Output installed packages in requirements format.
  list                        List installed packages.
  show                        Show information about installed packages.
  check                       Verify installed packages have compatible dependencies.
  config                      Manage local and global configuration.
  search                      Search PyPI for packages.
  wheel                       Build wheels from your requirements.
  hash                        Compute hashes of package archives.
  completion                  A helper command used for command completion.
  debug                       Show information useful for debugging.
  help                        Show help for commands.
```

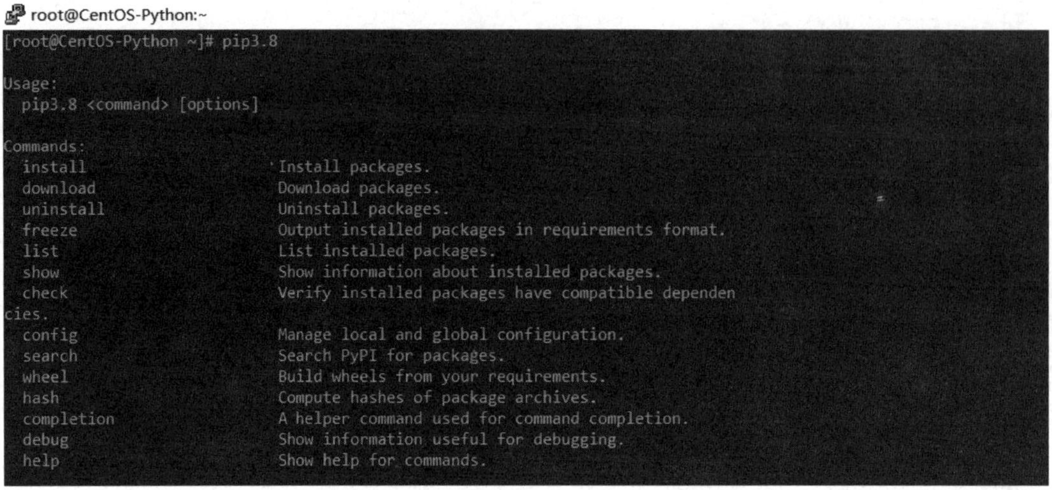

```
root@CentOS-Python:~
[root@CentOS-Python ~]# pip3.8

Usage:
  pip3.8 <command> [options]

Commands:
  install                     Install packages.
  download                    Download packages.
  uninstall                   Uninstall packages.
  freeze                      Output installed packages in requirements format.
  list                        List installed packages.
  show                        Show information about installed packages.
  check                       Verify installed packages have compatible dependen
cies.
  config                      Manage local and global configuration.
  search                      Search PyPI for packages.
  wheel                       Build wheels from your requirements.
  hash                        Compute hashes of package archives.
  completion                  A helper command used for command completion.
  debug                       Show information useful for debugging.
  help                        Show help for commands.
```

对于网络工程师来说，最常用的 Python 第三方模块无疑是用来 SSH 登录网络设备的 Paramiko 和 Netmiko。首先使用命令 pip3.10 install Paramiko 和 pip3.10 install Netmiko 来分别安装它们（如果你的 Python 不是 3.10.x 版本，而是其他 Python 3 版本，则将 pip3.10 替换成 pip3 即可）。在安装之前，请确认你的 CentOS 8 主机或虚拟机能够连上外网。

```
[root@CentOS-Python ~]#pip3.10 install Paramiko
[root@CentOS-Python ~]#pip3.10 install netmiko
```

注：如果你使用的是 Python 2，则需要使用 pip2 install Netmiko == 2.4.2 来安装 Netmiko，因为 2020 年 1 月后，所有通过 pip 安装的 Netmiko 都默认只支持 Python 3。

使用 pip 安装好 Paramiko 和 Netmiko 后，打开 Python 测试是否可以使用 import Paramiko 和 import Netmiko 来引用它，如果没报错，则说明安装成功。

```
>>> import paramiko
>>> import netmiko
Traceback (most recent call last):
  File "<stdin>", line 1, in <module>
ModuleNotFoundError: No module named 'Netmiko'
>>>
```

由上面的代码可以看到，import Paramiko 后没有任何问题，但是 import Netmiko 却收到解释器报错“ModuleNotFoundError: No module named 'Netmiko'”，这是因为为了效果演示，笔者故意只安装了 Paramiko 而没有安装 Netmiko。关于 Paramiko 模块的具体用法，后面将会详细介绍，这里仅做演示来说明如何在 Python 中通过 pip 安装第三方模块。

3.5.4　from…import…

对于带自定义函数的模块，除了 import 语句，我们还可以使用 from…import…来导入模块。**它的作用是省去了在调用模块函数时必须加上模块名的麻烦**，该语句具体的格式为 **from [模块名] import [函数名]**，举例如下。

将脚本 2 的代码修改如下，然后运行脚本 2。

```
[root@CentOS-Python ~]#cat script2.py
#coding=utf-8
from script1 import test
print ("这是脚本 2.")
test()

#运行 script2
[root@CentOS-Python ~]#python script2.py
这是脚本 2.
这是带函数的脚本 1.
```

因为我们在脚本 2 中使用了 from script1 import test，所以在调用脚本 1 的 test()函数时不再需要写成 script1.test()，直接用 test()即可。

3.5.5　if __name__ == '__main__':

在 3.5.1 节所举的例子中，脚本 2 在导入了模块脚本 1 后，立即就引用了脚本 1 的代码，即 print ("这是脚本 1.")。有时我们只是希望使用所导入模块中的部分函数，并不希望一次性全部引入模块中的代码内容，这时就可以用 **if __name__ == '__main__':**这个判断语句来达到目的。

在基本语法中已经讲过，Python 中前后带双下画线的变量叫作内置变量，如__name__，关于内置变量的内容已经超出了本书的范围。我们可以这么来理解：**如果一个 Python 脚本文件中用到了 if __name__ == '__main__':判断语句，则所有写在其下面的代码都将不会在该脚本被其他脚本用作模块导入时被执行。**

我们首先将脚本 1 和脚本 2 的代码分别修改如下，在脚本 1 中将 print ("这是脚本 1.")写在 if __name__ == '__main__':的下面。

```
[root@CentOS-Python ~]#cat script1.py
#coding=utf-8
if __name__ == '__main__':
   print ("这是脚本 1.")
[root@CentOS-Python ~]#

[root@CentOS-Python ~]#cat script2.py
#coding=utf-8
import script1
print ("这是脚本 2.")
[root@CentOS-Python ~]#
```

然后运行脚本 2，发现虽然脚本 2 导入了脚本 1，但是输出结果中却不再出现“这是脚本 1.”的输出结果。

```
[root@CentOS-Python ~]#python3.10 script2.py
这是脚本 2.
[root@CentOS-Python ~]#
```

3.6 正则表达式

在网络工程师的日常工作中，少不了要在路由器、交换机、防火墙等设备的命令行中使用各种 show 或者 display 命令来查询配置、设备信息或者进行排错，比如思科和 Arista 设备上最常见的 show run、show log、show interface 和 show ip int brief 等命令。通常这些命令输出的信息和回显内容过多，需要用管道符号|（Pipeline）配合 grep （Juniper 设备）或者 include/exclude/begin（思科设备）等命令来匹配或筛选我们所需要的信息。举个例子，要查询一台 24 口的思科 2960 交换机当前有多少个端口是 up 的，可以使用命令 show ip int brief | i up，代码如下。

```
2960#show ip int b | i up
FastEthernet0/2 unassigned YES unset up up
FastEthernet0/10  unassigned YES unset up up
FastEthernet0/11  unassigned YES unset up up
FastEthernet0/12  unassigned YES unset up up
FastEthernet0/15  unassigned YES unset up up
FastEthernet0/23  unassigned YES unset up up
FastEthernet0/24  unassigned YES unset up up
```

```
GigabitEthernet0/2  unassigned YES unset up up
Loopback0 unassigned YES NVRAM up up
```

管道符号后面的命令部分（i up）即本节将要讲的**正则表达式（Regular Expression）**。另外，如果你是 CCIE 或者是拥有数年从业经验、负责过大型企业网或运营商网络运维的资深网络工程师，那么应该不会对 BGP（Border Gateway Protocol，边界网关协议）中出现的自治域路径访问控制列表（as-path access-list）感到陌生。as-path access-list 是正则表达式在计算机网络中最典型的应用，比如下面的 as-path access-list 1 中的^4$就是一个标准的正则表达式。

```
ip as-path access-list 1 permit ^4$
```

3.6.1　什么是正则表达式

根据维基百科的解释：

正则表达式（Regular Expression，在代码中常简写为 regex、regexp 或 RE），又称**正规表示式**、**正规表示法**、**正规表达式**、**规则表达式**、**常规表示法**，是计算机科学的一个概念。正则表达式使用单个字符串来描述、匹配一系列匹配某个句法规则的字符串。在很多文本编辑器中，正则表达式通常被用来检索、替换那些匹配某个模式的文本。

在 Python 中，我们使用正则表达式来对文本内容做解析（Parse），即从一组给定的字符串中通过给定的字符来搜索和匹配我们需要的“模式”（Pattern，**可以把“模式”理解成我们要搜索和匹配的“关键词”**）。正则表达式本身并不是 Python 的一部分，它拥有自己独特的语法及独立的处理引擎，效率上可能不如字符串自带的各种方法，但它的功能远比字符串自带的方法强大得多，因为它同时支持**精确匹配（Exact Match）**和**模糊匹配（Wildcard Match）**。

3.6.2　正则表达式的验证

怎么知道自己的正则表达式是否写正确了呢？方法很简单，可以通过在线正则表达式模拟器来校验自己写的正则表达式是否正确。在线正则表达式模拟器很多，通过搜索引擎很容易找到。在线正则表达式模拟器的使用也很简单，通常只需要提供文本内容（即字符串内容），以及用来匹配模式的正则表达式即可。下图的 regex101 就是笔者常用的在线正则表达式模拟器，后面在举例讲解正则表达式时，笔者会提供在该模拟器上验证的截图来佐证，如下图所示。

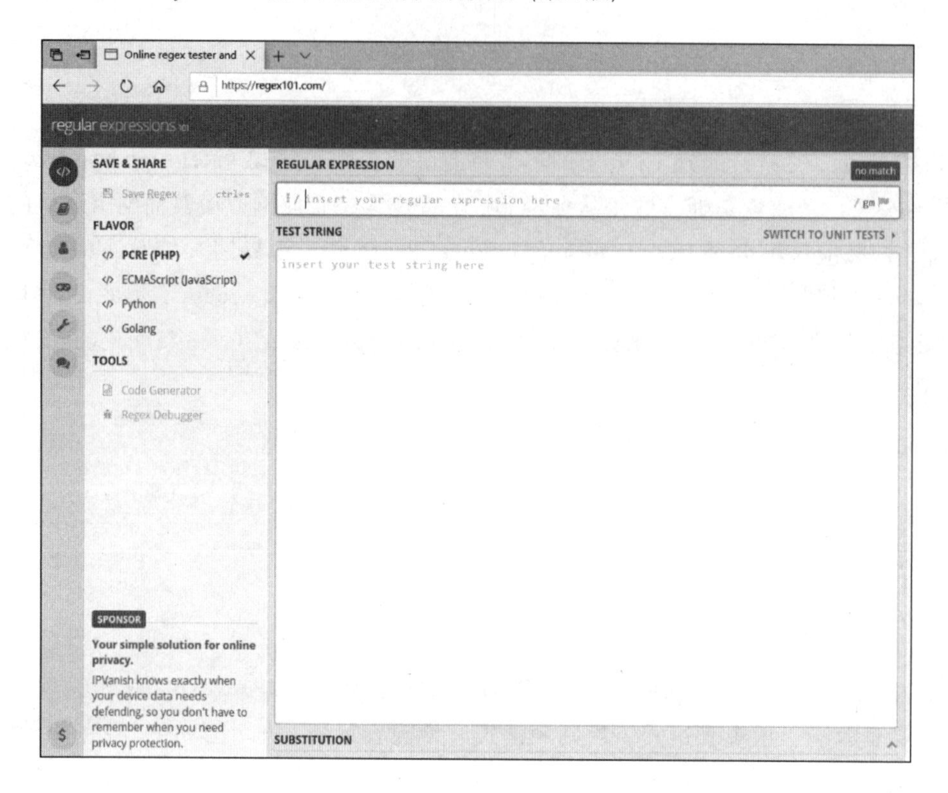

3.6.3 正则表达式的规则

正则表达式是一套十分强大但也十分复杂的字符匹配工具，本节只选取其中部分网络运维中常用并且适合网络工程师学习的知识点，然后配合案例讲解。首先来了解正则表达式的**精确匹配和模糊匹配**，其中模糊匹配又包括**匹配符号（Matching Characters）**和**特殊序列（Special Sequence）**。

1. 精确匹配

精确匹配即明文给出我们想要匹配的模式。比如上面讲到的在 24 口的思科 2960 交换机里查找 up 的端口，我们就在管道符号|后面明文给出模式 up。又比如我们想在下面的交换机日志中找出所有日志类型为%LINK-3-UPDOWN 的日志，那我们按照要求明文给出模式%LINK-3-UPDOWN 即可，即命令 show logging | i %LINK-3- UPDOWN。

```
2960#show logging | i %LINK-3-UPDOWN
000458: Feb 17 17:10:31.906: %LINK-3-UPDOWN: Interface FastEthernet0/2, changed state
to down
```

但是如果同时有多种日志类型需要匹配，代码如下。

```
000459: Feb 17 17:10:35.202: %LINK-3-UPDOWN: Interface FastEthernet0/2, changed state to up
000460: Feb 17 17:10:36.209: %LINEPROTO-5-UPDOWN: Line protocol on Interface FastEthernet0/2, changed state to up
000461: Feb 17 22:39:26.464: %SSH-5-SSH2_SESSION: SSH2 Session request from 10.1.1.1 (tty = 0) using crypto cipher 'aes128-cbc', hmac 'hmac-sha1' Succeeded
000462: Feb 17 22:39:27.748: %SSH-5-SSH2_USERAUTH: User 'test' authentication for SSH2 Session from 10.1.1.1 (tty = 0) using crypto cipher 'aes128-cbc', hmac 'hmac-sha1' Succeeded
```

这时，再用精确匹配就显得很笨拙，因为要通过命令 show logging | i %LINK-3-UPDOWN|%LINEPROTO-5-UPDOWN|%SSH-5-SSH2_SESSION|%SSH-5-SSH2_USERAUTH 把所有感兴趣的日志类型都明文列出来，这里只有 4 种日志类型还比较容易，如果有几十上百种日志类型去匹配，再进行精确匹配的工作量就会很大，这时我们需要借助模糊匹配来完成这项任务。

2. 模糊匹配

模糊匹配包括匹配符号和特殊序列，下面分别讲解。

正则表达式中常见的匹配符号如下表所示。

匹配符号	用　法
.	匹配除换行符外的所有字符，匹配次数为 1 次
*	用来匹配紧靠该符号左边的符号，匹配次数为 0 次或多次
+	用来匹配紧靠该符号左边的符号，匹配次数为 1 次或多次
?	用来匹配紧靠该符号左边的符号，匹配次数为 0 次或 1 次
{m}	用来匹配紧靠该符号左边的符号，指定匹配次数为 *m* 次。例如字符串'abbbbcccd'，使用 ab{2}将匹配到 abb，使用 bc{3}d 将匹配到 bcccd
{m,n}	用来匹配紧靠该符号左边的符号，指定匹配次数为最少 *m* 次，最多 *n* 次。例如，如果字符串为'abbbbcccd'，使用 ab{2,3}将只能匹配到 abbb；如果字符串为'abbbcccdabbccd'，使用 ab{2,3}将能同时匹配到 abbb 和 abb；如果字符串内容为'abcd'，使用 ab{2,3}将匹配不到任何东西
{m,}	用来匹配紧靠该符号左边的符号，指定匹配次数最少为 *m* 次，最多无限次
{,n}	用来匹配紧靠该符号左边的符号，指定匹配次数最少为 0 次，最多为 *n* 次

续表

匹配符号	用　法
\	转义字符，用来匹配上述“匹配符号”。例如字符串内容中出现了问号？，而又想精确匹配这个问号，那就要用\？来进行匹配。除此之外，\也用来表示一个特殊序列，特殊序列将在下节讲到
[]	表示字符集合，用来精确匹配。例如要精确匹配一个数字，可以使用[0-9]；如果要精确匹配一个小写字母，可以用[a-z]；如果要精确匹配一个大写字母，可以用[A-Z]；如果要匹配一个数字、字母或者下画线，可以用[0-9a-zA-Z_]。另外，在[]中加^表示取非，比如[^0-9]表示匹配一个非数字的字符，[^a-z]表示匹配一个非字母的字符，依此类推
|	表示“或匹配”（两项中匹配其中任意一项），比如要匹配 FastEthernet 和 GigabitEthernet 这两种端口名，可以写作 Fa|Gi
(…)	组合，匹配括号内的任意正则表达式，并表示组合的开始和结尾，例如(b|cd)ef 表示匹配 bef 或 cdef

3. 贪婪匹配

*、+、?、{m}、{m，}和{m，n}这 6 种匹配符号默认都是贪婪匹配的，即会尽可能多地去匹配符合条件的内容。

假设给定的一组字符串为“xxzyxzyz”，我们使用正则表达式模式 x.*y 来做匹配（注：**精确匹配和模糊匹配可以混用**）。在匹配到第一个“x”后，开始匹配.*，**因为.和*默认是贪婪匹配的，它会一直往后匹配，直到匹配到最后一个**“y”，因此匹配结果为“xxzyxzy”，可以在 regex101 上验证，如下图所示。

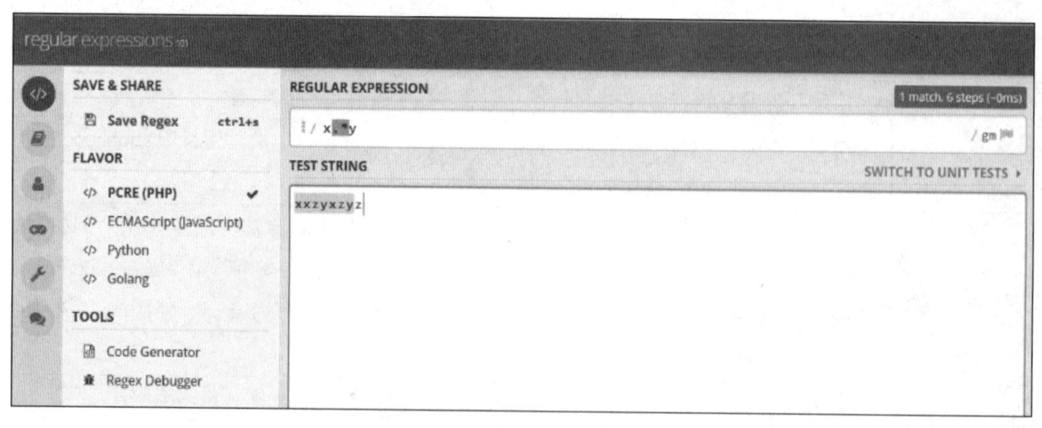

又假设给定的字符串依然为“xxzyxzyz”，我们使用正则表达式模式 xz.*y 来做匹配，在匹配到第一个“xz”后，开始匹配“贪婪”的.*，这里将会一直往后匹配，直到最后一个“y”，因此匹配结果为“xzyxzy”，在 regex101 上验证，如下图所示。

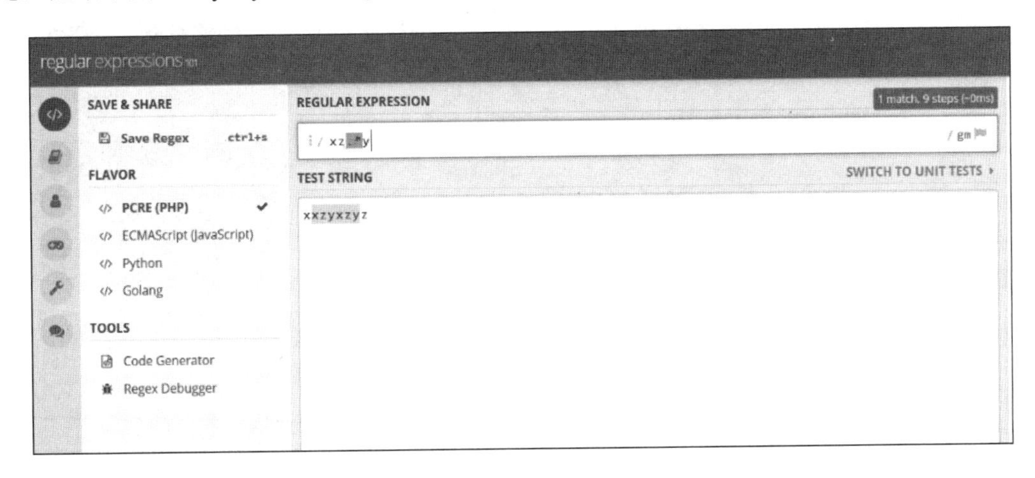

4. 非贪婪匹配

要实现非贪婪匹配很简单，就是在上述 6 种贪婪匹配符号后面加上问号?即可，即*?、+?、??、{m}?、{m，}?和{m，n}?。

假设给定的另一组字符串为“xxzyzyz”（注意不是之前的“xxzyxzyz”），我们使用正则表达式模式 x.*?y 来做匹配。在匹配到第一个“x”后，开始匹配.*?，**因为.*?是非贪婪匹配的，它在匹配到第一个“y”后便随即停止**，因此匹配结果为“xxzy”，在 regex101 上验证，如下图所示。

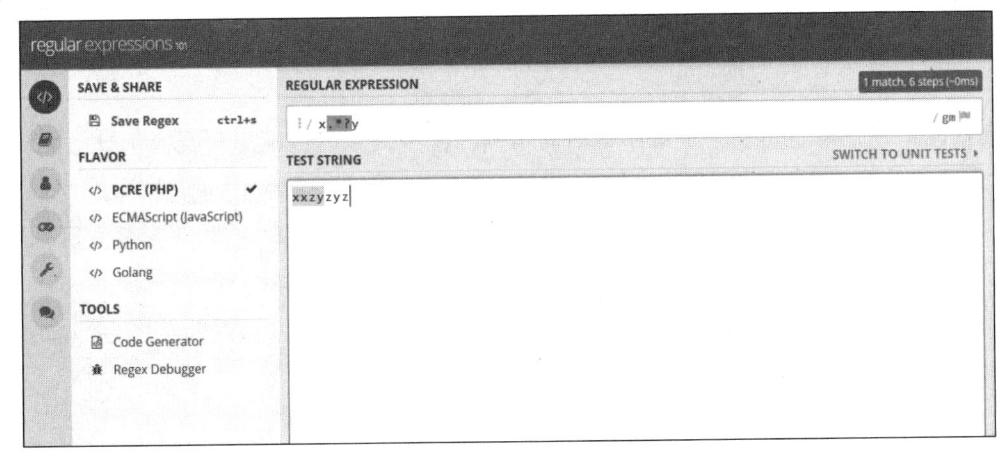

又假设给定的字符串依然为“xxzyzyz”，我们使用正则表达式模式 xz.*?y 来做匹配，在匹配到第一个“xz”后，开始匹配.*?，它在匹配到第一个“y”后便随即停止，因此匹配结果为“xzy”，在 regex101 上验证，如下图所示。

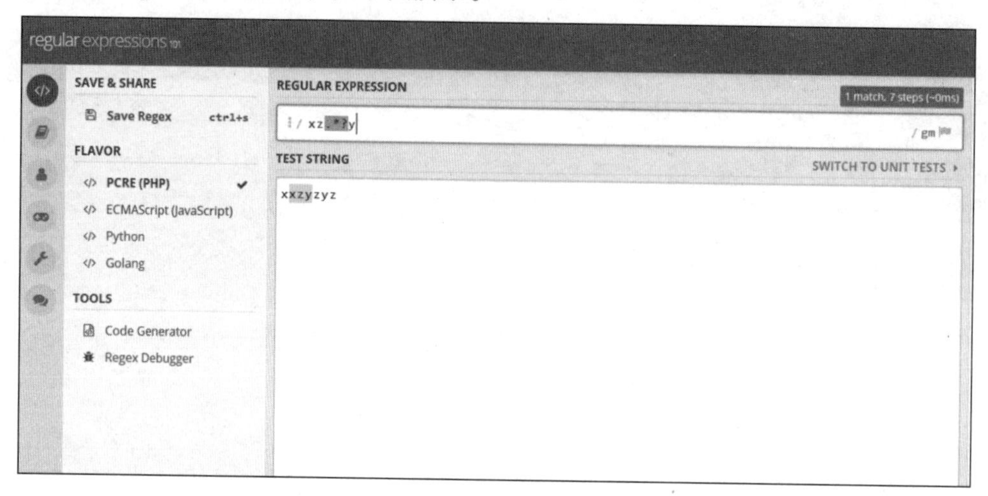

◎　正则表达式中常见的特殊序列

特殊序列由转义符号\和一个字符组成，常见的特殊序列及其用法如下表所示。

特殊序列	用　法
\d	匹配任意一个十进制数字，等价于[0-9]
\D	\d 取非，匹配任意一个非十进制数字，等价于[^0-9]
\w	匹配任意一个字母、十进制数字及下画线，等价于[a-zA-Z0-9]
\W	\w 取非，等价于[^a-zA-Z0-9]
\s	匹配任意一个空白字符，包括空格、换行符\n
\S	\s 取非，匹配任意一个非空白字符

模糊匹配在正则表达式中很常用，前面精确匹配中提到的匹配思科交换机日志类型的例子可以用模糊匹配来处理，比如我们要在下面的日志中同时匹配 %LINK-3-UPDOWN、%LINEPROTO-5-UPDOWN、%SSH-5-SSH2_SESSION 和 %SSH-5-SSH2_USERAUTH 4 种日志类型，用正则表达式**%\w{3,9}-\d-\w{6,13}**即可完全匹配。

```
000459: Feb 17 17:10:35.202: %LINK-3-UPDOWN: Interface FastEthernet0/2, changed state to up
000460: Feb 17 17:10:36.209: %LINEPROTO-5-UPDOWN: Line protocol on Interface FastEthernet0/2, changed state to up
```

```
000461: Feb 17 22:39:26.464: %SSH-5-SSH2_SESSION: SSH2 Session request from 10.1.1.1
(tty = 0) using crypto cipher 'aes128-cbc',  hmac 'hmac-sha1' Succeeded
000462: Feb 17 22:39:27.748: %SSH-5-SSH2_USERAUTH: User 'test' authentication for SSH2
Session from 10.1.1.1 (tty = 0) using crypto cipher 'aes128-cbc',  hmac 'hmac-sha1'
Succeeded
```

关于该正则表达式%\w{3,9}-\d-\w{4,13}是如何完整匹配上述 4 种日志类型的讲解如下。

首先将%\w{3,9}-\d-\w{4,13}拆分为 6 部分。

第 1 部分：%用来精确匹配百分号“%”（4 种日志全部以“%”开头），如下图所示。

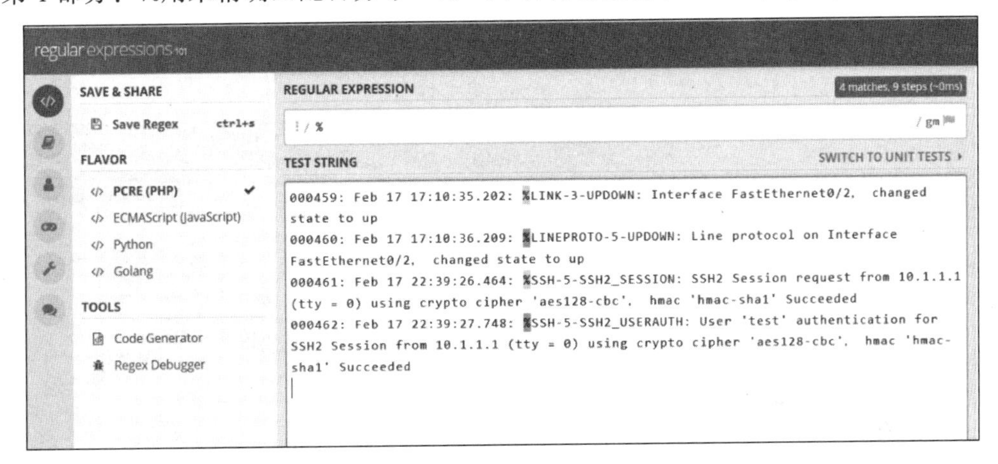

第 2 部分：\w{3,9}用来匹配“SSH”和“LINEPROTO”，如下图所示。

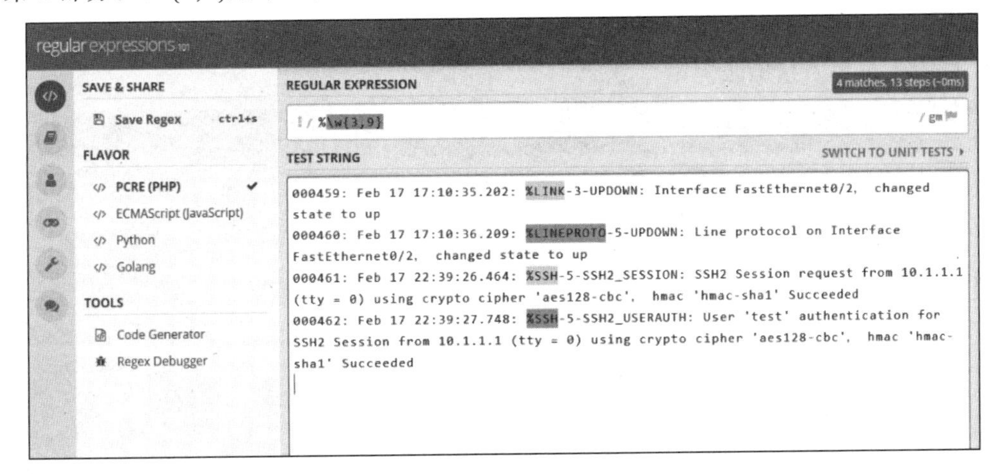

第 3 部分：-用来精确匹配第一个“-”，如下图所示。

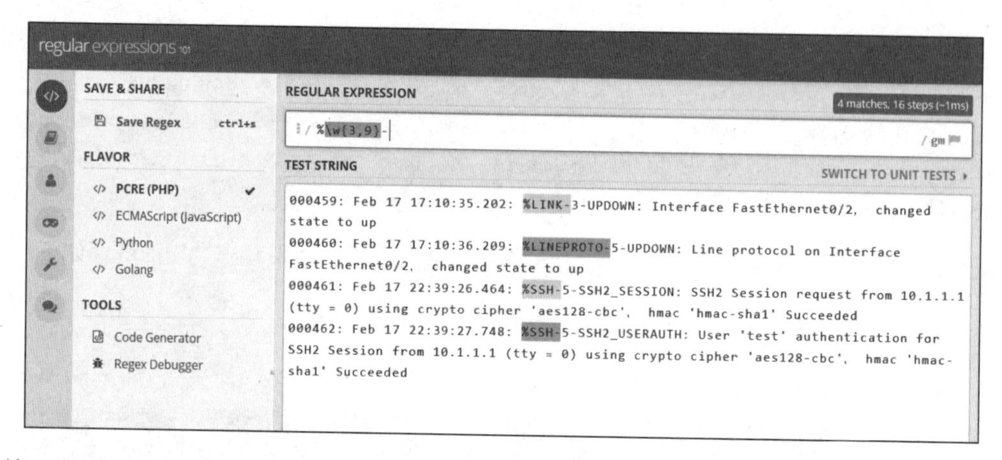

第 4 部分：\d 用来匹配数字“3”或“5”，如下图所示。

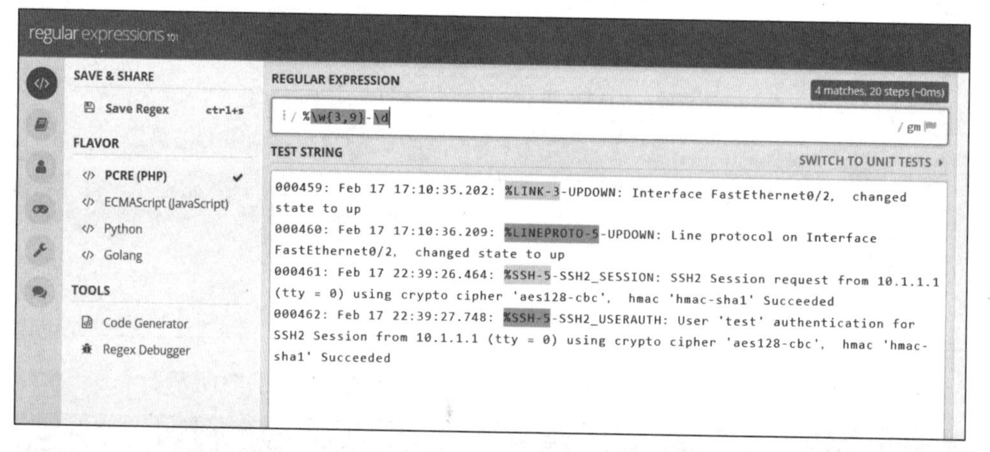

第 5 部分：-用来精确匹配第二个“-”，如下图所示。

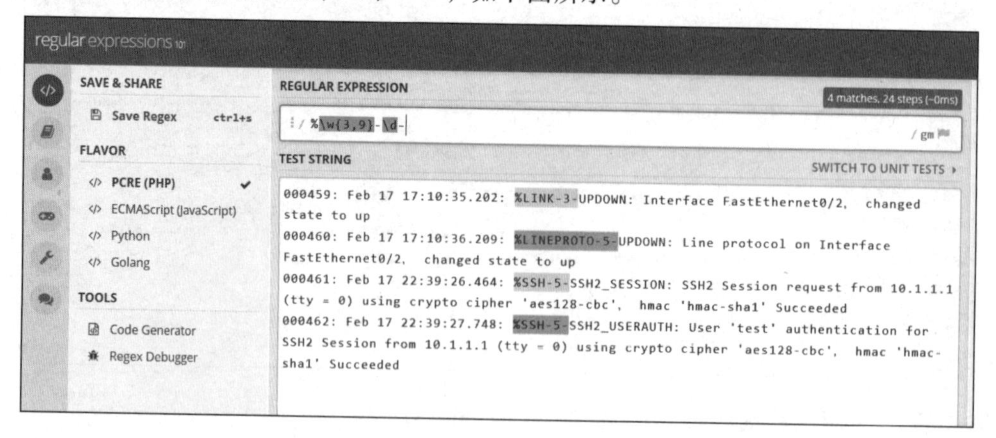

第 6 部分：由上图可知，至此还剩下“UPDOWN”“SSH2_SESSION”“SSH2_USERAUTH”3 部分需要匹配，因为\w 可以同时匹配数字、字母及下画线，因此，用\w{6,13}即可完整匹配最后这 3 部分，如下图所示。

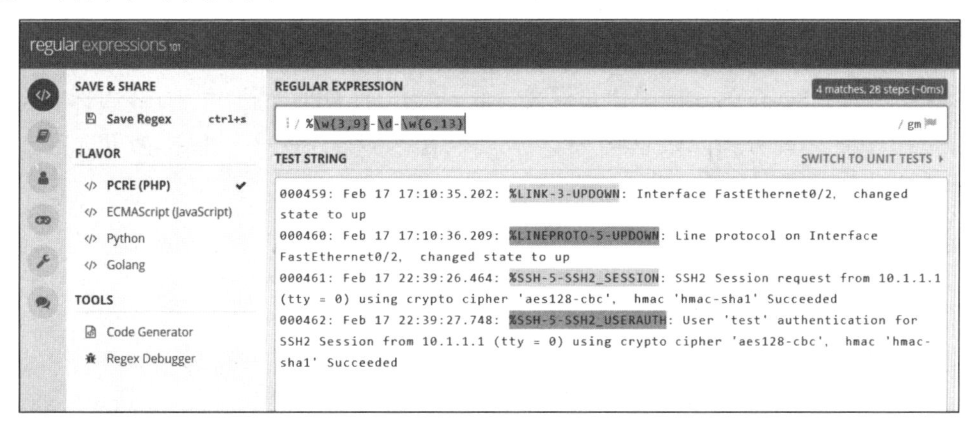

3.6.4 正则表达式在 Python 中的应用

在 Python 中，我们使用 **import re** 来导入正则表达式这个内建模块（无须使用 pip 来安装）。

```
import re
```

首先使用 dir()函数来看下 re 模块中有哪些内建函数和方法。

```
>>> import re
>>> dir(re)
['DEBUG', 'DOTALL', 'I', 'IGNORECASE', 'L', 'LOCALE', 'M', 'MULTILINE', 'S', 'Scanner',
'T', 'TEMPLATE', 'U', 'UNICODE', 'VERBOSE', 'X', '_MAXCACHE', '__all__',
'__builtins__', '__doc__', '__file__', '__name__', '__package__', '__version__',
'_alphanum', '_cache', '_cache_repl', '_compile', '_compile_repl', '_expand',
'_locale', '_pattern_type', '_pickle', '_subx', 'compile', 'copy_reg', 'error',
'escape', 'findall', 'finditer', 'match', 'purge', 'search', 'split', 'sre_compile',
'sre_parse', 'sub', 'subn', 'sys', 'template']
```

其中，网络工程师较常用的 Python 正则表达式的函数主要有 4 种，分别为 re.match()、re.search()、re.findall()和 re.sub()，下面对它们分别进行说明。

1. re.match()

re.match()函数用来在字符串的起始位置匹配指定的模式，如果匹配成功，则 **re.match()**

的返回值为匹配到的对象。如果想查看匹配到的对象的具体值，则还要对该对象调用 group() 函数。如果匹配到的模式不在字符串的起始位置，则 re.match()将返回空值（None）。

re.match()函数的语法如下。

```
re.match(pattern, string, flags=0)
```

pattern 即我们要匹配的正则表达式模式。string 为要匹配的字符串。flags 为标志位，用来控制正则表达式的匹配方式，如是否区分大小写、是否多行匹配等。flags 为可选项，不是很常用。

举例如下。

```
[root@CentOS-Python ~]#python3.10
Python 3.10.6 (default, Apr 27 2020, 23:06:10)
[GCC 8.3.1 20190507 (Red Hat 8.3.1-4)] on linux
Type "help", "copyright", "credits" or "license" for more information.
>>> import re
>>> test = 'Test match() function of regular expression.'
>>> a = re.match(r'Test', test)
>>> print (a)
<re.Match object; span=(0, 4), match='Test'>
>>> print (a.group())
Test
>>>
```

我们使用 re.match()函数，从字符串“Test match() function of regular expression.”里精确匹配模式“Test”，因为“Test”位于该段字符串的起始位置，所以匹配成功，并且返回一个匹配到的对象<re.Match object; span=(0, 4), match='Test'>（即用 print(a)看到的内容），为了查看该对象的具体值，我们可以对该对象调用 group()方法，得到具体值“Test”（即用 print (a.group()) 看到的内容），**该值的数据类型为字符串**，group()函数在 Python 的正则表达式中很常用，务必熟练使用。

如果我们不从字符串的起始位置去匹配，而是去匹配中间或末尾的字符串内容，则 re.match()将匹配不到任何内容，从而返回空值（None）。比如我们尝试匹配“function”这个词，因为“function”不在“Test match() function of regular expression.”的开头，所以 re.match() 的返回值为 None。

```
>>> import re
>>> test = 'Test match() function of regular expression.'
>>> a = re.match(r'function', test)
```

```
>>> print (a)
None
>>>
```

在上面的两个例子中，我们分别在模式“Test”和“function”的前面加上了一个 r，**这个 r 代表原始字符串（Raw String）**。在 Python 中，原始字符串主要用来处理特殊字符所产生的歧义，比如前面讲到的转义字符\就是一种特殊字符，它会产生很多不必要的歧义。举个例子，假设你要在 Windows 中用 Python 的 open()函数打开一个文件，代码如下。

```
>>> f = open('C:\Program Files\test.txt', 'r')
IOError: [Errno 2] No such file or directory: 'C:\\Program Files\test.txt'
>>>
```

这时，你会发现该文件打不开了，Python 返回了一个“IOError: [Errno 2] No such file or directory: 'C:\\Program Files\test.txt'”错误，这是因为“\t”被当作了不属于文件名的特殊符号。解决的办法也很简单，就是在代表文件所在路径的字符串前面使用原始字符串。

```
>>> f = open(r'C:\Program Files\test.txt', 'r')
>>> print f
<open file 'C:\\Program Files\\test.txt', mode 'r' at 0x0000000002A01150>
>>>
```

在正则表达式中使用原始字符串也是同样的道理，只是还有更多其他原因，对这些原因的解释已经超出了本书的范围。网络工程师只需记住：**在正则表达式中，建议使用原始字符串**。

2. re.search()

re.search()函数和 re.match()一样，返回值为**字符串，但是它比 re.match()更灵活，因为它允许在字符串的任意位置匹配指定的模式**。

re.search()函数的语法如下。

```
re.search(pattern, string, flags=0)
```

前面我们用 re.match()在“Test match() function of regular expression.”中尝试匹配“function”不成功，因为“function”不在该字符串的起始位置。我们改用 re.search()来匹配。

```
>>>import re
>>> test ='Test search() function of regular expression.'
>>> a = re.search(r'function', test)
>>> print (a)
```

```
<re.Match object; span=(14, 22), match='function'>
>>> print (a.group())
function
>>>
```

虽然 re.search()可以在字符串的任意位置匹配模式，但是它和 re.match()一样**一次只能匹配一个字符串内容**，比如下面是某台路由器上 show ip int brief 命令的回显内容，我们希望用正则表达式来匹配该回显内容中出现的所有 IPv4 地址。

```
Router#show ip int b
Interface IP-Address OK? Method Status Protocol
GigabitEthernet1/1 192.168.121.181 YES NVRAM up up
GigabitEthernet1/2 192.168.110.2  YES NVRAM up up
GigabitEthernet2/1 10.254.254.1 YES NVRAM up up
GigabitEthernet2/2 10.254.254.5 YES NVRAM up up
```

我们尝试用 re.search()来匹配。

```
>>> test = '''Router#show ip int b
Interface IP-Address OK? Method Status Protocol
GigabitEthernet1/1 192.168.121.181 YES NVRAM up up
GigabitEthernet1/2 192.168.110.2  YES NVRAM up up
GigabitEthernet2/1 10.254.254.1 YES NVRAM up up
GigabitEthernet2/2 10.254.254.5 YES NVRAM up up '''
>>> ip_address = re.search(r'\d{1, 3}\.\d{1, 3}\.\d{1, 3}\.\d{1, 3}', test)
>>> print (ip_address.group())
192.168.121.181
```

我们用正则表达式\d{1，3}\.\d{1，3}\.\d{1，3}\.\d{1，3}作为模式来匹配任意 IPv4 地址，注意我们在分割每段 IP 地址的“.”前面都加了转义符号\，如果不加\，写成\d{1，3}.\d{1，3}.\d{1，3}.\d{1，3}，则将会匹配到 GigabitEthernet1/1 中的“1/1”，原因请读者自行思考。

print (a.group())后可以看到只匹配到了 192.168.121.181 这一个 IPv4 地址，这就是 re.search()的短板。如果想匹配其他所有 IPv4 地址（192.168.110.2，10.254.254.1，10.254.254.5），则必须用下面要讲的 re.findall()。

3. re.findall()

如果字符串中有多个能被模式匹配到的关键词，并且我们希望把它们全部匹配出来，则要使用 re.findall()。同 re.match()和 re.search()不一样，**re.findall()的返回值为列表，每个被模**

式匹配到的字符串内容分别是该列表中的元素之一。

re.findall()函数的语法如下。

```
re.findall(pattern, string, flags=0)
```

还是以上面尝试匹配 show ip int brief 命令的回显内容中所有 IPv4 地址的例子为例。

```
>>> test = '''Router#show ip int b
Interface IP-Address OK? Method Status Protocol
GigabitEthernet1/1 192.168.121.181 YES NVRAM up up
GigabitEthernet1/2 192.168.110.2  YES NVRAM up up
GigabitEthernet2/1 10.254.254.1 YES NVRAM up up
GigabitEthernet2/2 10.254.254.5 YES NVRAM up up '''
>>> ip_address = re.findall(r'\d{1, 3}\.\d{1, 3}\.\d{1, 3}\.\d{1, 3}', test)
>>> type(ip_address)
<type 'list'>
>>> print (ip_address)
['192.168.121.181', '192.168.110.2', '10.254.254.1', '10.254.254.5']
>>>
```

这里成功通过 re.findall()匹配到所有的 4 个 IPv4 地址，每个 IPv4 地址分别为 re.findall()返回的列表中的一个元素。

4. re.sub()

最后要讲的 **re.sub()函数是用来替换字符串里被匹配到的字符串内容的**，类似 Word 的替换功能。除此之外，它还可以定义最大替换数（Maximum Number of Replacement）来指定 sub()函数所能替换的字符串内容的数量，默认状态下为全部替换。**re.sub()的返回值是字符串**。

re.sub()函数的语法如下。

```
a = re.sub(pattern, replacement, string, optional flags)
```

re.sub()函数的语法与前面讲到的 re.match()、re.search()、re.findall() 3 个函数略有不同，re.sub()函数里多了一个 replacement 参数，它表示被替换后的字符串内容。optional flags 可以用来指定所替换的字符串内容的数量，如果只想替换其中 1 个字符串内容，则可以将 optional flags 位设为 1；如果想替换其中的前两个字符串内容，则设为 2；依此类推。如果 optional flags 位空缺，则默认状态下为全部替换。

下面以某台路由器上的 ARP 表的输出内容为例，我们用 re.sub()来将其中所有的 MAC 地址全部替换为 1234.56ab.cdef。

```
>>> test = '''
... Router#show ip arp
... Protocol Address Age (min) Hardware Addr Type Interface
... Internet 10.1.21.1     -    b4a9.5aff.c845 ARPA  TenGigabitEthernet2/1
... Internet 10.11.22.1    51   b4a9.5a35.aa84 ARPA  TenGigabitEthernet2/2
... Internet 10.201.13.17  -    b4a9.5abe.4345 ARPA  TenGigabitEthernet2/3'''
>>> a = re.sub(r'\w{4}\.\w{4}\.\w{4}', '1234.56ab.cdef', test)
>>> print (a)

Router#show ip arp
Protocol  Address     Age (min)  Hardware Addr   Type   Interface
Internet  10.1.21.1       -      1234.56ab.cdef  ARPA   TenGigabitEthernet2/1
Internet  10.11.22.1     51      1234.56ab.cdef  ARPA   TenGigabitEthernet2/2
Internet  10.201.13.17    -      1234.56ab.cdef  ARPA   TenGigabitEthernet2/3
```

因为 optional flags 位空缺，所以默认将 3 个 MAC 地址全部替换成 1234.56ab.cdef。如果将 optional flags 位设为 1，则只会替换第一个 MAC 地址，效果如下。

```
>>> a = re.sub(r'\w{4}\.\w{4}\.\w{4}', '1234.56ab.cdef', test, 1)
>>> print (a)

Router#show ip arp
Protocol  Address     Age (min)  Hardware Addr   Type   Interface
Internet  10.1.21.1       -      1234.56ab.cdef  ARPA   TenGigabitEthernet2/1
Internet  10.11.22.1     51      b4a9.5a35.aa84  ARPA   TenGigabitEthernet2/2
Internet  10.201.13.17    -      b4a9.5abe.4345  ARPA   TenGigabitEthernet2/3
```

如果只希望替换某一个特定的 MAC 要怎么做呢？方法也很简单，这个问题留给读者朋友自行思考和实验。

3.7 异常处理

异常处理（Exception Handling）是 Python 中很常用的知识点。通常在写完代码第一次运行脚本时，我们难免会遇到一些代码错误。Python 中有两种代码错误：**语法错误（Syntax Errors）**和**异常（Exceptions）**。比如下面这种忘了在 if 语句末尾加冒号:就是一种典型的语法错误，Python 会回复一个“SyntaxError: invalid syntax”的报错信息。

```
>>> if True
  File "<stdin>",  line 1,  in ?
    if True
          ^
SyntaxError: invalid syntax
>>>
```

有时一条语句在语法上是正确的，但是执行代码后依然会引发错误，这类错误叫作异常。异常的种类很多，比如把零当作除数的“零除错误”（ZeroDivisonError）、变量还没创建就被调用的“命名错误”（NameError）、数据类型使用有误的“类型错误”（TypeError），以及尝试打开不存在的文件时会遇到的“I/O 错误”（IOError）等都是很常见的异常。这些都是 Python 中常见的内置异常，也就是在没有导入第三方模块的情况下会遇到的异常，举例如下。

```
#零除错误
>>> 100 / 0
Traceback (most recent call last):
  File "<stdin>", line 1, in <module>
ZeroDivisionError: division by zero
>>>

#命名错误
>>> print name
Traceback (most recent call last):
  File "<stdin>",  line 1,  in ?
NameError: name 'name' is not defined
>>>

#类型错误
>>> a = 10
>>> print ("There are " + a + "books.")
Traceback (most recent call last):
  File "<stdin>", line 1, in <module>
TypeError: can only concatenate str (not "int") to str
>>>

#I/O 错误
>>> f = open('abc.txt')
Traceback (most recent call last):
  File "<stdin>", line 1, in <module>
```

```
FileNotFoundError: [Errno 2] No such file or directory: 'abc.txt'
>>>
```

除了这些常见的 Python 内置异常，从第三方导入的模块也有自己独有的异常，比如后面实验部分将会重点讲的 Paramiko 就有与 SSH 用户名、密码错误相关的“AuthenticationException 异常”，以及网络设备 IP 地址不可达导致的 Socket 模块的“socket.error 异常”。关于这些第三方模块的使用及它们的异常处理在第 4 章的实验 3 将会重点介绍。

使用异常处理能提高代码的鲁棒性，帮助程序员快速修复代码中出现的错误。在 Python 中，我们使用 try…except…语句来做异常处理，举例如下。

```
>>> for i in range(1, 6):
...     print (i/0)
...
Traceback (most recent call last):
  File "<stdin>", line 2, in <module>
ZeroDivisionError: integer division or modulo by zero
>>>

>>> for i in range(1, 6):
...     try:
...        print (i/0)
...     except ZeroDivisionError:
...        print "Division by 0 is not allowed"
...
Division by 0 is not allowed
Division by 0 is not allowed
Division by 0 is not allowed
Division by 0 is not allowed
Division by 0 is not allowed
>>>
```

我们故意尝试触发零除错误，在没有做异常处理时，如果 Python 解释器发现 range(1, 6) 返回的第一个数字 1 试图去和 0 相除，会马上返回一个 ZeroDivisonError，**并且程序就此终止**。在使用 try…except…做异常处理时，我们使用 except 去主动捕获零除错误这个异常类型（except ZeroDivisonError:），并告诉 Python 解释器：在遇到零除错误时，**不要马上终止程序**，而是打印出“Division by 0 is not allowed”这个信息，告知用户代码具体出了什么问题，**然后继续执行剩下的代码，直至完成**。正因如此，在遇到零除错误后，程序并没有终止，而是接连打印出了 5 个“Division by 0 is not allowed”（因为 range(1,6)返回了 1、2、3、4、5 共 5 个整数）。

为了更清楚地展示 try…except…带来的当执行代码遇到语法错误和异常时，Python 解释器“**不会马上终止程序**”的好处，我们把上述代码稍作修改。

```
>>> for i in range(1, 6):
...     if i == 1:
...         print (i/0)
...     else:
...         print (i/1)
...
Traceback (most recent call last):
 File "<stdin>", line 3, in <module>
ZeroDivisionError: integer division or modulo by zero
>>>

>>> for i in range (1, 6):
...     try:
...         if i == 1:
...             print (i/0)
...         else:
...             print (i/1)
...     except ZeroDivisionError:
...         print 'Division by 0 is not allowed'
...
Division by 0 is not allowed
2
3
4
5
>>>
```

我们在原先的 for 循环语句里加入了 if 判断语句，只有当 range(1, 6)创建的整数 1 ~ 5 的数字为 1 时，才会触发零除错误。当整数列表里的数字不为 1 时，我们把它和 1 相除并打印出结果。在没有使用异常处理的情况下，Python 解释器在返回 ZeroDivisionError: integer division or modulo by zero 后马上终止了程序。使用异常处理后，在触发零除错误时，程序并没有被中断，而是打印“Division by 0 is not allowed”后继续执行。

另外，在上面的两个例子中，我们都在 except 语句后面加入了 ZeroDivisionError，这种提前捕获异常类型的做法是因为我们知道自己的程序将会触发零除错误，如果这时代码里出现了另外一种异常，则程序还是会被中断，因为 except 并没有捕获该异常。一般来说，我们很难记住 Python 中所有可能出现的异常类型，如果代码复杂，则无法预测脚本中将会出现什么样的异常类型，当出现这种情况时，我们要怎么做异常处理呢？有两种方法：一是在 except 语句后面不接任何异常类型，直接写成“except:”。二是通过 Exceptions 捕获所有异常。下面分别举例讲解。

```
>>> try:
...     10 / 0
... except:
...     print ("There's an error.")
...
There's an error.
>>>

>>> try:
...     10/0
... except Exception as e:
...     print (e)
...
integer division or modulo by zero
>>>
```

当单独使用 except、不捕获任何异常时，可以确保程序在遭遇任何异常时都不会被中断，并返回自定义打印出来的错误信息（比如上面例子中的 There's an error.）来代替所有可能出现的异常类型。这么做的优点是省事，但缺点也很明显：我们无法知道代码中具体出现了什么类型的异常，导致代码排错困难。

而使用 Exceptions 时（except Exception as e:），不仅可以捕获除 SystemExit、KeyboardInterrupt、GeneratorExit 外的所有异常，还能让 Python 告诉我们具体的错误原因（比如上面例子中出现的零除错误），方便对代码进行排错。如果要捕获 SystemExit、KeyboardInterrupt 和 GeneratorExit 这 3 种异常，则只需要把 Exceptions 替换成 BaseExceptions 即可。

3.8 类

作为一个面向对象编程（object-oriented programming）的语言，我们可以把 Python 中的类（Class）理解为一个模板。我们可以将自己定义好的类（也就是模板）实例化（instantiate）给一个对象（Object），所有被同一个类实例化的对象都继承了该类下所有的方法（即我们在该类下自定义的函数），唯一的区别是它们的初始属性（attribute）会因为根据我们在进行实例化初始配置时所传入参数的不同而不同。

举个例子，全世界有很多网络设备厂商都在生产或曾经生产过路由器和交换机，虽然它们的性能或各自使用的部分网络协议存在差异，但是它们都具备一些相同的术语和属性，比如 IP 地址、MAC 地址、VLAN、MTU、设备型号、操作系统版本等，正因为有这些相同的特征和属性，我们可以创建一个叫作 Router 的类或者叫作 Switch 的类来描述全天下所有的路由器和交换机，每一台路由器或交换机就是将要被我们创建的 Router 类和 Switch 类所实例化的对象。

3.8.1　怎么创建类

以创建一个 Switch 类为例，代码如下。

```
class Switch:
    def __init__(self, model, os_version, ip_add):
        self.model = model
        self.os_version = os_version
        self.ip_add = ip_add


SW1 = Switch('Cisco 9300', '16.12.04', '192.168.2.11')
```

代码讲解如下。

这里通过 class Switch:定义了一个叫作 Switch 的类，如果你曾经自学过和类相关的知识，肯定会注意到这里的 class Switch:也可以写成 class Switch(object):，在类名后加上（xxxxx）的做法叫作"继承"（inheritance，继承的概念我们后面会讲到），在 Python 2 中定义类的时候是否使用（object）来继承 object 这个 Python 自带的类是有很大区别的，具体的区别大家有兴趣的可以自行去扩展阅读，这里就不详述了，因为 Python 2 已经不是我们需要重点关注的内容了。大家只需要记住：在 Python 3 中，在类名后不管加不加（object）都没有任何区别，一般图省事的都可以不加。

```
class Switch:
```

接下来我们创建一个特殊函数，叫作__init__()。注意，在 Python 中，函数名前后各带两个下画线__的函数叫作魔法函数（Magic Methods）（关于魔法函数的讲解已经超出了本节的范围）。这里的__init__()就是一个典型的魔法函数，它的作用是在我们将类实例化给一个对象后，立即就要执行该函数让该对象完成初始化配置，__init__()中的参数 self 代表实例本身，而后面的参数 model、os_version、ip_add 等则是我们要手动赋值给对象本身的初始参数。

```
def__init__(self, model, os_version, ip_add):
    self.model = model
    self.os_version = os_version
    self.ip_add = ip_add
```

定义好 Switch 类和它的__init__()函数后，我们就可以将 Switch 类实例化给一个叫作 SW1 的对象。注意，在调用__init__()函数向对象传入参数时，参数 self 本身不用被传，我们只需要传入 model、os_version 及 ip_add3 个参数即可，这 3 个参数即成为 SW1 对象的初始属性。

```
SW1 = Switch('Cisco 9300','16.12.04','192.168.2.11')
```

此时，我们使用 print (SW1.model)、print (SW1.os_version)及 print (SW1.ip_add)即可打印出对象 SW1 的这 3 个初始属性，如下图所示。

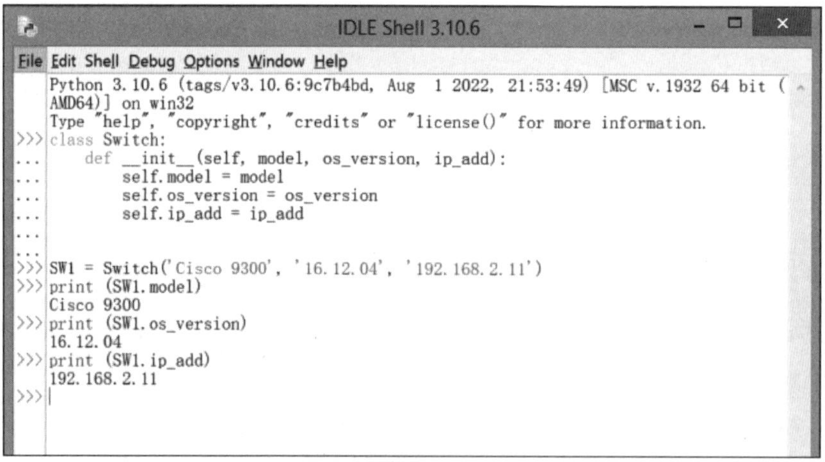

```
IDLE Shell 3.10.6
File Edit Shell Debug Options Window Help
Python 3.10.6 (tags/v3.10.6:9c7b4bd, Aug  1 2022, 21:53:49) [MSC v.1932 64 bit (
AMD64)] on win32
Type "help", "copyright", "credits" or "license()" for more information.
>>> class Switch:
...     def __init__(self, model, os_version, ip_add):
...         self.model = model
...         self.os_version = os_version
...         self.ip_add = ip_add
...
...
>>> SW1 = Switch('Cisco 9300', '16.12.04', '192.168.2.11')
>>> print (SW1.model)
Cisco 9300
>>> print (SW1.os_version)
16.12.04
>>> print (SW1.ip_add)
192.168.2.11
>>>
```

同样的道理，定义好 Switch 类后，我们还可以借助这个“模板”将 Switch 类实例化给第二个对象 SW2。

```
SW2 = Switch('Cisco 9300','16.12.04','192.168.2.12')
```

可以看到，对象 SW1 和 SW2 的设备型号及 OS 版本都是一样的，唯一不同的是 IP 地址，就像世界上可以随意找出两个性别和身高一模一样但名字不同的两个人（对象）一样。

3.8.2　方法

除了使用__init__()魔法函数来定义对象的初始属性，我们还可以在类下面自定义方法。比如可以定义一个 description()方法来获取 SW1 和 SW2 的 3 个初始属性，代码如下。

```
class Switch:
  def __init__(self, model, os_version, ip_add):
    self.model = model
    self.os_version = os_version
    self.ip_add = ip_add

  def description(self):
    description = f'Model  : {self.model}\n'\
                  f' OS Version : {self.os_version}\n'\
                  f' IP Address : {self.ip_add}\n'
    return description

SW1 = Switch('Cisco 9300', '16.12.04', '192.168.2.11')
SW2 = Switch('Cisco 9300', '16.12.04', '192.168.2.12')

print (' SW1\n', SW1.description())
print (' SW2\n', SW2.description())print (' SW2\n', SW2.description())
```

运行脚本看效果。

```
IDLE Shell 3.10.6
File Edit Shell Debug Options Window Help
Python 3.10.6 (tags/v3.10.6:9c7b4bd, Aug  1 2022, 21:53:49) [MSC v.1932 64 bit (
AMD64)] on win32
Type "help", "copyright", "credits" or "license()" for more information.
>>>
==================== RESTART: C:\Users\WANGYOL\Desktop\1.py ====================
 SW1
 Model  : Cisco 9300
 OS Version : 16.12.04
 IP Address : 192.168.2.11

 SW2
 Model  : Cisco 9300
 OS Version : 16.12.04
 IP Address : 192.168.2.12

>>>
```

3.8.3 继承

前面讲到，在类名后加上一个(xxxxx)的做法叫作“继承（Inheritance)”，xxxxx 就是要继承的类的名称。在 Python 中，继承的一方叫作“子类”，被继承的一方叫作“父类”。子类会继承父类所有的属性(比如上面 class Switch 中的 model、os_version 和 ip_add)、函数(__init__())及方法（description()），不用重新写一遍。

在前面的例子中，我们已经创建了 Switch 类，现在如果我们要创建 Router 类，有两个办法：一是从头再写一个 class Router；二是直接用 class Router(Switch)来继承已有的 Switch 类。显然第二种方法更省事。

来看下面的代码。

```
class Switch:
  def __init__(self, model, os_version, ip_add):
    self.model = model
    self.os_version = os_version
    self.ip_add = ip_add

  def description(self):
    description = f'Model  : {self.model}\n'\
                f' OS Version : {self.os_version}\n'\
                f' IP Address : {self.ip_add}\n'
    return description

class Router(Switch):
    pass

SW1 = Switch('Cisco 9300', '16.12.04', '192.168.2.11')
SW2 = Switch('Cisco 9300', '16.12.04', '192.168.2.12')
Router1 = Router ('Cisco ISR 4400', '16.10.05', '172.16.1.100')
Router2 = Router ('Cisco ISR 4400', '16.10.05', '172.16.1.101')

print (' SW1\n', SW1.description())
print (' SW2\n', SW2.description())
print (' Router1\n', Router1.description())
print (' Router2\n', Router2.description())
```

通过 class Router(Switch)来创建 Router 类，让它继承 Switch 类，此时 Switch 类为父类，Router 类为子类（注意，在代码里，父类必须写在子类的前面（或者说上面））。创建 Router 子类后，因为它继承了 Switch 父类的所有初始属性、函数和方法，因此我们可以什么事都不做，在 class Router(Switch):下面只写一个 pass 即可。然后直接创建 Router1 和 Router2 两个对象，将 Router 类直接实例化给它们，最后调用 description()方法将 Router1 和 Router2 的型号、OS 版本及 IP 地址打印出来。

```
class Router(Switch):
    pass

Router1 = Router ('Cisco ISR 4400', '16.10.05', '172.16.1.100')
Router2 = Router ('Cisco ISR 4400', '16.10.05', '172.16.1.101')

print (' Router1\n', Router1.description())
print (' Router2\n', Router2.description())
```

运行脚本效果如下图所示。

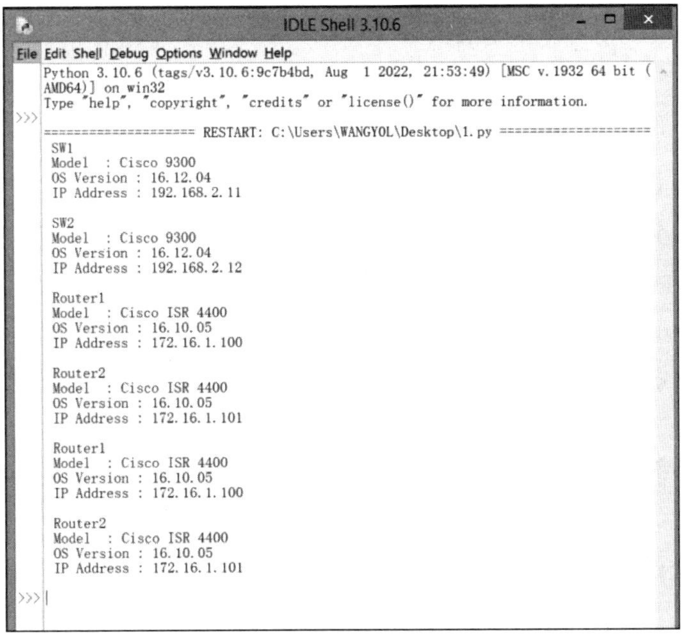

```
IDLE Shell 3.10.6
File Edit Shell Debug Options Window Help
Python 3.10.6 (tags/v3.10.6:9c7b4bd, Aug  1 2022, 21:53:49) [MSC v.1932 64 bit (
AMD64)] on win32
Type "help", "copyright", "credits" or "license()" for more information.
>>>
==================== RESTART: C:\Users\WANGYOL\Desktop\1.py ====================
 SW1
 Model   : Cisco 9300
 OS Version : 16.12.04
 IP Address : 192.168.2.11

 SW2
 Model   : Cisco 9300
 OS Version : 16.12.04
 IP Address : 192.168.2.12

 Router1
 Model   : Cisco ISR 4400
 OS Version : 16.10.05
 IP Address : 172.16.1.100

 Router2
 Model   : Cisco ISR 4400
 OS Version : 16.10.05
 IP Address : 172.16.1.101

 Router1
 Model   : Cisco ISR 4400
 OS Version : 16.10.05
 IP Address : 172.16.1.100

 Router2
 Model   : Cisco ISR 4400
 OS Version : 16.10.05
 IP Address : 172.16.1.101

>>>
```

第 4 章

Python 网络运维实验（网络模拟器）

在第 2 章和第 3 章详细介绍了 Python 的基本语法和进阶语法后，本章和第 5~9 章将分别以实验和实战的形式讲解 Python 的 Paramiko、Netmiko、Nornir 等模块和框架在网络运维中的具体应用。除介绍模块的若干引例，本章设置 4 个实验，难度循序渐进。本章和后面所有章节里的每个 Python 脚本代码都将提供详细的分段讲解，并且提供脚本运行前、脚本运行中、脚本运行后的截图，帮助读者清晰、直观地了解 Python 是如何把繁杂、单调、耗时的传统网络运维工作实现自动化的。思科实验部分已经逐行代码讲解，限于篇幅，华为实验部分将力求简洁，凸显代码迁移注意事项和适配心得感悟。

所有实验都将针对思科和华为的设备，分别在 GNS3 和 ENSP 模拟器上演示。考虑到华为 ENSP 官方已停更，所有实验脚本也同步在真机设备上进行了测试。关于网络模拟器，除了 GNS3，ENSP，HCL，目前市面上还有 EVE，PNET 等 x86 架构平台，书中实验脚本也都有尝试成功过。在 Python 网络自动化践行方面更多是注重联机交互，而非注重模拟器功能或性能。换句话说，不同模拟机（或真机）联机思路都是类似的。我们也期待华为下一代模拟器（据说叫 ENSP-NG）早日问世。

另外，为了兼顾使用不同操作系统的读者，大部分和思科设备相关的脚本将在 CentOS 上运行和演示，大部分和华为设备相关的脚本将在 Windows 10 上运行和演示。本书实验部分更倾向于引领读者通过若干实验练习后能掌握迁移方法，从而有效适配自己真实网络环境（不限于这些平台和厂商），提高生产效能。

4.1　实验运行环境

4.1.1　实验操作系统

主机操作系统：CentOS 8、Windows 10

网络设备：GNS3 模拟器上运行的思科三层交换机、ENSP 模拟器上运行的华为三层交换机

网络设备 OS 版本：思科 IOS（vios_12-ADVENTERPRISEK9-M）、华为 VRP 模拟器 Version 5.110 (S5700 V200R001C00)

Python 版本：3.10.6

4.1.2　思科实验网络拓扑

思科实验网络拓扑如下图所示。

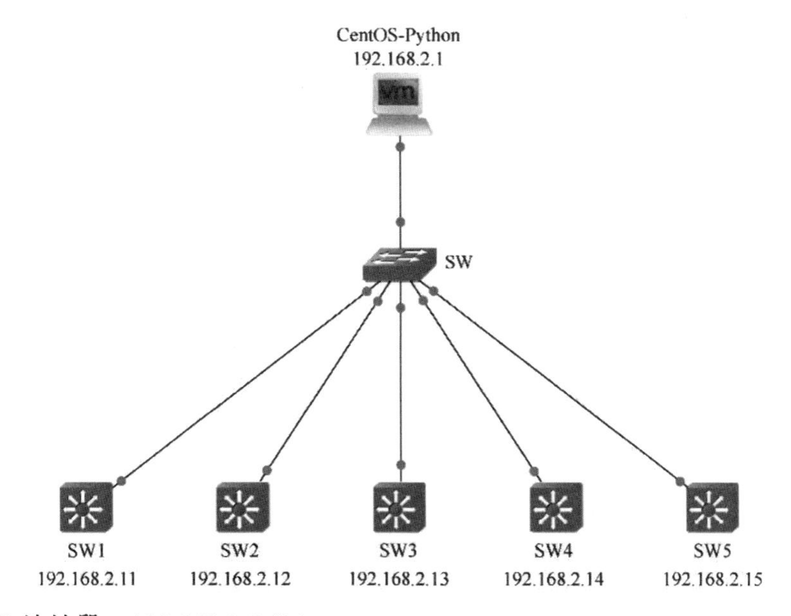

局域网 IP 地址段：192.168.2.0 /24

运行 Python 的 CentOS 主机：192.168.2.1

SW1：192.168.2.11

SW2：192.168.2.12

SW3：192.168.2.13

SW4：192.168.2.14

SW5：192.168.2.15

4.1.3　华为实验网络拓扑

华为实验网络拓扑如下图所示。

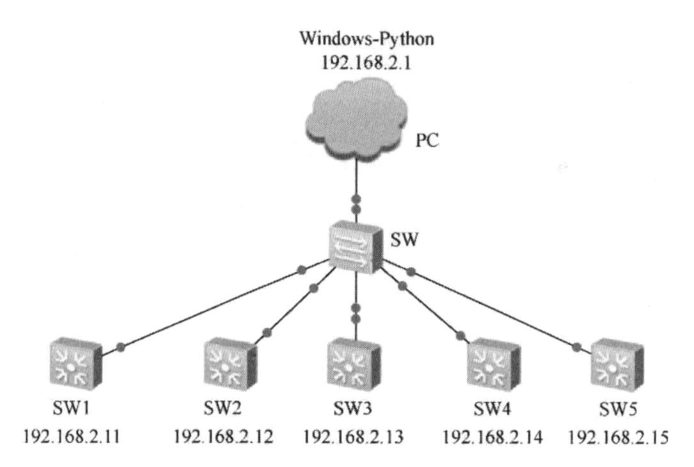

华为实验网络拓扑与思科实验网络拓扑完全一致，只是将 CentOS 主机改为 Windows 主机。

对于上述思科实验网络拓扑和华为实验网络拓扑，所有交换机都已经预配好了 SSH 和 Telnet，用户名为 python，密码为 123，用户权限为 15 级，SSH 或 Telnet 登录后直接进入特权模式，无须输入 enable 密码。模拟器或真机的不同版本有可能会因账号安全规则不允许使用“123”等简单规则文本作为密码，根据实际情况进行相应调整即可。同样地，还有用户权限问题，有些设备只设置了 1 ~ 3 级。

4.2　Python 中的 Telnet 和 SSH 模块

在 Python 中，支持 Telnet/SSH 远程登录访问网络设备的模块很多，常见的有 Telnetlib、Ciscolib、Paramiko、Netmiko 和 Pexpect。其中，Telnetlib 和 Ciscolib 对应 Telnet 协议，后面 3 个对应 SSH 协议。

因为篇幅有限，Ciscolib 和 Pexpect 不在本书的讨论范围之内，本书将简单介绍 Telnetlib，重点介绍 Paramiko（第 4 章）和 Netmiko（第 9 章），后面的实验脚本也都将基于 Paramiko 和 Netmiko 来完成。关于这 3 种模块的使用方法将通过下面的实验来介绍。

实验目的：通过 Telnetlib、Netmiko 和 Paramiko 模块，分别登录交换机 SW1（192.168.2.11）、SW2（192.168.2.12）、SW3（192.168.2.13），给 SW1 的 loopback 1 端口配置 IP 地址 1.1.1.1/32，

给 SW2 的 loopback 1 端口配置 IP 地址 2.2.2.2/32，给 SW3 的 loopback 1 端口配置 IP 地址 3.3.3.3/32。

4.2.1　Telnetlib

在 Python 中，我们使用 Telnetlib 模块来 Telnet 远程登录网络设备，Telnetlib 为 Python 内建模块，不需要 pip 下载安装就能直接使用。鉴于 Telnet 的安全性，通常不建议在生产网络中使用 Telnet，这里只举一例来讲解 Telnetlib 模块的使用方法。

Telnetlib 在 Python 2 和 Python 3 中有非常大的区别，虽然本书内容基于 Python 3.10，但是鉴于部分读者对 Telnet 还有需求，并且可能使用过 Python 2，这里将分别介绍 Telnetlib 在 Python 2 和 Python 3 中的使用方法。

1. Telnetlib 在 Python 2 中的应用（思科设备）

首先手动登录 SW1，确认它此时没有 loopback 1 这个端口，再开启 debug telnet 便于后面运行代码时进行验证。

```
SW1#
SW1#show run int loop 1
                      ^
% Invalid input detected at '^' marker.

SW1#debug telnet
Incoming Telnet debugging is on
SW1#
```

然后在运行 Python 的主机上（后面统称“主机”）创建下面的脚本，将其命名为 telnet.py。

```
[root@CentOS-Python ~]#cat telnet.py
import telnetlib

host = "192.168.2.11"
user = "python"
password = "123"

tn = telnetlib.Telnet(host)
tn.read_until("Username: ")
tn.write(user + "\n")
tn.read_until("Password: ")
tn.write(password + "\n")
```

```
tn.write("conf t\n")
tn.write("int loopback 1\n")
tn.write("ip address 1.1.1.1 255.255.255.255\n")
tn.write("end\n")
tn.write("exit\n")

print (tn.read_all())
```

代码分段讲解如下。

（1）首先通过 import 语句导入 telnetlib 模块。

```
import telnetlib
```

（2）然后创建 host、user、password 3 个变量，分别对应 SW1 的管理 IP 地址、Telnet 用户名和密码，注意这 3 个变量的数据类型均为字符串。

```
host = "192.168.2.11"
user = "python"
password = "123"
```

（3）调用 telnetlib 的 Telnet()函数，将它赋值给变量 tn，尝试以 Telnet 方式登录 192.168.2.11。

```
tn = telnetlib.Telnet(host)
```

（4）通过 Telnet 登录思科交换机时，终端输出最下面提示的信息始终为“Username:”，如下图所示。

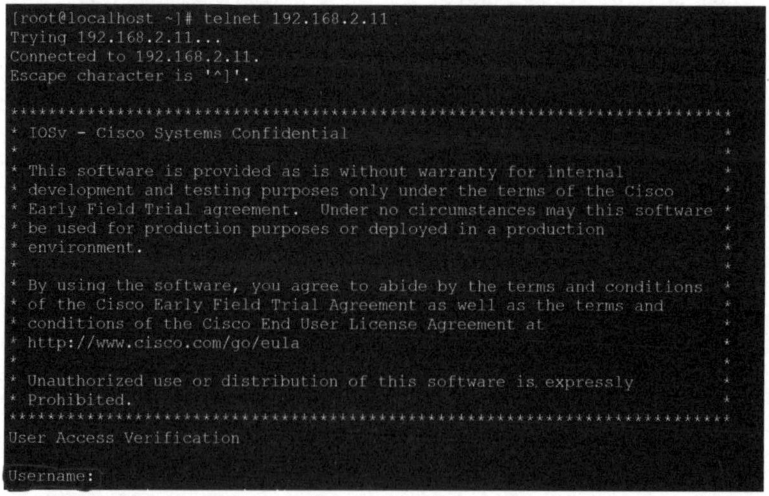

```
[root@localhost ~]# telnet 192.168.2.11
Trying 192.168.2.11...
Connected to 192.168.2.11.
Escape character is '^]'.

**************************************************************************
* IOSv - Cisco Systems Confidential                                      *
*                                                                        *
* This software is provided as is without warranty for internal        *
* development and testing purposes only under the terms of the Cisco   *
* Early Field Trial agreement.  Under no circumstances may this software *
* be used for production purposes or deployed in a production          *
* environment.                                                           *
*                                                                        *
* By using the software, you agree to abide by the terms and conditions *
* of the Cisco Early Field Trial Agreement as well as the terms and    *
* conditions of the Cisco End User License Agreement at                *
* http://www.cisco.com/go/eula                                           *
*                                                                        *
* Unauthorized use or distribution of this software is expressly       *
* Prohibited.                                                            *
**************************************************************************
User Access Verification

Username:
```

我们通过 tn.read_until("Username:") 函数来告诉 Python：如果在终端信息里读到 "Username:" 字样，则使用 tn.write(user + "\n")函数来输入 Telnet 用户名并回车。

```
tn.read_until("Username: ")
tn.write(user + "\n")
```

（5）同理，在输入 Username 后，接下来终端将显示 "Password:"，如下图所示。

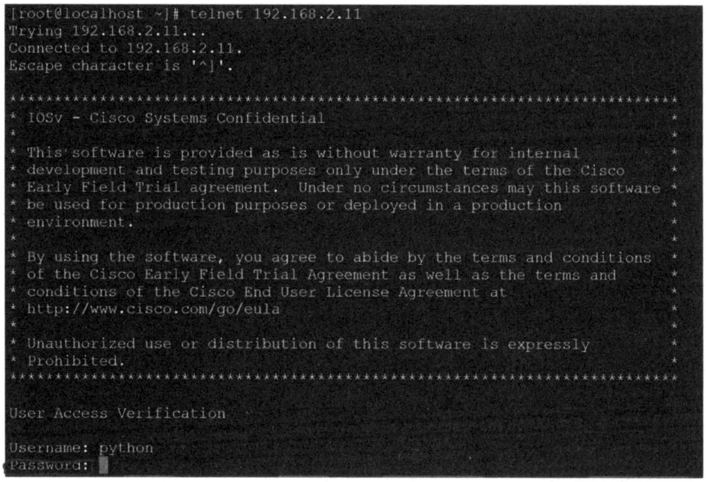

```
[root@localhost ~]# telnet 192.168.2.11
Trying 192.168.2.11...
Connected to 192.168.2.11.
Escape character is '^]'.

**************************************************************************
* IOSv - Cisco Systems Confidential                                      *
*                                                                        *
* This software is provided as is without warranty for internal         *
* development and testing purposes only under the terms of the Cisco    *
* Early Field Trial agreement.  Under no circumstances may this software *
* be used for production purposes or deployed in a production           *
* environment.                                                           *
*                                                                        *
* By using the software, you agree to abide by the terms and conditions  *
* of the Cisco Early Field Trial Agreement as well as the terms and      *
* conditions of the Cisco End User License Agreement at                  *
* http://www.cisco.com/go/eula                                           *
*                                                                        *
* Unauthorized use or distribution of this software is expressly        *
* Prohibited.                                                            *
**************************************************************************

User Access Verification

Username: python
Password:
```

我们通过 tn.read_until("Password: ")函数来告诉 Python：如果在终端信息里读到"Password:"字样，则使用 tn.write(password + "\n")函数来输入 Telnet 密码并回车。

```
tn.read_until("Password: ")
tn.write(password + "\n")
```

（6）至此，我们已经让 Python 成功通过 Telnet 登录 SW1 了。接下来就是继续用 Telnetlib 的 write()函数在 SW1 上输入各种配置命令，给 loopback 1 端口配置 1.1.1.1/32 这个 IP 地址，这是网络工程师最熟悉的部分，就不再解释了。这里只提一点，必须通过 tn.write ("exit\n")来退出 Telnet，否则位于脚本末尾的 print tn.read_all()将失效，原因在下一步中解释。

```
tn.write("conf t\n")
tn.write("int loopback 1\n")
tn.write("ip address 1.1.1.1 255.255.255.255\n")
tn.write("end\n")
tn.write("exit\n")
```

（7）最后我们通过 Telnetlib 的 read_all()方法将登录 SW1 后执行命令的所有过程都记录下来，通过 print (tn.read_all())将其打印，这样就能清楚地看到 Python 对 SW1 做了什么。注：

read_all()方法只有在退出 Telnet 后才会生效，因此我们必须在其之前通过 tn.write ("exit\n")退出 Telnet。

```
print (tn.read_all())
```

运行代码后看结果、做验证。

（1）在主机上输入 python2 telnet.py，用 Python 2 来运行代码，如下图所示。

```
[root@CentOS-Python ~]# python2 telnet.py

**************************************************************************
* IOSv is strictly limited to use for evaluation, demonstration and IOS  *
* education. IOSv is provided as-is and is not supported by Cisco's      *
* Technical Advisory Center. Any use or disclosure, in whole or in part, *
* of the IOSv Software or Documentation to any third party for any       *
* purposes is expressly prohibited except as otherwise authorized by     *
* Cisco in writing.                                                      *
**************************************************************************
SW1#conf t
Enter configuration commands, one per line.  End with CNTL/Z.
SW1(config)#int loopback 1
SW1(config-if)#ip address 1.1.1.1 255.255.255.255
SW1(config-if)#end
SW1#exit

[root@CentOS-Python ~]#
```

（2）同一时间在 SW1 上可以看到 Python 通过 Telnet 登录 SW 的详细 debug 日志（我们已经在 SW1 上开启了 debug telnet），这里能够看到“*May　2 14:00: 10.023:%LINK-3- UPDOWN: Interface Loopback1, changed state to up”“*May　2 14:00: 10.668: %SYS-5- CONFIG_I: Configured from console by python on vty0 (192.168.2.1)”和“*May 2 14:00: 11.060: %LINEPROTO-5-UPDOWN: Line protocol on Interface Loopback1, changed state to up”3 条日志，表明 loopback 1 端口已经被运行 Python 的主机 192.168.2.1 创建，并且端口已经被开启了。

```
SW1#
*May  2 14:00:07.396: Telnet2: 1 1 251 1
*May  2 14:00:07.397: TCP2: Telnet sent WILL ECHO (1)
*May  2 14:00:07.397: Telnet2: 2 2 251 3
*May  2 14:00:07.398: TCP2: Telnet sent WILL SUPPRESS-GA (3)
*May  2 14:00:07.398: Telnet2: 80000 80000 253 24
*May  2 14:00:07.399: TCP2: Telnet sent DO TTY-TYPE (24)
*May  2 14:00:07.399: Telnet2: 10000000 10000000 253 31
*May  2 14:00:07.400: TCP2: Telnet sent DO WINDOW-SIZE (31)
*May  2 14:00:07.466: TCP2: Telnet received DONT ECHO (1)
*May  2 14:00:07.467: TCP2: Telnet sent WONT ECHO (1)
*May  2 14:00:07.479: TCP2: Telnet received DONT SUPPRESS-GA (3)
*May  2 14:00:07.479: TCP2: Telnet sent WONT SUPPRESS-GA (3)
SW1#
*May  2 14:00:07.481: TCP2: Telnet received WONT TTY-TYPE (24)
*May  2 14:00:07.482: TCP2: Telnet sent DONT TTY-TYPE (24)
*May  2 14:00:07.484: TCP2: Telnet received WONT WINDOW-SIZE (31)
*May  2 14:00:07.486: TCP2: Telnet sent DONT WINDOW-SIZE (31)
*May  2 14:00:07.663: TCP2: Telnet received DONT ECHO (1)
*May  2 14:00:07.664: TCP2: Telnet received DONT SUPPRESS-GA (3)
*May  2 14:00:07.664: TCP2: Telnet received WONT TTY-TYPE (24)
*May  2 14:00:07.664: TCP2: Telnet received WONT WINDOW-SIZE (31)
SW1#
*May  2 14:00:10.023: %LINK-3-UPDOWN: Interface Loopback1, changed state to up
SW1#
*May  2 14:00:10.668: %SYS-5-CONFIG_I: Configured from console by python on vty0 (192.168.2.1)
*May  2 14:00:11.060: %LINEPROTO-5-UPDOWN: Line protocol on Interface Loopback1, changed state to up
SW1#
```

（3）最后在 SW1 上输入 show run int loop 1 做验证，确认 loopback 1 端口的 IP 地址配置正确，通过 Telnetlib 模块登录交换机修改配置的实验成功，如下图所示。

```
SW1#show run int loop 1
Building configuration...

Current configuration : 63 bytes
!
interface Loopback1
 ip address 1.1.1.1 255.255.255.255
end

SW1#
```

2. Telnetlib 在 Python 3 中的应用（思科设备）

在 Python 2 中，Telnetlib 模块下所有函数的返回值均为字符串，比如前面给出的代码中的 tn.read_until("Username: ")，这里的“Username:”即一个字符串。同理，tn.write(user + "\n")中的 user 虽然是一个变量，但是该变量的类型依然为字符串，后面的 tn.read_until ("Password: ")和 tn.write(password + "\n")，以及 tn.write('conf t\n')等都是同样的原理。

而在 Python 3 中，Telnetlib 模块下所有函数的返回值都变成字节型字符串(Byte Strings)。因此，在 Python 3 中使用 Telnetlib 需要注意以下几点。

- 在字符串的前面需要加上一个 b。
- 在变量和 Telnetlib 函数后面需要加上.encode('ascii')函数。
- 在 read_all()函数后面需要加上 decode('ascii')函数。

下面是在 Python 3 或 Python 3.10 中使用的 Telnetlib 脚本。

```
import telnetlib

host = "192.168.2.11"
user = "python"
password = "123"

tn = telnetlib.Telnet(host)
tn.read_until(b"Username: ")
tn.write(user.encode('ascii') + b"\n")
tn.read_until(b"Password: ")
tn.write(password.encode('ascii') + b"\n")
```

```
tn.write(b"conf t\n")
tn.write(b"int loopback 1\n")
tn.write(b"ip address 1.1.1.1 255.255.255.255\n")
tn.write(b"end\n")
tn.write(b"exit\n")

print (tn.read_all().decode('ascii'))
```

值得一提的是，这段 Python 3 的 Telnetlib 脚本在 Python 2 中依然可以正确使用。但是 Python 2 的那段脚本在 Python 3 中却无法兼容，感兴趣的读者可以自行在 Python 3 或者 Python 3.10 中试验一下，如果不使用 b 和.encode('ascii')函数来做字节型字符串会收到什么类型的错误。

3. Telnetlib 在 Python 3 中的应用（华为设备）

当下的普遍观点都认为 Telnet 协议是明文的，并不安全，大趋势都会往 SSH 协议改造。但是，Telnet 协议在现网中仍被大量使用。除历史遗留问题外，安全范畴可分为设备安全和通道安全，如果通道安全可控，如内网严格实行物理隔离，则 Telnet 协议依然有其用武之地。下面仅通过 Python 3（已成主流）演示如何从上述思科脚本迁移成华为脚本，Python 2 也是同样的思路。

```
import telnetlib
#引入 time 模块做延迟等待，休眠
import time

host = "192.168.2.11"
user = "python"
password = "123"

tn = telnetlib.Telnet(host)
#不同设备厂家，等待关键字有所差异
tn.read_until(b"Username:")
tn.write(user.encode('ascii') + b"\n")
tn.read_until(b"Password:")
tn.write(password.encode('ascii') + b"\n")
```

```
tn.write(b"sys\n")
tn.write(b"interface LoopBack 0\n")
tn.write(b"ip address 1.1.1.1 255.255.255.255\n")

#延迟等待
time.sleep(0.5)
tn.write(b"quit\n")

#华为设备，Telnetlib 库的 read_all 不可用
#print (tn.read_all().decode('ascii'))
print(tn.read_very_eager().decode('ascii'))

tn.close()
```

在 IDLE 脚本模式下，按“F5”键运行脚本后，我们检查一下设备端口配置，如下图所示。

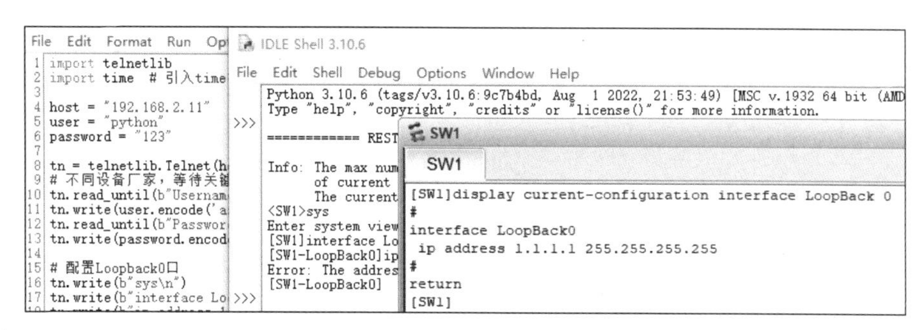

首先，迁移思路的共同点在于我们对设备操控流程的把握上，比如监听等待的关键字是否首字符大写、是否尾随空格符、指令结构翻译、指令操作后的延迟时长等。其次，迁移思路的特点还在于对应的 Python 库对不同设备的支持程度上。针对 Telnetlib 库，在操控华为设备的 Python 代码中，使用 Telnetlib 库的 read_all 会报错。

我们在代码适配过程中，通过查阅手册、搜索交流、大胆尝试等手段，经过对好几种 read 方法进行尝试，最后确定 read_very_eager 方法适合华为设备。当满足生产后，我们可以继续对比几种 read 方法的差别，尝试定位差异和寻找原因。

4.2.2　Paramiko 和 Netmiko

Python 中支持 SSH 协议实现远程连接设备的模块主要有 Paramiko 和 Netmiko 两种。Paramiko 是 Python 中一个非常著名的开源 SSHv2 项目，基于 Python 2.7 和 Python 3.4+开发，

于 2013 年 1 月发布，最早的作者是 Jeff Forcier。Paramiko 同时支持 SSH 的服务端和客户端，源码可以在 GitHub 上下载。

Netmiko 是另一个 SSH 开源项目，于 2014 年年底发布，作者是 Kirk Byers。根据作者的介绍，Netmiko 是基于 Paramiko 项目开发的，在 Paramiko 的基础上主要做了支持多厂商设备、简化命令 show 的执行和回显内容的读取、简化网络设备的配置命令等改进，具体的改进细节在后面会提到。

目前，NetDevOps 界的主流意见是 Netmiko 比 Paramiko 好用，事实上也的确如此。但是根据笔者学习和使用 Python 的经验来看，Netmiko 将太多东西简化、优化，反而不利于初学者学习。初学者如果从 Paramiko 上手，则能更直观地理解使用 Python 来 SSH 远程登录、管理设备时需要注意什么问题，更全面地锻炼自己的编程能力，并且在需要时能更顺利地在短时间内上手 Netmiko，让自己的技术更全面。反之则基本是不可逆的（可以想象成从一开始就学手动挡汽车和从一开始就学自动挡汽车的两位司机的区别）。

基于上述考虑，本章和第 5 章的所有实验都将以 Paramiko 模块实现，在第 7 章中会重点讲解 Netmiko 模块，这里只以一个简单的实验来讲解如何使用 Netmiko 登录 SW2，并为它的 loopback 1 端口配置 IP 地址 2.2.2.2/32。

1. Netmiko 实验举例（思科设备）

Netmiko 为 Python 第三方模块，需要使用 pip 来下载安装，因为实验基于 Python 3.10，这里用命令 pip3.10 install Netmiko 来下载安装 Netmiko，如下图所示。

```
[root@localhost ~]# pip3.10 install netmiko
Collecting netmiko
  Downloading netmiko-4.1.2-py3-none-any.whl (196 kB)
     ━━━━━━━━━━━━━━━━━━━━━━━━━━━━━━━━━━━━━━━━ 196.8/196.8 kB 1.1 MB/s eta 0:00:00
Requirement already satisfied: pyserial in /usr/local/lib/python3.10/site-packag
es (from netmiko) (3.5)
Requirement already satisfied: tenacity in /usr/local/lib/python3.10/site-packag
es (from netmiko) (8.0.1)
Requirement already satisfied: ntc-templates>=2.0.0 in /usr/local/lib/python3.10
/site-packages (from netmiko) (3.0.0)
Requirement already satisfied: setuptools>=38.4.0 in /usr/local/lib/python3.10/s
ite-packages (from netmiko) (63.2.0)
Requirement already satisfied: scp>=0.13.3 in /usr/local/lib/python3.10/site-pac
kages (from netmiko) (0.14.4)
Requirement already satisfied: pyyaml>=5.3 in /usr/local/lib/python3.10/site-pac
kages (from netmiko) (6.0)
Collecting textfsm==1.1.2
  Downloading textfsm-1.1.2-py2.py3-none-any.whl (44 kB)
     ━━━━━━━━━━━━━━━━━━━━━━━━━━━━━━━━━━━━━━━━ 44.7/44.7 kB 6.7 MB/s eta 0:00:00
```

```
Requirement already satisfied: paramiko>=2.7.2 in /usr/local/lib/python3.10/site
-packages (from netmiko) (2.11.0)
Requirement already satisfied: six in /usr/local/lib/python3.10/site-packages (f
rom textfsm==1.1.2->netmiko) (1.16.0)
Requirement already satisfied: future in /usr/local/lib/python3.10/site-packages
 (from textfsm==1.1.2->netmiko) (0.18.2)
Requirement already satisfied: bcrypt>=3.1.3 in /usr/local/lib/python3.10/site-p
ackages (from paramiko>=2.7.2->netmiko) (3.2.2)
Requirement already satisfied: pynacl>=1.0.1 in /usr/local/lib/python3.10/site-p
ackages (from paramiko>=2.7.2->netmiko) (1.5.0)
Requirement already satisfied: cryptography>=2.5 in /usr/local/lib/python3.10/si
te-packages (from paramiko>=2.7.2->netmiko) (37.0.4)
Requirement already satisfied: cffi>=1.1 in /usr/local/lib/python3.10/site-packa
ges (from bcrypt>=3.1.3->paramiko>=2.7.2->netmiko) (1.15.1)
Requirement already satisfied: pycparser in /usr/local/lib/python3.10/site-packa
ges (from cffi>=1.1->bcrypt>=3.1.3->paramiko>=2.7.2->netmiko) (2.21)
Installing collected packages: textfsm, netmiko
  Attempting uninstall: textfsm
    Found existing installation: textfsm 1.1.3
    Uninstalling textfsm-1.1.3:
      Successfully uninstalled textfsm-1.1.3
Successfully installed netmiko-4.1.2 textfsm-1.1.2
WARNING: Running pip as the 'root' user can result in broken permissions and con
flicting behaviour with the system package manager. It is recommended to use a v
irtual environment instead: https://pip.pypa.io/warnings/venv

[      ] A new release of pip available: 22.2.1 -> 22.2.2
[      ] To update, run: pip install --upgrade pip
[root@localhost ~]#
```

下载完毕后，进入 Python 3.10 解释器，如果 import netmiko 没有报错，则说明 Netmiko 安装成功，如下图所示。

```
root@localhost:~
[root@localhost ~]# python3.10
Python 3.10.6 (main, Aug  6 2022, 13:49:15) [GCC 8.5.0 20210514 (Red Hat 8.5.0-15)] on linux
Type "help", "copyright", "credits" or "license" for more information.
>>> import netmiko
>>>
```

与 Telnetlib 的实验步骤一样，首先登录 SW2，确认 loopback 1 端口当前不存在，然后启用 debug ip ssh 来监督验证下面创建的 Netmiko 脚本是否以 SSH 协议登录访问 SW2，如下图所示。

```
SW2#show run int loop1
                      ^
% Invalid input detected at '^' marker.

SW2#debug ip ssh
Incoming SSH debugging is on
SW2#
```

接下来在主机上创建一个名为 ssh_netmiko.py 的 Python 3.10 脚本，内容如下。

```
[root@CentOS-Python ~]#cat ssh_netmiko.py
from netmiko import ConnectHandler

SW2 = {
    'device_type': 'cisco_ios',
    'ip': '192.168.2.12',
    'username': 'python',
    'password': '123',
}

connect = ConnectHandler(**SW2)
print ("Successfully connected to " + SW2['ip'])
config_commands = ['int loop 1', 'ip address 2.2.2.2 255.255.255.255']
output = connect.send_config_set(config_commands)
print (output)
result = connect.send_command('show run int loop 1')
print (result)
```

代码分段讲解如下。

（1）首先通过 import 语句从 Netmiko 模块导入它的链接库函数 ConnectHandler()。该函数用来实现 SSH 登录网络设备，是 Netmiko 最重要的函数。

```
from netmiko import ConnectHandler
```

（2）创建一个名为 SW2 的字典，该字典包含“device_type”“ip”“username”“password”4 个必选的键，其中后面 3 个键的意思很好理解，这里主要说下“device_type”。前面提到支持多厂商的设备是 Netmiko 的优势之一，截至 2019 年 1 月，Netmiko 支持 Arista、Cisco、HP、Juniper、Alcatel、Huawei 和 Extreme 等绝大多数主流厂商的设备。除此之外，Netmiko 同样支持拥有多种不同 OS 类型的厂商的设备，比如针对 Cisco 的设备，Netmiko 能同时支持 Cisco ASA、Cisco IOS、Cisco IOS-XE、Cisco IOS-XR、Cisco NX-OS 和 Cisco SG300 共 6 种不同 OS 类型的设备。由于不同厂商的设备登录 SSH 后命令行界面和特性不尽相同，因此我们必须通过“device_type”来指定需要登录的设备的类型。因为实验里我们用到的是 Cisco IOS 设备，因此“deivce_type”的键值为“cisco_ios”。

```
SW2 = {
    'device_type': 'cisco_ios',
```

```
    'ip': '192.168.2.12',
    'username': 'python',
    'password': '123',
}
```

（3）调用 ConnectHandler()函数，用已经创建的字典 SW2 进行 SSH 连接，将它赋值给 connect 变量，注意 SW2 前面的**为关键字参数 kwargs（key word arguments）。在函数里，关键字参数表示传入函数的参数为字典格式。

```
connect = ConnectHandler(**SW2)
```

（4）如果 SSH 登录设备成功，则提示用户并告知所登录的交换机的 IP 地址。

```
print ("Successfully connected to " + SW2['ip'])
```

（5）创建一个名为 config_commands 的列表，其元素为需要在交换机上依次执行的命令。

```
config_commands = ['int loop 1', 'ip address 2.2.2.2 255.255.255.255']
```

（6）然后以刚刚创建的 config_commands 列表为参数，调用 ConnectHandler()的 send_config_set()函数来使用上述命令对 SW2 做配置，并将配置过程打印出来。

```
output = connect.send_config_set(config_commands)
print (output)
```

（7）最后调用 ConnectHandler()的 send_command()函数，对交换机输入命令 show run int loop 1 并将回显内容打印出来。需要注意的是，send_command()一次只能向设备输入一个命令，而 send_config_set()则可向设备一次输入多个命令。

```
result = connect.send_command('show run int loop 1')
print (result)
```

运行代码后看结果、做验证。

（1）在主机上输入 python3.10 ssh_netmiko.py 运行代码，可以看到除了我们在代码里写的 int loop 1 和 ip address 2.2.2.2 255.255.255.255 两个命令，**Netmiko 额外替我们输入了 3 个命令，一个是 config term，一个是 end，还有一个是 write memory**。最后我们能看到 show run int loop 1 命令的回显内容，证实 loopback1 端口配置成功，如下图所示。

```
root@localhost:~
[root@localhost ~]# python3.10 ssh_netmiko.py
Successfully connected to 192.168.2.12
configure terminal
Enter configuration commands, one per line.  End with CNTL/Z.
SW2(config)#int loop 1
SW2(config-if)#ip address 2.2.2.2 255.255.255.255
SW2(config-if)#end
SW2#
Building configuration...

Current configuration : 63 bytes
!
interface Loopback1
 ip address 2.2.2.2 255.255.255.255
end

[root@localhost ~]#
```

（2）同一时间在 SW2 上可以看到通过 Netmiko 来 SSH 登录 SW2 的详细日志（我们已经在 SW2 上开启了 debug ip ssh），如下图所示。

```
SW2#
*May  3 05:25:13.262: SSH0: starting SSH control process
*May  3 05:25:13.262: SSH0: sent protocol version id SSH-2.0-Cisco-1.25
*May  3 05:25:13.275: SSH0: protocol version id is - SSH-2.0-paramiko_2.7.1
*May  3 05:25:13.275: SSH2 0: kexinit sent: hostkey algo = ssh-rsa
*May  3 05:25:13.276: SSH2 0: kexinit sent: encryption algo = aes128-ctr,aes192-ctr,aes256-ctr,aes128-cbc,3des-cbc,aes192
-cbc,aes256-cbc
*May  3 05:25:13.292: SSH2 0: kexinit sent: mac algo = hmac-sha1,hmac-sha1-96
*May  3 05:25:13.293: SSH2 0: send:packet of  length 368 (length also includes padlen of 5)
*May  3 05:25:13.297: SSH2 0: SSH2_MSG_KEXINIT sent
*May  3 05:25:13.298: SSH2 0: ssh_receive: 880 bytes received
*May  3 05:25:13.298: SSH2 0: input: total packet length of 880 bytes
*May  3 05:25:13.299: SSH2 0: partial packet length(block size)8 bytes,needed 872 bytes,
               maclen 0
```

（3）最后我们在 SW2 上输入 show run int loop 1 做验证，确认 loopback 1 端口的 IP 地址配置正确，通过 Netmiko 模块登录交换机修改配置的实验成功，如下图所示。

```
SW2#show run int loop1
Building configuration...

Current configuration : 63 bytes
!
interface Loopback1
 ip address 2.2.2.2 255.255.255.255
end

SW2#
```

注：在使用 Netmiko 的 ConnectHandler 的脚本中，不能将脚本命名为 netmiko.py，否则会遇到“ImportError: cannot import name 'ConnectHandler' from partially initialized module 'Netmiko' (most likely due to a circular import) (/root/Netmiko.py)”这个错误。Paramiko 也一样，脚本名字不能为 paramiko.py，如下图所示。

root@localhost:~

```
[root@localhost ~]# python3.10 netmiko.py
Traceback (most recent call last):
  File "/root/netmiko.py", line 1, in <module>
    from netmiko import ConnectHandler
  File "/root/netmiko.py", line 1, in <module>
    from netmiko import ConnectHandler
ImportError: cannot import name 'ConnectHandler' from partially initialized module 'netmiko'
[root@localhost ~]#
```

2. Netmiko 实验举例（华为设备）

我们尝试把思科脚本“翻译”成华为脚本。Netmiko 原生就支持华为设备，但区分软件平台，如 V5、V8 使用的 device_type 字段是不同的，我们需要根据不同平台设定不同的 device_type 值。此外，它还支持华为的 OLT 设备（OLT 是 GPON 领域的，严格来说并非传统数通领域）。

```
from netmiko import ConnectHandler

SW2 = {
    #设备类型，指令结构进行修改适配
    'device_type': 'huawei',
    'ip': '192.168.2.12',
    'username': 'python',
    'password': '123',
}

connect = ConnectHandler(**SW2)
print ("Successfully connected to " + SW2['ip'])
config_commands = ['int loop 1', 'ip address 2.2.2.2 255.255.255.255']
output = connect.send_config_set(config_commands)
print (output)
result = connect.send_command('disp cur int loop 1')
print (result)
```

在 Telnetlib 模块实验中，我们使用 IDLE 脚本模式，按“F5”键执行脚本。这次我们尝试用 CMD 方式来执行脚本。此时执行脚本的回显“截屏”就在 CMD 上直接打印出来了。同样地，我们可以回到设备上查看相应接口的配置，看是否符合预期。

这几种方式操作完后，不知道读者朋友们是否找到一些感觉了？可以自行尝试修改 Python 脚本，实现把设备上对应的测试配置删掉功能，运行后再人工检查。

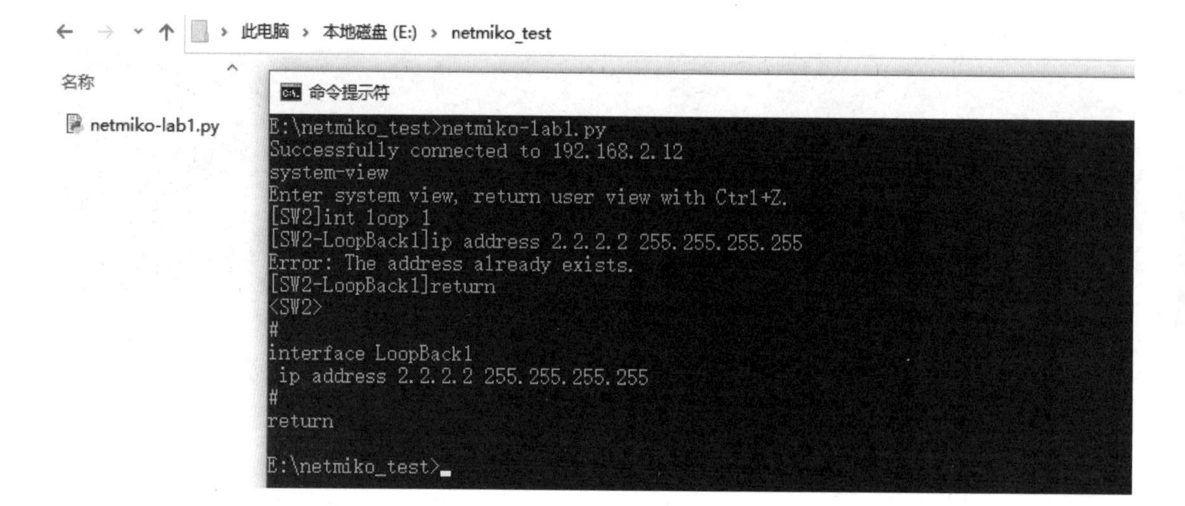

3. Paramiko 实验举例（思科设备）

与 Netmiko 不同，Paramiko 不会在做配置的时候替我们自动加上 config term、end 和 write memory 等命令，也不会在执行各种 show 命令后自动保存该命令的回显内容，一切都需要我们手动搞定。另外，Python 不像人类，后者在手动输入每个命令后会间隔一定时间，再输入下一个命令。Python 是一次性执行所有脚本里的命令的，中间没有间隔时间。当你要一次性输入很多个命令时，便经常会发生 SSH 终端跟不上速度，导致某些命令缺失没有被输入的问题（用传统的"复制、粘贴"方法给网络设备做配置的人应该遇到过这个问题）。同样，在用 print()函数输入回显内容或者用 open.write()将回显内容写入文档进行保存时，也会因为缺乏间隔时间而导致 Python "截屏"不完整，从而导致回显内容不完整。Netmiko 自动帮我们解决了这个问题，也就是说，不管上面所举的 Netmiko 例子中 config_commands 列表中的元素（命令）有多少个，都不会出现因为间隔时间不足而导致配置命令缺失的问题。而在 Paramiko 中，我们必须导入 time 模块，使用该模块的 sleep()方法来解决这个问题，关于 time 模块和 sleep()方法的使用将在下面的实验中讲到。

下面我们用 Paramiko 来完成最后一个实验：用 Paramiko 来登录 SW3 并为它的 loopback 1 端口配置 IP 地址 3.3.3.3/32。

与 Netmiko 一样，Paramiko 也是第三方模块，需要使用 pip 来下载安装，方法如下图所示。

```
[root@localhost ~]# pip3.10 install paramiko
Collecting paramiko
  Using cached paramiko-2.11.0-py2.py3-none-any.whl (212 kB)
Requirement already satisfied: cryptography>=2.5 in /usr/local/lib/python3.10/si
te-packages (from paramiko) (37.0.4)
Requirement already satisfied: pynacl>=1.0.1 in /usr/local/lib/python3.10/site-p
ackages (from paramiko) (1.5.0)
Requirement already satisfied: six in /usr/local/lib/python3.10/site-packages (f
rom paramiko) (1.16.0)
Requirement already satisfied: bcrypt>=3.1.3 in /usr/local/lib/python3.10/site-p
ackages (from paramiko) (3.2.2)
Requirement already satisfied: cffi>=1.1 in /usr/local/lib/python3.10/site-packa
ges (from bcrypt>=3.1.3->paramiko) (1.15.1)
Requirement already satisfied: pycparser in /usr/local/lib/python3.10/site-packa
ges (from cffi>=1.1->bcrypt>=3.1.3->paramiko) (2.21)
Installing collected packages: paramiko
Successfully installed paramiko-2.11.0
WARNING: Running pip as the 'root' user can result in broken permissions and con
flicting behaviour with the system package manager. It is recommended to use a v
irtual environment instead: https://pip.pypa.io/warnings/venv

[      ] A new release of pip available: 22.2.1 -> 22.2.2
[      ] To update, run: pip install --upgrade pip
[root@localhost ~]#
```

下载完毕后，进入 Python 3.10 解释器，如果 import paramiko 没有报错，则说明 Paramiko 安装成功，如下图所示。

```
[root@localhost ~]# python3.10
Python 3.10.6 (main, Aug  6 2022, 13:49:15) [GCC 8.5.0 20210514 (Red Hat 8.5.0-1
5)] on linux
Type "help", "copyright", "credits" or "license" for more information.
>>> import paramiko
>>>
```

与前面两个实验一样，首先登录 SW3，确认 loopback 1 端口目前并不存在，然后启用 debug ip ssh 来监督验证 Paramiko 脚本是否以 SSH 协议登录访问 SW3，如下图所示。

```
SW3#show run int loop1
                     ^
% Invalid input detected at '^' marker.

SW3#debug ip ssh
Incoming SSH debugging is on
SW3#
```

接下来在主机上创建一个名为 ssh_paramiko.py 的 Python 3.10 脚本，内容如下。

```
import paramiko
import time

ip = "192.168.2.13"
username = "python"
password = "123"
```

```
ssh_client = paramiko.SSHClient()
ssh_client.set_missing_host_key_policy(paramiko.AutoAddPolicy())
ssh_client.connect(hostname=ip, username=username, password=password)

print ("Successfully connected to ", ip)
command = ssh_client.invoke_shell()
command.send("configure terminal\n")
command.send("int loop 1\n")
command.send("ip address 3.3.3.3 255.255.255.255\n")
command.send("end\n")
command.send("wr mem\n")

time.sleep(2)
output = command.recv(65535)
print (output.decode("ascii"))

ssh_client.close
```

代码分段讲解如下。

（1）首先通过 import 语句导入 paramiko 和 time 两个模块。

```
import paramiko
import time
```

（2）创建 3 个变量：ip、username 和 password，分别对应我们要登录的交换机（SW3）的管理 IP 地址、SSH 用户名和密码。

```
ip = "192.168.2.13"
username = "python"
password = "123"
```

（3）调用 Paramiko 的 SSHClient()方法，将其赋值给变量 ssh_client。顾名思义，CentOS 主机做 SSH 客户端，而 SSH 服务端则是我们要登录的 SW3（192.168.2.13）。

```
ssh_client = paramiko.SSHClient()
```

（4）在默认情况下，Paramiko 会拒绝任何未知的 SSH 公钥（publickey），这里我们需要使用 ssh_client.set_missing_host_key_policy(paramiko.AutoAddPolicy()) 来让 Paramiko 接受 SSH 服务端（也就是 SW3）提供的公钥，这是任何时候使用 Paramiko 都要用到的标准配置。

```
ssh_client.set_missing_host_key_policy(paramiko.AutoAddPolicy())
```

（5）在做完 Paramiko 关于 SSH 公钥相关的配置后，调用 Paramiko.SSHClient()的 connect()函数进行 SSH 登录，该函数包含 3 个必选的参数 hostname、username 和 password，分别对应我们创建的 ip、username 和 password 3 个变量，也就是远程登录的设备的主机名/IP 地址、SSH 用户名和密码。如果 SSH 登录设备成功，则提示用户并告知所登录的交换机的管理 IP 地址。

```
ssh_client.connect (hostname=ip, username=username, password=password)
print ("Successfully connected to ", ip)
```

（6）SSH 连接成功后，需要调用 Paramiko.SSHClient()的 invoke_shell()方法来唤醒 shell，也就是思科设备 IOS 命令行，并将它赋值给变量 command。

```
command = ssh_client.invoke_shell()
```

（7）之后便可以调用 invoke_shell()的 command()函数来向 SW3 “发号施令”了。这里注意需要手动在代码中输入 configure terminal、end 和 wr mem 3 个命令，这一点与 Netmiko 不同。

```
command.send("configure terminal\n")
command.send("int loop 1\n")
command.send("ip address 3.3.3.3 255.255.255.255\n")
command.send("end\n")
command.send("wr mem\n")
```

（8）前面讲到，Python 是一次性执行所有脚本里的命令的，中间没有间隔时间，这样会导致某些命令遗漏和回显内容不完整的问题。在用 Paramiko 的 recv()函数将回显结果保存之前，我们需要调用 time 模块下的 sleep()函数手动让 Python 休眠 2s，这样回显内容才能被完整打印出来（sleep()中参数的单位为 s）。这里的 command.recv(65535)中的 65535 代表截取 65535 个字符的回显内容，这也是 Paramiko 一次能截取的最大回显内容数。另外，与 Telnetlib 类似，在 Python 3 中，Paramiko 截取的回显内容格式为字节型字符串，需要用 decode("ascii")将其解析为 ASCII 编码，否则打印出来的 output 的内容格式会很难看（下面验证部分会给出不带 decode("ascii")的回显内容供参考）。

```
time.sleep(2)
output = command.recv(65535)
print (output.decode("ascii"))
```

（9）配置完毕后，使用 close 方法退出 SSH。

```
ssh_client.close
```

运行代码后看结果、做验证。

（1）在主机上输入 python ssh_paramiko.py 来运行代码，可以看到脚本提示登录 192.168.2.13（SW3）成功，然后给 loopback1 端口配置 IP 地址 3.3.3.3，最后保存配置。代码运行如下图所示。

```
[root@localhost ~]# python3.10 ssh_paramiko.py
Successfully connected to  192.168.2.13

****************************************************************************
* IOSv - Cisco Systems Confidential                                         *
*                                                                          *
* This software is provided as is without warranty for internal            *
* development and testing purposes only under the terms of the Cisco       *
* Early Field Trial agreement.  Under no circumstances may this software *
* be used for production purposes or deployed in a production              *
* environment.                                                             *
*                                                                          *
* By using the software, you agree to abide by the terms and conditions  *
* of the Cisco Early Field Trial Agreement as well as the terms and        *
* conditions of the Cisco End User License Agreement at                    *
* http://www.cisco.com/go/eula                                             *
*                                                                          *
* Unauthorized use or distribution of this software is expressly          *
* Prohibited.                                                              *
****************************************************************************SW3#co
nfigure terminal
Enter configuration commands, one per line.  End with CNTL/Z.
SW3(config)#int loop 1
SW3(config-if)#ip address 3.3.3.3 255.255.255.255
SW3(config-if)#end
SW3#wr mem
Building configuration...

[root@localhost ~]#
```

上面提到，如果我们不在 print (output)后面加上 decode("ascii")会怎样呢？如下图所示。

```
[root@localhost ~]# python3.10 ssh_paramiko.py
Successfully connected to  192.168.2.13
b'\r\n***************************************************************************
                                              *\r\n* This software is provided a
he Cisco     *\r\n* Early Field Trial agreement.  Under no circumstances may thi
                                                          *\r\n*
he terms and conditions  *\r\n* of the Cisco Early Field Trial Agreement as well
ttp://www.cisco.com/go/eula                                           *\r\n*
f this software is expressly         *\r\n* Prohibited.
*****\r\nSW3#configure terminal\r\nEnter configuration commands, one per line.
-if)#end\r\nSW3#wr mem\r\nBuilding configuration...\r\n'
[root@localhost ~]#
```

是不是很乱？根本不知道脚本在做什么。

注：如果在 CentOS 中运行代码时遇到 SSH 用户名和密码都正确，但是一直出现“Authentication failed”验证失败的异常提示（如下图所示），很可能是主机之前用 ssh-keygen 命令生成过 RSA 密钥对用来做 SSH 免密码登录（比如在使用 Gitlab 时可以用到）。一旦在 CentOS 上生成了本地 RSA 密钥对，Paramiko 就会一直尝试使用该密钥对来登录设备，原因

是默认情况下 Paramiko.SSHClient().connect()中的 look_for_keys 可选参数默认为 True。通常情况下，我们登录交换机等网络设备时不是 SSH 免密登录，没有用到 RSA 密钥对，但是 Paramiko 又一直尝试使用该密钥对来登录设备，导致“Authentication failed”验证失败。

```
[root@localhost ~]# python ssh_paramiko.py
Traceback (most recent call last):
  File "ssh_paramiko.py", line 10, in <module>
    ssh_client.connect(hostname=ip,username=username,password=password,look_for_keys=True)
  File "/usr/lib/python2.7/site-packages/paramiko/client.py", line 437, in connect
    passphrase,
  File "/usr/lib/python2.7/site-packages/paramiko/client.py", line 749, in _auth
    raise saved_exception
paramiko.ssh_exception.AuthenticationException: Authentication failed.
[root@localhost ~]#
```

在交换机上开启 debug ip ssh 后，能看到用户名 python 的公钥缺失，导致验证失败的日志记录，如下图所示。

```
*Mar 25 06:52:18.152: SSH2 0: Using method = publickey
*Mar 25 06:52:18.153: S
S3#SH2 0: Publickey for 'python' not found
*Mar 25 06:52:18.153: SSH2 0: Pubkey Authentication failed for user 'python'
*Mar 25 06:52:18.153: SSH0: password authentication failed for python
*Mar 25 06:52:20.153: SSH2 0: Authentications that can continue = publickey,keyboard-interactive,password
*Mar 25 06:52:20.153: SSH2 0: send:packet of  length 64 (length also includes padlen of 14)
*Mar 25 06:52:20.153: SSH2 0: computed MAC for sequence no.#6 type 51
*Mar 25 06:52:20.169: SSH2 0: ssh_receive: 52 bytes received
*Mar 25 06:52:20.169: SSH2 0: input: total packet length of 32 bytes
*Mar 25 06:52:20.169: SSH2 0: partial packet length(block size)16 bytes,needed 16 bytes,
            maclen 20
*Mar 25 06:52:20.170: SSH2 0: MAC compared for #6 :ok
*Mar 25 06:52:20.170: SSH2 0: input: padlength 10 bytes
S3#
*Mar 25 06:52:20.171: SSH2 0: send:packet of  length 80 (length also includes padlen of 16)
*Mar 25 06:52:20.171: SSH2 0: computed MAC for sequence no.#7 type 1
*Mar 25 06:52:20.274: SSH0: Session disconnected - error 0x07
S3#
```

解决办法也很简单，将 look_for_keys 参数修改为 False 即可。

```
ssh_client.connect(hostname=ip,  username=username, password=password,
look_for_keys=False)
```

（2）同一时间在 SW3 上可以看到我们通过 Python 来 SSH 登录 SW3 的详细日志（之前我们已经在 SW3 开启了 debug ip ssh），如下图所示。

```
SW3#debug ip ssh
Incoming SSH debugging is on
SW3#
*May  3 05:48:26.936: SSH0: starting SSH control process
*May  3 05:48:26.937: SSH0: sent protocol version id SSH-2.0-Cisco-1.25
*May  3 05:48:26.973: SSH0: protocol version id is - SSH-2.0-paramiko_2.7.1
*May  3 05:48:26.975: SSH2 0: kexinit sent: hostkey algo = ssh-rsa
*May  3 05:48:26.976: SSH2 0: kexinit sent: encryption algo = aes128-ctr,aes192-ctr,aes256-ctr,aes128-cbc,3des-cbc,aes192
-cbc,aes256-cbc
*May  3 05:48:26.977: SSH2 0: kexinit sent: mac algo = hmac-sha1,hmac-sha1-96
*May  3 05:48:26.979: SSH2 0: send:packet of  length 368 (length also includes padlen of 5)
*May  3 05:48:26.989: SSH2 0: SSH2_MSG_KEXINIT sent
*May  3 05:48:26.991: SSH2 0: ssh_receive: 880 bytes received
*May  3 05:48:26.992: SSH2 0: input: total packet length of 880 bytes
*May  3 05:48:26.993: SSH2 0: partial packet length(block size)8 bytes,needed 872 bytes,
            maclen 0
*May  3 05:48:26.993: SSH2 0: input: padlength 10 bytes
```

（3）最后我们在 SW3 上输入 show run int loop 1 做验证，确认 loopback 1 端口的 IP 地址配置正确，通过 Paramiko 模块登录交换机修改配置的实验成功，如下图所示。

```
SW3#show run int loop1
Building configuration...

Current configuration : 63 bytes
!
interface Loopback1
 ip address 3.3.3.3 255.255.255.255
end

SW3#
```

4. Paramiko 实验举例（华为设备）

Netmiko 是基于 Paramiko 的二次开发。一般来说，Netmiko 对用户比较友好，容易上手，但有时候会增加额外开销，如内置主动延迟等。通常来说，某些第三方库或框架对一些国产设备在功能方面支持不足。因此，如果我们多掌握些“朴素”的 Paramiko 技能，则 Python 自动化实施起来会更加灵活。

```
import paramiko
import time

ip = "192.168.2.13"
username = "python"
password = "123"

ssh_client = paramiko.SSHClient()
ssh_client.set_missing_host_key_policy(paramiko.AutoAddPolicy())
ssh_client.connect(hostname=ip,username=username,password=password)

print ("Successfully connected to ",ip)
command = ssh_client.invoke_shell()
command.send("sys\n")
command.send("int loop 1\n")
command.send("ip address 3.3.3.3 255.255.255.255\n")
command.send("return\n")
command.send("save\n")
command.send("Y\n")

time.sleep(2)
```

```
output = command.recv(65535)
print (output.decode("ascii"))

ssh_client.close
```

如果跟着前面的实验“敲”过来，相信读者们多少掌握了一些适配华为等国产设备的技巧了。当我们有了一定基础后，就可以参考别人的运维脚本，逐步“翻译”成适配自己实际网络的有效脚本。

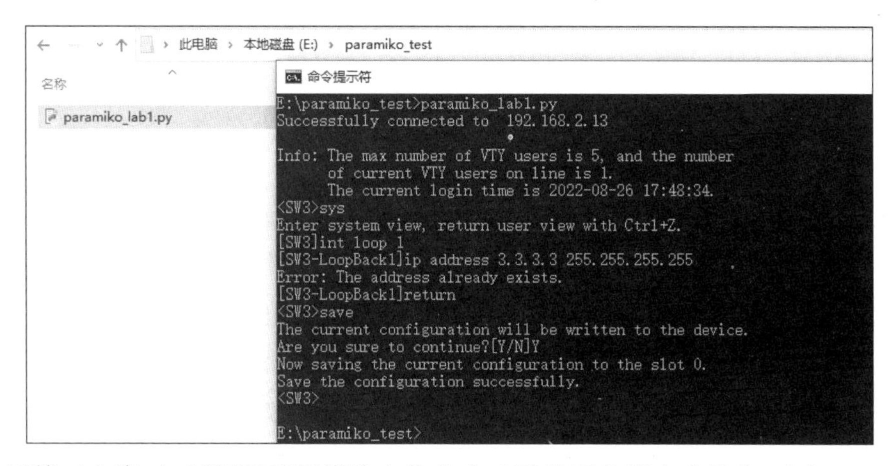

在使用第三方库时，切记不要用模块名作为自己要编写的脚本文件名，比如 paramiko.py。这种情况初学者很容易“踩坑”，需要多多留意。可以用 paramiko_lab1.py 或 paramiko20220808.py 等，这样可避免因 Python 脚本名与模块名相同而带来的麻烦。

在举例介绍了 Telnetlib、Netmiko 和 Paramiko 的基本使用方法后，下面将正式进入实验部分。每个实验都将以网络运维中常见的需求为实验背景，讲解怎样用 Python 来实现这些需求的自动化，所有实验都将使用 Paramiko 实现，并分别在思科和华为交换机上演示。

4.3　实验 1　input()函数和 getpass 模块（思科设备）

在针对 Telnetlib、Netmiko 和 Paramiko 模块的基础知识讲解中，我们都将 SSH 登录的用户名和密码明文写在了脚本里，这种做法在实验练习中可以使用，但是在生产环境中是不够安全的。在生产环境中，正确的做法是使用 input()函数和 getpass 模块来分别提示用户手动输入 SSH 用户名和密码，这是本实验将重点讲解的部分。

4.3.1　实验目的

- 使用 input()函数和 getpass 模块实现交互式的 SSH 用户名和密码输入。
- 通过 for 循环同时为 5 台交换机 SW1 ~ SW5 配置 VLAN 10 ~ VLAN 20。

4.3.2　实验准备

（1）运行代码前，首先检查 5 台交换机的配置，确认它们都没有 VLAN 10 ~ VLAN 20。

运行代码前，SW1 的配置如下图所示。

```
SW1#show vlan b

VLAN Name                             Status    Ports
---- -------------------------------- --------- -------------------------------
1    default                          active    Gi0/0, Gi0/1, Gi0/2, Gi0/3
                                                Gi1/0, Gi1/1, Gi1/2, Gi1/3
                                                Gi2/0, Gi2/1, Gi2/2, Gi2/3
                                                Gi3/0, Gi3/1, Gi3/2, Gi3/3
1002 fddi-default                     act/unsup
1003 token-ring-default               act/unsup
1004 fddinet-default                  act/unsup
1005 trnet-default                    act/unsup
SW1#
```

运行代码前，SW2 的配置如下图所示。

```
SW2#show vlan b

VLAN Name                             Status    Ports
---- -------------------------------- --------- -------------------------------
1    default                          active    Gi0/0, Gi0/1, Gi0/2, Gi0/3
                                                Gi1/0, Gi1/1, Gi1/2, Gi1/3
                                                Gi2/0, Gi2/1, Gi2/2, Gi2/3
                                                Gi3/0, Gi3/1, Gi3/2, Gi3/3
1002 fddi-default                     act/unsup
1003 token-ring-default               act/unsup
1004 fddinet-default                  act/unsup
1005 trnet-default                    act/unsup
SW2#
```

运行代码前，SW3 的配置如下图所示。

```
SW3#show vlan b

VLAN Name                             Status    Ports
---- -------------------------------- --------- -------------------------------
1    default                          active    Gi0/0, Gi0/1, Gi0/2, Gi0/3
                                                Gi1/0, Gi1/1, Gi1/2, Gi1/3
                                                Gi2/0, Gi2/1, Gi2/2, Gi2/3
                                                Gi3/0, Gi3/1, Gi3/2, Gi3/3
1002 fddi-default                     act/unsup
1003 token-ring-default               act/unsup
1004 fddinet-default                  act/unsup
1005 trnet-default                    act/unsup
SW3#
```

运行代码前，SW4 的配置如下图所示。

```
SW4#show vlan b

VLAN Name                             Status    Ports
---- -------------------------------- --------- -------------------------------
1    default                          active    Gi0/0, Gi0/1, Gi0/2, Gi0/3
                                                Gi1/0, Gi1/1, Gi1/2, Gi1/3
                                                Gi2/0, Gi2/1, Gi2/2, Gi2/3
                                                Gi3/0, Gi3/1, Gi3/2, Gi3/3
1002 fddi-default                     act/unsup
1003 token-ring-default               act/unsup
1004 fddinet-default                  act/unsup
1005 trnet-default                    act/unsup
SW4#
```

运行代码前，SW5 的配置如下图所示。

```
SW5#show vlan b

VLAN Name                             Status    Ports
---- -------------------------------- --------- -------------------------------
1    default                          active    Gi0/0, Gi0/1, Gi0/2, Gi0/3
                                                Gi1/0, Gi1/1, Gi1/2, Gi1/3
                                                Gi2/0, Gi2/1, Gi2/2, Gi2/3
                                                Gi3/0, Gi3/1, Gi3/2, Gi3/3
1002 fddi-default                     act/unsup
1003 token-ring-default               act/unsup
1004 fddinet-default                  act/unsup
1005 trnet-default                    act/unsup
SW5#
```

（2）在主机上创建实验 1 的 Python 脚本，将其命名为 lab1.py，如下图所示。

```
[root@CentOS-Python ~]# vi lab1.py
```

4.3.3　实验代码

将下列代码写入脚本 lab1.py。

```
import paramiko
import time
import getpass

username = input('Username: ')
password = getpass.getpass('Password: ')

for i in range(11,16):
    ip = "192.168.2." + str(i)
    ssh_client = paramiko.SSHClient()
    ssh_client.set_missing_host_key_policy(paramiko.AutoAddPolicy())
    ssh_client.connect(hostname=ip,username=username,password=password,
look_for_keys=False)
```

```
    print ("Successfully connect to ", ip)
    command = ssh_client.invoke_shell()
    command.send("configure terminal\n")
    for n in range (10,21):
        print ("Creating VLAN " + str(n))
        command.send("vlan " + str(n) +  "\n")
        command.send("name Python_VLAN " + str(n) +  "\n")
        time.sleep(1)

    command.send("end\n")
    command.send("wr mem\n")
    time.sleep(2)
    output = command.recv(65535)
    print (output.decode('ascii'))

ssh_client.close
```

4.3.4　代码分段讲解

（1）首先导入 paramiko、time 和 getpass 3 种模块。前两种模块的用法已经讲过，这里讲下 getpass 模块。getpass 是 Python 的内建模块，无须通过 pip 下载安装即可使用。它和 input()函数一样，都是 Python 的交互式功能，用来提示用户输入密码，区别是如果用 input()输入密码，用户输入的密码是明文可见的，如果你身边坐了其他人，密码就这么暴露了。而通过 getpass 输入的密码则是隐藏不可见的，安全性很高，所以强烈建议使用 getpass 来输入密码，使用 input()来输入用户名。注：getpass 在 Windows 中有 bug，输出的密码依然明文可见，但是不影响脚本的运行。

```
import paramiko
import time
import getpass

username = input('Username: ')
password = getpass.getpass('Password: ')
```

（2）因为 5 个交换机 SW1 ~ SW5 的 IP 地址是连续的（192.168.2.11-15），我们可以配合 for i in range(11,16)做一个简单的 for 循环来遍历 11 ~ 15 的迭代值（在 Python 3 中，range()函数的返回值不再是列表，而是一组迭代值），然后以此配合下一行代码 ip = "192.168.2." + str(i)来实现循环（批量）登录交换机 SW1 ~ SW5。注意：这里的 i 是整数，整数不能和字符串做拼接，所以要用 str(i)先将 i 转化成字符串。

```
for i in range(11,16):
    ip = "192.168.2." + str(i)
    ssh_client = Paramiko.SSHClient()
    ssh_client.set_missing_host_key_policy(Paramiko.AutoAddPolicy())
    ssh_client.connect(hostname=ip, username=username, password=password,
look_for_keys=False)
    print ("Successfully connect to ", ip)
    command = ssh_client.invoke_shell()
    command.send("configure terminal\n")
```

（3）同样的道理，我们要创建 VLAN 10 ~ VLAN 20，这些 VLAN ID 是连续的，所以又可以配合一个简单的 for 循环 for n in range (10,21)来达到循环配置 VLAN 10 ~ VLAN 20 的目的，这里使用的是嵌套 for 循环，需要注意缩进。每创建一个 VLAN，都先打印内容"print ("Creating VLAN " + str(n))"来提示用户当前正在创建的 VLAN。每个 VLAN 的命名格式都是 Python_VLAN XX，比如 VLAN 10 的名字是 Python_VLAN 10，VLAN 11 的名字是 Python_VLAN 11，依此类推。每创建一个 VLAN 之间都需要 1s 的间隔。

```
    for n in range (10, 21):
    print ("Creating VLAN " + str(n))
    command.send("vlan " + str(n) + "\n")
    command.send("name Python_VLAN " + str(n) + "\n")
    time.sleep(1)
```

（4）最后保存配置，间隔 2s 后打印出回显内容，并关闭 SSH。

```
    command.send("end\n")
    command.send("wr mem\n")
    time.sleep(2)
    output = command.recv(65535)
    print (output)

ssh_client.close
```

4.3.5　验证

（1）因打印出的回显内容过长，这里只截取自动登录 SW1、SW2 和 SW3 做配置的部分代码。可以看到：当运行脚本后系统提示输入用户名和密码时，我们输入的用户名是可见的，密码是不可见的，原因就是输入用户名时我们使用的是 input()，输入密码时我们用的是 getpass.getpass()，如下图所示。

```
[root@localhost ~]# python3.10 lab1.py
Username: python
Password:
Successfully connect to  192.168.2.11
Creating VLAN 10
Creating VLAN 11
Creating VLAN 12
Creating VLAN 13
Creating VLAN 14
Creating VLAN 15
Creating VLAN 16
Creating VLAN 17
Creating VLAN 18
Creating VLAN 19
Creating VLAN 20

**************************************************************************
* IOSv - Cisco Systems Confidential                                      *
*                                                                        *
* This software is provided as is without warranty for internal         *
* development and testing purposes only under the terms of the Cisco    *
* Early Field Trial agreement.  Under no circumstances may this software *
* be used for production purposes or deployed in a production           *
* environment.                                                           *
*                                                                        *
* By using the software, you agree to abide by the terms and conditions *
* of the Cisco Early Field Trial Agreement as well as the terms and     *
* conditions of the Cisco End User License Agreement at                 *
* http://www.cisco.com/go/eula                                           *
*                                                                        *
* Unauthorized use or distribution of this software is expressly        *
* Prohibited.                                                            *
**************************************************************************
SW1#configure terminal
Enter configuration commands, one per line.  End with CNTL/Z.
SW1(config)#vlan 10
SW1(config-vlan)#name Python_VLAN 10
SW1(config-vlan)#vlan 11
SW1(config-vlan)#name Python_VLAN 11
SW1(config-vlan)#vlan 12
SW1(config-vlan)#name Python_VLAN 12
SW1(config-vlan)#vlan 13
SW1(config-vlan)#name Python_VLAN 13
SW1(config-vlan)#vlan 14
SW1(config-vlan)#name Python_VLAN 14
SW1(config-vlan)#vlan 15
SW1(config-vlan)#name Python_VLAN 15
SW1(config-vlan)#vlan 16
SW1(config-vlan)#name Python_VLAN 16
SW1(config-vlan)#vlan 17
SW1(config-vlan)#name Python_VLAN 17
SW1(config-vlan)#vlan 18
SW1(config-vlan)#name Python_VLAN 18
SW1(config-vlan)#vlan 19
SW1(config-vlan)#name Python_VLAN 19
SW1(config-vlan)#vlan 20
SW1(config-vlan)#name Python_VLAN 20
SW1(config-vlan)#end
SW1#wr mem
```

```
Successfully connect to  192.168.2.12
Creating VLAN 10
Creating VLAN 11
Creating VLAN 12
Creating VLAN 13
Creating VLAN 14
Creating VLAN 15
Creating VLAN 16
Creating VLAN 17
Creating VLAN 18
Creating VLAN 19
Creating VLAN 20

**************************************************************************
* IOSv - Cisco Systems Confidential                                      *
*                                                                        *
* This software is provided as is without warranty for internal          *
* development and testing purposes only under the terms of the Cisco     *
* Early Field Trial agreement.  Under no circumstances may this software *
* be used for production purposes or deployed in a production            *
* environment.                                                           *
*                                                                        *
* By using the software, you agree to abide by the terms and conditions  *
* of the Cisco Early Field Trial Agreement as well as the terms and      *
* conditions of the Cisco End User License Agreement at                  *
* http://www.cisco.com/go/eula                                           *
*                                                                        *
* Unauthorized use or distribution of this software is expressly         *
* Prohibited.                                                            *
**************************************************************************
SW2#configure terminal
Enter configuration commands, one per line.  End with CNTL/Z.
SW2(config)#vlan 10
SW2(config-vlan)#name Python_VLAN 10
SW2(config-vlan)#vlan 11
SW2(config-vlan)#name Python_VLAN 11
SW2(config-vlan)#vlan 12
SW2(config-vlan)#name Python_VLAN 12
SW2(config-vlan)#vlan 13
SW2(config-vlan)#name Python_VLAN 13
SW2(config-vlan)#vlan 14
SW2(config-vlan)#name Python_VLAN 14
SW2(config-vlan)#vlan 15
SW2(config-vlan)#name Python_VLAN 15
SW2(config-vlan)#vlan 16
SW2(config-vlan)#name Python_VLAN 16
SW2(config-vlan)#vlan 17
SW2(config-vlan)#name Python_VLAN 17
SW2(config-vlan)#vlan 18
SW2(config-vlan)#name Python_VLAN 18
SW2(config-vlan)#vlan 19
SW2(config-vlan)#name Python_VLAN 19
SW2(config-vlan)#vlan 20
SW2(config-vlan)#name Python_VLAN 20
SW2(config-vlan)#end
SW2#wr mem
Building configuration...

Successfully connect to  192.168.2.13
```

```
Successfully connect to  192.168.2.13
Creating VLAN 10
Creating VLAN 11
Creating VLAN 12
Creating VLAN 13
Creating VLAN 14
Creating VLAN 15
Creating VLAN 16
Creating VLAN 17
Creating VLAN 18
Creating VLAN 19
Creating VLAN 20

**************************************************************************
* IOSv - Cisco Systems Confidential                                      *
*                                                                        *
* This software is provided as is without warranty for internal          *
* development and testing purposes only under the terms of the Cisco     *
* Early Field Trial agreement.  Under no circumstances may this software *
* be used for production purposes or deployed in a production            *
* environment.                                                           *
*                                                                        *
* By using the software, you agree to abide by the terms and conditions  *
* of the Cisco Early Field Trial Agreement as well as the terms and      *
* conditions of the Cisco End User License Agreement at                  *
* http://www.cisco.com/go/eula                                           *
*                                                                        *
* Unauthorized use or distribution of this software is expressly         *
* Prohibited.                                                            *
**************************************************************************
SW3#configure terminal
Enter configuration commands, one per line.  End with CNTL/Z.
SW3(config)#vlan 10
SW3(config-vlan)#name Python_VLAN 10
SW3(config-vlan)#vlan 11
SW3(config-vlan)#name Python_VLAN 11
SW3(config-vlan)#vlan 12
SW3(config-vlan)#name Python_VLAN 12
SW3(config-vlan)#vlan 13
SW3(config-vlan)#name Python_VLAN 13
SW3(config-vlan)#vlan 14
SW3(config-vlan)#name Python_VLAN 14
SW3(config-vlan)#vlan 15
SW3(config-vlan)#name Python_VLAN 15
SW3(config-vlan)#vlan 16
SW3(config-vlan)#name Python_VLAN 16
SW3(config-vlan)#vlan 17
SW3(config-vlan)#name Python_VLAN 17
SW3(config-vlan)#vlan 18
SW3(config-vlan)#name Python_VLAN 18
SW3(config-vlan)#vlan 19
SW3(config-vlan)#name Python_VLAN 19
SW3(config-vlan)#vlan 20
SW3(config-vlan)#name Python_VLAN 20
SW3(config-vlan)#end
SW3#wr mem
Building configuration...

Successfully connect to  192.168.2.14
```

（2）依次登录 5 个交换机验证配置。

运行代码后，SW1 的配置如下图所示。

```
SW1#show vlan b

VLAN Name                             Status    Ports
---- -------------------------------- --------- -------------------------------
1    default                          active    Gi0/0, Gi0/1, Gi0/2, Gi0/3
                                                Gi1/0, Gi1/1, Gi1/2, Gi1/3
                                                Gi2/0, Gi2/1, Gi2/2, Gi2/3
                                                Gi3/0, Gi3/1, Gi3/2, Gi3/3
10   Python_VLAN 10                   active
11   Python_VLAN 11                   active
12   Python_VLAN 12                   active
13   Python_VLAN 13                   active
14   Python_VLAN 14                   active
15   Python_VLAN 15                   active
16   Python_VLAN 16                   active
17   Python_VLAN 17                   active
18   Python_VLAN 18                   active
19   Python_VLAN 19                   active
20   Python_VLAN 20                   active
1002 fddi-default                     act/unsup
1003 token-ring-default               act/unsup
1004 fddinet-default                  act/unsup
1005 trnet-default                    act/unsup
SW1#
```

运行代码后，SW2 的配置如下图所示。

```
SW2#show vlan b

VLAN Name                             Status    Ports
---- -------------------------------- --------- -------------------------------
1    default                          active    Gi0/0, Gi0/1, Gi0/2, Gi0/3
                                                Gi1/0, Gi1/1, Gi1/2, Gi1/3
                                                Gi2/0, Gi2/1, Gi2/2, Gi2/3
                                                Gi3/0, Gi3/1, Gi3/2, Gi3/3
10   Python_VLAN 10                   active
11   Python_VLAN 11                   active
12   Python_VLAN 12                   active
13   Python_VLAN 13                   active
14   Python_VLAN 14                   active
15   Python_VLAN 15                   active
16   Python_VLAN 16                   active
17   Python_VLAN 17                   active
18   Python_VLAN 18                   active
19   Python_VLAN 19                   active
20   Python_VLAN 20                   active
1002 fddi-default                     act/unsup
1003 token-ring-default               act/unsup
1004 fddinet-default                  act/unsup
1005 trnet-default                    act/unsup
SW2#
```

运行代码后，SW3 的配置如下图所示。

```
SW3#show vlan b

VLAN Name                             Status    Ports
---- -------------------------------- --------- -------------------------------
1    default                          active    Gi0/0, Gi0/1, Gi0/2, Gi0/3
                                                Gi1/0, Gi1/1, Gi1/2, Gi1/3
                                                Gi2/0, Gi2/1, Gi2/2, Gi2/3
                                                Gi3/0, Gi3/1, Gi3/2, Gi3/3
10   Python_VLAN 10                   active
11   Python_VLAN 11                   active
12   Python_VLAN 12                   active
13   Python_VLAN 13                   active
14   Python_VLAN 14                   active
15   Python_VLAN 15                   active
16   Python_VLAN 16                   active
17   Python_VLAN 17                   active
18   Python_VLAN 18                   active
19   Python_VLAN 19                   active
20   Python_VLAN 20                   active
1002 fddi-default                     act/unsup
1003 token-ring-default               act/unsup
1004 fddinet-default                  act/unsup
1005 trnet-default                    act/unsup
SW3#
```

运行代码后，SW4 的配置如下图所示。

```
SW4#show vlan b

VLAN Name                             Status    Ports
---- -------------------------------- --------- -------------------------------
1    default                          active    Gi0/0, Gi0/1, Gi0/2, Gi0/3
                                                Gi1/0, Gi1/1, Gi1/2, Gi1/3
                                                Gi2/0, Gi2/1, Gi2/2, Gi2/3
                                                Gi3/0, Gi3/1, Gi3/2, Gi3/3
10   Python_VLAN 10                   active
11   Python_VLAN 11                   active
12   Python_VLAN 12                   active
13   Python_VLAN 13                   active
14   Python_VLAN 14                   active
15   Python_VLAN 15                   active
16   Python_VLAN 16                   active
17   Python_VLAN 17                   active
18   Python_VLAN 18                   active
19   Python_VLAN 19                   active
20   Python_VLAN 20                   active
1002 fddi-default                     act/unsup
1003 token-ring-default               act/unsup
1004 fddinet-default                  act/unsup
1005 trnet-default                    act/unsup
SW4#
```

运行代码后，SW5 的配置如下图所示。

```
SW5#show vlan b

VLAN Name                             Status    Ports
---- -------------------------------- --------- -------------------------------
1    default                          active    Gi0/0, Gi0/1, Gi0/2, Gi0/3
                                                Gi1/0, Gi1/1, Gi1/2, Gi1/3
                                                Gi2/0, Gi2/1, Gi2/2, Gi2/3
                                                Gi3/0, Gi3/1, Gi3/2, Gi3/3
10   Python_VLAN 10                   active
11   Python_VLAN 11                   active
12   Python_VLAN 12                   active
13   Python_VLAN 13                   active
14   Python_VLAN 14                   active
15   Python_VLAN 15                   active
16   Python_VLAN 16                   active
17   Python_VLAN 17                   active
18   Python_VLAN 18                   active
19   Python_VLAN 19                   active
20   Python_VLAN 20                   active
1002 fddi-default                     act/unsup
1003 token-ring-default               act/unsup
1004 fddinet-default                  act/unsup
1005 trnet-default                    act/unsup
SW5#
```

4.4　实验 1　input()函数和 getpass 模块（华为设备）

以下 Python 代码由上述思科实验代码移植而成，命令推送的逻辑基本上是一致的，只是指令结构有些许变化而已。

```
import paramiko
import time
import getpass

username = input('Username: ')
password = getpass.getpass('Password: ')
for i in range(11,16):
    ip = "192.168.2." + str(i)

    ssh_client = paramiko.SSHClient()
    ssh_client.set_missing_host_key_policy(paramiko.AutoAddPolicy())
    ssh_client.connect(hostname=ip,username=username,password=password,look_for_
keys=False)
    print ("Successfully connect to ", ip)
    command = ssh_client.invoke_shell()
```

```
    #关闭分屏功能
    command.send('screen-length 0 temporary\n')
    #进入系统视图
    command.send("sys\n")

    #以下命令推送全部转换成华为指令
    for n in range(10,21):
        print("Creating VLAN " + str(n))
        command.send("vlan " + str(n) +  "\n")
        command.send("desc Python_VLAN " + str(n) +  "\n")
        time.sleep(1)

    command.send('return\n')
    command.send('save\n')
    command.send('Y\n')
    #在 output 前，一定记得休眠等待，否则极可能回显抓取不全
    time.sleep(2)
    output = command.recv(65535)
    print(output.decode('ASCII'))

ssh_client.close
```

在 Windows 上直接用 IDLE 脚本模式运行，虽然整个流程能走得通，但是会出现一个问题，即代码在运行到 getpass.getpass('Password: ') 时，输入密码会明文显示，且前后会出现警告提示，如下图所示。

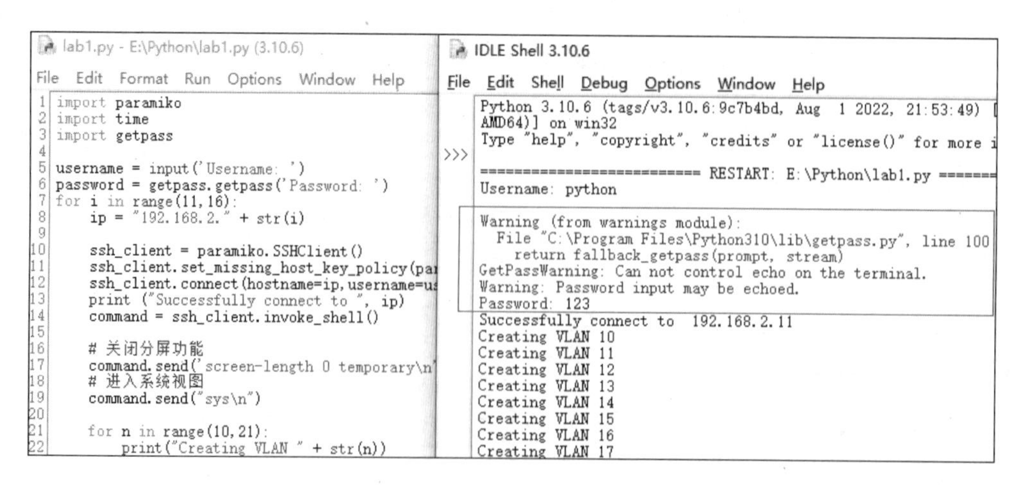

此时，改用 CMD 执行脚本，就没有这个问题了。代码执行完成后，我们手工登录到设备上，就可以看到 Python 脚本已经替我们完成这一系列“烦琐”的配置工作，如下图所示。

```
E:\Python>lab1.py
Username: python
Password:
Successfully connect to
Creating VLAN 10
Creating VLAN 11
Creating VLAN 12
Creating VLAN 13
Creating VLAN 14
Creating VLAN 15
Creating VLAN 16
Creating VLAN 17
```

```
SW5
---------------------------------------------------------------------------
1    enable  default     enable  disable    VLAN 0001
10   enable  default     enable  disable    Python_VLAN 10
11   enable  default     enable  disable    Python_VLAN 11
12   enable  default     enable  disable    Python_VLAN 12
13   enable  default     enable  disable    Python_VLAN 13
14   enable  default     enable  disable    Python_VLAN 14
15   enable  default     enable  disable    Python_VLAN 15
16   enable  default     enable  disable    Python_VLAN 16
```

如果待操作设备多或待下发的配置多，你应该已经能感受到通过 Python 脚本来释放人工劳动的魅力了。请慢慢来，其魅力还远不止于此。Python 代码具有累积效应，下次遇到同样或类似的场景时，我们稍微改一改代码，又可以适配到新场景中去了。

作为初学者，关于 getpass.getpass 在 IDLE 脚本模式下直接运行为什么会出现提示和显示明文密码，这类知识可以适当“后置”，后续留意，有时候突然就明白了。目前不必过度深挖，学习要沿着知识主线进行编织。

4.5　实验 2　批量登录地址不连续的交换机（思科设备）

在生产环境中，交换机的管理 IP 地址基本不可能像实验 1 中那样是连续的，有些交换机的管理 IP 地址甚至在不同的网段。在这种情况下，我们就不能简单地用 for 循环来登录交换机了。我们要额外建立一个文本文件，把需要登录的交换机的管理 IP 地址全部写进去，然后用 for 循环配合 open()函数来读取该文档中的管理 IP 地址，从而达到批量登录交换机的目的。

4.5.1　实验目的

- 通过 Python 脚本批量登录所有交换机，并在每个交换机上都开启 EIGRP。

4.5.2　实验准备

（1）把 SW5 的管理地址从 192.168.2.15 改成 192.168.2.55，如下图所示。

```
S5#conf t
Enter configuration commands, one per line.  End with CNTL/Z.
S5(config)#int vlan 1
S5(config-if)#ip add 192.168.2.55 255.255.255.0
S5(config-if)#end
S5#
*Mar 26 07:03:04.886: %SYS-5-CONFIG_I: Configured from console by console
S5#
```

（2）在 CentOS 上创建一个名为 ip_list 的 TXT 文件，把所有交换机的管理 IP 地址都放进去，注意该文件和等下要创建的脚本位于同一个文件夹下，如下图所示。

```
[root@CentOS-Python ~]# cat ip_list.txt
192.168.2.11
192.168.2.12
192.168.2.13
192.168.2.14
192.168.2.55
[root@CentOS-Python ~]#
```

（3）在运行代码前，检查 5 台交换机的配置，确认它们都没有开启 EIGRP。

运行代码前，SW1 的配置如下图所示。

```
SW1#show run | s router eigrp
SW1#
```

运行代码前，SW2 的配置如下图所示。

```
SW2#show run | s router eigrp
SW2#
```

运行代码前，SW3 的配置如下图所示。

```
SW3#show run | s router eigrp
SW3#
```

运行代码前，SW4 的配置如下图所示。

```
SW4#show run | s router eigrp
SW4#
```

运行代码前，SW5 的配置如下图所示。

```
SW5#show run | s router eigrp
SW5#
```

（4）在主机上创建实验 2 的脚本，将其命名为 lab2.py，如下图所示。

```
[root@CentOS-Python ~]# vi lab2.py
```

4.5.3　实验代码

将下列代码写入脚本 lab2.py。

```
import paramiko
import time
```

```
from getpass import getpass

username = input('Username: ')
password = getpass('password: ')

f = open("ip_list.txt", "r")
for line in f.readlines():
    ip = line.strip()
    ssh_client = paramiko.SSHClient()
    ssh_client.set_missing_host_key_policy(paramiko.AutoAddPolicy())
    ssh_client.connect(hostname=ip, username=username, password=password,
                       look_for_keys=False)
    print ("Successfully connect to ", ip)
    remote_connection = ssh_client.invoke_shell()
    remote_connection.send("conf t\n")
    remote_connection.send("router eigrp 1\n")
    remote_connection.send("end\n")
    remote_connection.send("wr mem\n")
    time.sleep(1)
    output = remote_connection.recv(65535)
    print (output.decode("ascii"))

f.close()
ssh_client.close
```

4.5.4　代码分段讲解

（1）和实验 1 稍有不同，我们在导入 getpass 模块时用的是 from getpass import getpass，因此我们可以把 getpass.getpass('password: ')简写成 getpass('password: ')。

```
import paramiko
import time
from getpass import getpass

username = input('Username: ')
password = getpass('password: ')
```

（2）用 open()函数打开之前创建好的包含 5 个交换机的管理 IP 地址的文档（ip_list.txt），通过 for 循环来依次遍历 readlines()方法返回的列表中的每个元素（即每个交换机的管理 IP 地址），即可达到批量依次登录 SW1 ~ SW5 的目的，即使这 5 个交换机的管理 IP 地址不是连续的。

```
f = open("ip_list.txt", "r")
for line in f.readlines():
    ip = line.strip()
    ssh_client = paramiko.SSHClient()
    ssh_client.set_missing_host_key_policy(paramiko.AutoAddPolicy())
    ssh_client.connect(hostname=ip, username=username, password=password,
                       look_for_keys=False)
    print ("Successfully connect to ", ip)
```

（3）登录每台交换机后配置 EIGRP，将回显内容打印出来。

```
    remote_connection = ssh_client.invoke_shell()
    remote_connection.send("conf t\n")
    remote_connection.send("router eigrp 1\n")
    remote_connection.send("end\n")
    remote_connection.send("wr mem\n")
    time.sleep(1)
    output = remote_connection.recv(65535)
    print (output.decode("ascii"))
```

（4）文件有开有关，脚本结束前用 close()关掉 ip_list.txt 文档，并且关闭 SSH 链接。

```
f.close()
ssh_client.close
```

4.5.5　验证

（1）因打印出的回显内容过长，这里只截取自动登录 SW1、SW2 做配置的部分代码，可以看见代码中自动登录了每个交换机开启 EIGRP 并保存配置，随后退出，如下图所示。

```
[root@localhost ~]# python3.10 lab2.py
Username: python
password:
Successfully connect to  192.168.2.11

**************************************************************************
* IOSv - Cisco Systems Confidential                                      *
*                                                                        *
* This software is provided as is without warranty for internal         *
* development and testing purposes only under the terms of the Cisco     *
* Early Field Trial agreement.  Under no circumstances may this software *
* be used for production purposes or deployed in a production            *
* environment.                                                           *
*                                                                        *
* By using the software, you agree to abide by the terms and conditions  *
* of the Cisco Early Field Trial Agreement as well as the terms and      *
* conditions of the Cisco End User License Agreement at                  *
* http://www.cisco.com/go/eula                                           *
*                                                                        *
* Unauthorized use or distribution of this software is expressly         *
* Prohibited.                                                            *
**************************************************************************
SW1#conf t
Enter configuration commands, one per line.  End with CNTL/Z.
SW1(config)#router eigrp 1
SW1(config-router)#end
SW1#wr mem
Building configuration...

Successfully connect to  192.168.2.12

**************************************************************************
* IOSv - Cisco Systems Confidential                                      *
*                                                                        *
* This software is provided as is without warranty for internal         *
* development and testing purposes only under the terms of the Cisco     *
* Early Field Trial agreement.  Under no circumstances may this software *
* be used for production purposes or deployed in a production            *
* environment.                                                           *
*                                                                        *
* By using the software, you agree to abide by the terms and conditions  *
* of the Cisco Early Field Trial Agreement as well as the terms and      *
* conditions of the Cisco End User License Agreement at                  *
* http://www.cisco.com/go/eula                                           *
*                                                                        *
* Unauthorized use or distribution of this software is expressly         *
* Prohibited.                                                            *
**************************************************************************
SW2#conf t
Enter configuration commands, one per line.  End with CNTL/Z.
SW2(config)#router eigrp 1
SW2(config-router)#end
SW2#wr mem
Building configuration...

Successfully connect to  192.168.2.13
```

（2）依次登录 5 个交换机验证配置。

运行代码后，SW1 的配置如下图所示。

```
SW1#show run | s router eigrp
router eigrp 1
SW1#
```

运行代码后，SW2 的配置如下图所示。

```
SW2#show run | s router eigrp
router eigrp 1
SW2#
```

运行代码后，SW3 的配置如下图所示。

```
SW3#show run | s router eigrp
router eigrp 1
SW3#
```

运行代码后，SW4 的配置如下图所示。

```
SW4#show run | s router eigrp
router eigrp 1
SW4#
```

运行代码后，SW5 的配置如下图所示。

```
SW5#show run | s router eigrp
router eigrp 1
SW5#
```

4.6　实验 2　批量登录地址不连续的交换机（华为设备）

与思科实验保持一致，先将 SW5 的管理地址修改一下，制作地址不连续的效果，如下图所示。

```
[SW5-Vlanif1]dis this
#
interface Vlanif1
 ip address 192.168.2.15 255.255.255.0
#
return
[SW5-Vlanif1]
[SW5-Vlanif1]
[SW5-Vlanif1] ip address 192.168.2.55 24
[SW5-Vlanif1]dis this
#
interface Vlanif1
 ip address 192.168.2.55 255.255.255.0
#
return
[SW5-Vlanif1]
```

准备 ip_list.txt 文件，存放交换机的 IP 地址列表，如下图所示。

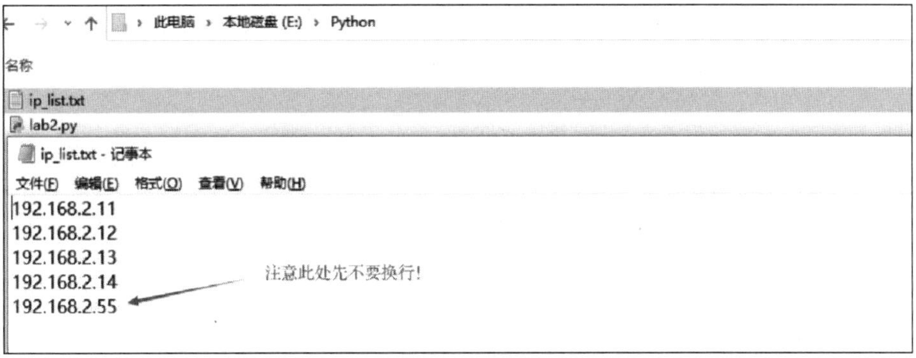

← → ˅ ↑ › 此电脑 › 本地磁盘 (E:) › Python

名称

ip_list.txt

lab2.py

ip_list.txt - 记事本

文件(F) 编辑(E) 格式(O) 查看(V) 帮助(H)

```
192.168.2.11
192.168.2.12
192.168.2.13
192.168.2.14
192.168.2.55
```

注意此处先不要换行！

Eigrp 是思科设备特有的协议，而华为设备并没有。我们换一个功能，将华为交换机默认开启的 stp 关闭。实验前可先检查一下，以便前后做对比，如下图所示。

```
[SW1]display stp bri
 MSTID  Port                        Role  STP State     Protection
   0    GigabitEthernet0/0/1        DESI  FORWARDING      NONE
```

```
import paramiko
import time
from getpass import getpass

username = input('Username: ')
password = getpass('password: ')

f = open("ip_list.txt", "r")
for line in f.readlines():
    ip = line.strip()
    ssh_client = paramiko.SSHClient()
    ssh_client.set_missing_host_key_policy(paramiko.AutoAddPolicy())
    ssh_client.connect(hostname=ip,username=username,
                  password=password,look_for_keys=False)
    print ("Successfully connect to ", ip)
    remote_connection = ssh_client.invoke_shell()
    #关闭分屏功能
    remote_connection.send('screen-length 0 temporary\n')
    #进入系统视图
    remote_connection.send('sys\n')
    #关闭消息通知（防止 log 信息刷屏）
    remote_connection.send('undo info-center enable\n')
```

```
    #将交换机 stp 关闭
    remote_connection.send('undo stp enable\n')
    remote_connection.send('Y\n')
    time.sleep(2)

    #返回用户视图
    remote_connection.send('return\n')
    #执行保存
    remote_connection.send('save\n')
    remote_connection.send('Y\n')
    time.sleep(1)
    output = remote_connection.recv(65535)
    print (output.decode("ascii"))
f.close()
ssh_client.close
```

在脚本运行完成后，我们回到设备上进行检查，发现 stp 功能已经被禁用，如下图所示。

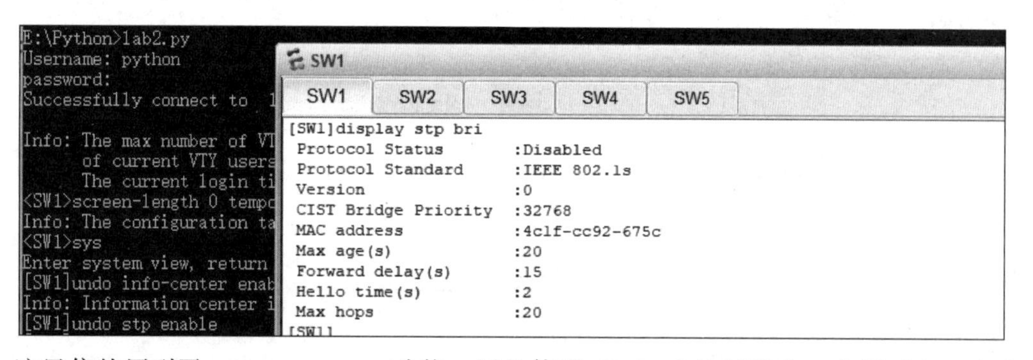

这里依然用到了 getpass.getpass 功能，因此使用 CMD 来执行脚本。有别于上一个实验，我们在 import 的时候做了些调整，适当体验 import 的不同处理细节，持续编织我们的知识体系。对于交换机，默认 stp 功能是开启的，30 秒左右的收敛时长有时候会影响我们的脚本执行，实验环境下可以关掉 stp，以便调测。真实生产环境下则不可以随意关掉 stp 功能，否则可能会造成网络环路引发广播风暴。

另外，从 ip_list.txt 文件截图中，我们特别强调最后一行不要回车换行，那么如果“不小心”回车换行了呢？会出现什么情况？Python 脚本要怎么处理这种情况呢？请读者自行尝试解决。

4.7　实验 3　异常处理的应用（思科设备）

在网络设备数量超过千台甚至上万台的大型企业网中，难免会遇到某些设备的管理 IP 地址不通、SSH 连接失败的情况，设备数量越多，这种情况发生的概率越高。这个时候如果你想用 Python 批量配置所有的设备，就一定要注意这种情况，很可能你的脚本运行了还不到一半就因为中间某一个连接不通的设备而停止了。比如你有 5000 台交换机需要统一更改本地用户名和密码，前 500 台交换机的连通性都没有问题，第 501 台交换机因为某个网络问题导致管理 IP 地址不可达，SSH 连不上，此时 Python 会返回一个"socket.error: [Errno 10060] A connection attempt failed because the connected party did not properly respond after a period of time，or established connection failed because connected host has failed to respond"错误，然后脚本就此停住！脚本不会再对剩下的 4500 台交换机做配置，也就意味着"挂机"失败！如下图所示。

```
Traceback (most recent call last):
  File "Verify configuration violation.py", line 23, in <module>
    ssh_client.connect(hostname=ip,username=username,password=password,look_for_
keys=False)
  File "C:\Python27\lib\site-packages\paramiko\client.py", line 338, in connect
    retry_on_signal(lambda: sock.connect(addr))
  File "C:\Python27\lib\site-packages\paramiko\util.py", line 279, in retry_on_s
ignal
    return function()
  File "C:\Python27\lib\site-packages\paramiko\client.py", line 338, in <lambda>

    retry_on_signal(lambda: sock.connect(addr))
  File "C:\Python27\lib\socket.py", line 228, in meth
    return getattr(self._sock,name)(*args)
socket.error: [Errno 10060] A connection attempt failed because the connected pa
rty did not properly respond after a period of time, or established connection f
ailed because connected host has failed to respond
```

同样的问题也会发生在当你输入了错误的交换机用户名和密码时，或者某些交换机和其他大部分交换机用户名和密码不一致时（因为我们只能输入一次用户名和密码，用户名和密码不一致会导致无法登录个别交换机的情况发生），也许你会问大型企业网不都是统一配置 AAA 配合 TACACS 或者 RADIUS 做用户访问管理吗？怎么还会出现登录账号和密码不一致的问题？这个现象就发生在笔者目前所任职的沙特阿卜杜拉国王科技大学，学校里的 TACACS 服务器（思科 ACS）已经服役 9 年，当前的问题是每天早晨 8 点左右该 ACS 会"失效"，必须手动重启 ACS 所在的服务器才能解决问题，在 ACS 无法正常工作期间，我们只能通过网络设备的本地账号和密码登录。鉴于此，我们已经部署了思科的 ISE 来替代 ACS 做 TACACS

服务器，但由于学校网络过于庞大，迁徙过程漫长，就导致了部分设备已经迁徙，使用了 ISE 配置的账号和密码；而另一部分还没有迁徙的设备在 ACS 出问题的时候只能用本地的账号和密码，这就出现了两套账号和密码的情况，后果就是使用 Paramiko 来 SSH 登录网络设备的 Python 会返回“Paramiko.ssh_exception.AuthenticationException: Authentication failed”的错误，如下图所示，导致脚本戛然而止，无法继续运行。

```
Traceback (most recent call last):
  File "Verify configuration violation.py", line 25, in <module>
    ssh_client.connect(hostname=ip,username=username,password=password,look_for_
keys=False)
  File "C:\Python27\lib\site-packages\paramiko\client.py", line 424, in connect
    passphrase,
  File "C:\Python27\lib\site-packages\paramiko\client.py", line 714, in _auth
    raise saved_exception
paramiko.ssh_exception.AuthenticationException: Authentication failed.
```

解决上述两个问题的方法也许你已经想到了，就是使用我们在第 3 章中讲到的异常处理。下面我们就用实验来演示异常处理在上述两种情况中的应用。

4.7.1　实验目的

- 使用实验 1 和实验 2 中的网络拓扑，将交换机 SW3（192.168.2.13）用户名 python 的密码从 123 改为 456，并将 SW4（192.168.2.14）的 Gi0/0 端口断掉。
- 创建一个带有 try…except…异常处理语句的脚本来批量在交换机 SW1 ~ SW5 上执行 show clock 命令，让脚本在 SW3、SW4 分别因为用户名和密码不匹配，以及连通性出现故障的情况下，依然可以不受干扰，进而完成剩余的配置。

4.7.2　实验准备

（1）首先将 SW3 用户名 python 的密码从 123 改为 456，如下图所示。

```
SW3#conf t
Enter configuration commands, one per line.  End with CNTL/Z.
SW3(config)#username python password 456
SW3(config)#end
```

（2）然后将 SW4 的端口 Gi0/0 关掉，如下图所示。

```
SW4#conf t
Enter configuration commands, one per line.  End with CNTL/Z.
SW4(config)#int gi0/0
SW4(config-if)#shutdown
SW4(config-if)#end
SW4#
*May  4 08:31:15.664: %SYS-5-CONFIG_I: Configured from console by console
SW4#
```

（3）在主机上创建一个名为 ip_list.txt 的文本文件，内含 SW1 ~ SW5 的管理 IP 地址。注意，在实验 2 中，我们已经将 SW5 的管理 IP 地址从 192.168.2.15 改成了 192.168.2.55，如下图所示。

```
[root@CentOS-Python ~]# cat ip_list.txt
192.168.2.11
192.168.2.12
192.168.2.13
192.168.2.14
192.168.2.55
[root@CentOS-Python ~]#
```

（4）延续实验 2 的思路，我们在主机上创建一个名为 cmd.txt 的文本文件，写入我们要在 SW1 ~ SW5 上执行的命令：show clock，如下图所示。

```
[root@CentOS-Python ~]#
[root@CentOS-Python ~]# cat cmd.txt
show clock
[root@CentOS-Python ~]#
```

（5）最后创建实验 3 的脚本文件 lab3.py，如下图所示。

```
[root@CentOS-Python ~]#
[root@CentOS-Python ~]# vi lab3.py
```

4.7.3　实验代码

将下列代码写入脚本 lab3.py。

```
import paramiko
import time
import getpass
import sys
import socket

username = input('Username: ')
password = getpass.getpass('password: ')
ip_file = sys.argv[1]
cmd_file = sys.argv[2]

switch_with_authentication_issue = []
switch_not_reachable = []
```

```
iplist = open(ip_file, 'r')
for line in iplist.readlines():
    try:
        ip = line.strip()
        ssh_client = paramiko.SSHClient()
        ssh_client.set_missing_host_key_policy(paramiko.AutoAddPolicy())
        ssh_client.connect(hostname=ip, username=username, password=password,
                    look_for_keys=False)
        print ("You have successfully connect to ", ip)
        command = ssh_client.invoke_shell()
        cmdlist = open(cmd_file, 'r')
        cmdlist.seek(0)
        for line in cmdlist.readlines():
            command.send(line + "\n")
        time.sleep(2)
        cmdlist.close()
        output = command.recv(65535)
        print (output.decode('ascii'))
    except paramiko.ssh_exception.AuthenticationException:
        print ("User authentication failed for " + ip + ".")
        switch_with_authentication_issue.append(ip)
    except socket.error:
        print (ip +  " is not reachable.")
        switch_not_reachable.append(ip)

iplist.close()
ssh_client.close

print ('\nUser authentication failed for below switches: ')
for i in switch_with_authentication_issue:
    print (i)

print ('\nBelow switches are not reachable: ')
for i in switch_not_reachable:
    print (i)
```

4.7.4　代码分段讲解

（1）为了使用异常处理来应对网络设备不可达引起的 socket.error，必须导入 Python 的内建模块 socket。

```
import paramiko
import time
import getpass
import sys
import socket

username = input('Username: ')
password = getpass.getpass('password: ')
ip_file = sys.argv[1]
cmd_file = sys.argv[2]
```

（2）创建两个空列表，分别命名为 switch_with_authentication_issue 和 switch_not_ reachable，它们的作用是在脚本最后配合 for 循环来统计哪些设备是因为认证问题而无法登录的，哪些设备是因为设备本身不可达而无法登录的。

```
switch_with_authentication_issue = []
switch_not_reachable = []
```

（3）在 for 循环下使用 try…except…异常处理语句。当 SSH 登录交换机时，如果用户名和密码不正确，Python 会报错“Paramiko.ssh_exception.AuthenticationException”，因此我们用 except Paramiko.ssh_exception.AuthenticationException:来应对该异常，一旦有交换机出现用户名和密码不正确的情况，打印出“User authentication failed for [交换机 IP]”来取代前面的“Paramiko.ssh_exception.AuthenticationException”错误信息，然后将出现该异常的交换机的管理 IP 地址用列表的 append()方法放入 switch_with_authentication_issue 列表中。同理，用 except socket.error:来应对交换机不可达时返回的错误“socket.error: [Errno 10060] A connection attempt failed because the connected party did not properly respond after a period of time, or established connection failed because connected host has failed to respond”，并打印出“[交换机 IP] is not reachable”来取代上述错误信息，然后将出现该错误的交换机的管理 IP 地址用列表的 append()方法放入 switch_not_reachable 列表中。

```
iplist = open(ip_file, 'r')
for line in iplist.readlines():
```

```
    try:
        ip = line.strip()
        ssh_client = paramiko.SSHClient()
        ssh_client.set_missing_host_key_policy(paramiko.AutoAddPolicy())
        ssh_client.connect(hostname=ip, username=username, password=password,
                    look_for_keys=False)
        print ("You have successfully connect to ", ip)
        command = ssh_client.invoke_shell()
        cmdlist = open(cmd_file, 'r')
        cmdlist.seek(0)
        for line in cmdlist.readlines():
            command.send(line + "\n")
        time.sleep(2)
        cmdlist.close()
        output = command.recv(65535)
        print (output.decode('ascii'))
    except paramiko.ssh_exception.AuthenticationException:
        print ("User authentication failed for " + ip + ".")
        switch_with_authentication_issue.append(ip)
    except socket.error:
        print (ip +  " is not reachable.")
        switch_not_reachable.append(ip)

iplist.close()
ssh_client.close
```

（4）最后使用 for 循环，打印出 switch_with_authentication_issue 和 switch_not_reachable 两个列表中的元素，这样就能清楚看到有哪些交换机的用户名和密码验证失败，哪些交换机的管理 IP 地址不可达。

```
print ('\nUser authentication failed for below switches: ')
for i in switch_with_authentication_issue:
    print (i)

print ('\nBelow switches are not reachable: ')
for i in switch_not_reachable:
    print (i)
```

4.7.5　验证

如下图所示，重点部分已经用线标注。

```
[root@localhost ~]# python3 lab3.py ip_list.txt cmd.txt
Username: python
password:
You have successfully connect to  192.168.2.11

CCCC
# # # # # # # # # # # # # # # # # # # # # # # # # #
#                                                 #
#   Access restricted to Authorised users only    #
#   Unauthorised access will result in penalties  #
#                                                 #
# # # # # # # # # # # # # # # # # # # # # # # # # #

[root@localhost ~]#
S1#show clock
*06:07:20.072 PDT Sat Mar 30 2019
S1#
S1#
You have successfully connect to  192.168.2.12

**************************************************************************
* IOSv - Cisco Systems Confidential                                      *
*                                                                        *
* This software is provided as is without warranty for internal          *
* development and testing purposes only under the terms of the Cisco     *
* Early Field Trial agreement.  Under no circumstances may this software *
* be used for production purposes or deployed in a production            *
* environment.                                                           *
*                                                                        *
* By using the software, you agree to abide by the terms and conditions  *
* of the Cisco Early Field Trial Agreement as well as the terms and      *
* conditions of the Cisco End User License Agreement at                  *
* http://www.cisco.com/go/eula                                           *
*                                                                        *
* Unauthorized use or distribution of this software is expressly         *
* Prohibited.                                                            *
**************************************************************************
S2#show clock
*13:06:52.531 UTC Sat Mar 30 2019
S2#
S2#
User authentication failed for 192.168.2.13.
192.168.2.14 is not reachable.
You have successfully connect to  192.168.2.55

**************************************************************************
* IOSv - Cisco Systems Confidential                                      *
*                                                                        *
* This software is provided as is without warranty for internal          *
* development and testing purposes only under the terms of the Cisco     *
* Early Field Trial agreement.  Under no circumstances may this software *
* be used for production purposes or deployed in a production            *
* environment.                                                           *
*                                                                        *
* By using the software, you agree to abide by the terms and conditions  *
* of the Cisco Early Field Trial Agreement as well as the terms and      *
* conditions of the Cisco End User License Agreement at                  *
* http://www.cisco.com/go/eula                                           *
*                                                                        *
* Unauthorized use or distribution of this software is expressly         *
* Prohibited.                                                            *
**************************************************************************
S5#show clock
*13:08:41.053 UTC Sat Mar 30 2019
S5#
S5#

User authentication failed for below switches:
192.168.2.13

Below switches are not reachable:
192.168.2.14
```

（1）本应出现的“Paramiko.ssh_exception.AuthenticationException”错误已经被“User authentication failed for 192.168.2.13”取代，并且 Python 脚本并未就此停止运行，而是继续尝试登录下一个交换机 SW4，也就是 192.168.2.14。

（2）本应出现的“socket.error: [Errno 10060] A connection attempt failed because the connected party did not properly respond after a period of time, or established connection failed because connected host has failed to respond”错误已经被“192.168.2.14 is not reachable”取代，并且 Python 脚本并未就此停止运行，而是继续尝试登录下一个交换机 SW5，也就是 192.168.2.55。

（3）在脚本的最后，可以看到哪些交换机出现了用户名和密码认证失败的情况，哪些交换机出现了管理 IP 地址不可达的情况。

4.8　实验 3　异常处理的应用（华为设备）

我们先把 SW3、SW4 相应的操作在设备上执行同样的修改，将交换机 SW3（192.168.2.13）用户名 python 的密码从 123 改为 456，并将 SW4（192.168.2.14）的 Gi0/0/1 端口断掉。SW5 的管理 IP 地址依然是 192.168.2.55，如下图所示。

```
[SW3-aaa] local-user python password cipher 456
[SW3-aaa]
```

```
[SW4]int GigabitEthernet 0/0/1
[SW4-GigabitEthernet0/0/1]shut
[SW4-GigabitEthernet0/0/1]
```

接下来准备华为设备的指令文件 cmd.txt。此时，整个实验目录结构如下图所示。

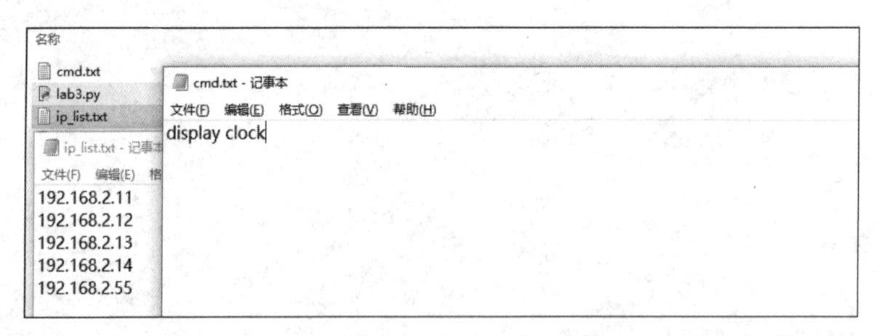

在“朴素”的 Paramiko 实验中，思科实验转华为实验的“翻译”过程主要集中在指令结构的调整上。这个实验只有一个查看设备时间的指令，且独立存放在 cmd.txt 文件中。你是否能想到，此时沿用思科实验的代码来直接操作即可呢？答案是可以来试试，直接沿用思科实验的脚本即可，如下图所示。

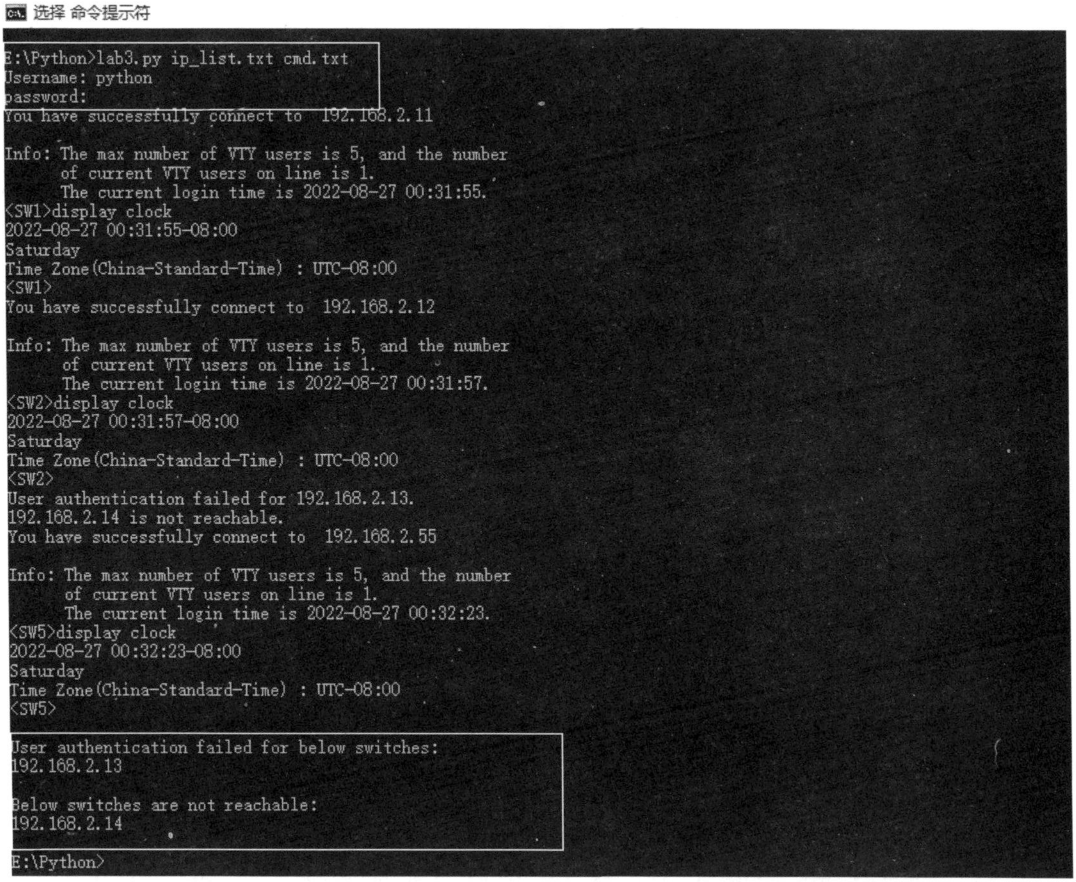

选择 命令提示符

```
E:\Python>lab3.py ip_list.txt cmd.txt
Username: python
password:
You have successfully connect to  192.168.2.11

Info: The max number of VTY users is 5, and the number
      of current VTY users on line is 1.
      The current login time is 2022-08-27 00:31:55.
<SW1>display clock
2022-08-27 00:31:55-08:00
Saturday
Time Zone(China-Standard-Time) : UTC-08:00
<SW1>
You have successfully connect to  192.168.2.12

Info: The max number of VTY users is 5, and the number
      of current VTY users on line is 1.
      The current login time is 2022-08-27 00:31:57.
<SW2>display clock
2022-08-27 00:31:57-08:00
Saturday
Time Zone(China-Standard-Time) : UTC-08:00
<SW2>
User authentication failed for 192.168.2.13.
192.168.2.14 is not reachable.
You have successfully connect to  192.168.2.55

Info: The max number of VTY users is 5, and the number
      of current VTY users on line is 1.
      The current login time is 2022-08-27 00:32:23.
<SW5>display clock
2022-08-27 00:32:23-08:00
Saturday
Time Zone(China-Standard-Time) : UTC-08:00
<SW5>

User authentication failed for below switches:
192.168.2.13

Below switches are not reachable:
192.168.2.14

E:\Python>
```

配合异常处理，可以让脚本更具有健壮性，因为此时遇到异常是捕获异常后继续往下执行，而非抛出异常后中断脚本。配合 sys.argv 操作，可以让脚本更具有“通用性”，因为此时我们可以把一些相同的操作放在统一的脚本文件中，而把差异化操作放在不同的独立文件中，相当于有“参数”可以控制。

当逐步理解 Python 基础知识及第三方库的运行逻辑后，我们就可以根据自己的需要，编写代码，参考移植代码，甚至在不同厂商设备间原封不动地运行同一份 Python 脚本。但是，这种“直接”情况的概率是很低的，往往都得调整适配，特别是遇到一些高级库，如 Netmiko、Nornir 等，都是区分厂商的，甚至同品牌还区分软件版本。因此，网络工程师在践行 Python 自动化过程中还得注重稳扎稳打，把基础打牢，万变不离其宗！仅仅想着要一份别人的代码，转手就在自己的网络中运行起来并得到自己想要的结果，实际上是很难的，除非恰巧遇到这种特例。

4.9　实验 4　用 Python 实现网络设备的配置备份（思科设备）

将网络设备的配置做备份是网络运维中必不可少的一项工作，根据公司的规模和要求不同，管理层可能会要求对网络设备的配置做月备、周备甚至日备。传统的备份思科交换机配置的办法是手动 SSH 远程登录设备，然后输入命令 term len 0 和 show run，将回显内容手动复制、粘贴到一个 TXT 或者 Word 文本文件中，效率十分低下，在有成百上千台设备需要备份的网络中尤为明显。实验 4 将演示如何用 Python 来解决这个困扰了传统网络工程师很多年的网络运维痛点问题。

4.9.1　实验目的

- 在 CentOS 8 主机上开启 FTP Server 服务（鉴于 TFTP 在 CentOS 8 上安装使用的复杂程度，这里用 FTP 替代），创建 Python 脚本，将 SW1 ~ SW5 的 running configuration 备份保存到 TFTP 服务器上。

4.9.2　实验准备

（1）将 SW5 的管理 IP 地址改回 192.168.2.15，将 SW3 的用户名 python 的密码从 456 改回 123，将 SW4 的 Gi0/0 端口重新开启，如下图所示。

```
SW5#
SW5#conf t
Enter configuration commands, one per line.  End with CNTL/Z.
SW5(config)#int vlan 1
SW5(config-if)#ip add 192.168.2.15 255.255.255.0
SW5(config-if)#end
SW5#wr mem
SW3#
SW3#conf t
Enter configuration commands, one per line.  End with CNTL/Z.
SW3(config)#username python password 123
SW3(config)#end
SW3#wr mem
Building configuration...
SW4#conf t
Enter configuration commands, one per line.  End with CNTL/Z.
SW4(config)#int gi0/0
SW4(config-if)#no shut
SW4(config-if)#end
SW4#
*May  8 10:05:55.505: %SYS-5-CONFIG_I: Configured from console by console
SW4#
*May  8 10:05:56.060: %LINK-3-UPDOWN: Interface GigabitEthernet0/0, changed state to up
*May  8 10:05:57.064: %LINEPROTO-5-UPDOWN: Line protocol on Interface GigabitEthernet0/0, changed state to up
SW4#
```

（2）将 CentOS 上的 ip_list.txt 里 SW5 的管理 IP 地址也改回 192.168.2.15，如下图所示。

```
[root@CentOS-Python ~]# cat ip_list.txt
192.168.2.11
192.168.2.12
192.168.2.13
192.168.2.14
192.168.2.15
[root@CentOS-Python ~]#
```

（3）在 CentOS 8 主机上输入下列命令下载安装 vsftpd（FTP 服务），安装前需要确认主机能否连通外网。

```
dnf install vsftpd -y
```

（4）安装完成后，在 CentOS 8 主机上输入下面两个命令分别让 CentOS 在当前和开机时启动 vsftpd 服务，如下图所示。

```
[root@CentOS-Python ~]# dnf install vsftpd -y
Last metadata expiration check: 0:06:21 ago on Fri 08 May 2020 04:22:58 AM EDT.
Dependencies resolved.
================================================================================
 Package                                    Architecture
================================================================================
Installing:
 vsftpd                                     x86_64

Transaction Summary
================================================================================
Install  1 Package

Total download size: 180 k
Installed size: 359 k
Downloading Packages:
vsftpd-3.0.3-28.el8.x86_64.rpm
--------------------------------------------------------------------------------
Total
Running transaction check
Transaction check succeeded.
Running transaction test
Transaction test succeeded.
Running transaction
  Preparing        :
  Installing       : vsftpd-3.0.3-28.el8.x86_64
  Running scriptlet: vsftpd-3.0.3-28.el8.x86_64
  Verifying        : vsftpd-3.0.3-28.el8.x86_64

Installed:
  vsftpd-3.0.3-28.el8.x86_64

Complete!
[root@CentOS-Python ~]#
```

```
systemctl start vsftpd
systemctl enable vsftpd
```

```
[root@CentOS-Python ~]# systemctl start vsftpd
[root@CentOS-Python ~]# systemctl enable vsftpd
Created symlink /etc/systemd/system/multi-user.target.wants/vsftpd.service → /usr/lib/s
ystemd/system/vsftpd.service.
[root@CentOS-Python ~]#
```

（5）输入下列命令确认 vsftpd 已经被启动运行，如下图所示。

```
systemctl status vsftpd
```

```
[root@CentOS-Python ~]# systemctl status vsftpd
● vsftpd.service - Vsftpd ftp daemon
   Loaded: loaded (/usr/lib/systemd/system/vsftpd.service; enabled; vendor preset: disabled)
   Active: active (running) since Fri 2020-05-08 04:35:25 EDT; 2min 58s ago
 Main PID: 33940 (vsftpd)
    Tasks: 1 (limit: 11343)
   Memory: 552.0K
   CGroup: /system.slice/vsftpd.service
           └─33940 /usr/sbin/vsftpd /etc/vsftpd/vsftpd.conf

May 08 04:35:25 CentOS-Python systemd[1]: Starting Vsftpd ftp daemon...
May 08 04:35:25 CentOS-Python systemd[1]: Started Vsftpd ftp daemon.
[root@CentOS-Python ~]#
```

（6）输入下列命令关闭 CentOS 8 的防火墙（仅用作实验演示，生产环境中建议修改防火墙策略）并验证，如下图所示。

```
systemctl stop firewalld
systemctl status firewalld
```

```
[root@CentOS-Python ~]# systemctl stop firewalld
[root@CentOS-Python ~]# systemctl status firewalld
● firewalld.service - firewalld - dynamic firewall daemon
   Loaded: loaded (/usr/lib/systemd/system/firewalld.service; enabled; vendor preset: enabled)
   Active: inactive (dead) since Fri 2020-05-08 05:12:23 EDT; 11s ago
     Docs: man:firewalld(1)
  Process: 1110 ExecStart=/usr/sbin/firewalld --nofork --nopid $FIREWALLD_ARGS (code=exited, status=0/SUCCESS)
 Main PID: 1110 (code=exited, status=0/SUCCESS)

May 08 04:22:20 CentOS-Python systemd[1]: Starting firewalld - dynamic firewall daemon...
May 08 04:22:20 CentOS-Python systemd[1]: Started firewalld - dynamic firewall daemon.
May 08 04:22:20 CentOS-Python firewalld[1110]: WARNING: AllowZoneDrifting is enabled. This is considered an insec
May 08 05:12:22 CentOS-Python systemd[1]: Stopping firewalld - dynamic firewall daemon...
May 08 05:12:23 CentOS-Python systemd[1]: Stopped firewalld - dynamic firewall daemon.
[root@CentOS-Python ~]#
```

（7）在 CentOS 上输入下列命令分别创建新的用户名 python，并根据提示设置密码 python，用户名 python 将稍后作为 FTP 的用户使用名，而交换机的 running configuration 也将被保存在 /home/python 文件夹下，如下图所示。

```
useradd -create-home python
passwd python
```

```
[root@CentOS-Python ~]# useradd --create-home python
[root@CentOS-Python ~]# passwd python
Changing password for user python.
New password:
BAD PASSWORD: The password is shorter than 8 characters
Retype new password:
passwd: all authentication tokens updated successfully.
[root@CentOS-Python ~]#
```

（8）在主机上创建实验 4 的脚本，将其命名为 lab4.py，如下图所示。

```
[root@CentOS-Python ~]#
[root@CentOS-Python ~]# vi lab4.py
```

4.9.3　实验代码

将下列代码写入脚本 lab4.py。

```
import paramiko
import time
import getpass

username = input('Username: ')
password = getpass.getpass('password: ')

f = open("ip_list.txt")

for line in f.readlines():
    ip_address = line.strip()
    ssh_client = paramiko.SSHClient()
    ssh_client.set_missing_host_key_policy(paramiko.AutoAddPolicy())
    ssh_client.connect(hostname=ip_address,username=username,password=password,
                       look_for_keys=False)
    print ("Successfully connect to ", ip_address)
    command = ssh_client.invoke_shell()
    command.send("configure terminal\n")
    command.send("ip ftp username python\n")
    command.send("ip ftp password python\n")
    command.send("file prompt quiet\n")
    command.send("end\n")
    command.send("copy running-config ftp://192.168.2.1\n")
    time.sleep(5)
```

```
    output = command.recv(65535)
    print (output.decode('ascii'))

f.close()
ssh_client.close
```

4.9.4　代码分段讲解

（1）实验 4 的代码难度不大，下面这段代码的作用在实验 2 中已经有详细解释，这里不再赘述。

```
import paramiko
import time
import getpass

username = input('Username: ')
password = getpass.getpass('password: ')

f = open("ip_list.txt")

for line in f.readlines():
    ip_address = line.strip()
    ssh_client = paramiko.SSHClient()
    ssh_client.set_missing_host_key_policy(paramiko.AutoAddPolicy())
    ssh_client.connect(hostname=ip_address,username=username,password=password,
                       look_for_keys=False)
    print ("Successfully connect to ", ip_address)
    command = ssh_client.invoke_shell()
```

（2）首先我们在每个交换机中通过命令 ip ftp username python 和 ip ftp password python 创建 FTP 用户名和密码，该用户名和密码同我们在 CentOS 主机上创建的一样。

```
    command.send("configure terminal\n")
    command.send("ip ftp username python\n")
    command.send("ip ftp password python\n")
```

（3）开启 file prompt quiet，然后将交换机的配置文件备份到 CentOS 主机。

```
    command.send("file prompt quiet\n")
```

```
command.send("end\n")
command.send("copy running-config ftp://192.168.2.1\n")
```

命令 file prompt 用来修改交换机文件操作的提醒方式有 alert、noisy 和 quiet 3 种模式，默认是 alert。该模式和 noisy 都会在用户进行文件操作时提示用户确认目标主机地址及目标文件名等参数，比如我们在交换机里输入命令 copy running-config ftp://192.168.2.1 将交换机的配置文件通过 FTP 备份到 CentOS 主机 192.168.2.1 后，如果使用 file prompt alert 或者 file prompt noisy，则交换机都会提醒你对目标主机地址及目标文件名进行确认，举例如下图所示。

```
SW1#conf t
Enter configuration commands, one per line.  End with CNTL/Z.
SW1(config)#file pr
SW1(config)#file prompt ?
  alert  Prompt only for destructive file operations
  noisy  Confirm all file operation parameters
  quiet  Seldom prompt for file operations
  <cr>

SW1(config)#file prompt alert
SW1(config)#end
SW1#
SW1#
SW1#
SW1#
*May  8 10:40:30.010: %SYS-5-CONFIG_I: Configured from console by console
SW1#copy running-config ftp://192.168.2.1
Address or name of remote host [192.168.2.1]?
Destination filename [sw1-confg]?
Writing swi-confg !
3822 bytes copied in 25.809 secs (148 bytes/sec)
SW1#
```

系统已经自动设置好目标文件名，其格式为“交换机的 hostname-config”，如果你对这个系统默认设置好的目标文件名没有问题，那么 alert 和 noisy 这两种 file prompt 模式不但对我们没有任何帮助，反而还会影响脚本的运行，因此我们使用 quiet 模式。关于 quiet 模式的效果举例如下图所示。

```
SW1#conf t
Enter configuration commands, one per line.  End with CNTL/Z.
SW1(config)#file prompt quiet
SW1(config)#do copy running-config ftp://192.168.2.1
Writing sw1-confg !
3840 bytes copied in 25.533 secs (150 bytes/sec)
SW1(config)#
```

（4）后面部分的代码都是在实验 1 和 2 中讲过的，不再赘述。

```
time.sleep(3)
output = command.recv(65535)
print (output.decode('ascii'))
```

```
f.close()
```

4.9.5　验证

（1）执行代码前，在 CentOS 主机上确认/home/python 文件夹下没有任何文件，如下图所示。

```
[root@localhost ~]# cd /home/python/
[root@localhost python]# ls
[root@localhost python]#
[root@localhost python]#
```

（2）回到脚本 lab4.py 所在的文件夹，执行脚本 lab4.py 后，输入交换机的 SSH 用户名和密码然后看效果，这里只截取脚本在 SW1 和 SW2 上运行后的回显内容，如下图所示。

```
[root@localhost ~]# python3.10 lab4.py
Username: python
password:
Successfully connect to  192.168.2.11

**************************************************************************
* IOSv - Cisco Systems Confidential                                      *
*                                                                        *
* This software is provided as is without warranty for internal          *
* development and testing purposes only under the terms of the Cisco     *
* Early Field Trial agreement.  Under no circumstances may this software *
* be used for production purposes or deployed in a production            *
* environment.                                                           *
*                                                                        *
* By using the software, you agree to abide by the terms and conditions  *
* of the Cisco Early Field Trial Agreement as well as the terms and      *
* conditions of the Cisco End User License Agreement at                  *
* http://www.cisco.com/go/eula                                           *
*                                                                        *
* Unauthorized use or distribution of this software is expressly         *
* Prohibited.                                                            *
**************************************************************************
SW1#configure terminal
Enter configuration commands, one per line.  End with CNTL/Z.
SW1(config)#ip ftp username python
SW1(config)#ip ftp password python
SW1(config)#file prompt quiet
SW1(config)#end
SW1#copy running-config ftp://192.168.2.1
Writing sw1-confg !
5487 bytes copied in 0.447 secs (12275 bytes/sec)
SW1#
Successfully connect to  192.168.2.12

**************************************************************************
* IOSv - Cisco Systems Confidential                                      *
*                                                                        *
* This software is provided as is without warranty for internal          *
* development and testing purposes only under the terms of the Cisco     *
* Early Field Trial agreement.  Under no circumstances may this software *
* be used for production purposes or deployed in a production            *
* environment.                                                           *
*                                                                        *
* By using the software, you agree to abide by the terms and conditions  *
* of the Cisco Early Field Trial Agreement as well as the terms and      *
* conditions of the Cisco End User License Agreement at                  *
* http://www.cisco.com/go/eula                                           *
*                                                                        *
* Unauthorized use or distribution of this software is expressly         *
* Prohibited.                                                            *
**************************************************************************
SW2#configure terminal
Enter configuration commands, one per line.  End with CNTL/Z.
SW2(config)#ip ftp username python
SW2(config)#ip ftp password python
SW2(config)#file prompt quiet
SW2(config)#end
SW2#copy running-config ftp://192.168.2.1
Writing sw2-confg !
```

（3）脚本运行完毕后，回到/home/python，此时可以看到 SW1～SW5 的 running config 都被成功备份到该文件夹下，如下图所示。

```
[root@localhost ~]# cd /home/python/
[root@localhost python]# ls
sw1-confg  sw2-confg  sw3-confg  sw4-confg  sw5-confg
[root@localhost python]# 
```

（4）用 cat 打开其中任意一个 config 文件，验证其内容，如下图所示。

```
[root@localhost python]# cat sw1-confg

!
! Last configuration change at 07:56:16 UTC Wed Aug 10 2022 by python
!
version 15.0
service timestamps debug datetime msec
service timestamps log datetime msec
no service password-encryption
service compress-config
!
hostname SW1
!
boot-start-marker
boot-end-marker
!
!
!
username python privilege 15 password 0 123
no aaa new-model
!
!
!
!
!
vtp domain CISCO-vIOS
vtp mode transparent
!
!
!
ip domain-name test
ip cef
no ipv6 cef
!
!
file prompt quiet
spanning-tree mode pvst
spanning-tree extend system-id
!
vlan internal allocation policy ascending
!
vlan 10
 name Python_VLAN 10
!
vlan 11
 name Python_VLAN 11
!
vlan 12
 name Python_VLAN 12
!
vlan 13
 name Python_VLAN 13
!
vlan 14
 name Python_VLAN 14
!
vlan 15
 name Python_VLAN 15
!
```

4.10　实验 4　用 Python 实现网络设备的配置备份（华为设备）

对于网络工程师来说，华为设备通常采用 CLI，即手工登录设备，取消分屏，执行 disp cur，抓取回显，保存成配置文件。通过 Python 基础知识及早前实验，或许你现在已经可以用 Python 脚本进行了。上述思科实验采用另一种方法，即类似于复制系统的备份文件到 FTP 服务器上。Windows 是 FTP 服务端，SW1 ~ SW5 是客户端。我们通过 Windows 上的 Python 脚本，登录到每台交换机上实施 FTP 操作，推送备份文件至服务器。

这次实验，我们先将 SW5 的管理 IP 地址改回 192.168.2.15，将 SW3 的用户名 python 的密码从 456 改回 123，将 SW4 的 Gi0/0/1 端口重新开启，ip_list.txt 里 SW5 的管理 IP 地址也改回 192.168.2.15。同时，我们在实验目录上新建一个 lab4 文件夹，用于后面放置备份的配置文件。

我们找一个很轻巧的 Windows10 环境下可运行的 FTP 服务端软件 3CDaemon（类似的软件有很多），配置 FTP 账号密码，登录目录，单击“应用”→“确定”，然后启动 FTP 服务器功能，如下图所示。

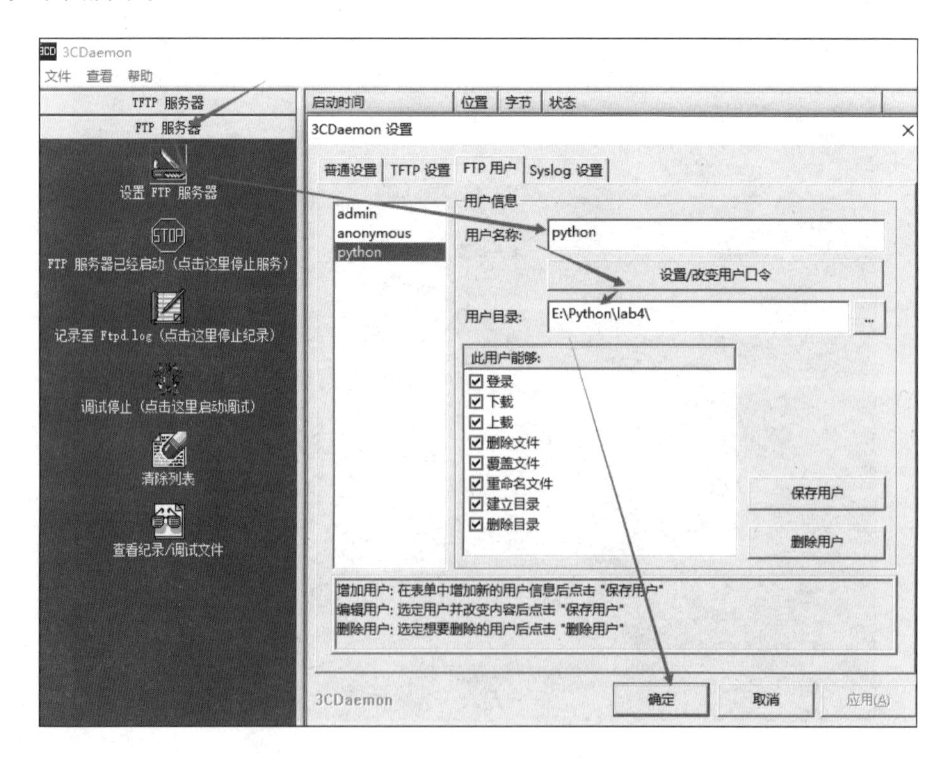

```
import paramiko
import time
import getpass

username = input('Username: ')
password = getpass.getpass('password: ')

f = open("ip_list.txt")

for line in f.readlines():
    ip_address = line.strip()
    ssh_client = paramiko.SSHClient()
    ssh_client.set_missing_host_key_policy(paramiko.AutoAddPolicy())
    ssh_client.connect(hostname=ip_address,username=username,
                       password=password,look_for_keys=False)
    print ("Successfully connect to ", ip_address)

    command = ssh_client.invoke_shell()

    #FTP 相关操作
    command.send('ftp 192.168.2.1\n')
    time.sleep(1)
    command.send('python\n')
    time.sleep(1)
    command.send('python\n')
    time.sleep(1)
    command.send('bin\n')
    command.send('put vrpcfg.zip '+ ip_address +'_vrpcfg.zip'+'\n')
    time.sleep(1)
    command.send('quit\n')
    time.sleep(1)

    #因 Windows 10 中文版本会出现中文，不能直接用'ascii'
    output = command.recv(65535)
    print(output.decode('GB2312'))

f.close()
ssh_client.close
```

运行 Python 脚本，输入交换机用户名和密码，随即 Python 就到 SW 上进行 FTP 复制配置文件，最终在 lab4 文件夹中就有了 5 台设备的配置文件，如下图所示。

有时候换个思路，解决问题可能就有新方法了。本实验 FTP 服务端和客户端与网络工程师的常规思维是相反的。这样从“下载备份”变成了“上传备份”，备份期间各网络设备并不需要开启 FTP 服务功能，也不需要进行其他额外配置。

另外，编解码相关知识是网络工程师践行 Python 自动化的一大“拦路虎”。如若初学，建议变量名、文件路径等都用纯英文形式，不要用中文。涉及中文，一定会有编解码的问题，后面随着主线学习不断深入，再逐步把编解码相关知识编织开来。

第 5 章

Python 网络运维实战（真机）

本书所有代码均在实际网络真机设备上测试运行过。由于现网设备厂商和类型繁多，笔者在前面引例及后续各第三方模块详解部分，采用网络模拟器进行标准化梳理，以便读者检索阅读及实验复现。本书旨在将各种常见的 Python 网络自动化知识模块进行拆解细化，注重传授思路，日常读者根据实际网络情况，组合使用各知识模块，实现不同功能，类似于“乐高”积木的组合搭建。不过，考虑到如果全书仅仅以网络模拟器为对象撰写，则可能略显单薄，因而特安排“真机实战”章节。本章将以笔者在工作中遇到的若干实际案例，演示 Python 在工作实战中的应用。

◎　真机运行环境

主机操作系统：CentOS 8、Windows 10

网络设备：思科 2960、2960S、2960X、3750、3850 若干，华为 Quidway 系列、CE 系列等

网络设备 OS 版本：思科 IOS、IOS-XE，华为 VRP

Python 版本：3.10.6

实验网络拓扑：与模拟器运行环境类似，不同点是运行 Python 的主机和网络设备处在不同网段中，并且网络设备的管理 IP 地址并不连续。

5.1　实验 1　大规模批量修改交换机 QoS 的配置（思科设备）

在第 4 章的实验 2 中提到了，要使用 Python 来批量连接管理 IP 地址不连续的网络设备，可以把设备的管理 IP 地址预先写入一个文本文件，然后在代码中使用 for 循环配合 open()函数和 readlines()函数逐行读取该文本文件里的管理 IP 地址，达到循环批量登录多台网络设备的目的。

在成功登录交换机后，我们可以配合 command.send()来对网络设备“发号施令”，但在前面的例子中我们都是将要输入的命令预先写在脚本里，如 command.send("conf t\n")、command.send("router eigrp 1\n")和 command.send("end\n")等。这种将配置命令预先写在脚本里的方法便于初学者理解和学习，在只有几台设备的实验环境中常用。但是在有成千上万台网络设备需要管理的生产环境中，这种方法显得很笨拙，缺乏灵活性。举例来说，假设生产环境中有不同型号、不同操作系统、不同命令格式的设备各 1 000 台，比如思科的 3750 和 3850 交换机，前者运行的是 IOS，后者运行的是 IOS-XE。

最近你接到任务，需要分别给这两种交换机批量修改 QoS 的配置，因为两者的命令格式差异巨大（一个是 MLS QoS，一个是 MQC QoS），必须反复修改 command.send()部分的代码。如果只是简单数条命令还好办，一旦遇到大规模的配置，那么这种方法的效率会很低。

解决这个问题的思路是分别创建两个文本文件，一个用来存放配置 3750 交换机要用的命令集，另一个用来存放配置 3850 交换机要用到的命令集，然后在 Python 脚本里同样通过 for 循环加 open()函数来读取两个文件里的内容，达到分别给所有 3750 和 3850 交换机做 QoS 配置的目的，这样做的好处是无须修改 command.send()部分的代码，因为所有的命令行已经在文本文件里预先设置好了。

但是新的问题又来了，每次配备不同型号的设备，都必须手动修改 open()函数所打开的配置文本文件及 IP 地址文件。如给 3750 交换机做配置时，需要 open('command_ 3750.txt')和 open('ip_3750.txt')；给 3850 交换机做配置时，又需要 open('command_3850.txt')和 open('ip_3850.txt')，这样一来二去修改配置脚本的做法大大缺乏灵活性。如果只有两种不同型号、不同命令格式的设备还能应付，那么当生产环境中同时使用 3750（IOS）、3850（IOS-XE）、Nexus 3k/5k/7k/9k（NX-OS）、CRS3/ASR9K（IOS-XR），甚至其他厂商的设备，而又要对所有这些设备同时修改某个共有的配置。例如网络新添加了某台 TACACS 服务器，要统一给所有设备修改它们的 AAA 配置；又或者网络新添加了某台 NMS 系统，要统一给所有设备修改 SNMP 配置。因为不同 OS 的设备的配置命令完全不同，这时就能体会到痛苦了。此时我们可以用下面实验中的 sys.argv 来解决这个问题。

5.1.1 实验背景

本实验将在真机上完成。

- 假设现在手边有 3 台管理 IP 地址在 192.168.100.x/24 网段的 3750 交换机和 3 台管理

IP 地址在 172.16.100.x/24 网段的 3850 交换机，它们的 hostname 和管理 IP 地址分别如下。

3750_1: 192.168.100.11
3750_2: 192.168.100.22
3750_3: 192.168.100.33

3850_1: 172.16.100.11
3850_2: 172.16.100.22
3850_3: 172.16.100.33

5.1.2　实验目的

- 修改所有 3750 和 3850 交换机的 QoS 配置，更改它们出队列（output queue）的队列参数集 2（queue-set 2）的缓存（buffers）配置，给队列 1、2、3 和 4 分别分配 15%、25%、40%和 20%的缓存（默认状况下是 25%、25%、25%和 25%）。

5.1.3　实验准备

（1）首先创建名为 command_3750.txt 和 ip_3750.txt 的两个文本文件，分别用来保存我们将要配置 3750 交换机的 QoS 命令，以及所有 3750 交换机的管理 IP 地址。

```
[root@CentOS-Python ~]#cat command_3750.txt
configure terminal
mls qos queue-set output 1 buffers 15 25 40 20
end
wr mem

[root@CentOS-Python ~]#cat ip_3750.txt
192.168.100.11
192.168.100.22
192.168.100.33
```

（2）同理，创建名为 command_3850.txt 和 ip_3850.txt 的两个文本文件，分别用来保存我们将要配置 3850 交换机的 QoS 命令，以及所有 3850 交换机的管理 IP 地址。

```
[root@CentOS-Python ~]#cat command_3850.txt
configure terminal
class-map match-any cos7
match cos 7
class-map match-any cos1
match cos 1
exit
policy-map queue-buffer
class cos7
bandwidth percent 10
queue-buffers ratio 15
class cos1
bandwidth percent 30
queue-buffers ratio 25
exit
exit
interface gi1/0/1
service-policy output queue-buffer
end
wr mem

[root@CentOS-Python ~]#cat ip_3850.txt
172.16.100.11
172.16.100.22
172.16.100.33
```

（3）在主机上创建实验 1 的脚本，将其命名为 lab1.py，如下图所示。

```
[root@CentOS-Python ~]#
[root@CentOS-Python ~]# vi lab1.py_
```

5.1.4　实验代码

将下列代码写入脚本 lab1.py。

```
import paramiko
import time
```

```
import getpass
import sys

username = input('username: ')
password = getpass.getpass('password: ')
ip_file = sys.argv[1]
cmd_file = sys.argv[2]

iplist = open(ip_file, 'r')
for line in iplist.readlines():
    ip = line.strip()
    ssh_client = paramiko.SSHClient()
    ssh_client.set_missing_host_key_policy(paramiko.AutoAddPolicy())
    ssh_client.connect(hostname=ip, username=username, password=password)
    print "You have successfully connect to ", ip
    command = ssh_client.invoke_shell()
    cmdlist = open(cmd_file, 'r')
    cmdlist.seek(0)
    for line in cmdlist.readlines():
        command.send(line + "\n")
        time.sleep(1)
    cmdlist.close()
    output = command.recv(65535)
    print (output.decode("ascii"))

iplist.close()
ssh_client.close
```

5.1.5　代码分段讲解

（1）因为要用到 sys.argv，所以我们导入了 sys 模块。sys 模块是 Python 中十分常用的内建模块。其余部分的代码都是在第 4 章中讲过的，不再赘述。

```
import paramiko
import time
import getpass
import sys
```

```
username = input('username: ')
password = getpass.getpass('password: ')
```

（2）创建两个变量：ip_file 和 cmd_file，分别对应 sys.argv[1]和 sys.argv[2]。

```
ip_file = sys.argv[1]
cmd_file = sys.argv[2]
```

argv 是 argument variable 参数变量的简写形式，这个变量的返回值是一个列表，该列表中的元素即我们在主机命令行里运行 Python 脚本时输入的命令。sys.argv[0] 一般是被调用的.py 脚本的文件名，从 sys.argv[1]开始就是为这个脚本添加的参数。举个例子，我们现在返回主机，输入下面这条命令。

```
[root@CentOS-Python ~]#python3.10 lab1.py ip_3750.txt cmd_3750.txt
```

那么，这时的 sys.argv 即含有 lab1.py、ip_3750.txt、cmd_3750.txt 3 个元素的列表。这时，sys.argv[0] = lab1.py，sys.argv[1] = ip_3750.txt，sys.argv[2] = cmd_3750.txt。相应地，代码里的 ip_file = sys.argv[1] 此时等同于 ip_file = ip_3750.txt，cmd_file = sys.argv[2]此时等同于 cmd_file = cmd_3750.txt。同理，如果这时我们在主机上执行如下命令。

```
[root@CentOS-Python ~]#python3.10 lab1.py ip_3850.txt cmd_3850.txt
```

则此时 ip_file = ip_3850.txt，cmd_file = cmd_3850.txt。由此可见，配合 sys.argv，我们可以很灵活地选用脚本需要调用的参数（文本文件），而无须反反复复地修改脚本代码。

（3）需要注意的是，在剩下的代码中，我们没有在脚本里预先写好具体的 QoS 配置命令，取而代之的是通过 cmd_file = sys.argv[2]配合 cmdlist = open(cmd_file, 'r')和 for line in cmdlist.readlines()来读取独立于脚本之外的配置命令文件，可以随意在命令行里选择我们想要的配置命令文件，也就是本实验中的 cmd_3750.txt 和 cmd_3850.txt。

```
iplist = open(ip_file, 'r')
for line in iplist.readlines():
    ip = line.strip()
    ssh_client = paramiko.SSHClient()
    ssh_client.set_missing_host_key_policy(paramiko.AutoAddPolicy())
    ssh_client.connect(hostname=ip, username=username, password=password)
    print "You have successfully connect to ", ip
    command = ssh_client.invoke_shell()
    cmdlist = open(cmd_file, 'r')
    cmdlist.seek(0)
    for line in cmdlist.readlines():
```

```
        command.send(line + "\n")
        time.sleep(1)
    cmdlist.close()
    output = command.recv(65535)
    print (output.decode("ascii"))

iplist.close()
ssh_client.close
```

5.1.6　验证

```
[root@CentOS-Python ~]#python3.10 lab1.py ip_3750.txt cmd_3750.txt
Username: python
password:
You have successfully connect to 192.168.100.11
3750_1#conf t
3750_1(config)#mls qos queue-set output 1 buffers 15 25 40 20
3750_1(config)#end
3750_1#wr mem
Building configuration...
[OK]

You have successfully connect to 192.168.100.22
3750_2#conf t
3750_2(config)#mls qos queue-set output 1 buffers 15 25 40 20
3750_2(config)#end
3750_2#wr mem
Building configuration...
[OK]

You have successfully connect to 192.168.100.33
3750_3#conf t
3750_3(config)#mls qos queue-set output 1 buffers 15 25 40 20
3750_3(config)#end
3750_3#wr mem
Building configuration...
[OK]
```

```
[root@CentOS-Python ~]#python lab1.py ip_3850.txt cmd_3850.txt
Username: python
password:

You have successfully connect to 172.16.100.11
3850_1#configure terminal
Enter configuration commands,  one per line.  End with CNTL/Z.
3850_1(config)#class-map match-any cos7
3850_1(config-cmap)#match cos 7
3850_1(config-cmap)#class-map match-any cos1
3850_1(config-cmap)#match cos 1
3850_1(config-cmap)#exit
3850_1(config)#policy-map queue-buffer
3850_1(config-pmap)#class cos7
3850_1(config-pmap-c)#bandwidth percent 10
3850_1(config-pmap-c)#queue-buffers ratio 15
3850_1(config-pmap-c)#class cos1
3850_1(config-pmap-c)#bandwidth percent 30
3850_1(config-pmap-c)#queue-buffers ratio 25
3850_1(config-pmap-c)#exit
3850_1(config-pmap)#exit
3850_1(config)#interface gi1/0/1
3850_1(config-if)#service-policy output queue-buffer
3850_1(config-if)#end
3850_1#wr mem
Building configuration...
Compressed configuration from 62654 bytes to 19670 bytes[OK]

You have successfully connect to 172.16.100.22
3850_2#configure terminal
Enter configuration commands,  one per line.  End with CNTL/Z.
3850_2(config)#class-map match-any cos7
3850_2(config-cmap)#match cos 7
3850_2(config-cmap)#class-map match-any cos1
3850_2(config-cmap)#match cos 1
3850_2(config-cmap)#exit
3850_2(config)#policy-map queue-buffer
3850_2(config-pmap)#class cos7
3850_2(config-pmap-c)#bandwidth percent 10
```

```
3850_2(config-pmap-c)#queue-buffers ratio 15
3850_2(config-pmap-c)#class cos1
3850_2(config-pmap-c)#bandwidth percent 30
3850_2(config-pmap-c)#queue-buffers ratio 25
3850_2(config-pmap-c)#exit
3850_2(config-pmap)#exit
3850_2(config)#interface gi1/0/1
3850_2(config-if)#service-policy output queue-buffer
3850_2(config-if)#end
3850_2#wr mem
Building configuration...
Compressed configuration from 62654 bytes to 19670 bytes[OK]

You have successfully connect to 172.16.100.33
3850_3#configure terminal
Enter configuration commands, one per line.  End with CNTL/Z.
3850_3(config)#class-map match-any cos7
3850_3(config-cmap)#match cos 7
3850_3(config-cmap)#class-map match-any cos1
3850_3(config-cmap)#match cos 1
3850_3(config-cmap)#exit
3850_3(config)#policy-map queue-buffer
3850_3(config-pmap)#class cos7
3850_3(config-pmap-c)#bandwidth percent 10
3850_3(config-pmap-c)#queue-buffers ratio 15
3850_3(config-pmap-c)#class cos1
3850_3(config-pmap-c)#bandwidth percent 30
3850_3(config-pmap-c)#queue-buffers ratio 25
3850_3(config-pmap-c)#exit
3850_3(config-pmap)#exit
3850_3(config)#interface gi1/0/1
3850_3(config-if)#service-policy output queue-buffer
3850_3(config-if)#end
3850_3#wr mem
Building configuration...
Compressed configuration from 62654 bytes to 19670 bytes[OK]
```

5.2　实验 2　pythonping 的使用方法（思科设备）

在第 3 章中，我们曾经提到过在 Python 中用来执行 ping 命令的模块有很多种，os、subprocess 及 pyping 都可以用来 ping 指定的 IP 地址或者 URL。三者的区别是 os 和 subprocess 在执行 ping 命令时脚本会将系统执行 ping 时的回显内容显示出来，有时这些回显内容并不是必要的，例如下面是用 subprocess 模块在 CentOS 主机上执行 ping -c 3 www.cisco.com 命令的脚本及执行脚本后的回显内容。

```
[root@CentOS-Python ~]#cat ping.py

import subprocess

target = 'www.cisco.com'
ping_result = subprocess.call(['ping','-c','3',target])
if ping_result == 0:
    print (target + ' is reachable.')
else:
    print (target + ' is not reachable.')


[root@CentOS-Python ~]#python3.10 ping.py
PING www.cisco.com (27.151.12.183) 56(84) bytes of data.
64 bytes from 27.151.12.183 (27.151.12.183): icmp_seq=1 ttl=55 time=19.8 ms
64 bytes from 27.151.12.183 (27.151.12.183): icmp_seq=2 ttl=55 time=19.9 ms
64 bytes from 27.151.12.183 (27.151.12.183): icmp_seq=3 ttl=55 time=19.9 ms

--- www.cisco.com ping statistics ---
3 packets transmitted, 3 received, 0% packet loss, time 5ms
rtt min/avg/max/mdev = 19.765/19.875/19.935/0.077 ms
www.cisco.com is reachable.
[root@CentOS-Python ~]#
```

由上面的代码可以明显看到，上述回显内容过多，在用脚本一次性 ping 成百上千个 IP 地址或者 URL 时非常影响美观和阅读，因为我们真正关心的其实是最后一句用 Python 打印出来的通知用户目标 IP 地址或者 URL 是否可达的内容“www.cisco.com is reachable”，而用 pyping 来执行 ping 命令则不会有回显内容过多的问题。很遗憾的是，pyping 只支持 Python 2，并且截至 2020 年 5 月，作者似乎也没有继续更新 pyping 来支持 Python 3 的意愿（虽然 pyping 依然能通过 pip3 下载 Python 3 的版本，但是使用时会报错“ModuleNotFoundError: No module named 'core'”）。

在 Python 3.10 里，我们可以使用 **pythonping** 作为 pyping 的替代品。本节实验将详细介绍 pythonping 的使用方法。

5.2.1　实验背景

本实验将在真机上完成。

- 某公司有 48 口的思科 3750 交换机共 1000 台，分别分布在 5 个子网掩码为/24 的 B 类网络子网下：172.16.0.x /24，172.16.1.x /24，172.16.2.x /24，172.16.3.x /24，172.16.4.x /24。

5.2.2　实验目的

- 在不借助任何第三方 NMS 软件或网络安全工具帮助的前提下，使用 Python 脚本依次 ping 所有交换机的管理 IP 地址，来确定当前（需要记录下运行脚本时的时间，要求精确到年月日和时分秒）有哪些交换机可达，并且统计当前每个交换机有多少用户端的物理端口是 up 的（级联端口不算），以及这 1000 台交换机所有 up 的用户端物理端口的总数，并统计网络里的端口使用率（也就是物理端口的 up 率）。

5.2.3　实验思路

- 根据实验目的，我们可以写两个脚本，脚本 1 通过导入 pythonping 模块来扫描这 5 个网段下所有交换机的管理 IP 地址，看哪些管理 IP 地址是可达的。因为子网掩码是 255.255.255.0，意味着每个网段下的管理 IP 地址的前三位都是固定不变的，只有最后一位会在 1 ~ 254 中变化。我们可以在第一个脚本中使用 for 循环来 ping .1 到.254，然后将所有该网段下可达的交换机管理 IP 地址都写入并保存在一个名为 reachable_ip.txt 的文本文件中。
- 因为这里有 5 个连续的/24 的网段需要扫描（从 172.16.0.x 到 172.16.4.x），我们可以在脚本 1 中再写一个 for 循环来连续 ping 这 5 个网段，然后把上一个 ping .1 到.254 的 for 循环嵌入这一个 for 循环，这样就能让 Python 一次性把 5 个/24 网段下总共 1270 个可用 IP 地址（254 × 5 = 1270）全部 ping 一遍。

- 在用脚本 1 生成 reachable_ip.txt 文件后，我们可以再写一个脚本 2 来读取该文本文件中所有可达的交换机的管理 IP 地址，依次登录所有这些可达的交换机，输入命令 show ip int brief | i up 查看哪些端口是 up 的，再配合正则表达式（re 模块），在回显内容中匹配我们所要的用户端物理端口号（Gix/x/x），统计它们的总数，即可得到当前一个交换机有多少个物理端口是 up 的。（注：因为 show ip int brief | i up 的回显内容里也会出现 10GB 的级联端口 Tex/x/x 及虚拟端口，比如 VLAN 或者 loopback 端口，所以这里强调的是用正则表达式来匹配用户端物理端口 Gix/x/x）。

5.2.4　实验准备——脚本 1

（1）pythonping 为第三方模块，使用前需要通过 pip 下载安装，安装完成后进入 Python 3.10 编辑器，如果 import pythonping 没有报错，则说明安装成功，如下图所示。

```
[root@localhost ~]# pip3.10 install pythonping
Collecting pythonping
  Using cached pythonping-1.1.2-py3-none-any.whl (15 kB)
Installing collected packages: pythonping
Successfully installed pythonping-1.1.2
WARNING: Running pip as the 'root' user can result in broken permissions and con
flicting behaviour with the system package manager. It is recommended to use a v
irtual environment instead: https://pip.pypa.io/warnings/venv

[      ] A new release of pip available: [illegible] -> 22.2.2
[      ] To update, run: pip install --upgrade pip
[root@localhost ~]# python3.10
Python 3.10.6 (main, Aug  6 2022, 13:49:15) [GCC 8.5.0 20210514 (Red Hat 8.5.0-1
5)] on linux
Type "help", "copyright", "credits" or "license" for more information.
>>> import pythonping
>>> 
```

（2）在主机上创建一个新的文件夹，取名为 lab2，在该文件夹下创建实验 2 的脚本 1 文件 lab2_1.py，如下图所示。

```
[root@CentOS-Python ~]# mkdir lab2
[root@CentOS-Python ~]# cd lab2
[root@CentOS-Python lab2]# vi lab2_1.py
```

5.2.5　实验代码——脚本 1

将下列代码写入脚本 lab2_1.py。

```
from pythonping import ping
import os

if os.path.exists('reachable_ip.txt'):
    os.remove('reachable_ip.txt')

third_octet = range(5)
last_octet = range(1, 255)

for ip3 in third_octet:
    for ip4 in last_octet:
        ip = '172.16.' + str(ip3) + '.' + str(ip4)
        ping_result = ping(ip)
        f = open('reachable_ip.txt', 'a')
        if 'Reply' in str(ping_result):
            print (ip + ' is reachable. ')
            f.write(ip + "\n")
        else:
            print (ip + ' is not reachable. ')

f.close()
```

5.2.6　脚本 1 代码分段讲解

（1）我们导入 pythonping 和 os 两个模块。在 pythonping 中，最核心的函数显然是 ping()，因为 pythonping 的模块名长度偏长，这里用 from...import...将 ping()函数导入后，后面调用时就能省去使用 pythonping.ping()完整函数路径的麻烦，体会到直接使用 ping()的便利。至于 os 模块的用法下面会讲到。

```
from pythonping import ping
import os
```

（2）每次我们运行脚本 1，都不希望保留上一次运行脚本时生成的 reachable_ip.txt 文件，因为在有成千上万台设备的大型网络里，每时每刻可达交换机的数量都有可能改变。这时可以用 os 模块下的 os.path.exists()方法来判断该文件是否存在，如果存在，则用 os.remove()方法将该文件删除。这样可以保证每次运行脚本 1 时，reachable_ip.txt 文件中只会包含本次运行脚本后所有可达的交换机管理 IP 地址。

```
if os.path.exists('reachable_ip.txt'):
    os.remove('reachable_ip.txt')
```

（3）实验思路中讲到，5 个/24 网段的管理 IP 地址是有规律可循的，头两位都不变，为 172.16，第三位为 0 ~ 4，第四位为 1 ~ 254，因此可以通过 range(5)和 range(1，255)分别创建两个整数列表来囊括管理 IP 地址的第三位和第四位，为后面的两个 for 循环做准备。

```
third_octet = range(5)
last_octet = range(1, 255)
```

（4）通过两个 for 循环做嵌套，依次从 172.16.0.1、172.16.0.2、172.16.0.3……一直遍历到 172.16.4.254，然后配合 ping()函数来依次 ping 所有这些管理 IP 地址（前面已经提到 ping()函数是通过 from pythonping import ping 从 pythonping 模块导入的）。

```
for ip3 in third_octet:
    for ip4 in last_octet:
        ip = '172.16.' + str(ip3) + '.' + str(ip4)
        ping_result = ping(ip)
```

（5）在实验 2 的脚本里，我们将所有可达的管理 IP 地址以追加模式（a）写入 reachable_ip.txt 文件。因为 pythonping 模块在运行过程中不显示任何回显内容，就无法知道脚本运行的进度，所以通过 print (ip + ' is reachable. ')和 print (ip + ' is not reachable. ') 分别打印出目标 IP 地址是否可达的信息。

```
        f = open('reachable_ip.txt', 'a')
        if 'Reply' in str(ping_result):
            print (ip + ' is reachable. ')
            f.write(ip + "\n")
        else:
            print (ip + ' is not reachable. ')
```

这里重点解释下 if 'Reply' in str(ping_result):的用法和原理。

在使用 os、subprocess 和 pyping 等模块做 ping 测试时，如果目标 IP 地址可达，则它们会返回整数 0；如果不可达，则返回非 0 的整数。而 pythonping 不同，在 pythonping 中，ping()函数默认对目标 IP 地址 ping 4 次，当目标 IP 地址可达时，ping()函数返回的是“Reply from x.x.x.x, x bytes in xx.xx ms”；如果不可达，则返回的是“Request timed out”，如下图所示。

```
[root@localhost ~]# python3.10
Python 3.10.6 (main, Aug  6 2022, 13:49:15) [GCC 8.5.0 20210514 (Red Hat 8.5.0-1
5)] on linux
Type "help", "copyright", "credits" or "license" for more information.
>>> from pythonping import ping
>>> ping('www.cisco.com')
Reply from 2.20.7.24, 29 bytes in 107.67ms
Reply from 2.20.7.24, 29 bytes in 107.79ms
Reply from 2.20.7.24, 29 bytes in 107.74ms
Reply from 2.20.7.24, 29 bytes in 107.83ms

Round Trip Times min/avg/max is 107.67/107.76/107.83 ms
>>> ping('10.1.1.1')
Request timed out
Request timed out
Request timed out
Request timed out

Round Trip Times min/avg/max is 2000/2000.0/2000 ms
>>>
```

也许你会问：既然 pythonping 的 ping()函数返回的不再是 0 或非 0 的整数，那么我们怎么将上面代表目标 IP 地址可达的“Reply from x.x.x.x, x bytes in xx.xx ms”和代表目标 IP 地址不可达的“Request timed out”通过 if 语句将可达的目标 IP 地址打印出来并写入 reachable_ip.txt 文件呢？也许你猜到用成员运算符 in 来判断返回值中是否有“Reply”和“Request”这两个字符串。如果有“Reply”，则说明目标可达；如果没有，则说明目标不可达，但前提是 ping()函数返回值的类型必须是字符串。而实际情况是 ping()函数返回值的类型是一个叫作 pythonping.executor.ResponseList 的特殊类型，如下图所示。

```
>>>
>>> from pythonping import ping
>>> ping_result = ping('www.cisco.com')
>>> type(ping_result)
<class 'pythonping.executor.ResponseList'>
>>>
```

因此，必须通过 str()函数将它转换成字符串后才能使用成员运算符 in 来作判断（即这里的 str(ping_result)，否则 **if 'Reply' in str(ping_result):**会永远返回布尔值 **False**，表示目标不可达。

（6）最后在退出程序前不要忘记关闭已经打开的 reachable_ip.txt 文件。

```
f.close()
```

5.2.7　脚本 1 验证

（1）运行脚本 1 前，确认/root/lab2 文件夹下只有 lab2_1.py 这一个文件，如下图所示。

```
[root@CentOS-Python lab2]# ls
lab2_1.py
[root@CentOS-Python lab2]#
```

（2）运行脚本 1，因为 1270 个管理 IP 地址实在太多，不方便截图演示，所以这里将 last_octet = range(1，255)改为 last_octet = range(1，4)，只 ping 每个网段下前 3 个管理 IP 地址，也就是总共 15 个管理 IP 地址。运行脚本后的回显内容如下图所示。

```
[root@CentOS-Python lab2]# python3.8 lab2_1.py
172.16.0.1 is reachable.
172.16.0.2 is not reachable.
172.16.0.3 is not reachable.
172.16.1.1 is not reachable.
172.16.1.2 is not reachable.
172.16.1.3 is not reachable.
172.16.2.1 is not reachable.
172.16.2.2 is not reachable.
172.16.2.3 is not reachable.
172.16.3.1 is not reachable.
172.16.3.2 is not reachable.
172.16.3.3 is not reachable.
```

（3）再次查看/root/lab2 文件夹，可以看到这时多出来一个 reachable_ip.txt 的文本文件，该文件正是脚本自动生成的，用来保存所有可达交换机的管理 IP 地址。可以看到只有 172.16.0.1 这一个管理 IP 地址被写入 reachable_ip.txt，至此证明脚本 1 运行成功，如下图所示。

```
[root@CentOS-Python lab2]# ls
lab2_1.py  reachable_ip.txt
[root@CentOS-Python lab2]# cat reachable_ip.txt
172.16.0.1
[root@CentOS-Python lab2]# _
```

5.2.8　实验准备——脚本 2

（1）讲解脚本 2 之前，先来看下在一个 48 口的思科 3750 交换机里输入命令 show ip int brief | i up 后能得到什么样的回显内容，如下图所示。

```
#show ip int b | i up
Vlan3999                                 YES NVRAM  up                    up
GigabitEthernet1/0/1   unassigned        YES unset  up                    up
GigabitEthernet1/0/3   unassigned        YES unset  up                    up
GigabitEthernet1/0/5   unassigned        YES unset  up                    up
GigabitEthernet1/0/6   unassigned        YES unset  up                    up
GigabitEthernet1/0/7   unassigned        YES unset  up                    up
GigabitEthernet1/0/9   unassigned        YES unset  up                    up
GigabitEthernet1/0/10  unassigned        YES unset  up                    up
GigabitEthernet1/0/11  unassigned        YES unset  up                    up
GigabitEthernet1/0/12  unassigned        YES unset  up                    up
GigabitEthernet1/0/15  unassigned        YES unset  up                    up
GigabitEthernet1/0/16  unassigned        YES unset  up                    up
GigabitEthernet1/0/17  unassigned        YES unset  up                    up
GigabitEthernet1/0/18  unassigned        YES unset  up                    up
GigabitEthernet1/0/20  unassigned        YES unset  up                    up
GigabitEthernet1/0/21  unassigned        YES unset  up                    up
GigabitEthernet1/0/22  unassigned        YES unset  up                    up
GigabitEthernet1/0/29  unassigned        YES unset  up                    up
GigabitEthernet1/0/31  unassigned        YES unset  up                    up
```

```
GigabitEthernet1/0/33  unassigned      YES unset  up                    up
GigabitEthernet1/0/34  unassigned      YES unset  up                    up
GigabitEthernet1/0/36  unassigned      YES unset  up                    up
GigabitEthernet1/0/37  unassigned      YES unset  up                    up
GigabitEthernet1/0/38  unassigned      YES unset  up                    up
GigabitEthernet1/0/39  unassigned      YES unset  up                    up
GigabitEthernet1/0/42  unassigned      YES unset  up                    up
GigabitEthernet1/0/45  unassigned      YES unset  up                    up
Te1/0/1                unassigned      YES unset  up                    up
Te1/0/2                unassigned      YES unset  up                    up
```

由上面可以看到，除了 GigabitEthernet 用户端物理端口，还有 VLAN 虚拟端口和两个万兆的级联端口 Te1/0/1 和 Te1/0/2。在实验目的中已经明确说明不考虑虚拟端口和级联端口，只统计共有多少个用户端物理端口是 up 的。

（2）创建实验 2 的第二个脚本，取名为 lab2_2.py，如下图所示。

```
[root@CentOS-Python lab2]#
[root@CentOS-Python lab2]# vi lab2_2.py
```

5.2.9　实验代码——脚本 2

将下列代码写入脚本 lab2_2.py。

```
import paramiko
import time
import re
from datetime import datetime
import socket
import getpass

username = input('Enter your SSH username: ')
password = getpass.getpass('Enter your SSH password: ')
now = datetime.now()
date = "%s-%s-%s" % (now.month, now.day, now.year)
time_now = "%s:%s:%s" % (now.hour, now.minute, now.second)

switch_with_tacacs_issue = []
switch_not_reachable = []
total_number_of_up_port = 0

iplist = open('reachable_ip.txt')
number_of_switch = len(iplist.readlines())
total_number_of_ports = number_of_switch * 48

iplist.seek(0)
for line in iplist.readlines():
```

```
    try:
        ip = line.strip()
        ssh_client = paramiko.SSHClient()
        ssh_client.set_missing_host_key_policy(paramiko.AutoAddPolicy())
        ssh_client.connect(hostname=ip, username=username, password=password)
        print ("\nYou have successfully connect to ", ip)
        command = ssh_client.invoke_shell()
        command.send('term len 0\n')
        command.send('show ip int b | i up\n')
        time.sleep(1)
        output = command.recv(65535)
        #print (output)
        search_up_port = re.findall(r'GigabitEthernet', output)
        number_of_up_port = len(search_up_port)
        print (ip + " has " + str(number_of_up_port) + " ports up.")
        total_number_of_up_port += number_of_up_port
    except Paramiko.ssh_exception.AuthenticationException:
        print ("TACACS is not working for " + ip + ".")
        switch_with_tacacs_issue.append(ip)
    except socket.error:
        print (ip + " is not reachable.")
        switch_not_reachable.append(ip)
iplist.close()

print ("\n")
print ("There are totally " + str(total_number_of_ports) + " ports available in the
network.")
print (str(total_number_of_up_port) + " ports are currently up.")
print ("Port up rate is %.2f%%" % (total_number_of_up_port / float(total_number_of_ports)
* 100))
print ('\nTACACS is not working for below switches: ')
for i in switch_with_tacacs_issue:
    print (i)
print ('\nBelow switches are not reachable: ')
for i in switch_not_reachable:
    print (i)
f = open(date + ".txt", "a+")
f.write('As of ' + date + " " + time_now)
f.write("\n\nThere are totally " + str(total_number_of_ports) + " ports available in
the network.")
f.write("\n" + str(total_number_of_up_port) + " ports are currently up.")
f.write("\nPort up rate is %.2f%%" % (total_number_of_up_port / float
```

```
(total_number_of_ports) * 100))
f.write("\n*********************************************************\n\n")
f.close()
```

5.2.10　脚本 2 代码分段讲解

（1）这里导入了 datetime 模块，这个 Python 内置模块可以用来显示运行脚本时的系统日期和时间，因为实验目的里提到需要记录下运行脚本时的时间（精确到年月日和时分秒）。其余模块的用法在前面已经都讲过，这里不再赘述。

```
import paramiko
import time
import re
from datetime import datetime
import socket
import getpass
```

（2）使用 input()和 getpass.getpass()来分别提示用户输入 SSH 登录交换机的用户名和密码。记录当前时间可以调用 datetime.now()方法，我们将它赋值给变量 now。datetime.now()方法下面又包含了.year()（年）、.month()（月）、.day()（日）、.hour()（时）、.minute()（分）、.second()（秒）几个子方法，这里将“月-日-年”赋值给变量 date，将“时:分:秒”赋值给变量 time_now。

```
username = input('Enter your SSH username: ')
password = getpass.getpass('Enter your SSH password: ')
now = datetime.now()
date = "%s-%s-%s" % (now.month, now.day, now.year)
time_now = "%s:%s:%s" % (now.hour, now.minute, now.second)
```

（3）创建 switch_with_tacacs_issue 和 switch_not_reachable 两个空列表来统计有哪些交换机的 TACACS 失效导致用户验证失败，哪些交换机的管理 IP 地址不可达。另外创建一个 total_number_of_up_port 变量，将其初始值设为 0，再用累加的方法统计所有可达的 3750 交换机上状态为 up 的端口的总数。

```
switch_with_tacacs_issue = []
switch_not_reachable = []
total_number_of_up_port = 0
```

（4）用 open()函数打开脚本 1 创建的 reachable_ip.txt 文件，用 readlines()将其内容以列表形式返回，再配合 len()函数得到可达交换机的数量，因为每个交换机都有 48 个端口，所以通过交换机数量×48 可以得到端口总数（无论端口状态是否为 up）。

```
iplist = open('reachable_ip.txt')
number_of_switch = len(iplist.readlines())
total_number_of_ports = number_of_switch * 48
```

（5）因为已经用 open()函数打开过一次 reachable_ip.txt 文件，所以要用 seek(0)回到文件的起始位置。这里需要注意用来应对交换机登录失败的问题，异常处理语句 try 要写在 for 循环的下面，剩下的代码是使用 Paramiko 配合 for 循环登录交换机并进入命令行的最基础的知识点，这里不再赘述。

```
iplist.seek(0)
for line in iplist.readlines():
    try:
        ip = line.strip()
        ssh_client = Paramiko.SSHClient()
        ssh_client.set_missing_host_key_policy(Paramiko.AutoAddPolicy())
        ssh_client.connect(hostname=ip, username=username, password=password)
        print ("\nYou have successfully connect to ", ip)
        command = ssh_client.invoke_shell()
```

（6）因为是 48 口的交换机，show ip int brief | i up 的回显内容会比较长，无法一次性完整地显示，所以首先要用命令 term len 0 完整地显示所有的回显内容。“sleep”1s 后再用 recv(65535)将所有的回显内容保存在变量 output 中，如果你想在脚本执行过程中查看完整的回显内容，可以选择 print (output)；如果不想看，则在其前面加上注释符号#。

```
        command.send('term len 0\n')
        command.send('show ip int b | i up\n')
        time.sleep(1)
        output = command.recv(65535)
        #print (output)
```

（7）因为我们只想统计有多少个用户端的物理端口（GigabitEthernet）是 up 的，所以可以用正则表达式的 findall()方法去精确匹配 GigabitEthernet，将 findall()返回的列表赋值给变量 search_up_port，然后通过 len(search_up_port)即可得到 up 的物理端口的数量，并将该数量赋值给变量 number_of_port。随后打印出每个交换机有多少个用户端物理端口是 up 的。因为前面已经定义了变量 total_number_of_up_port，并将整数 0 赋值给它，所以可以通过 total_number_of_up_port += number_of_up_port 方法将每个交换机 up 的物理端口数量累加起来，最后得到整个网络下 up 的物理端口的总数。后面的异常处理语句 except 和关闭文件的部分不再赘述。

```
        search_up_port = re.findall(r'GigabitEthernet', output)
        number_of_up_port = len(search_up_port)
        print (ip + " has " + str(number_of_up_port) + " ports up.")
        total_number_of_up_port += number_of_up_port
    except Paramiko.ssh_exception.AuthenticationException:
        print ("TACACS is not working for " + ip + ".")
        switch_with_tacacs_issue.append(ip)
    except socket.error:
        print (ip + " is not reachable.")
        switch_not_reachable.append(ip)
iplist.close()
```

（8）最后，除了将各种统计信息打印出来，我们还将另外创建一个文件，通过 f = open(date + ".txt", "a + ")将运行脚本时的日期用作该脚本的名字，将统计信息写入，方便以后调阅查看。注意写入的内容里有 f.write('As of ' + date + " " + time_now)，这样可以清晰直观地看到我们是在哪一天的几时几分几秒运行的脚本。为什么要用日期作为文件名呢？这样做的好处是一旦运行脚本时的日期不同，脚本就会自动创建一个新的文件，例如 2018 年 6 月 16 日运行了一次脚本，Python 创建了一个名为 6-16-2018.txt 的文件，如果第二天再运行一次脚本，Python 又会创建一个名为 6-17-2018.txt 的文件。如果在同一天里数次运行脚本，则多次运行的结果会以追加的形式写入同一个.txt 文件，不会创建新文件。这么做可以配合 Windows 的 Task Scheduler 或者 Linux 的 Crontab 来定期自动执行脚本，每天自动生成当天的端口使用量的统计情况，方便公司管理层随时观察网络里交换机的端口使用情况。

```
print ("\n")
print ("There are totally " + str(total_number_of_ports) + " ports available in the
network.")
print (str(total_number_of_up_port) + " ports are currently up.")
print ("Port up rate is %.2f%%" % (total_number_of_up_port / float(total_number_of_ports)
* 100))
print ('\nTACACS is not working for below switches: ')
for i in switch_with_tacacs_issue:
    print (i)
print ('\nBelow switches are not reachable: ')
for i in switch_not_reachable:
    print (i)
f = open(date + ".txt", "a+")
f.write('As of ' + date + " " + time_now)
f.write("\n\nThere are totally " + str(total_number_of_ports) + " ports available in
the network.")
f.write("\n" + str(total_number_of_up_port) + " ports are currently up.")
```

```
f.write("\nPort up rate is %.2f%%" % (total_number_of_up_port /
float(total_number_of_ports) * 100))
f.write("\n*************************************************************\n\n")
f.close()
```

5.2.11　脚本 2 验证

（1）移动到/root/lab2，运行脚本 lab2_2.py，让脚本读取 reachable_ip.txt 文件自动登录所有可达的交换机（出于演示的目的，这里将之前除 172.16.0.1 外所有不可达的交换机全部开启，并将它们的管理 IP 地址写入 reachable_ip.txt），并查看每个交换机当前有多少个端口是 up 的，最后给出统计数据，如下图所示。

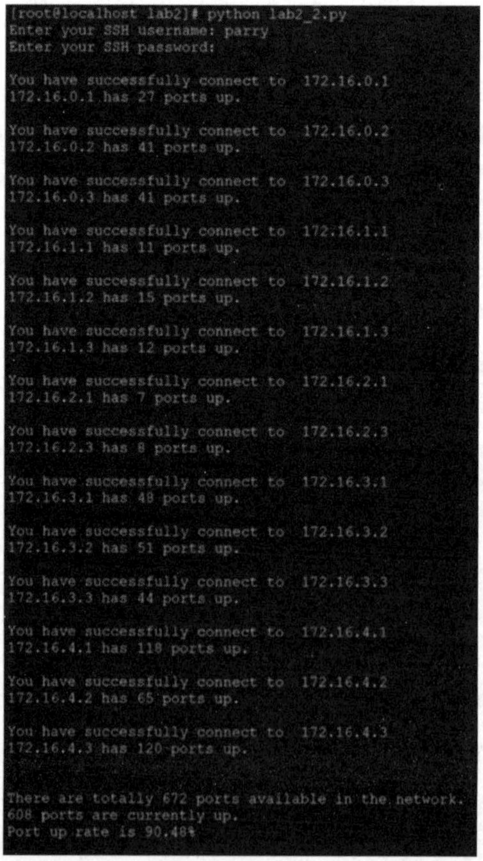

```
[root@localhost lab2]# python lab2_2.py
Enter your SSH username: parry
Enter your SSH password:

You have successfully connect to  172.16.0.1
172.16.0.1 has 27 ports up.

You have successfully connect to  172.16.0.2
172.16.0.2 has 41 ports up.

You have successfully connect to  172.16.0.3
172.16.0.3 has 41 ports up.

You have successfully connect to  172.16.1.1
172.16.1.1 has 11 ports up.

You have successfully connect to  172.16.1.2
172.16.1.2 has 15 ports up.

You have successfully connect to  172.16.1.3
172.16.1.3 has 12 ports up.

You have successfully connect to  172.16.2.1
172.16.2.1 has 7 ports up.

You have successfully connect to  172.16.2.3
172.16.2.3 has 8 ports up.

You have successfully connect to  172.16.3.1
172.16.3.1 has 48 ports up.

You have successfully connect to  172.16.3.2
172.16.3.2 has 51 ports up.

You have successfully connect to  172.16.3.3
172.16.3.3 has 44 ports up.

You have successfully connect to  172.16.4.1
172.16.4.1 has 118 ports up.

You have successfully connect to  172.16.4.2
172.16.4.2 has 65 ports up.

You have successfully connect to  172.16.4.3
172.16.4.3 has 120 ports up.


There are totally 672 ports available in the network.
608 ports are currently up.
Port up rate is 90.48%
```

（2）再次查看/root/lab2 文件夹，可以看到这时多出来一个名为 4-22-2019.txt 的文本文件，该文件是脚本 2 自动生成的，用来保存每次运行脚本后的统计信息。文件名即运行该脚本的日期。在 Linux 中配合 Crontab，在 Windows Server 中配合任务计划程序（Task Scheduler）定期每天运行该脚本，即可达到每天自动化监控交换机端口使用率的目的，如下图所示。

```
[root@localhost lab2]# ls -l
total 16
-rw-r--r-- 1 root root  195 Apr 22 08:40 4-22-2019.txt
-rw-r--r-- 1 root root  513 Apr  6 17:38 lab2_1.py
-rw-r--r-- 1 root root 2504 Apr 15 12:56 lab2_2.py
-rw-r--r-- 1 root root  154 Apr 22 08:39 reachable_ip.txt
[root@localhost lab2]# cat 4-22-2019.txt
As of 4-22-2019 8:39:44

There are totally 672 ports available in the network.
608 ports are currently up.
Port up rate is 90.48%
************************************************************

[root@localhost lab2]#
```

5.3　实验 3　利用 Python 脚本检查交换机的配置（思科设备）

前面几个实验（包括第 4 章）已经由浅入深地讲解了 Python 在网络运维中的实际应用技巧，并给出了相应的范例。这里将继续讨论如何使用 Python 来解决在大型生产网络中常见的一个网络运维自动化的需求。

5.3.1　实验背景

本实验将在真机上完成。

- 某公司有 1000 台思科 2960、100 台思科 2960S、300 台思科 2960X 交换机，均为 24 口，它们的型号和 IOS 版本分别如下表所示。

类　别	型　号	IOS 版本
2960	WS-C2960-24PC-L	c2960-lanbasek9-mz.122-55.SE5
2960S	WS-C2960S-F24PS-L	c2960s-universalk9-mz.150-2.SE5
2960X	WS-C2960X-24PS-L	c2960x-universalk9-mz.152-2.E5

最近公司决定将所有上述交换机的 IOS 版本升级，修补漏洞，消除 bug。升级后的 IOS 版本分别如下。

2960：c2960-lanbasek9-mz.122-55.SE12

2960S：c2960s-universalk9-mz.150-2.SE11

2960X：c2960x-universalk9-mz.152-2.E8

5.3.2 实验目的

- 所有交换机的新 IOS 版本已经手动完成上传，并且 boot system 的路径也改成了新的 IOS 版本，为确保 IOS 版本升级顺利完成，需要创建 Python 脚本检查所有交换机的配置是否正确，避免升级 IOS 版本的过程中出现人为的错误。

5.3.3 实验思路

升级交换机 IOS 版本的验证部分可以分为以下两个步骤。

1. 重启交换机前的验证步骤

将新版 IOS 上传到 flash 后到重启交换机前可能会遇到以下 3 种人为错误。

（1）IOS 版本和交换机型号货不对板，比如把 2960 交换机的 IOS 版本上传给了 2960X 交换机。

（2）boot system 忘记修改或者修改错误。

（3）重启交换机前，忘记执行 write memory 保存配置。

因此，在重启所有交换机前，需要检验交换机的 3 处配置。

（1）show inventory | i PID: WS。

（2）show flash: | i c2960。

上述两步的目的是查看交换机型号和闪存的 IOS 版本，避免前者和后者货不对板的情况发生。

（3）show boot | i BOOT path。

这一步的目的是检查 boot system 配置是否修改正确。

2. 重启交换机后的验证步骤

重启交换机后的验证步骤较为简单，只需要用 show ver | b SW Version 验证交换机的 IOS 版本是否已经成功升级到新版本即可。

综上所述，我们可以分别写两个脚本来对应上面两个验证步骤。脚本 1 用作重启交换机前的验证，脚本 2 用作重启交换机后的验证。

5.3.4 实验准备——脚本 1

（1）在主机上创建一个新的文件夹，取名为 lab3，在该文件夹下创建实验 3 的脚本 1 文件 lab3_1.py，如下图所示。

```
[root@CentOS-Python ~]#
[root@CentOS-Python ~]# mkdir lab3
[root@CentOS-Python ~]# cd lab3
[root@CentOS-Python lab3]# vi lab3_1.py
```

（2）在主机上创建一个 ip_list.txt 文件，该文件用来保存我们要登录验证 IOS 版本是否升级成功的所有交换机的管理 IP 地址，如下图所示。

```
[root@CentOS-Python lab3]# cat ip_list.txt
172.16.206.39
172.16.206.40
172.16.206.41
172.16.206.42
172.16.206.43
172.16.206.119
172.16.206.128
172.16.206.129
172.16.206.145
172.16.206.146
[root@CentOS-Python lab3]#
```

5.3.5 实验代码——脚本 1

将下列代码写入脚本 lab3_1.py。

```
import paramiko
import time
import getpass
import sys
import re
import socket

username = input("Username: ")
password = getpass.getpass("Password: ")
iplist = open('ip_list.txt', 'r+')
```

```
switch_upgraded = []
switch_not_upgraded = []
switch_with_tacacs_issue = []
switch_not_reachable = []

for line in iplist.readlines():
    try:
        ip_address = line.strip()
        ssh_client = paramiko.SSHClient()
        ssh_client.set_missing_host_key_policy(paramiko.AutoAddPolicy())
        ssh_client.connect(hostname=ip_address, username=username, password=password)
        print ("Successfully connect to ", ip_address)
        command = ssh_client.invoke_shell(width=300)
        command.send("show inventory | i PID: WS\n")
        time.sleep(0.5)
        command.send("show flash: | i c2960\n")
        time.sleep(0.5)
        command.send("show boot | i BOOT path\n")
        time.sleep(0.5)
        output = command.recv(65535)
        command.send("wr mem\n")
        switch_model = re.search(r'WS-C2960\w?-\w{4, 5}-L', output)
        ios_version = re.search(r'c2960\w?-\w{8, 10}\d?-mz.\d{3}-\d{1, 2}.\w{2, 4}
(.bin)?', output)
        boot_system = re.search(r'flash:.+mz.\d{3}-\d{1, 2}\.\w{2, 4}\.bin', output)
        if switch_model.group() == "WS-C2960-24PC-L" and ios_version.group() ==
"c2960-lanbasek9-mz.122-55.SE12.bin" and boot_system.group() ==
'flash:c2960-lanbasek9-mz.122-55.SE12.bin' or boot_system.group() ==
'flash:/c2960-lanbasek9-mz.122-55.SE12.bin':
            switch_upgraded.append(ip_address)
        elif switch_model.group() == "WS-C2960S-F24PS-L" and ios_version.group() ==
"c2960s-universalk9-mz.150-2.SE11.bin" and boot_system.group() ==
'flash:c2960s-universalk9-mz.150-2.SE11.bin' or boot_system.group() ==
'flash:/c2960s-universalk9-mz.150-2.SE11.bin':
            switch_upgraded.append(ip_address)
```

```
        elif switch_model.group() == "WS-C2960X-24PS-L" and ios_version.group() ==
"c2960x-universalk9-mz.152-2.E8.bin" and boot_system.group() ==
'flash:c2960x-universalk9-mz.152-2.E8.bin' or boot_system.group() ==
'flash:/c2960x-universalk9-mz.152-2.E8.bin':
            switch_upgraded.append(ip_address)
        else:
            switch_not_upgraded.append(ip_address)
    except paramiko.ssh_exception.AuthenticationException:
        print ("TACACS is not working for " + ip_address + ".")
        switch_with_tacacs_issue.append(ip_address)
    except socket.error:
        print (ip_address +  " is not reachable.")
        switch_not_reachable.append(ip_address)

iplist.close()
ssh_client.close

print ('\nTACACS is not working for below switches: ')
for i in switch_with_tacacs_issue:
    print (i)

print ('\nBelow switches are not reachable: ')
for i in switch_not_reachable:
    print (i)

print ('\nBelow switches IOS version are up-to-date: ')
for i in switch_upgraded:
    print (i)

print ('\nBelow switches IOS version are not updated yet: ')
for i in switch_not_upgraded:
    print (i)
```

5.3.6　脚本 1 代码分段讲解

（1）在代码开始的部分导入模块，创建 4 个空列表，用来统计有多少台交换机已经成功升级 IOS 版本，有多少台交换机没有升级 IOS 版本，有哪些交换机因为 TACACS 或者管理 IP 地址不可达而无法登录，以及配合 for loop 和 Paramiko 依次循环 SSH 登录所有交换机，这些部分都是老生常谈的话题，这里不再赘述。

```
import paramiko
import time
import getpass
import sys
import re
import socket

username = input("Username: ")
password = getpass.getpass("Password: ")
iplist = open('ip_list.txt', 'r+')

switch_upgraded = []
switch_not_upgraded = []
switch_with_tacacs_issue = []
switch_not_reachable = []

for line in iplist.readlines():
    try:
        ip_address = line.strip()
        ssh_client = paramiko.SSHClient()
        ssh_client.set_missing_host_key_policy(paramiko.AutoAddPolicy())
        ssh_client.connect(hostname=ip_address, username=username, password=password)
        print ("Successfully connect to ", ip_address)
```

（2）因为后面要用 show flash: | i c2960 查看闪存下面的 IOS 文件，而 IOS 文件名通常很长，如果不调整宽度，则会导致后面截取的 output 不完整，从而影响正则表达式对关键词的匹配。所以可以用 command = ssh_client.invoke_shell(width=300)来调整 Paramiko 回显内容的宽度（默认为 100）。

```
command = ssh_client.invoke_shell(width=300)
```

（3）command.send("show inventory | i PID: WS\n")、command.send("show flash: | i c2960\n")和 command.send("show boot | i BOOT path\n")这 3 个命令的作用前面已经讲过，这里给大家看下在生产网络下的 2960、2960S 和 2960X 交换机中执行这 3 条命令得到的输出结果（均为 IOS 上传完毕，重启交换机前的输出结果），如下图所示。

2960 交换机

```
AS-2960-EC3-G-3630#show inventory | i PID: WS
PID: WS-C2960-24PC-L    , VID: V02   ,
AS-2960-EC3-G-3630#
AS-2960-EC3-G-3630#
AS-2960-EC3-G-3630#show flash: | i c2960
    2  -rwx     9827106   Mar 4 2018 09:32:04 +03:00  c2960-lanbasek9-mz.122-55.SE12.bin
    4  drwx         512   Mar 1 1993 03:08:37 +03:00  c2960-lanbase-mz.122-44.SE2
AS-2960-EC3-G-3630#
AS-2960-EC3-G-3630#
AS-2960-EC3-G-3630#show boot | i BOOT path
BOOT path-list      : flash:c2960-lanbasek9-mz.122-55.SE12.bin
AS-2960-EC3-G-3630#
```


2960S 交换机

```
AS-2960-EC3-G-3748#show inventory | i PID: WS
PID: WS-C2960S-F24PS-L , VID: V01   ,
AS-2960-EC3-G-3748#
AS-2960-EC3-G-3748#
AS-2960-EC3-G-3748#show flash: | i c2960
    4  -rwx    14572032  May 10 2018 14:48:49 +03:00  c2960s-universalk9-mz.150-2.SE11.bin
    7  drwx         512   Mar 1 1993 03:16:11 +03:00  c2960s-universalk9-mz.150-2.SE5
AS-2960-EC3-G-3748#
AS-2960-EC3-G-3748#
AS-2960-EC3-G-3748#show boot | i BOOT path
BOOT path-list      : flash:c2960s-universalk9-mz.150-2.SE11.bin
```


2960X 交换机

```
AS-2960-EC3-G-3622#show inventory | i PID: WS
PID: WS-C2960X-24PS-L   , VID: V04
AS-2960-EC3-G-3622#
AS-2960-EC3-G-3622#
AS-2960-EC3-G-3622#show flash: | i c2960
    5  -rwx    21287936   Apr 9 2018 15:40:23 +03:00  c2960x-universalk9-mz.152-2.E8.bin
AS-2960-EC3-G-3622#
AS-2960-EC3-G-3622#
AS-2960-EC3-G-3622#show boot | i BOOT path
BOOT path-list      : flash:/c2960x-universalk9-mz.152-2.E8.bin
AS-2960-EC3-G-3622#
```


从上面的 show inventory | i PID: WS 的回显内容可以看到，3 种交换机的型号分别为 WS-C2960-24PC-L、WS-C2960S-F24PS-L 和 WS-C2960X-24PS-L，可以用正则表达式 switch_model = re.search(r'WS-C2960\w?-\w{4，5}-L'， output)来匹配。

再来看 show flash: | i c2960 的输出结果，我们得到了对应 3 种交换机型号的 IOS 的.bin 文件名分别为 c2960-lanbasek9-mz.122-55.SE12.bin、c2960s-universalk9-mz.150-2. SE11.bin 和 c2960x-universalk9-mz.152-2.E8.bin，可以用正则表达式 ios_version = re.search(r'c2960\w?-\w{8，10}\d?-mz.\d{3}-\d{1，2}.\w{2，4}(.bin)?'，output)来匹配。

同理，最后的 show boot | i BOOT path 分别得到了 flash:c2960-lanbasek9-mz.122-55. SE12.bin、flash:c2960s-universalk9-mz.150-2.SE11.bin 和 flash:/c2960x-universalk9-mz.152-2. E8.bin 3 种不同的 boot_system 引导路径，用正则表达式 boot_system = re.search(r'flash:/?c2960\w?-\w{9，11}-mz.\d{3}-\d{1，2}.\w{2，4}.bin'，output)来匹配。如果你足够细心，则会发现 boot_system 路径可以写成 flash:，也可以写成 flash:/，有些网络工程师喜欢加/，有些网络工程师不喜欢加，所以在正则表达式中我们用 flash:/?将两种情况都匹配到了。

```
        command.send("show inventory | i PID: WS\n")
        time.sleep(0.5)
        command.send("show flash: | i c2960\n")
        time.sleep(0.5)
        command.send("show boot | i BOOT path\n")
        time.sleep(0.5)
        output = command.recv(65535)
        command.send("wr mem\n")
        switch_model = re.search(r'WS-C2960\w?-\w{4, 5}-L',  output)
        ios_version = re.search(r'c2960\w?-\w{8, 10}\d?-mz.\d{3}-\d{1, 2}.\w{2,
4}(.bin)?',  output)
        boot_system = re.search(r'flash:.+mz.\d{3}-\d{1, 2}\.\w{2, 4}\.bin',  output)
```

（4）用正则表达式匹配所有可能出现的输出结果后，接下来就可以用 if 语句配合 and 和 or 两个布尔逻辑运算来作判断，先看第一条匹配 2960 交换机的 if 语句。

```
        if switch_model.group() == "WS-C2960-24PC-L" and ios_version.group() ==
"c2960-lanbasek9-mz.122-55.SE12.bin" and boot_system.group() ==
'flash:c2960-lanbasek9-mz.122-55.SE12.bin' or boot_system.group() ==
'flash:/c2960-lanbasek9-mz.122-55.SE12.bin':
            switch_upgraded.append(ip_address)
```

如果交换机型号为 WS-2960-24PC-L 且交换机的 IOS 版本为 c2960-lanbasek9-mz.122-55.SE12.bin，并且 boot_system 的引导路径为 flash:c2960-lanbasek9-mz.122-55.SE12.bin（不加/）或者 flash:/c2960-lanbasek9-mz.122-55.SE12.bin（加/），则将该交换机的管理 IP 地址加入 switch_upgraded 列表。

同理，再来看第二条匹配 2960S 交换机的 elif 语句。

```
        elif switch_model.group() == "WS-C2960X-24PS-L" and ios_version.group() ==
"c2960x-universalk9-mz.152-2.E8.bin" and boot_system.group() ==
'flash:c2960x-universalk9-mz.152-2.E8.bin' or boot_system.group() ==
'flash:/c2960x-universalk9-mz.152-2.E8.bin':
            switch_upgraded.append(ip_address)
```

如果交换机型号为 WS-C2960S-F24PS-L 且交换机的 IOS 版本为 c2960s-universalk9-mz.150-2.SE11.bin，并且 flash:c2960s-universalk9-mz.150-2.SE11.bin（不加/）或者 flash:/c2960s-universalk9-mz.150-2.SE11.bin（加/），则将该交换机的管理 IP 地址加入 switch_upgraded 列表。

第三条匹配 2960X 交换机的同理，这里略过不讲。

如果任何交换机都不满足以上 3 条匹配条件，则将它的管理 IP 地址加入 switch_not_upgraded 列表。

```
        elif switch_model.group() == "WS-C2960X-24PS-L" and ios_version.group() ==
"c2960x-universalk9-mz.152-2.E8.bin" and boot_system.group() ==
'flash:c2960x-universalk9-mz.152-2.E8.bin' or boot_system.group() ==
'flash:/c2960x-universalk9-mz.152-2.E8.bin':
            switch_upgraded.append(ip_address)
        else:
            switch_not_upgraded.append(ip_address)
```

（5）最后将 switch_upgraded 和 switch_not_upgraded 两个列表的所有元素都打印出来，这样就能清楚地看到哪些交换机已经成功升级，哪些还没有升级，以及它们的地址。其余的代码内容略过不讲。

```
    except Paramiko.ssh_exception.AuthenticationException:
        print ("TACACS is not working for " + ip_address + ".")
        switch_with_tacacs_issue.append(ip_address)
    except socket.error:
        print (ip_address + " is not reachable.")
        switch_not_reachable.append(ip_address)

iplist.close()
ssh_client.close

print ('\nTACACS is not working for below switches: ')
for i in switch_with_tacacs_issue:
    print (i)

print ('\nBelow switches are not reachable: ')
for i in switch_not_reachable:
    print (i)

print ('\nBelow switches IOS version are up-to-date: ')
for i in switch_upgraded:
    print (i)
```

```
print ('\nBelow switches IOS version are not updated yet: ')
for i in switch_not_upgraded:
    print (i)
```

5.3.7　脚本 1 验证

移动到/root/lab3，执行脚本 1（lab3_1.py），脚本自动登录 ip_list.txt 中所有交换机的管理 IP 地址，然后执行重启交换机前的验证步骤，可以看到 172.16.206.119 和 172.16.206.146 两个交换机重启前的 IOS 版本升级配置有误，如下图所示。

```
[root@localhost lab3]# python lab3_1.py
Username: parry
Password:
Successfully connect to  172.16.206.39
Successfully connect to  172.16.206.40
Successfully connect to  172.16.206.41
Successfully connect to  172.16.206.42
Successfully connect to  172.16.206.43
Successfully connect to  172.16.206.119
Successfully connect to  172.16.206.128
Successfully connect to  172.16.206.129
Successfully connect to  172.16.206.145
Successfully connect to  172.16.206.146

TACACS is not working for below switches:

Below switches are not reachable:

Below switches IOS version are up-to-date:
172.16.206.39
172.16.206.40
172.16.206.41
172.16.206.42
172.16.206.43
172.16.206.128
172.16.206.129
172.16.206.145

Below switches IOS version are not updated yet:
172.16.206.119
172.16.206.146
```

5.3.8　实验准备——脚本 2

我们来看脚本 2，也就是重启交换机后，验证其 IOS 版本是否成功升级的脚本。

首先移动到 lab3 文件夹下，创建 lab3_2.py 脚本，如下图所示。

```
[root@CentOS-Python ~]# cd lab3
[root@CentOS-Python lab3]# vi lab3_2.py
```

然后在维护窗口时段重启所有交换机，相信读到这里的所有读者都有能力独自写一个脚本来批量重启交换机了。

5.3.9　实验代码——脚本 2

将下列代码写入脚本 lab3_2.py。

```
import paramiko
import time
import getpass
import sys
import re
import socket

username = input("Username: ")
password = getpass.getpass("Password: ")
iplist = open('ip_list.txt', 'r+')

switch_upgraded = []
switch_not_upgraded = []
switch_with_tacacs_issue = []
switch_not_reachable = []

for line in iplist.readlines():
    try:
        ip_address = line.strip()
        ssh_client = paramiko.SSHClient()
        ssh_client.set_missing_host_key_policy(paramiko.AutoAddPolicy())
        ssh_client.connect(hostname=ip_address, username=username, password=password)
        print ("Successfully connect to ", ip_address)
        command = ssh_client.invoke_shell(width=300)
        command.send("show ver | b SW Version\n")
        time.sleep(0.5)
        output = command.recv(65535)
        print (output)
        ios_version = re.search(r'\d{2}.\d\(\d{1, 2}\)\w{2, 4}', output)
        if ios_version.group() == '12.2(55)SE12':
            switch_upgraded.append(ip_address)
        elif ios_version.group() == '15.2(2)E8':
            switch_upgraded.append(ip_address)
        elif ios_version.group() == '15.0(2)SE11':
```

```
            switch_upgraded.append(ip_address)
        else:
            switch_not_upgraded.append(ip_address)
    except paramiko.ssh_exception.AuthenticationException:
        print ("TACACS is not working for " + ip_address + ".")
        switch_with_tacacs_issue.append(ip_address)
    except socket.error:
        print (ip_address + " is not reachable." )
        switch_not_reachable.append(ip_address)

iplist.close()
ssh_client.close

print ('\nTACACS is not working for below switches: ')
for i in switch_with_tacacs_issue:
    print (i)

print ('\nBelow switches are not reachable: ')
for i in switch_not_reachable:
    print (i)

print ('\nBelow switches IOS version are up-to-date: ')
for i in switch_upgraded:
    print (i)

print ('\nBelow switches IOS version are not updated yet: ')
for i in switch_not_upgraded:
    print (i)
```

5.3.10　脚本 2 代码分段讲解

（1）这段代码和脚本 1 基本相同，我们同样创建了 switch_upgraded 和 switch_not_ upgraded 两个空列表，用来统计有哪些交换机在重启后 IOS 版本升级成功，有哪些交换机在重启后 IOS 版本升级不成功。

```
import paramiko
import time
import getpass
```

```
import sys
import re
import socket

username = input("Username: ")
password = getpass.getpass("Password: ")
iplist = open('ip_list.txt', 'r+')

switch_upgraded = []
switch_not_upgraded = []
switch_with_tacacs_issue = []
switch_not_reachable = []

for line in iplist.readlines():
    try:
        ip_address = line.strip()
        ssh_client = paramiko.SSHClient()
        ssh_client.set_missing_host_key_policy(paramiko.AutoAddPolicy())
        ssh_client.connect(hostname=ip_address, username=username, password=password)
        print ("Successfully connect to ", ip_address)
        command = ssh_client.invoke_shell(width=300)
        command.send("show ver | b SW Version\n")
        time.sleep(0.5)
        output = command.recv(65535)
        print (output)
```

（2）重启交换机后直接输入 show ver | b SW Version 查看当前的 IOS 版本，再用正则表达式 ios_version = re.search(r'\d{2}.\d\(\d{1，2}\)\w{2，4}', output)来匹配，只要能匹配到 12.2(55)SE12、15.2(2)E8、15.0(2)SE11 这 3 种 IOS 版本中的任意一种，就将交换机的管理 IP 地址添加到 switch_upgraded 列表。反之，如果上述 3 种 IOS 版本都匹配不到，则将交换机的管理 IP 地址添加到 switch_not_reachable 列表。最后用 for 循环将 switch_ upgraded 和 switch_not_reachable 中的元素一一打印出来，输出内容的格式和脚本 1 并无二致。

```
        ios_version = re.search(r'\d{2}.\d\(\d{1, 2}\)\w{2, 4}', output)
        if ios_version.group() == '12.2(55)SE12':
            switch_upgraded.append(ip_address)
        elif ios_version.group() == '15.2(2)E8':
            switch_upgraded.append(ip_address)
```

```
        elif ios_version.group() == '15.0(2)SE11':
            switch_upgraded.append(ip_address)
        else:
            switch_not_upgraded.append(ip_address)
    except Paramiko.ssh_exception.AuthenticationException:
        print ("TACACS is not working for " + ip_address + ".")
        switch_with_tacacs_issue.append(ip_address)
    except socket.error:
        print (ip_address +  " is not reachable." )
        switch_not_reachable.append(ip_address)

iplist.close()
ssh_client.close

print ('\nTACACS is not working for below switches: ')
for i in switch_with_tacacs_issue:
    print (i)

print ('\nBelow switches are not reachable: ')
for i in switch_not_reachable:
    print (i)

print ('\nBelow switches IOS version are up-to-date: ')
for i in switch_upgraded:
    print (i)

print ('\nBelow switches IOS version are not updated yet: ')
for i in switch_not_upgraded:
    print (i)
```

5.3.11　脚本 2 验证

移动至/root/lab3，执行脚本 lab3_2.py，可以看到在重启后，所有交换机的 IOS 版本都已升级成功，包括在运行脚本 lab3_1.py 时出问题的 172.16.206.119 和 172.16.206.146 两台交换机，如下图所示。

```
[root@localhost lab3]# python lab3_2.py
Username: parry
Password:
Successfully connect to  172.16.206.39
Successfully connect to  172.16.206.40
Successfully connect to  172.16.206.41
Successfully connect to  172.16.206.42
Successfully connect to  172.16.206.43
Successfully connect to  172.16.206.119
Successfully connect to  172.16.206.128
Successfully connect to  172.16.206.129
Successfully connect to  172.16.206.145
Successfully connect to  172.16.206.146

TACACS is not working for below switches:

Below switches are not reachable:

Below switches IOS version are up-to-date:
172.16.206.39
172.16.206.40
172.16.206.41
172.16.206.42
172.16.206.43
172.16.206.119
172.16.206.128
172.16.206.129
172.16.206.145
172.16.206.146

Below switches IOS version are not updated yet:
```

5.4　实验 4　现网超长命令回显处理（华为设备）

5.4.1　实验背景

在前面的实验中，设备往往一联机就马上执行分页取消操作，华为设备命令如下。

```
screen-length 0 temporary
```

但是，在现网环境中总有场景需要配合回显分页功能来实现回显完整抓取。比如某一账号开通时没有授权分屏指令、某一台防火墙设备配置几万行代码等。这种实际情况通常会影响到 CLI 命令回显抓取的完整性，甚至有时候回显等待太久而造成读取超时等，从而联机失败。

现有一台已安装 Python 的 Windows 运维终端，其 IP 地址（192.168.2.101）等经过网管专用网络（会运行 OSPF、BGP、MPLS 等）在经由跳板机和防火墙做安全 NAT 策略等转换成另一 IP 地址（172.25.1.231），接入一台交换机，其 IP 地址为 172.25.1.234。在逻辑上经过抽象处理后，可简单理解为：运维终端 PC（172.25.1.231）直连真机 Quidway S5328C-EI

（172.25.1.234），在同一个局域网中。它们所在网段的信息为 172.25.1.224/27，网关均为 172.25.1.225。

现在假设这台交换机的配置很长，把它理解成一台行为管控防火墙也可以。那么，如何通过 Python 脚本与真机设备分页符进行互动，而非联机后直接取消分页？如何通过这种方式把现网设备中的超长回显逐步提取出来？

```
#
  ---- More ----
```

为了更好地控制实验，避免篇幅太长，这里介绍一个命令，用来控制回显长度。

```
#每次回显 10 行，随即分页
screen-length 10 temporary
```

命令实施效果如下图所示。

```
<Quidway-S5300-SW1-HW>disp cur
!Software Version V200R005C00SPC500
#
sysname Quidway-S5300-SW1-HW
#
undo info-center enable
#
vlan batch 201
#
undo authentication unified-mode
#
  ---- More ----
```

这样，现网的实际场景就变成了可手工控制的实验场景。

5.4.2　实验目的

遇到比较“棘手”的问题时，通常可以回归最“朴素”的 Paramiko 模块。

（1）Python 脚本 Paramiko 登录一台交换机（如 Layer3Switch-1），设置回显分页，参数为 10。

（2）回显信息如出现分页符 More 字样，则 Paramiko 继续发送空格指令，直到分页结束。

（3）处理回显信息，将分页符等处理掉。

5.4.3　实验思路

实施拆解动作步骤时，先完成 Python 脚本怎么在带分页的回显上发送空格指令，并把每个分页都抓取出来，且抓取完整，再考虑怎么把 More 字样做规范化处理。

5.4.4　实验准备——脚本 1

将下列代码写入脚本 lab4_1.py。

```
import paramiko
import time

ip = "172.25.1.234"
username = "python"
password = "1234abcd"

ssh_client = paramiko.SSHClient()
ssh_client.set_missing_host_key_policy(paramiko.AutoAddPolicy())
ssh_client.connect(hostname=ip, username=username,
                    password=password, look_for_keys=False)
print("Successfully connected to ",ip)

command = ssh_client.invoke_shell()
command.send("screen-length 10 temporary \n")
command.send("disp cur \n")
time.sleep(0.5)

#定义一个字符串变量来装分页回显，对回显进行拼接
output = ""

#出现分页符 More 则发送空格命令，直到分页结束
while True:
    page = command.recv(65535).decode("ASCII")
    output += page
    time.sleep(0.1)
    if page.endswith('>') or page.endswith(']'):
    #另一种等效判断方式
    #if page[-1] in ['>',']']:
        break
    if "  ---- More ----" in page:
        command.send(" ")

print(output)
ssh_client.close
```

脚本代码讲解已直接标注在脚本中。这里有一个小技巧，如果在 Python 脚本中，有太多 or 结构逻辑，则我们可以尝试用 in 结构来做等效替换。

5.4.5　脚本 1 验证

运行脚本 lab4_1.py，如下图所示，此时已经能完整获取 Python 脚本与真机设备互动的指令回显。这里的 More 相关字为分页标识，后面的一些“乱码”为发送空格键等符号显示内容。

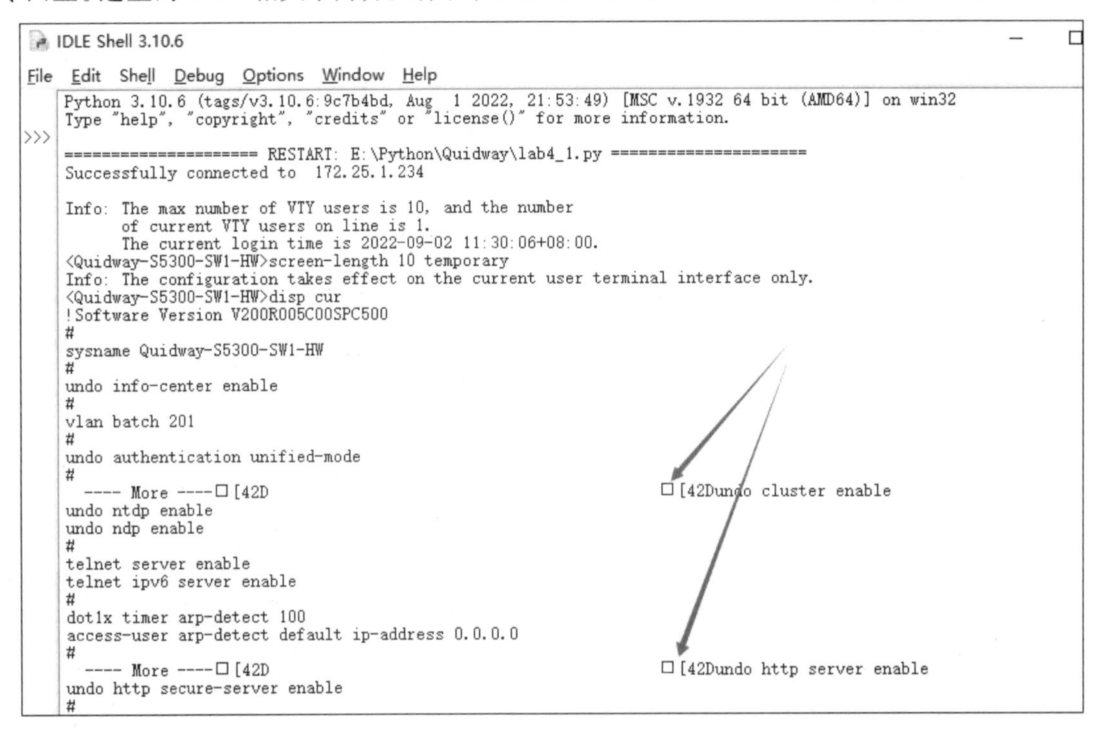

经过 Paramiko 模块检测 More 符号，发送空格键，再检测 More 符号，一点点就可以把全部命令回显都抓取完整。

5.4.6　实验准备——脚本 2

现在回显已成功抓取，只是有一些文本干扰项。处理文本有一个“祖传”工具——正则表达式。于是对上述代码调优，将下列代码写入脚本 lab4_2.py。

```
import paramiko
import time
```

```
import re

ip = "172.25.1.234"
username = "python"
password = "1234abcd"

ssh_client = paramiko.SSHClient()
ssh_client.set_missing_host_key_policy(paramiko.AutoAddPolicy())
ssh_client.connect(hostname=ip, username=username, password=password,
look_for_keys=False)
print("Successfully connected to ",ip)

command = ssh_client.invoke_shell()
command.send("screen-length 10 temporary \n")
command.send("disp cur \n")
time.sleep(0.5)

output = ""

while True:
    page = command.recv(65535)
    page = page.decode("ASCII")
    output += page
    time.sleep(0.1)
    if page.endswith('>') or page.endswith(']'):
    #if page[-1] in ['>',']']:
        break
    if "  ---- More ----" in page:
        command.send(" ")

#sub 替换的内容可以是正则表达式
output = re.sub(r"  ---- More ----.*42D", "", output)
print(output)
ssh_client.close
```

脚本代码讲解已直接标注在脚本中。理论上这里可以用字符串的 replace 方法，但是从 lab4_1.py 运行的结果来看，回显出现一个乱码（方框），这种情况不好把控，因此改用 re 模块的 sub 来处理。

5.4.7　脚本 2 验证

运行脚本 lab4_2.py，如下图所示，此时 More 相关字为分页标识，后面的一些“乱码”已全部处理完毕，还原成标准命令回显。

```
lab4_2.py - E:\Python\Quidway\lab4_2.py (3.10.6)
File Edit Format Run Options Window Help
import paramiko
import time
import re

ip = "172.25.1.234"
username = "python"
password = "1234abcd"

ssh_client = paramiko.SSHClient()
ssh_client.set_missing_host_key_policy(parami
ssh_client.connect(hostname=ip, username=user
print("Successfully connected to ",ip)

command = ssh_client.invoke_shell()
command.send("screen-length 10 temporary \n")
command.send("disp cur \n")
time.sleep(0.5)

output = ""

while True:
    page = command.recv(65535)
    page = page.decode("ASCII")
    output += page
    time.sleep(0.1)
    if page.endswith('>') or page.endswith(']
    #if page[-1] in ['>',']']:
        break
    if "  ---- More ----" in page:
        command.send(" ")

# sub 此时要替换的内容可以是正则表达式，我们
output = re.sub(r"  ---- More ----.*42D", "",
print(output)
ssh_client.close
```

```
IDLE Shell 3.10.6
File Edit Shell Debug Options Window Help
Python 3.10.6 (tags/v3.10.6:9c7b4bd, Aug  1 2022, 21:53:49) [
AMD64)] on win32
Type "help", "copyright", "credits" or "license()" for more i
>>>
===================== RESTART: E:\Python\Quidway\lab4_2.py ==
Successfully connected to  172.25.1.234

Info: The max number of VTY users is 10, and the number
      of current VTY users on line is 1.
      The current login time is 2022-09-02 11:39:49+08:00.
<Quidway-S5300-SW1-HW>screen-length 10 temporary
Info: The configuration takes effect on the current user term
.
<Quidway-S5300-SW1-HW>disp cur
!Software Version V200R005C00SPC500
#
sysname Quidway-S5300-SW1-HW
#
undo info-center enable
#
vlan batch 201
#
undo authentication unified-mode
#
undo cluster enable
undo ntdp enable
undo ndp enable
#
telnet server enable
telnet ipv6 server enable
#
dot1x timer arp-detect 100
access-user arp-detect default ip-address 0.0.0.0
#
undo http server enable
undo http secure-server enable
```

通常来说，在现网实际生产环境中，这些实战应用是综合的。联机、交互、抓取、保存、解析、入库、呈现……等一系列操作，我们可分解成一个个小目标、小任务来完成，逐步积累经验。

5.5　实验 5　自定义 ping 工具及 exe 打包（华为设备）

在交流互动中，读者们经常提及在 Windows 上对 Python 脚本打包成 exe 文件。

5.5.1　实验背景

日常需要打包成 exe 的场景大体有如下几种。

（1）Python 脚本需要 Windows 跳板机才能接入运维网络来执行，而跳板机不允许安装 Python 或不允许随意安装第三方模块。

（2）Python 脚本完成后，供完全不懂 Python 的人员使用。

（3）日常简化操作，非必要不打开脚本，防止误操作修改代码。

（4）程序打包后配合一些界面交互功能，工作内容显性化效果更佳。

5.5.2　实验目的

- 初步了解 pyinstaller 模块。
- 做一个 Python 脚本，将其打包成 exe，测试运行。
- 将这个 exe 包放到其他人电脑上，测试运行。

5.5.3　实验思路

将 Python 脚本打包成 exe，常用为 pyinstaller 模块，我们可以到官网看看如何快速切入使用，如下图所示。

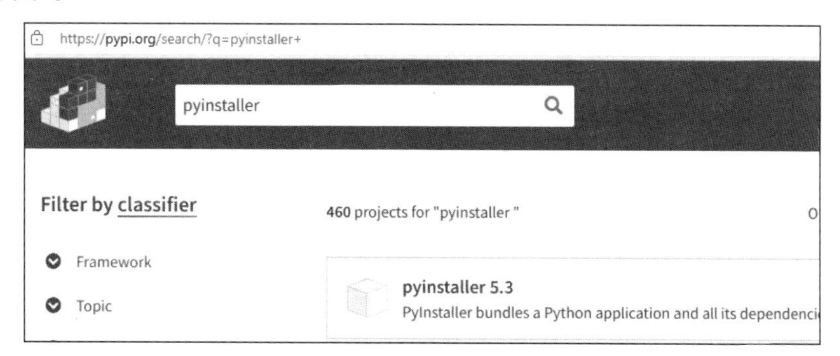

与其他模块类似，使用前需进行 pip 安装。安装以后可以使用 pip show 命令进行检查。

```
pip3 install pyinstaller
pip3 show pyinstaller
```

前面 pythonping 模块已详细介绍了用 Python 进行 ping 操作。为了丰富本书内容，这里再演示使用另一个实现 ping 功能的第三方模块——ping3。

```
pip3 install ping3
pip3 show ping3
```

综上，两个库的版本号如下。

```
Name: pyinstaller
Version: 5.3
```

```
-----
Name: ping3
Version: 4.0.3
```

实验思路：打包好的 Python 程序读取文件 toping_list.txt 存放待 ping 的 IP 地址或域名，批量进行 ping 测试，每次以追加形式将结果写入 toping_result.txt。

5.5.4 实验准备——脚本

```
from ping3 import ping
from datetime import datetime

def fun_ping3(ih):
    #ih 表示 IP 地址或域名
    result = ping(ih)
    now = datetime.now().strftime('%Y-%m-%d %H:%M:%S')
    if result is None:
        print(ih + '  不通')
        with open('toping_result.txt',mode='a') as f:
            f.write(f'{now}   {ih},不通\n')

    elif result is False:
        print(ih + '  域名解析失败')
        with open('toping_result.txt',mode='a') as f:
            f.write(f'{now}   {ih},域名解析失败\n')

    else:
        print(ih + '  可达')
        with open('toping_result.txt',mode='a') as f:
            f.write(f'{now}   {ih},可达\n')

if __name__ == '__main__':
    #toping_list.txt 存放 IP 地址或域名，一项一行
    with open('toping_list.txt') as f:
        toping_list = f.readlines()

    #ping 测试结果文件采用追加加入模式
    with open('toping_result.txt',mode='a') as f:
        f.write('\n')
```

```
for each in toping_list:
    fun_ping3(each.strip())
```

脚本代码讲解已直接标注在脚本中。通过在脚本中封装函数、加入__name__等内容，让整个代码更规范。

5.5.5　脚本验证

执行脚本之前，我们需要把 IP 地址存入 toping_list.txt 文件中。在现网实际生产环境中，这些 IP 地址或域名的数量可能很多。

```
192.168.2.1
172.25.1.234
www.baidu.com
www.huawei.com
1.1.1.1
www.butong1212.com
```

实验文件夹如下图所示。

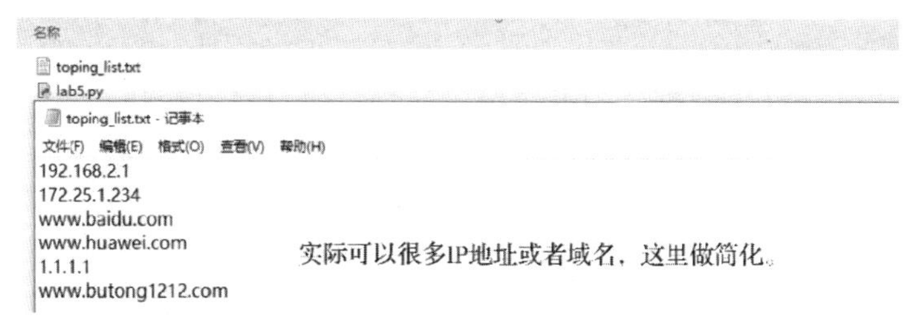

执行 Python 脚本，如下图所示，代码执行完会生成一个 toping_result.txt 文件，用来存放 ping 测试结果。

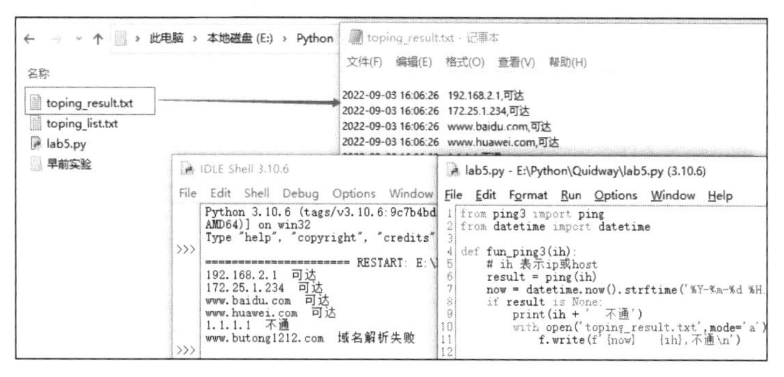

5.5.6 脚本打包

Python 脚本 lab5.py 测试没问题后，即可打包成 exe，如下图所示。

```
#CMD 先进入对应脚本的文件夹中
pyinstaller -F lab_pyinstaller.py
```

```
E:\Python\Quidway>pyinstaller -F lab5.py
376 INFO: PyInstaller: 5.3
376 INFO: Python: 3.10.6
386 INFO: Platform: Windows-10-10.0.19043-SP0
389 INFO: wrote E:\Python\Quidway\lab5.spec
392 INFO: UPX is not available.
393 INFO: Extending PYTHONPATH with paths
['E:\\Python\\Quidway']
861 INFO: checking Analysis
862 INFO: Building Analysis because Analysis-00.toc is non existent
```

在打包过程中，屏幕会持续滚动显示一些信息。我们静静等待最后提示“completed successfully”即可，如下图所示。

```
15159 INFO: Updating manifest in E:\Python\Quidway\dist\lab5.exe.notanexecutable
15164 INFO: Updating resource type 24 name 1 language 0
15204 INFO: Appending PKG archive to EXE
15251 INFO: Fixing EXE headers
15475 INFO: Building EXE from EXE-00.toc completed successfully.

E:\Python\Quidway>_
```

5.5.7 打包验证

打包完成后，文件目录结构如下图所示。

顺着目录结构，即可找到一个名为 lab5.exe 的可执行文件。如果直接双击 lab5.exe，黑色窗口会一闪而过。此时，需要把 toping_list.txt 文件放在 exe 的同路径的文件夹中，然后双击即可，如下图所示。

至此，脚本打包测试已完成。关于 ping3 库和 pyinstaller 模块还有其他参数和功能，甚至可以更换程序图标等，请读者自行尝试。

在 Python 中使用 pyinstaller 模块对一般脚本程序进行打包，还是比较简单的。然而，因为 pyinstaller 打包时需要配套把依赖项一并纳入，生成的 exe 及相关文件会显得很笨重，所占空间不小。如果待打包的代码的依赖关系本身就比较复杂，打包过程容易出现一些意想不到的情况。另外，由于不同 Windows 之间版本或者环境差异，A 主机打包后成功运行的 exe 文件，可能在 B 主机上无法执行。

第6章

Python 内置模块与第三方模块详解

在读完前面 5 章内容并动手实操后，相信读者已经对 Python 的基础知识和 Python 在网络运维中的应用有了一定程度的了解。在此基础上，本章将扩展讨论 Python 在网络运维自动化领域中十分重要的两个话题：“不支持 API 的传统设备”和“Python 默认同步、单线程效率低下”的问题，并将举例给出解决这些问题的方案及其原理。针对“不支持 API 的传统设备”的话题将在 6.1 ~ 6.4 节讨论，而在 6.5 和 6.6 两节将讨论如何应对第二个问题，在 6.7 节中也将介绍 CSV 和 Jinja2 这两个对网络运维工作很有帮助的内置和第三方模块。

1. 不支持 API 的传统设备

众所周知，NETCONF（RFC 6241）这个基于 XML、用来替代 CLI 和 SNMP 的网络配置和网络管理协议最早诞生于 2006 年 12 月，由 IETF（The Internet Engineering Task Force，互联网工程任务组）以 RFC 4741 发布。在进行一番修订后，IETF 又于 2011 年 6 月以 RFC 6241 将其再次发布。在此之前的一些“上了年纪”的网络设备，比如思科经典的基于 IOS 操作系统的 Catalyst 系列交换机（2960、3560、3750、4500、6500）是不支持 NETCONF 的，更别提对正在学习 NetDevOps 的网络工程师来说耳熟能详的 JSON、XML、YAML 这些数据类型，以及 YANG 这个数据模型语言了。简而言之，这些“古董级”的设备是没有 API 的。API 这类“时髦”的东西是思科为了顺应 SDN 时代潮流，进而推出新的 IOS-XE 操作系统后逐渐融进思科设备的，比较有代表性的使用 IOS-XE 的设备有思科的 3850、9200、9300 等新一代 Catalyst 系列交换机。

本书第 10 章、第 11 章将介绍 NETCONF、RESTCONF 相关内容，除思科设备外，也会涉及国产设备。

由于思科在 21 世纪初期在数据通信行业占据“霸主”的地位，加上这些古老的基于 IOS 的设备确实经典耐用、长盛不衰，因此如今我们仍然能在很多企业和公司的现网里看到它们的存在，就像今天仍然有一批 Windows XP 和 Windows 7 的忠实用户在坚守阵地一样。2016 年 5 月在美国拉斯维加斯举办的第 30 届 Interop 上（Interop 是由英国 Informa 会展公司每年举办一次的 IT 技术峰会），大会嘉宾，*Network Programmability and Automation: Skills for the Next-Generation Network Engineer* 一书的作者之一，网络运维自动化的先驱 Jason Edelman 在他的演讲中提到，目前市场上支持 NETCONF、API 的网络设备只占市场份额的 15%~20%，剩下的 80%~85%还都是只支持 SSH 访问命令行的传统设备。

那么问题来了：既然 IOS 设备当前还占据着极大的市场份额，那么是否意味着 JSON、XML、YAML 等在这些“古董级”的设备里真的毫无用武之地呢？答案是否定的，本章的 6.1~6.5 节将分别介绍 TextFSM、ntc-template、NAPALM、pyntc 这些 Python 的第三方模块和模板是如何将 JSON 和这些“老古董”巧妙结合在一起的，以及它到底能帮我们在这些没有 API 的设备上做什么。在此之前，我们会先介绍一些关于 JSON 的基础知识。

2. Python 默认同步、单线程效率低下的问题

对网络工程师来说，我们通常必须借助 Telnetlib、Paramiko 或 Netmiko 这些第三方开源模块才能通过 Telnet 或 SSH 来登录、操作和管理各种网络设备。因为 Python 默认的运行方式是同步、单线程的，也就意味着在不对脚本内容做额外调整的前提下，运行 Python 脚本的主机只能一台一台地登录设备执行代码。假设一个脚本登录一台交换机执行配置平均耗时 5s，那么在拥有 1000 台交换机的大型企业网里就要耗时 5000s 才能执行完脚本，效率太低。在 6.6 节将介绍如何通过使用 Netmiko 实现多线程来解决这个痛点问题，提升工作效率。

6.1　JSON

JSON 诞生于 1999 年 12 月，是 JavaScript Programming Language（Standard ECMA-262 3rd Edition）的一个子集合，是一种轻量级的数据交换格式。虽然 JSON 基于 JavaScript 开发，但它是一种“语言无关”（Language Independent）的文本格式，并且采用 C 语言家族，如 C、C++、C#、Java、Python 和 Perl 等语言的用法习惯，成为一门理想的数据交换语言（Data-Interchange Language）。

6.1.1　JSON 基础知识

JSON 的数据结构具有易读的特点，它由键值对（Collection）和对象（Object）组成，在

结构上非常类似 Python 的字典（Dictionary），如下是一个典型的 JSON 数据格式。

```
{
"intf":"Gigabitethernet0/0",
"status":"up"
}
```

与 Python 的字典一样，JSON 的键值对也由冒号分开，冒号左边的“intf”和“status”即键值对的键（Name），冒号右边的“Gigabitethernet0/0”和“up”即键值对的值（Value），每组键值对之间都用逗号隔开。

JSON 与字典不一样的地方如下。

（1）**JSON 里键的数据类型必须为字符串**，而在字典里字符串、常数、浮点数或者元组等都能作为键的数据类型。

（2）**JSON 里键的字符串内容必须使用双引号括起来**，不像字典里既可以用单引号，又可以用双引号来表示字符串。

JSON 里键值对的值又分为两种形式，一种形式是简单的值，包括字符串、整数等，比如上面的“Gigabitethernet0/0”和“up”就是一种简单的值。另一种形式被称为对象（Object），对象内容用大括号{}表示，对象中的键值对之间用逗号分开，**它们是无序的**，举例如下。

```
{"Vendor":"Cisco", "Model":"2960"}
```

当有多组对象存在时，我们将其称为 **JSON 阵列**（JSON Array），阵列以中括号[]表示，**阵列中的元素（即各个对象）是有序的（可以把它理解为列表）**，举例如下。

```
{
   "devices":[
       {"Vendor":"Cisco", "Model":"2960"},
       {"Vendor":"Cisco", "Model":"3560"},
       {"Vendor":"Cisco", "Model":"4500"}
       ]
}
```

6.1.2　JSON 在 Python 中的使用

Python 中已经内置了 JSON 模块，使用时只需 import json 即可，如下图所示。

```
[root@localhost ~]# python3.10
Python 3.10.6 (main, Aug  6 2022, 13:49:15) [GCC 8.5.0 20210514 (Red Hat 8.5.0-1
5)] on linux
Type "help", "copyright", "credits" or "license" for more information.
>>> import json
>>>
```

JSON 模块主要有两种函数:json.dumps()和 json.loads()。前者是 JSON 的编码器(Encoder),用来将 Python 中的对象转换成 JSON 格式的字符串，如下图所示。

```
>>> import json
>>>
>>> a = json.dumps('parry')
>>> type(a)
<class 'str'>
>>> type(json.dumps({"c": 0, "b": 0, "a": 0}, sort_keys=True))
<class 'str'>
>>>
>>> type({"c": 0, "b": 0, "a": 0})
<class 'dict'>
>>> type(json.dumps([1,2,3]))
<class 'str'>
```

由此可以看到，我们用 json.dumps()将 Python 中 3 种类型的对象：字符串（'parry'）、字典（{"c": 0, "b": 0, "a": 0}）、列表（[1,2,3]），转换成 JSON 格式的字符串，并用 type()函数进行了验证。

而 json.loads()的用法则是将 JSON 格式的字符串转换成 Python 的对象，如下图所示。

```
>>> json_list = '[1,2,3]'
>>> type(json_list)
<class 'str'>
>>> python_list = json.loads(json_list)
>>> print (python_list)
[1, 2, 3]
>>> type(python_list)
<class 'list'>
>>>
>>> json_dictionary = '{"vendor":"Cisco", "model":"2960"}'
>>> python_dictionary = json.loads(json_dictionary)
>>> print (python_dictionary)
{'vendor': 'Cisco', 'model': '2960'}
>>> type(python_dictionary)
<class 'dict'>
>>>
```

我们将两个 JSON 格式的字符串：“[1,2,3]”和“{"vendor":"Cisco", "model":"2960"}”，用 json.loads()转换成它们各自对应的 Python 对象：列表[1,2,3]和字典{"vendor":"Cisco", "model":"2960"}。需要注意的是，在创建对应 Python 字典类型的 JSON 字符串时，如果对键使用了单引号，则 Python 会返回“SyntaxError: invalid sytanx”，提示语法错误，如下图所示。

```
>>>
>>> json_dictionary = '{'vendor':'Cisco', 'model':'2960'}'
  File "<stdin>", line 1
    json_dictionary = '{'vendor':'Cisco', 'model':'2960'}'
                        ^
SyntaxError: invalid syntax
>>> exit()
```

原因就是：**JSON 里键的字符串内容必须使用双引号括起来**，不像字典里既可以用单引号，又可以用双引号来表示字符串，这点请务必注意。

6.2　正则表达式的痛点问题

作为本书的重点内容，我们知道，在 Python 中可以使用正则表达式来对字符串格式的文本内容做解析（Parse），从而匹配到我们感兴趣的文本内容，**但是这么做有一定限制**，比如我们需要用正则表达式从如下图所示的一台 2960 交换机 show ip int brief 命令的回显内容中找出哪些**物理端口**是 up 的。

```
SW1#show ip int brief
Interface              IP-Address      OK? Method Status                Protocol
GigabitEthernet0/0     unassigned      YES unset  up                    up
GigabitEthernet0/1     unassigned      YES unset  down                  down
GigabitEthernet0/2     unassigned      YES unset  down                  down
GigabitEthernet0/3     unassigned      YES unset  down                  down
GigabitEthernet1/0     unassigned      YES unset  down                  down
GigabitEthernet1/1     unassigned      YES unset  down                  down
GigabitEthernet1/2     unassigned      YES unset  down                  down
GigabitEthernet1/3     unassigned      YES unset  down                  down
GigabitEthernet2/0     unassigned      YES unset  down                  down
GigabitEthernet2/1     unassigned      YES unset  down                  down
GigabitEthernet2/2     unassigned      YES unset  down                  down
GigabitEthernet2/3     unassigned      YES unset  down                  down
GigabitEthernet3/0     unassigned      YES unset  down                  down
GigabitEthernet3/1     unassigned      YES unset  down                  down
GigabitEthernet3/2     unassigned      YES unset  down                  down
GigabitEthernet3/3     unassigned      YES unset  down                  down
Vlan1                  192.168.2.11    YES NVRAM  up                    up
```

如果用常规思维的正则表达式 GigabitEthernet\d\/\d 来匹配，则会将除 VLAN1 外的所有物理端口都匹配上，显然这种做法是错误的，因为当前只有 GigabitEthernet0/0 这一个物理端口的状态是 up 的，如下图所示。

```
REGULAR EXPRESSION                                          16 matches, 304 steps (~0ms)
/ GigabitEthernet\d\/\d                                                            / gm
TEST STRING                                                     SWITCH TO UNIT TESTS ›
SW1#show ip int brief
Interface              IP-Address      OK? Method Status                Protocol
GigabitEthernet0/0     unassigned      YES unset  up                    up
GigabitEthernet0/1     unassigned      YES unset  down                  down
GigabitEthernet0/2     unassigned      YES unset  down                  down
GigabitEthernet0/3     unassigned      YES unset  down                  down
GigabitEthernet1/0     unassigned      YES unset  down                  down
GigabitEthernet1/1     unassigned      YES unset  down                  down
GigabitEthernet1/2     unassigned      YES unset  down                  down
GigabitEthernet1/3     unassigned      YES unset  down                  down
GigabitEthernet2/0     unassigned      YES unset  down                  down
GigabitEthernet2/1     unassigned      YES unset  down                  down
GigabitEthernet2/2     unassigned      YES unset  down                  down
GigabitEthernet2/3     unassigned      YES unset  down                  down
GigabitEthernet3/0     unassigned      YES unset  down                  down
GigabitEthernet3/1     unassigned      YES unset  down                  down
GigabitEthernet3/2     unassigned      YES unset  down                  down
GigabitEthernet3/3     unassigned      YES unset  down                  down
Vlan1                  192.168.2.11    YES NVRAM  up                    up
```

大家也许此时想到了另一种方法，那就是在交换机的 show ip int brief 后面加上| i up 提前做好过滤，再用正则表达式 GigabitEthernet\d\/\d 来匹配，比如下图这样。

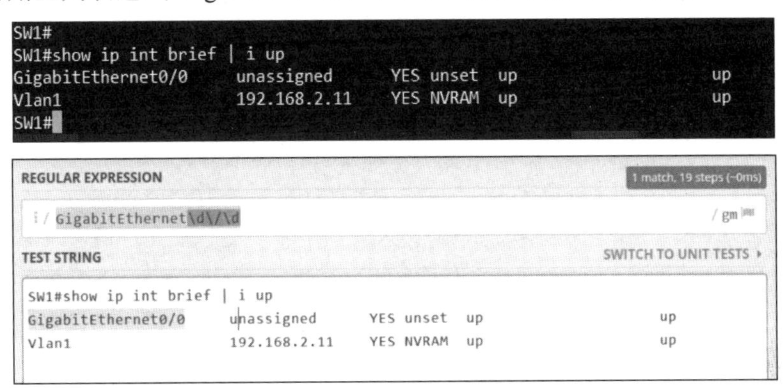

这种方法确实可以匹配出我们想要的结果，但是如果这时把要求变一变：找出当前交换机下所有 up 的端口（注意，现在不再只是找出所有 up 的物理端口，虚拟端口也要算进去），并且同时给出它们的端口号及 IP 地址。

这个要求意味着不仅要用正则表达式匹配上面文本里的 GigabitEthernet0/0、Vlan1、unassigned、192.168.2.11 这 4 项内容，还要保证正则表达式没有匹配到其他诸如 YES、unset、NVRAM、up 等文本内容，怎么样？这个难度是不是立即上升了几个等级？

为了解决类似这样的正则表达式的痛点问题，我们必须搬出“救兵”——**TextFSM** 了。

6.3　TextFSM 和 ntc-templates

本节做模块串讲，关于 TextFSM 基础语法入门将在第 7 章专门进行详解。

TextFSM 最早是由 Google 开发的一个 Python 开源模块，它能使用自定义的变量和规则设计出一个模板（Template），然后用该模板来处理文本内容，将这些**无规律的文本内容**按照自己打造的模板将它们整合成想要的**有序的数据格式**。

举个例子，在 6.2 节的例子中，如果有办法把 show ip int brief 的回显内容转换成 JSON 格式，将它们以 JSON 阵列的数据格式列出来，是不是会很方便配合 for 循环匹配出我们想要的东西呢？比如看到下面这样的 JSON 阵列后，读者是不是想到些什么呢？

```
[
  {
    "intf": "GigabitEthernet0/0",
```

```
    "ipaddr": "unassigned",
    "status": "up",
    "proto": "up"
  },
  {
    "intf": "GigabitEthernet0/1",
    "ipaddr": "unassigned",
    "status": "down",
    "proto": "down"
  },
  {
    "intf": "GigabitEthernet0/2",
    "ipaddr": "unassigned",
    "status": "down",
    "proto": "down"
  },
  {
    "intf": "GigabitEthernet0/3",
    "ipaddr": "unassigned",
    "status": "down",
    "proto": "down"
  },
]
```

由上面的代码可以看到，JSON 阵列格式在 Python 中实际上是一个数据类型为列表的对象（以 [开头，以]结尾），既然是列表，那么我们就能很方便地使用 for 语句遍历列表中的每个元素（这里所有的元素均为字典），并配合 if 语句，将端口状态为“up”（“status”键对应的值）的端口号（“intf”键对应的值）和 IP 地址（“ipaddr”键对应的值）一一打印出来即可（具体的代码将在 6.4 节中给出）。

下面介绍如何在 Python 中使用 TextFSM 创建我们需要的模板。

6.3.1　TextFSM 的安装

作为 Python 的第三方模块，TextFSM 有两种安装方法。

第一种方法是使用 pip 安装，如下图所示。

```
[root@localhost ~]# pip3.10 install textfsm
Requirement already satisfied: textfsm in /usr/local/lib/python3.10/site-package
s (1.1.2)
Requirement already satisfied: six in /usr/local/lib/python3.10/site-packages (f
rom textfsm) (1.16.0)
Requirement already satisfied: future in /usr/local/lib/python3.10/site-packages
 (from textfsm) (0.18.2)
WARNING: Running pip as the 'root' user can result in broken permissions and con
flicting behaviour with the system package manager. It is recommended to use a v
irtual environment instead: https://pip.pypa.io/warnings/venv

[      ] A new release of pip available:        -> 22.2.2
[      ] To update, run: pip install --upgrade pip
[root@localhost ~]#
```

pip 安装完毕后，进入 Python 并输入 import textfsm，如果没有报错，则说明安装成功，如下图所示。

```
[root@localhost ~]#
[root@localhost ~]# python3.10
Python 3.10.6 (main, Aug  6 2022, 13:49:15) [GCC 8.5.0 20210514 (Red Hat 8.5.0-1
5)] on linux
Type "help", "copyright", "credits" or "license" for more information.
>>> import textfsm
>>>
```

第二种方法是使用 git clone 命令从 GitHub 下载 TextFSM 的源码，如果你的 CentOS 8 主机没有安装 Git，系统会提醒你一并安装，具体如下图所示。

```
git clone https://github.com/google/textfsm.git
```

```
[root@localhost ~]# git clone https://github.com/google/textfsm.git
Cloning into 'textfsm'...
remote: Enumerating objects: 372, done.
remote: Counting objects: 100% (20/20), done.
remote: Compressing objects: 100% (16/16), done.
remote: Total 372 (delta 8), reused 6 (delta 4), pack-reused 352
Receiving objects: 100% (372/372), 179.09 KiB | 150.00 KiB/s, done.
Resolving deltas: 100% (218/218), done.
```

源码下载后，会看到当前目录下多出一个 textfsm 文件夹。进入该文件夹后，会看到如下图所示的 setup.py 文件。

```
[root@localhost ~]# ls
anaconda-ks.cfg                       ip_list.txt        Pictures          Templates
centos-gpg-keys-8-3.el8.noarch.rpm    lab1.py            Public            textfsm
custome_ansible_modules               lab2.py            __pycache__       Videos
Desktop                               lab4.py            Python-3.10.6
Documents                             Music              ssh_netmiko.py
Downloads                             original-ks.cfg    ssh_paramiko.py
[root@localhost ~]# cd textfsm
[root@localhost textfsm]# ls
COPYING   LICENSE.txt  README.md  setup.py  tests
examples  MANIFEST.in  setup.cfg  testdata  textfsm
[root@localhost textfsm]#
```

然后输入下面的命令执行该 py 文件进行安装，如下图所示。

```
python3.10 setup.py install
```

```
[root@localhost textfsm]# python3.10 setup.py install
/usr/local/lib/python3.10/site-packages/setuptools/dist.py:771: UserWarning: Usa
ge of dash-separated 'description-file' will not be supported in future versions
. Please use the underscore name 'description_file' instead
  warnings.warn(
running install
/usr/local/lib/python3.10/site-packages/setuptools/command/install.py:34: Setupt
oolsDeprecationWarning: setup.py install is deprecated. Use build and pip and ot
her standards-based tools.
  warnings.warn(
/usr/local/lib/python3.10/site-packages/setuptools/command/easy_install.py:144:
EasyInstallDeprecationWarning: easy_install command is deprecated. Use build and
 pip and other standards-based tools.
  warnings.warn(
running bdist_egg
running egg_info
creating textfsm.egg-info
```

6.3.2 TextFSM 模板的创建和应用

TextFSM 的语法本身并不难，但是使用者必须熟练掌握正则表达式。下面我们以思科 Nexus 7000 交换机上 show vlan 命令的回显内容为例，来看如何用 TextFSM 创建模板，以及如何使用模板将该 show vlan 命令的回显内容整理成我们想要的数据格式。

```
N7K#show vlan

VLAN Name                             Status    Ports
---- -------------------------------- --------- -------------------------------
1    default                          active    Eth1/1, Eth1/2, Eth1/3
                                                Eth1/5, Eth1/6, Eth1/7
2    VLAN0002                         active    Po100, Eth1/49, Eth1/50
3    VLAN0003                         active    Po100, Eth1/49, Eth1/50
4    VLAN0004                         active    Po100, Eth1/49, Eth1/50
5    VLAN0005                         active    Po100, Eth1/49, Eth1/50
6    VLAN0006                         active    Po100, Eth1/49, Eth1/50
7    VLAN0007                         active    Po100, Eth1/49, Eth1/50
8    VLAN0008                         active    Po100, Eth1/49, Eth1/50
```

我们创建一个 TextFSM 模板，模板内容如下。

```
Value VLAN_ID (\d+)
Value NAME (\w+)
Value STATUS (\w+)

Start
  ^${VLAN_ID}\s+${NAME}\s+${STATUS}\s+ -> Record
```

（1）在 TextFSM 中，我们使用 Value 语句来定义变量，这里定义了 3 个变量，分别是 VLAN_ID、NAME 和 STATUS。

```
Value VLAN_ID (\d+)
Value NAME (\w+)
Value STATUS (\w+)
```

（2）每个变量后面都有它自己对应的正则表达式模式（Pattern），这些模式写在括号()中。比如变量 VLAN_ID 顾名思义是要去匹配 VLAN 的 ID 的，所以它后面的正则表达式模式写为（\d+）。在 3.6 节中讲过，\d 这个特殊序列用来匹配数字，后面的+用来做贪婪匹配。同理，变量 NAME 是用来匹配 VLAN 的名称的，因为这里 VLAN 的名称掺杂了字母和数字，比如 VLAN0002，所以它的正则表达式模式写为（\w+）。\w 这个特殊序列用来匹配字母或数字，后面的+用来做贪婪匹配。同理，变量 STATUS（\w+）用来匹配 VLAN 状态，VLAN 状态会有 active 和 inactive 之分。

（3）在定义好变量后，我们使用 Start 语句来定义匹配规则，匹配规则由正则表达式的模式及变量名组成。

```
Start
  ^${VLAN_ID}\s+${NAME}\s+${STATUS}\s+ -> Record
```

（4）**Start 语句后面必须以正则表达式^开头**。^是正则表达式中的一种特殊字符，用于匹配输入字符串的开始位置，**注意紧随其后的$不是正则表达式里的$，它的作用不是用来匹配输入字符串的结尾位置，而是用来调用我们之前设置好的 VLAN_ID 并匹配该变量**。注意，在 TextFSM 中调用变量时可以用大括号{}，写成${VLAN_ID}，也可以不用，写成$VLAN_ID，但是 TextFSM 官方推荐使用大括号。VLAN_ID 对应的正则表达式恰巧是\d+，这样就匹配到了 1、2、3、4、5、6、7、8 这些 VLAN_ID，而后面的\s+$则表示匹配 1、2、3、4、5、6、7、8 后面的空白字符（\s 这个特殊序列用来匹配空白字符）。

```
^${VLAN_ID}\s+
```

（5）同理，我们调用变量 NAME，它对应的正则表达式模式为\w+，该特殊序列用来匹配 show vlan 命令回显内容中的 default、VLAN0002、VLAN0003、…、VLAN0008 等内容，而后面的\s 则用来匹配之后所有的空白字符。

```
${NAME}\s+
```

（6）变量{STATUS}也一样，它对应的\w+用来匹配 active 和 inactive 这两种 VLAN 状态（例子中给出的 show vlan 的回显内容中没有 inactive，但是不影响理解）以及后面的空白字符。

最后用-> Record 来结束 TextFSM 的匹配规则。

```
${STATUS}\s+ -> Record
```

在了解了 TextFSM 的语法基础后，接下来看怎么在 Python 中使用 TextFSM。首先将上面的 TextFSM 模板文件以文件名 show_vlan.template 保存，如下图所示。

```
[root@localhost textfsm]# cat show_vlan.template
Value VLAN_ID (\d+)
Value NAME (\w+)
Value STATUS (\w+)

Start
  ^${VLAN_ID}\s+${NAME}\s+${STATUS}\s+ -> Record
[root@localhost textfsm]#
```

然后在相同的文件夹下创建一个名叫 textfsm_demo.py 的 Python 脚本。

```
[root@localhost textfsm]# vi textfsm_demo.py
[root@localhost textfsm]#
```

将下面的代码写入该脚本。

```
from textfsm import TextFSM

output = '''
N7K#show vlan

VLAN Name                             Status    Ports
---- --------------------------------------------------------------------------
1    default                          active    Eth1/1, Eth1/2, Eth1/3
                                                Eth1/5, Eth1/6, Eth1/7
2    VLAN0002                         active    Po100, Eth1/49, Eth1/50
3    VLAN0003                         active    Po100, Eth1/49, Eth1/50
4    VLAN0004                         active    Po100, Eth1/49, Eth1/50
5    VLAN0005                         active    Po100, Eth1/49, Eth1/50
6    VLAN0006                         active    Po100, Eth1/49, Eth1/50
7    VLAN0007                         active    Po100, Eth1/49, Eth1/50
8    VLAN0008                         active    Po100, Eth1/49, Eth1/50
'''

f = open('show_vlan.template')
template = TextFSM(f)

print (template.ParseText(output))
```

代码分段讲解如下。

（1）首先我们用 from textfsm import TextFSM 引入 TextFSM 模块的 TextFSM 函数，该函数为 TextFSM 模块下最核心的类。

```
from textfsm import TextFSM
```

（2）将 show vlan 的回显内容以三引号字符串的形式赋值给变量 output。

```
output = '''
N7K#show vlan

VLAN Name                             Status    Ports
--------------------------------------------------------------------------------
1    default                          active    Eth1/1, Eth1/2, Eth1/3
                                                Eth1/5, Eth1/6, Eth1/7
2    VLAN0002                         active    Po100, Eth1/49, Eth1/50
3    VLAN0003                         active    Po100, Eth1/49, Eth1/50
4    VLAN0004                         active    Po100, Eth1/49, Eth1/50
5    VLAN0005                         active    Po100, Eth1/49, Eth1/50
6    VLAN0006                         active    Po100, Eth1/49, Eth1/50
7    VLAN0007                         active    Po100, Eth1/49, Eth1/50
8    VLAN0008                         active    Po100, Eth1/49, Eth1/50
'''
```

（3）打开之前创建好的模板文件 show_vlan.template，调用 TextFSM()函数将它赋值给变量 template，最后调用 template 下的 ParseText()函数对文本内容进行解析，ParseText()函数中的参数 output 即 show vlan 命令的回显内容，最后用 print()函数将被模板解析后的回显内容打印出来，看看是什么样的内容。

```
f = open('show_vlan.template')
template = TextFSM(f)

print (template.ParseText(output))
```

一切就绪后，执行脚本看效果，如下图所示。

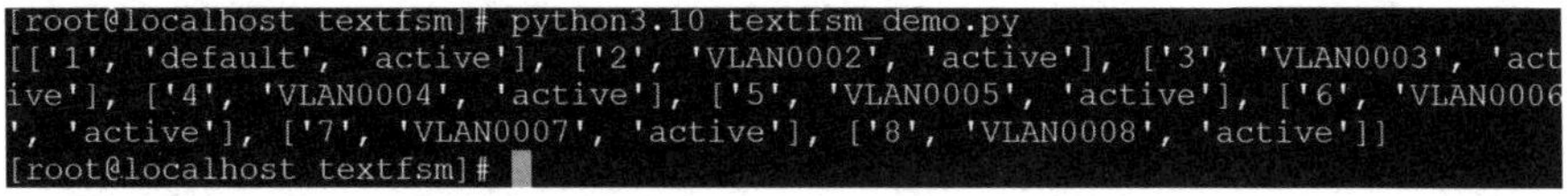

```
[root@localhost textfsm]# python3.10 textfsm_demo.py
[['1', 'default', 'active'], ['2', 'VLAN0002', 'active'], ['3', 'VLAN0003', 'act
ive'], ['4', 'VLAN0004', 'active'], ['5', 'VLAN0005', 'active'], ['6', 'VLAN0006
', 'active'], ['7', 'VLAN0007', 'active'], ['8', 'VLAN0008', 'active']]
[root@localhost textfsm]#
```

由此可以看到，之前无序的纯字符串文本内容被 TextFSM 模板解析后，已经被有序的嵌套列表替代，方便我们配合 for 循环做很多事情。

TextFSM 是一个十分强大也比较复杂的文字解析工具，更多关于 TextFSM 在网络运维自动化中的使用技巧会在第 7 章里详细讲解。

6.3.3　ntc-templates

用 TextFSM 制作的模板很好用，但是缺点也很明显：每个 TextFSM 模板都只能对应一条 show 或者 display 命令的回显内容，而目前每家知名厂商的网络设备都有上百种 show 或 display 命令，并且每家厂商的回显内容和格式都完全不同，有些厂商还有多种不同的操作系统，比如思科就有 IOS、IOS-XE、IOS-XR、NX-OS、ASA、WLC 等多种 OS 版本，这些版本又有各自特有的 show 命令，难道我们必须自己动手造轮子，每种操作系统的每条命令都要靠自己手动写一个对应的模板吗？不用担心，已经有前人帮我们造好了轮子，这就是 **ntc-templates**。

ntc-templates 是由 Network To Code 团队用 TextFSM 花费了无数心血开发出来的一套模板集，支持 Cisco IOS、Cisco ASA、Cisco NX-OS、Cisco IOS-XR、Arista、Avaya、Brocade、Checkpoint、Fortinet、Dell、Huawei、Palo Alto 等绝大多数主流厂商的设备，将在这些设备里输入 show 和 display 命令后得到的各式各样的回显文本内容整合成 JSON、XML、YAML 等格式。举例来说，针对思科的 IOS 设备，ntc-templates 提供了 show ip int brief、show cdp neighbor、show access-list 等常用的 show 命令对应的 TextFSM 模板（还有数百种思科和其他厂商的 TextFSM 模板等待读者自行去挖掘），如下图所示。

```
-rw-r--r--. 1 root root  481 May  8 10:02 cisco_ios_dir.textfsm
-rw-r--r--. 1 root root 1470 May  8 10:02 cisco_ios_show_access-list.textfsm
-rw-r--r--. 1 root root  953 May  8 10:02 cisco_ios_show_adjacency_detail.textfsm
-rw-r--r--. 1 root root  310 May  8 10:02 cisco_ios_show_aliases.textfsm
-rw-r--r--. 1 root root  180 May  8 10:02 cisco_ios_show_archive.textfsm
-rw-r--r--. 1 root root  403 May  8 10:02 cisco_ios_show_authentication_sessions.textfsm
-rw-r--r--. 1 root root 1096 May  8 10:02 cisco_ios_show_boot.textfsm
-rw-r--r--. 1 root root  264 May  8 10:02 cisco_ios_show_capability_feature_routing.textfsm
-rw-r--r--. 1 root root  937 May  8 10:02 cisco_ios_show_cdp_neighbors_detail.textfsm
-rw-r--r--. 1 root root  546 May  8 10:02 cisco_ios_show_cdp_neighbors.textfsm
-rw-r--r--. 1 root root  317 May  8 10:02 cisco_ios_show_clock.textfsm
-rw-r--r--. 1 root root 2327 May  8 10:02 cisco_ios_show_controller_t1.textfsm
-rw-r--r--. 1 root root  491 May  8 10:02 cisco_ios_show_dmvpn.textfsm
-rw-r--r--. 1 root root 1367 May  8 10:02 cisco_ios_show_dot1x_all.textfsm
-rw-r--r--. 1 root root  413 May  8 10:02 cisco_ios_show_environment_power_all.textfsm
-rw-r--r--. 1 root root 1603 May  8 10:02 cisco_ios_show_environment_temperature.textfsm
-rw-r--r--. 1 root root 1085 May  8 10:02 cisco_ios_show_etherchannel_summary.textfsm
-rw-r--r--. 1 root root  454 May  8 10:02 cisco_ios_show_hosts_summary.textfsm
-rw-r--r--. 1 root root  401 May  8 10:02 cisco_ios_show_interfaces_description.textfsm
-rw-r--r--. 1 root root  484 May  8 10:02 cisco_ios_show_interfaces_status.textfsm
-rw-r--r--. 1 root root 1181 May  8 10:02 cisco_ios_show_interfaces_switchport.textfsm
-rw-r--r--. 1 root root 2082 May  8 10:02 cisco_ios_show_interfaces.textfsm
-rw-r--r--. 1 root root  312 May  8 10:02 cisco_ios_show_interface_transceiver.textfsm
-rw-r--r--. 1 root root  570 May  8 10:02 cisco_ios_show_inventory.textfsm
-rw-r--r--. 1 root root 2000 May  8 10:02 cisco_ios_show_ip_access-lists.textfsm
-rw-r--r--. 1 root root  525 May  8 10:02 cisco_ios_show_ip_arp.textfsm
-rw-r--r--. 1 root root 1272 May  8 10:02 cisco_ios_show_ip_bgp_neighbors.textfsm
-rw-r--r--. 1 root root  502 May  8 10:02 cisco_ios_show_ip_bgp_summary.textfsm
-rw-r--r--. 1 root root 1954 May  8 10:02 cisco_ios_show_ip_bgp.textfsm
-rw-r--r--. 1 root root  711 May  8 10:02 cisco_ios_show_ip_device_tracking_all.textfsm
-rw-r--r--. 1 root root  663 May  8 10:02 cisco_ios_show_ip_eigrp_neighbors.textfsm
-rw-r--r--. 1 root root 2407 May  8 10:02 cisco_ios_show_ip_eigrp_topology.textfsm
-rwxr-xr-x. 1 root root  459 May  8 10:02 cisco_ios_show_ip_flow_toptalkers.textfsm
-rw-r--r--. 1 root root  288 May  8 10:02 cisco_ios_show_ip_interface_brief.textfsm
-rw-r--r--. 1 root root 1778 May  8 10:02 cisco_ios_show_ip_interface.textfsm
```

我们可以通过第 4 章的 Netmiko 模块来调用 TextFSM 和 ntc-tempalte。接下来就以实验的形式演示如何使用 ntc-template。

（1）首先确认主机的 Python 里安装了 Netmiko 模块，如下图所示。

```
[root@localhost ~]# python3.10
Python 3.10.6 (main, Aug  6 2022, 13:49:15) [GCC 8.5.0 20210514 (Red Hat 8.5.0-1
5)] on linux
Type "help", "copyright", "credits" or "license" for more information.
>>> import netmiko
>>>
```

（2）在根目录下创建一个名为 ntc-template 的文件夹，如下图所示。

```
[root@localhost ~]#
[root@localhost ~]# mkdir /ntc-template
```

（3）移动到该文件夹下，如下图所示。

```
[root@localhost ~]# cd /ntc-template/
[root@localhost ntc-template]#
```

（4）然后用 git clone 命令下载 ntc-templates.git 文件，如下图所示。

```
git clone https://github.com/networktocode/ntc-templates.git
```

```
[root@localhost ntc-template]# git clone https://github.com/networktocode/ntc-te
mplates.git
Cloning into 'ntc-templates'...
remote: Enumerating objects: 11136, done.
remote: Counting objects: 100% (16/16), done.
remote: Compressing objects: 100% (16/16), done.
remote: Total 11136 (delta 3), reused 5 (delta 0), pack-reused 11120
Receiving objects: 100% (11136/11136), 2.99 MiB | 51.00 KiB/s, done.
Resolving deltas: 100% (5833/5833), done.
[root@localhost ntc-template]#
```

（5）在安装完成后，可以用 ls 命令看到在/ntc-template 下面多出了另一个 ntc-templates 文件夹，再依次输入命令 cd ntc-templates/ntc_templates/templates/和 ls，就能看到全部的 ntc-templates 模板集了，如下图所示。

```
[root@localhost templates]# pwd
/ntc-template/ntc-templates/ntc_templates/templates
[root@localhost templates]# ls
alcatel_aos_show_vlan.textfsm
alcatel_sros_oam_mac-ping.textfsm
alcatel_sros_show_lag.textfsm
alcatel_sros_show_port.textfsm
alcatel_sros_show_router_bgp_routes_vpn-ipv4.textfsm
alcatel_sros_show_router_bgp_summary_family.textfsm
alcatel_sros_show_router_interface.textfsm
alcatel_sros_show_router_isis_adjacency.textfsm
alcatel_sros_show_router_isis_interface.textfsm
alcatel_sros_show_router_ldp_interface.textfsm
alcatel_sros_show_router_mpls_interface.textfsm
alcatel_sros_show_router_mpls_lsp.textfsm
alcatel_sros_show_router_ospf_interface.textfsm
alcatel_sros_show_router_pim_interface.textfsm
alcatel_sros_show_router_rsvp_interface.textfsm
alcatel_sros_show_service_id_base.textfsm
alcatel_sros_show_service_sap-using.textfsm
alcatel_sros_show_service_sdp.textfsm
alcatel_sros_show_service_sdp-using.textfsm
alcatel_sros_show_system_cpu.textfsm
arista_eos_bash_df_-h.textfsm
arista_eos_dir_flash.textfsm
arista_eos_show_boot-config.textfsm
arista_eos_show_clock.textfsm
arista_eos_show_environment_cooling.textfsm
arista_eos_show_environment_temperature.textfsm
arista_eos_show_hostname.textfsm
arista_eos_show_interfaces_description.textfsm
arista_eos_show_interfaces_status.textfsm
arista_eos_show_interfaces.textfsm
arista_eos_show_interfaces_transceiver_detail.textfsm
arista_eos_show_interfaces_transceiver.textfsm
cisco_ios_show_switch_detail_stack_ports.textfsm
cisco_ios_show_switch_detail.textfsm
cisco_ios_show_tacacs.textfsm
cisco_ios_show_version.textfsm
cisco_ios_show_vlan.textfsm
cisco_ios_show_vrf.textfsm
cisco_ios_show_vrrp_all.textfsm
cisco_ios_show_vrrp_brief.textfsm
cisco_ios_show_vtp_status.textfsm
cisco_ios_traceroute.textfsm
cisco_nxos_show_access-lists.textfsm
cisco_nxos_show_cdp_neighbors_detail.textfsm
cisco_nxos_show_cdp_neighbors.textfsm
cisco_nxos_show_clock.textfsm
cisco_nxos_show_configuration_session_summary.textfsm
cisco_nxos_show_cts_interface_all.textfsm
cisco_nxos_show_cts_interface_brief.textfsm
cisco_nxos_show_environment_temperature.textfsm
cisco_nxos_show_environment.textfsm
cisco_nxos_show_feature.textfsm
cisco_nxos_show_fex_id.textfsm
cisco_nxos_show_fex.textfsm
cisco_nxos_show_flogi_database.textfsm
cisco_nxos_show_forwarding_adjacency.textfsm
cisco_nxos_show_forwarding_ipv4_route.textfsm
cisco_nxos_show_hostname.textfsm
cisco_nxos_show_hsrp_all.textfsm
cisco_nxos_show_interface_brief.textfsm
cisco_nxos_show_interface_description.textfsm
cisco_nxos_show_interfaces_switchport.textfsm
cisco_nxos_show_interface_status.textfsm
cisco_nxos_show_interface.textfsm
```

（6）因为在 Netmiko 里是用变量 NET_TEXTFSM 来调用 ntc-templates 模板集的，所以还要用 export 来设置相应的环境变量，将 ntc-templates 模板集的完整路径/ntc-template/ntc-templates/templates 赋值给变量 NET_TEXTFSM，然后用 echo 命令检验，如下图所示。

```
export NET_TEXTFSM='/ntc-template/ntc-templates/ntc_templates/templates'
echo $NET_TEXTFSM
```

```
[root@localhost ~]# export NET_TEXTFSM='/ntc-template/ntc-templates/ntc_templates/templates'
[root@localhost ~]# echo $NET_TEXTFSM
/ntc-template/ntc-templates/ntc_templates/templates
[root@localhost ~]#
```

注：export 命令只在当前生效，下次 CentOS 重启后用 export 命令设置的环境变量就会失效，可以使用 cd ~ 命令回到当前用户的家目录，然后用 vi 编辑家目录下的.bashrc 文件，如下图所示。

```
[root@localhost ~]#
[root@localhost ~]# cd ~
[root@localhost ~]# vi .bashrc
```

将 export NET_TEXTFSM='/ntc-template/ntc-templates/ntc_templates/templates'写在.bashrc 的最下面并保存，这样下次 CentOS 重启后的环境变量设置仍然有效，如下图所示。

```
# .bashrc

# User specific aliases and functions

alias rm='rm -i'
alias cp='cp -i'
alias mv='mv -i'

# Source global definitions
if [ -f /etc/bashrc ]; then
        . /etc/bashrc
fi

export NET_TEXTFSM='/ntc-template/ntc-templates/ntc_templates/templates'
```

（7）在一切准备就绪后，创建一个 Python 3 的脚本，脚本取名为 test.py，脚本内容如下。

```
from netmiko import connectHandler
import json

SW1 = {
   'device_type': 'cisco_ios',
   'ip': '192.168.2.11',
   'username': 'python',
   'password': '123',
}
```

```
connect = connectHandler(**SW1)
print ("Successfully connected to " + SW1['ip'])
interfaces = connect.send_command('show ip int brief', use_textfsm=True)
print (json.dumps(interfaces, indent=2))
```

由上面的代码可以看到，我们导入 Netmiko，以便配合使用 TextFSM 和 ntc-template（导入 Netmiko 后就不用再导入 textfsm 模块了）。在 6.1 节中讲到，JSON 是 Python 的内置脚本，可以直接导入，关于 Netmiko 的其他基础用法请参考 4.2.2 节，这里不再赘述。

注意，我们在 Netmiko 的 send_command()函数中使用了 use_textfsm 参数，并将其设为 True，表示调用 TextFSM，而 TextFSM 调用的 ntc-templates 模板这个步骤我们已经在第 6 步通过改变环境变量做好了，这里不用操心。最后用 JSON 的 dumps()函数将解析后的 show ip int brief 的回显文本内容打印出来，注意后面设置缩进的 indent = 2 不能省掉，如果写成 print (json.dumps(result))，则 Python 会报错。

（8）运行脚本看效果，完美地得到我们预期想要的效果，如下图所示。

```
[root@localhost ~]# python3.10 test.py
Successfully connected to 192.168.2.11
[
  {
    "intf": "GigabitEthernet0/1",
    "ipaddr": "unassigned",
    "status": "down",
    "proto": "down"
  },
  {
    "intf": "GigabitEthernet0/2",
    "ipaddr": "unassigned",
    "status": "down",
    "proto": "down"
  },
  {
    "intf": "GigabitEthernet0/3",
    "ipaddr": "unassigned",
    "status": "down",
    "proto": "down"
  },
  {
    "intf": "GigabitEthernet0/0",
    "ipaddr": "192.168.2.11",
    "status": "up",
    "proto": "up"
  }
]
[root@localhost ~]#
```

（9）将之前的 Python 脚本修改如下。

```
from netmiko import ConnectHandler
import json
```

```
SW1 = {
    'device_type': 'cisco_ios',
    'ip': '192.168.2.11',
    'username': 'python',
    'password': '123',
}

connect = ConnectHandler(**SW1)
print ("Sucessfully connected to " + SW1['ip'])
interfaces = connect.send_command('show ip int brief', use_textfsm=True)
for interface in interfaces:
    if interface["status"] == 'up':
        print (f'{interface["intf"]} is up!  IP address: {interface["ipaddr"]}')
```

前面讲过的代码内容不再重复。需要注意的是，**在 Netmiko 调用 TextFSM 后，send_command()的返回值不再是字符串，而是列表。**所以这里的变量 interfaces 的数据类型为列表，列表里的每个元素又如上一步的 JSON 阵列一样为 JSON 对象，因此可以配合 for 循环和 if 语句将所有端口状态为 up 的端口及对应的 IP 地址都打印出来。

（10）最后再运行一次脚本，效果如下图所示。

```
[root@localhost ~]# python3.10 test.py
Sucessfully connected to 192.168.2.11
GigabitEthernet0/0 is up!  IP address: 192.168.2.11
[root@localhost ~]#
```

可以看到，在 ntc-templates 的协助下，我们成功地将 show ip int brief 的回显内容转换成了 JSON 阵列格式，并使用 for 循环和 if 语句找出了交换机 192.168.2.11 下所有 up 的端口，以及对应的 IP 地址信息，完美地解决了正则表达式的痛点问题。

6.4　NAPALM

前面举例讲解了 JSON 的基础知识及用法，以及如何在 Netmiko 中使用 TextFSM 和 ntc-templates 来对旧型 IOS 设备的各种 show 和 display 命令的回显内容进行解析，接下来介绍另一个十分强大的第三方开源模块：NAPALM，看看 NAPALM 是如何在对旧型 IOS 设备做网络运维时提供各种便利的。

6.4.1　什么是 NAPALM

NAPALM 全称为 Network Automation and Programmability Abstraction Layer with Multivendor support（在英语中，Napalm 是凝固汽油弹的意思，因此 NAPALM 的图标是一团火焰包围着一个串行接口，如下图所示）。顾名思义，NAPALM 是一种为多厂商的网络设备提供统一 API 的 Python 库，其源码可以在 GitHub 上下载。

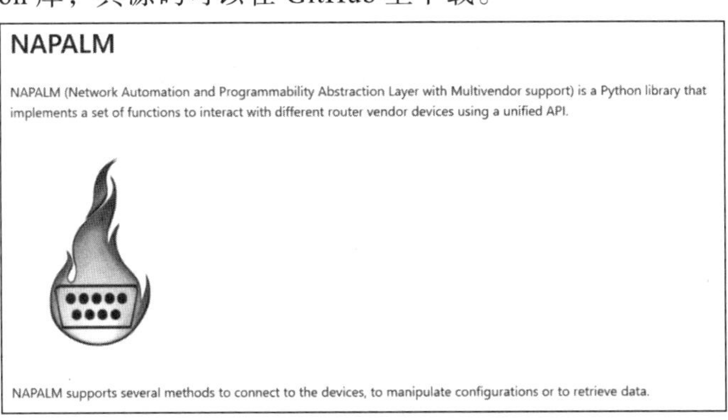

截至 2022 年 8 月，NAPALM 支持 Cisco、Arista、Juniper 3 家主流网络设备厂商的 5 种操作系统。

- Cisco IOS
- Cisco IOS-XR
- Cisco NX-OS
- Arista EOS
- Juniper JunOS

从上面来看，NAPALM 官方目前并没有支持华为设备，不过现已有热心人士基于 NAPALM 开发出 napalm-huawei-vrp 模块，旨在支持华为园区网交换机设备。本书将在 Nornir 详解章节实验 1（华为设备）中介绍 napalm-huawei-vrp 这个模块。

6.4.2　NAPALM 的优点

NAPALM 简单易用，首先它依赖 Netmiko，但是又不需要导入 Netmiko 便可独立使用（只需要保证运行脚本的主机安装了 Netmiko 即可），因此在脚本里节省了很多行代码，**让整个脚**

本更加简捷易懂。

我们知道 TextFSM 可以配合 ntc-templates，将在网络设备中输入各种 show 和 display 命令后得到的无序的纯字符串类型的回显内容整合成有序的数据结构（列表类型），方便我们使用 for 循环和 if 语句对这些文本内容做解析，从而解决 6.2 节提到的问题。而 NAPALM 提供的各种 API 也可以帮助我们在网络设备上得到我们感兴趣的设备信息和参数等内容，**并且 NAPALM 返回的数据类型也是列表**。目前 NAPALM 提供的 API 如下图所示，基本覆盖了网络运维中需要经常关注的各种网络信息和参数。

	EOS	IOS	IOSXR	JUNOS	NXOS	NXOS_SSH
get_arp_table	☑	☑	✘	✘	✘	☑
get_bgp_config	☑	☑	☑	☑	✘	✘
get_bgp_neighbors	☑	☑	☑	☑	☑	☑
get_bgp_neighbors_detail	☑	☑	☑	☑	✘	✘
get_config	☑	☑	☑	☑	☑	☑
get_environment	☑	☑	☑	☑	☑	☑
get_facts	☑	☑	☑	☑	☑	☑
get_firewall_policies	✘	✘	✘	✘	✘	✘
get_interfaces	☑	☑	☑	☑	☑	☑
get_interfaces_counters	☑	☑	☑	☑	✘	✘
get_interfaces_ip	☑	☑	☑	☑	☑	☑
get_ipv6_neighbors_table	✘	☑	✘	☑	✘	✘
get_lldp_neighbors	☑	☑	☑	☑	☑	☑
get_lldp_neighbors_detail	☑	☑	☑	☑	☑	☑
get_mac_address_table	☑	☑	☑	☑	☑	☑
get_network_instances	☑	☑	✘	☑	☑	✘
get_ntp_peers	✘	☑	☑	☑	☑	☑
get_ntp_servers	☑	☑	☑	☑	☑	☑
get_ntp_stats	☑	☑	☑	☑	☑	✘
get_optics	☑	☑	✘	☑	✘	✘
get_probes_config	✘	☑	☑	☑	✘	✘
get_probes_results	✘	✘	☑	☑	✘	✘
get_route_to	☑	☑	☑	☑	✘	☑
get_snmp_information	☑	☑	☑	☑	☑	☑
get_users	☑	☑	☑	☑	☑	☑
is_alive	☑	☑	☑	☑	☑	☑
ping	☑	☑	✘	☑	☑	☑
traceroute	☑	☑	☑	☑	☑	☑

NAPALM 的 API 分为 Getter 类和 Configuration 类。上面所列获取设备参数的 API 即 Getter 类，而 Configuration 类则支持对设备的配置做替换（Config.replace）、合并（Config.merge）、

比对（Compare Config）、原子更换（Atomic Changes）、回滚（Rollback）等操作，功能比 TextFSM 和 ntc-templates 更强大。

6.4.3　NAPALM 的缺点

前面讲到 ntc-templates 支持 Cisco IOS、Cisco ASA、Cisco NX-OS、Cisco IOS-XR、Arista、Avaya、Brocade、Checkpoint、Fortinet、Dell、Huawei、Palo Alto 等绝大多数主流厂商的设备。而相较于 ntc-templates，NAPALM 目前支持的厂商数还比较少。

由于 NAPALM 是被其他开发者按照他们自己的习惯和喜好提前造好的轮子，因此 NAPALM 的 API 返回的数据格式是统一、固定的，很难满足有个性化需求的使用者，比如在 6.3.3 节的最后，我们用 TextFSM 配合 ntc-templates 找出了一个 24 口的思科 2960 交换机中当前有哪些端口是 up 的，以及这些端口对应的 IP 地址，如下图所示。

```
[root@localhost ~]# python3.10 test.py
Sucessfully connected to 192.168.2.11
GigabitEthernet0/0 is up!  IP address: 192.168.2.11
[root@localhost ~]#
```

从运行脚本后打印出的内容可以看到，这里 GigabitEthernet0/0 是 up 的，其 IP 地址为 unassigned（二层端口）。另外，Vlan1 端口也是 up 的，其 IP 地址是 192.168.2.11。但是 NAPALM 的 get_interfaces_ip()只能给出交换机里配置了 IP 地址的端口，那些端口状态为 up、但是 IP 地址为 unassigned 的二层端口是不会被解析进去的，满足不了我们的需求。对同一台交换机使用 NAPALM 的 get_interfaces_ip()后的输出结果如下图所示（缺失了 GigabitEthernet0/0）。

```
[root@CentOS-Python python]# python3 naplm_test.py
{
  "Vlan1": {
    "ipv4": {
      "192.168.2.11": {
        "prefix_length": 24
      }
    }
  }
}
[root@CentOS-Python python]# _
```

也许你会说 ntc-templates 里的各种模板也是提前造好的轮子，返回的数据格式也是统一固定的。是的，所以当我们对回显内容的解析有个性化需求时，推荐使用 TextFSM 来编写自己需要的模板。

6.4.4　NAPALM 的安装

首先确认你的系统中已经安装了 Netmiko（可以使用命令 pip freeze | grep Netmiko，也可以进入 Python 解释器使用 import Netmiko 来确认），然后通过 pip 下载安装 NAPALM，如下图所示。

```
pip install napalm
```

```
[root@CentOS-Python python]# pip3.8 freeze | grep netmiko
netmiko==3.1.0
[root@CentOS-Python python]# pip3.8 install napalm
Requirement already satisfied: napalm in /usr/local/lib/python3.8/site-packages (3.0.0)
Requirement already satisfied: junos-eznc>=2.2.1 in /usr/local/lib/python3.8/site-packages (from napalm) (2.4.1)
Requirement already satisfied: netaddr in /usr/local/lib/python3.8/site-packages (from napalm) (0.7.19)
Requirement already satisfied: pyeapi>=0.8.2 in /usr/local/lib/python3.8/site-packages (from napalm) (0.8.3)
Requirement already satisfied: netmiko>=3.1.0 in /usr/local/lib/python3.8/site-packages (from napalm) (3.1.0)
Requirement already satisfied: textfsm in /usr/local/lib/python3.8/site-packages (from napalm) (1.1.0)
Requirement already satisfied: cffi>=1.11.3 in /usr/local/lib/python3.8/site-packages (from napalm) (1.14.0)
Requirement already satisfied: scp in /usr/local/lib/python3.8/site-packages (from napalm) (0.13.2)
Requirement already satisfied: future in /usr/local/lib/python3.8/site-packages (from napalm) (0.18.2)
Requirement already satisfied: ciscoconfparse in /usr/local/lib/python3.8/site-packages (from napalm) (1.5.4)
Requirement already satisfied: paramiko>=2.6.0 in /usr/local/lib/python3.8/site-packages (from napalm) (2.7.1)
Requirement already satisfied: pyYAML in /usr/local/lib/python3.8/site-packages (from napalm) (5.3.1)
Requirement already satisfied: requests>=2.7.0 in /usr/local/lib/python3.8/site-packages (from napalm) (2.23.0)
Requirement already satisfied: setuptools>=38.4.0 in /usr/local/lib/python3.8/site-packages (from napalm) (41.2.0)
Requirement already satisfied: jinja2 in /usr/local/lib/python3.8/site-packages (from napalm) (2.11.2)
Requirement already satisfied: lxml>=4.3.0 in /usr/local/lib/python3.8/site-packages (from napalm) (4.5.0)
Requirement already satisfied: six in /usr/local/lib/python3.8/site-packages (from junos-eznc>=2.2.1->napalm) (1.14.0)
Requirement already satisfied: ncclient>=0.6.3 in /usr/local/lib/python3.8/site-packages (from junos-eznc>=2.2.1->napalm) (0.6.7)
Requirement already satisfied: pyserial in /usr/local/lib/python3.8/site-packages (from junos-eznc>=2.2.1->napalm) (3.4)
Requirement already satisfied: ntc-templates in /usr/local/lib/python3.8/site-packages (from junos-eznc>=2.2.1->napalm) (1.4.1)
Requirement already satisfied: transitions in /usr/local/lib/python3.8/site-packages (from junos-eznc>=2.2.1->napalm) (0.8.1)
Requirement already satisfied: pyparsing in /usr/local/lib/python3.8/site-packages (from junos-eznc>=2.2.1->napalm) (2.4.7)
Requirement already satisfied: yamlordereddictloader in /usr/local/lib/python3.8/site-packages (from junos-eznc>=2.2.1->napalm) (0.4.0)
Requirement already satisfied: pycparser in /usr/local/lib/python3.8/site-packages (from cffi>=1.11.3->napalm) (2.20)
Requirement already satisfied: passlib in /usr/local/lib/python3.8/site-packages (from ciscoconfparse->napalm) (1.7.2)
Requirement already satisfied: dnspython in /usr/local/lib/python3.8/site-packages (from ciscoconfparse->napalm) (1.16.0)
Requirement already satisfied: colorama in /usr/local/lib/python3.8/site-packages (from ciscoconfparse->napalm) (0.4.3)
Requirement already satisfied: bcrypt>=3.1.3 in /usr/local/lib/python3.8/site-packages (from paramiko>=2.6.0->napalm) (3.1.7)
Requirement already satisfied: pynacl>=1.0.1 in /usr/local/lib/python3.8/site-packages (from paramiko>=2.6.0->napalm) (1.3.0)
Requirement already satisfied: cryptography>=2.5 in /usr/local/lib/python3.8/site-packages (from paramiko>=2.6.0->napalm) (2.9.2)
Requirement already satisfied: urllib3!=1.25.0,!=1.25.1,<1.26,>=1.21.1 in /usr/local/lib/python3.8/site-packages (from requests>=2.7.0->napalm) (1.25.9)
Requirement already satisfied: idna<3,>=2.5 in /usr/local/lib/python3.8/site-packages (from requests>=2.7.0->napalm) (2.9)
Requirement already satisfied: chardet<4,>=3.0.2 in /usr/local/lib/python3.8/site-packages (from requests>=2.7.0->napalm) (3.0.4)
Requirement already satisfied: certifi>=2017.4.17 in /usr/local/lib/python3.8/site-packages (from requests>=2.7.0->napalm) (2020.4.5.1)
Requirement already satisfied: MarkupSafe>=0.23 in /usr/local/lib/python3.8/site-packages (from jinja2->napalm) (1.1.1)
Requirement already satisfied: terminal in /usr/local/lib/python3.8/site-packages (from ntc-templates->junos-eznc>=2.2.1->napalm) (0.4.0)
WARNING: You are using pip version 19.2.3, however version 20.1 is available.
You should consider upgrading via the 'pip install --upgrade pip' command.
[root@CentOS-Python python]# _
```

进入 Python，如果输入 import napalm 没有报错，则说明安装成功，如下图所示。

```
[root@localhost ~]# python3.10
Python 3.10.6 (main, Aug  6 2022, 13:49:15) [GCC 8.5.0 20210514 (Red Hat 8.5.0-1
5)] on linux
Type "help", "copyright", "credits" or "license" for more information.
>>> import napalm
>>> 
```

6.4.5　NAPALM 的应用

我们以实验的形式举两个例子来分别演示如何使用 NAPALM 的 Getter 类和 Configuration 类的 API。

首先创建一个名为 napalm1.py 的 Python 脚本，如下图所示。

```
[root@CentOS-Python python]# vi napalm1.py_
```

然后在脚本里写入如下代码。

```
from napalm import get_network_driver

driver = get_network_driver('ios')
SW1 = driver('192.168.2.11','python','123')
SW1.open()

output = SW1.get_arp_table()
print (output)
```

由上面的代码可以看到，我们用来做实验的交换机是一台 IP 地址为 192.168.2.11、用户名为 python、密码为 123 的思科 2960 交换机。因为 NAPALM 支持多厂商设备不同的操作系统，所以首先需要导入 NAPALM 的 get_network_driver 类，用来指定该交换机对应的操作系统 IOS 版本。然后调用 NAPALM 的 open()方法，即完成了 SSH 远程登录交换机的操作。

接着对交换机调用 NAPALM 的 get_arp_table()方法，将它赋值给变量 output，最后将输出结果打印出来。

运行代码，效果如下图所示。

```
[root@CentOS-Python python]# python3.8 napalm1.py
[{'interface': 'Vlan1', 'mac': '00:0C:29:9E:A6:8A', 'ip': '192.168.2.1', 'age':
51.0}, {'interface': 'Vlan1', 'mac': '0C:BA:9D:CE:80:01', 'ip': '192.168.2.11',
'age': 0.0}]
[root@CentOS-Python python]# _
```

可以看到，我们**仅使用 6 行代码**就将交换机 show ip arp 命令的回显内容转换成了一个有序的列表形式的数据结构，**NAPALM 是不是非常简单易用**？

如果你觉得输出结果不易读，那么我们也可以配合 JSON 模块在脚本里添加两行代码，将输出内容转换成 JSON 阵列，代码如下。

```
from napalm import get_network_driver
import json

driver = get_network_driver('ios')
SW1 = driver('192.168.2.11','python','123')
SW1.open()
```

```
output = SW1.get_arp_table()
print (json.dumps(output, indent=2))
```

再次运行脚本看效果，如下图所示。

```
[root@localhost ~]# python3.10 napalm1.py
[
  {
    "interface": "GigabitEthernet0/0",
    "mac": "00:0C:29:F4:A5:EE",
    "ip": "192.168.2.1",
    "age": 63.0
  },
  {
    "interface": "GigabitEthernet0/0",
    "mac": "0C:1A:91:B4:48:00",
    "ip": "192.168.2.11",
    "age": -1.0
  }
]
[root@localhost ~]#
```

有关 NAPALM 的 Getter 类 API 就介绍到这里，其他比较常用的 Getter 类 API 包括 get_facts()、get_config()、get_interaces()、get_bgp_config()、get_bgp_interfaces()等，就留给读者自行去尝试和使用吧。

接下来演示如何使用 NAPALM 的 Configuration 类 API。前面提到，Configuration 类 API 包括替换（Config.replace）、合并（Config.merge）、比对（Compare Config）、原子更换（Atomic Changes）、回滚（Rollback）5 项操作，这里举例说明最常用的合并和比对的用法。

Configuration 类的合并实际就是给设备做配置，它的方法是首先创建一个扩展名为.cfg 的配置文件，将配置命令写入该配置文件；然后在脚本里使用 NAPALM 的 load_merge_candidate()函数读取配置文件中的命令并上传到目标设备；接着通过 commit_config()将这些命令在设备上配置执行。load_merge_candidate()函数是基于 SCP 协议向目标设备传送配置命令的，因此在使用之前，需要先在设备上开启 scp server，否则 NAPALM 会返回“napalm.base.exceptions.CommandErrorException: SCP file transfers are not enabled. Configure 'ip scp server enable' on the device.”的错误，如下图所示。

```
[root@CentOS-Python python]# python3 napalm2.py
Traceback (most recent call last):
  File "napalm2.py", line 7, in <module>
    iosvl2.load_merge_candidate(filename='napalm_config.cfg')
  File "/usr/local/lib/python3.6/site-packages/napalm/ios/ios.py", line 319, in load_merge_candidate
    file_system=self.dest_file_system,
  File "/usr/local/lib/python3.6/site-packages/napalm/ios/ios.py", line 285, in _load_candidate_wrapper
    file_system=file_system,
  File "/usr/local/lib/python3.6/site-packages/napalm/ios/ios.py", line 624, in _scp_file
    TransferClass=FileTransfer,
  File "/usr/local/lib/python3.6/site-packages/napalm/ios/ios.py", line 689, in _xfer_file
    raise CommandErrorException(msg)
napalm.base.exceptions.CommandErrorException: SCP file transfers are not enabled. Configure 'ip scp server enable' on the device.
```

```
SW1#conf t
Enter configuration commands, one per line.  End with CNTL/Z.
SW1(config)#ip scp server enable
```

在了解原理后，首先创建一个名为 napalm_config.cfg 的配置文件，将用来对交换机 VTY 的第 5～15 条线做基本配置的命令集写进该配置文件并保存，如下图所示。

```
[root@CentOS-Python python]# cat napalm_config.cfg
line vty 5 15
transport input ssh
transport output ssh
login local
```

然后创建一个名为 napalm2.py 的脚本，如下图所示。

```
[root@CentOS-Python python]# vi napalm2.py_
```

将下面的代码写入该脚本。

```
from napalm import get_network_driver

driver = get_network_driver('ios')
SW1 = driver('192.168.2.11', 'python', '123')
SW1.open()

SW1.load_merge_candidate(filename='napalm_config.cfg')
SW1.commit_config()
```

由上面的代码可知，我们用 load_merge_candidate(filename='napalm_config.cfg')加载之前创建好的配置文件 napalm_config.cfg，然后配合 NAPALM 的 commit_config()方法执行该配置文件的配置命令，即可完成对交换机的配置任务。

在执行代码前，我们先检查一遍交换机 192.168.2.11 的配置，确认当前交换机没有 line vty 5 15 的配置，如下图所示。

```
SW1#show run | s line vty
line vty 0 4
 login local
 transport input ssh
 transport output ssh
```

然后执行脚本（因为代码里没有使用任何 print()函数，所以执行代码后没有任何回显内容），如下图所示。

```
[root@CentOS-Python python]# python3 napalm2.py
```

接着回到交换机检查配置，如下图所示。

```
SW1#show run | s line vty
line vty 0 4
 login local
 transport input ssh
 transport output ssh
line vty 5 15
 login local
 transport input ssh
 transport output ssh
SW1#
SW1#show start | s line vty
line vty 0 4
 login local
 transport input ssh
 transport output ssh
line vty 5 15
 login local
 transport input ssh
 transport output ssh
SW1#
```

可以看到，在执行脚本后，NAPALM 已经完成了 VTY 5 ~ 15 线的配置，并且 show start | s line vty 命令的回显内容证实了在完成配置的同时，NAPALM 帮我们保存了该配置。

下面来看如何使用 Configuration 类的比对。首先我们手动将 line vty 5 15 下面的 login local、transport input ssh、transport output ssh 3 条命令都从交换机里移除，移除之后再输入 show run | s line vty 命令，可以看到之前的 3 条命令已分别被 no login、transport input none 和 transport output none 替代，如下图所示。

```
SW1(config)#line vty 5 15
SW1(config-line)#no login local
SW1(config-line)#no transport input
SW1(config-line)#no transport output
SW1(config-line)#end
SW1#
*May 15 10:22:05.047: %SYS-5-CONFIG_I: Configured from console by console
SW1#show run | s line vty
line vty 0 4
 login local
 transport input ssh
 transport output ssh
line vty 5 15
 no login
 transport input none
 transport output none
SW1#
```

然后创建一个名为 napalm3.py 的脚本，如下图所示。

```
[root@CentOS-Python python]# vi napalm3.py
```

将下面的代码写入该脚本。

```
from napalm import get_network_driver

driver = get_network_driver('ios')
SW1 = driver('192.168.2.11', 'python', '123')
SW1.open()

SW1.load_merge_candidate(filename='napalm_config.cfg')

differences = SW1.compare_config()
if len(differences) > 0:
   print(differences)
   SW1.commit_config()
else:
   print('No changes needed.')
   SW1.discard_config()
```

由上面的代码可知，我们照例用 load_merge_candidate(filename='napalm_config.cfg')加载之前创建好的配置文件 napalm_config.cfg，然后调用 NAPALM 的 compare_config()方法。顾名思义，compare_config()是 NAPALM 用来将 napalm_config.cfg 文件里的配置和交换机当前的配置做比对用的，它的返回值（即比对的结果）的数据类型为字符串。我们将该比对结果赋值给变量 difference，然后用 if 语句配合 len()函数对比对的结果 difference 进行分析，如果 len()返回的整数大于 0，则说明 difference 的内容为非空，也说明配置文件的命令和交换机当前的配置有区别，我们将比对的结果打印出来，再调用 commit_config()将漏掉的命令补上。反之，如果 len()返回的整数为 0，则说明配置文件里的命令和交换机当前的配置一致，打印 “No changes needed.” 提醒用户不需要对交换机配置做任何更改，并调用 NAPALM 的 discard_config()方法来放弃之前通过 load_merge_candidate(filename='napalm_ config.cfg')从配置文件 napalm_config.cfg 里加载好的配置命令。

执行脚本看效果，如下图所示。

```
[root@CentOS-Python python]# python3 napalm3.py
+transport input ssh
+transport output ssh
+login local
[root@CentOS-Python python]# _
```

大家注意到 3 条命令前的 3 个加号了吗？它代表比对的结果变量 difference 的具体内容，表示 compare_config()方法在将配置文件里的配置命令和交换机现有的配置做比对后，发现交换机还缺少这 3 条命令，然后通过 SW1.commit_config()将这 3 条命令在交换机里补上。我们登录交换机来进行验证，如下图所示。

```
SW1#show run | s line vty
line vty 0 4
 login local
 transport input ssh
 transport output ssh
line vty 5 15
 login local
 transport input ssh
 transport output ssh
SW1#
```

可以发现，NAPALM 将交换机缺少的 login local、transport input ssh 和 transport output ssh 这 3 条命令又补回来了。

6.5 asyncio

asyncio 是 Python 内置的异步模块，它可以大幅提升 Python 的工作效率。在讲解它的用法前，首先需要知道什么是**同步**（**Synchronous**），什么是**异步**（**Asynchronous**），以及为什么使用异步能够提升日常网络运维的工作效率。

注：本书第 1 版中介绍的基于 asyncio 的 netdev 模块由于作者 Sergey Yakovlev 不再对其更新和维护，在 Python 3.10 中已经无法使用，因此本书将不再介绍其使用方法。

6.5.1 同步与异步

所谓同步，可以理解为每当系统执行完一段代码或者函数后，系统都将一直等待该段代码或函数的返回值或消息，直到系统接收返回值或消息后才继续执行下一段代码或函数，**在等待返回值或消息期间，程序处于阻塞状态，系统将不做任何事情**。

在本书前面所有涉及管理多个设备的实验中，我们都是将设备的 IP 地址预先写入一个名为 ip_list.txt 的文本文件，然后在脚本里使用 open()函数将其打开，调用 readlines()函数并配合 for 循环读取每个设备的 IP 地址，再通过 Paramiko 或者 Netmiko 一台接一台设备地完成 SSH 登录。像这样 Python 一次只能登录一台设备，只有完成一台设备的配置后才能登录下一台设备继续配置的方式就是一种典型的“同步”。

而异步则恰恰相反，系统在执行完一段代码或者函数后，不用阻塞性地等待返回值或消息，而是继续执行下一段代码或函数，**在同一时间段里执行多个任务（而不是傻傻地等着一**

件事情做完并且结果出来后才去做下一件事情），将多个任务并行，从而提高程序的执行效率。如果你读过数学家华罗庚的《统筹方法》，那么对其中所举的例子一定不会感到陌生：同样是沏茶的步骤，因为烧水需要一段时间，你不用等水煮沸了以后才来洗茶杯、倒茶叶（类似“同步”），而是在等待烧水的过程中就把茶杯洗好、把茶叶倒好，等水烧开了就能直接泡茶喝了，这里烧水、洗茶杯、倒茶叶 3 个任务是在同一时间段内并行完成的。这就是一种典型的“异步”。

同步和异步有一个相同点：它们都是单线程下的概念。关于单线程和多线程的比较会在 6.6 节中讨论。

6.5.2　异步在 Python 中的应用

自从 Python 在 3.4.x 版本起开始支持异步后，关于异步的 Python 语法几经更改，在 Python 3.4、Python 3.5、Python 3.7 中的实现方式有很大不同，本书后面的例子都将基于 Python 3.10.6 来讲解异步的使用。

要了解异步在 Python 中的应用，必须知道什么是**协程（Coroutine）**、什么是**任务（Task）**、什么是**可等待对象（Awaitable Object）**。

我们可以把协程理解为线程的优化，看成一种微线程。它是一种比线程更节省资源、效率更高的系统调度机制。异步就是基于协程实现的。在 Python 中，实现协程的模块主要有 asyncio、gevent 和 tornado，使用较多的是 asyncio。首先来看下面的例子。

```
#coding=utf-8
import asyncio
import time

async def main():
    print('hello')
    await asyncio.sleep(1)
    print('world')

print (f"程序于 {time.strftime('%X')} 开始执行")
asyncio.run(main())
print (f"程序于 {time.strftime('%X')} 执行结束")
```

- 在 Python 中，我们通过在 def 语句前加上 async 语句将一个函数定义为协程函数，在上面的例子中，main()函数被定义为**协程函数**。

- 这里的 await asyncio.sleep(1)表示**临时中断**当前的函数 1s。如果程序中还有其他函数，则继续执行下一个函数，直到下一个函数执行完毕，再返回来执行 main()函数。因为除了一个 main()函数，就没有其他函数了，所以在 print('hello')后，main()函数休眠了 1s，然后继续 print('world')。
- 协程函数不是普通的函数，不能直接用 main()函数来调用，需要使用 asyncio.run(main()) 才能执行该协程函数。
- 我们配合 time 模块的 strftime()函数来记录程序开始前的时间和程序结束后的时间，可以看到总共耗时确实是 1s，如下图所示。

```
[root@localhost ~]# python3.10 asyncio_1.py
程序于 11:12:29 开始执行
hello
world
程序于 11:12:30 执行结束
[root@localhost ~]#
```

需要注意的是，不要把 await asyncio.sleep(1)和 time.sleep(1)弄混，后者是在同步中使用的休眠操作，前者是在异步中使用的，因为只有一个 main()函数需要执行，所以暂时感受不到这两者的区别，不用着急，继续看下面的两个例子。

```
#coding=utf-8
import asyncio
import time

async def say_after(what, delay):
    print(what)
    await asyncio.sleep(delay)

async def main():
    print (f"程序于 {time.strftime('%X')} 执行结束")
    await say_after('hello',1)
    await say_after('world',2)
    print (f"程序于 {time.strftime('%X')} 执行结束")

asyncio.run(main())
```

- 我们在协程函数 main()的基础上加入了另一个函数 say_after()。同样地，我们用 async 将它定义为协程函数。

- 我们在 main()函数中两次调用 say_after()函数，因为 say_after()函数是一个协程函数，因此在调用它时，前面必须加上 await。
- 当 main()函数第一次调用 say_after()函数时，首先打印出“hello”，然后休眠 1s；第二次调用 say_after()函数时，打印出“world”，再休眠 2s。两次调用总共花费 3s 来运行完整个程序，如下图所示。

```
[root@localhost ~]# python3.10 asyncio_2.py
程序于 11:16:17 执行结束
hello
world
程序于 11:16:20 执行结束
[root@localhost ~]#
```

这时你会说，第一次花费 1s，第二次花费 2s，总共 3s 时间，这没节省时间啊，两次调用的 say_after()函数并没有被**并行**啊，这和同步有什么区别？别急，继续往下看。

```
#coding=utf-8
import asyncio
import time

async def say_after(what, delay):
    await asyncio.sleep(delay)
    print(what)

async def main():
    task1 = asyncio.create_task(say_after('hello',1))
    task2 = asyncio.create_task(say_after('world',2))
    print (f"程序于 {time.strftime('%X')} 开始执行")
    await task1
    await task2
    print (f"程序于 {time.strftime('%X')} 执行结束")

asyncio.run(main())
```

- 要实现异步并行，需要将协程函数打包成一个**任务**，这里使用 asyncio 的 create_task()函数将 say_after()函数打包了两次，并分别赋值给 task1 和 task2 两个变量。然后使用 await 来调用 task1 和 task2 两个任务。
- 运行脚本后可以看到，因为 task1 和 task2 是并行执行的，所以程序总共耗时 2s 即告完成（11:22:04—11:22:06），如下图所示。

```
[root@localhost ~]# python3.10 asyncio_3.py
程序于 11:22:04 开始执行
hello
world
程序于 11:22:06 执行结束
[root@localhost ~]#
```

最后来看什么是**可等待对象**。可等待对象的定义很简单：如果一个对象可以在 await 语句中使用，那么它就是**可等待对象**。可等待对象主要有 3 种类型：协程、任务和 Future。协程和任务前面已经讲过，Future 不在本书的讨论范围内，感兴趣的读者可以自己参阅其他资料深入学习。

6.6 多线程

除了 6.5 节讲到的单线程异步并行，我们也可以使用多线程（Multithreading）来提升 Python 脚本的工作效率。第 4 章已经介绍了 Netmiko 的来历及安装和使用方法，本节主要介绍如何使用 Netmiko 配合 Python 的内置模块 threading 来实现多线程执行 Python 脚本。

6.6.1 单线程与多线程

在 6.5.1 节，我们引用了数学家华罗庚的《统筹方法》中的例子来说明单线程中同步和异步的区别。其实我们也可以引用同样的例子来说明单线程和多线程的区别。在《统筹方法》中讲到的沏茶的这个例子中，如果只有一个人来完成烧水、洗茶杯、倒茶叶 3 项任务，此时只有一个劳动力，我们就可以把它看成是单线程的。**假设我们能找来 3 个人分别负责烧水、洗茶杯、倒茶叶，并且保证 3 个人同时开工干活，那么我们就可以把它看成是多线程的，每一个劳动力都代表一个线程。**

在计算机的世界中也是一样的，一个程序可以启用多个线程同时完成多个任务，如果一个线程阻塞，其他线程并不受影响。现在的 CPU 都是多核的，单线程只用到其中的一核，这其实是对硬件资源的一种浪费（当然不可否认的是，随着时代的进步，现在的 CPU 已经足够强大，即使只用单核也能同时应付多个任务，这也是后来 Python 支持异步的原因）。如果使用多线程来运行 Python 脚本，则不仅能极大地提高脚本的运行速度，提高工作效率，并且还能充分利用主机的硬件资源。下面来看如何在 Python 中使用多线程。

6.6.2　多线程在 Python 中的应用

Python 3 已经内置了_thread 和 threading 两个模块来实现多线程。相较于_thread，threading 提供的方法更多而且更常用，因此我们将举例讲解 threading 模块的用法。首先来看下面这段代码。

```
import threading
import time

def say_after(what, delay):
    print (what)
    time.sleep(delay)
    print (what)

t = threading.Thread(target = say_after, args = ('hello',3))
t.start()
```

- 我们导入 Python 内置模块 threading 来实现多线程。之后定义一个 say_after(what, delay) 函数，该函数包含 what 和 delay 两个参数，分别用来表示打印的内容及 time.sleep() 休眠的时间。
- 随后使用 threading 的 Thread()函数为 say_after(what, delay)函数创建一个线程并将它赋值给变量 t，注意 Thread()函数的 target 参数对应的是函数名称（即 say_after），args 对应的是该 say_after 函数的参数，等同于 what = 'hello'和 delay = 3。
- 最后调用 threading 中的 start()来启动刚刚创建的线程。

运行代码看效果，如下图所示。

```
[root@localhost ~]# python3.10 threading1.py
hello
hello
[root@localhost ~]# 
```

在打印出第一个“hello”后，程序因为 time.sleep(3)休眠了 3s，之后打印出了第二个“hello”。因为这时我们只运行了 say_after(what, delay)这一个函数，并且只运行了一次，因此即使我们启用了多线程，也感受不到它和单线程的区别。接下来将代码修改如下。

```
#coding=utf-8
import threading
```

```
import time

def say_after(what, delay):
    print (what)
    time.sleep(delay)
    print (what)

t = threading.Thread(target = say_after, args = ('hello',3))

print (f"程序于 {time.strftime('%X')} 开始执行")
t.start()
print (f"程序于 {time.strftime('%X')} 执行结束")
```

- 这一次调用 time.strftime()来尝试记录程序执行前和执行后的时间，看看有什么“意想不到”的结果。

运行代码看效果，如下图所示。

```
[root@localhost ~]# python3.10 threading2.py
程序于 11:47:20 开始执行
hello
程序于 11:47:20 执行结束
hello
[root@localhost ~]#
```

这里你肯定会问为什么程序在 11:47:20 开始执行，又在同一时间结束？难道不是该休眠 3s 吗？为什么明明 print (f"程序于 {time.strftime('%X')} 开始执行")和 print (f"程序于 {time.strftime('%X')} 执行结束")分别写在 t.start()的上面和下面，但是不等第二个“hello”被打印出来，print (f"程序于 {time.strftime('%X')} 执行结束")就被执行了？

这是因为除了 threading.Thread()为 say_after()函数创建的**用户线程**，print(f"程序于{time.strftime('%X')}开始执行")和print(f"程序于{time.strftime('%X')}执行结束")两个print()函数也共同占用了公用的**内核线程**。也就是说，该脚本实际上调用了两个线程：一个是用户线程，一个是内核线程，也就构成了一个多线程的环境。**因为分属不同的线程，say_after()函数和函数之外的两个 print 语句是同时运行的，互不干涉**，所以 print(f"程序于 {time.strftime('%X')} 执行结束")不会像在单线程中那样等到 t.start()执行完才被执行，而是在 print (f"程序于 {time.strftime('%X')} 开始执行")被执行后就马上跟着执行。这也就解释了为什么你会看到原本需要休眠 3s 的脚本会在 11:47:20 同时开始和结束。

如果想要正确捕捉 say_after(what, delay)函数开始和结束的时间，需要额外使用 threading 模块的 join()方法，代码如下。

```
#coding=utf-8
import threading
import time

def say_after(what, delay):
    print (what)
    time.sleep(delay)
    print (what)

t = threading.Thread(target = say_after, args = ('hello',3))

print (f"程序于 {time.strftime('%X')} 开始执行")
t.start()
t.join()
print (f"程序于 {time.strftime('%X')} 执行结束")
```

- 我们只修改了代码的一处地方，即在 t.start()下面添加了一个 t.join()。**join()方法的作用是强制阻塞调用它的线程，直到该线程运行完毕或者终止（类似单线程同步）。**因为调用 join()方法的变量 t 正是用 threading.Thread()为 say_after(what, delay)函数创建的用户线程，所以使用内核线程的 print (f"程序于 {time.strftime('%X')} 执行结束")必须等待该用户线程执行完毕后才能继续执行，因此脚本在执行时的效果会让你觉得整体还是以单线程同步的方式运行的。

运行代码看效果，如下图所示。

```
[root@localhost ~]# python3.10 threading3.py
程序于 11:49:15 开始执行
hello
hello
程序于 11:49:18 执行结束
[root@localhost ~]#
```

可以看到，因为调用了 join()方法，在内核线程上运行的 print (f"程序于 {time.strftime('%X')}执行结束")必须等待在用户线程上运行的 say_after(what, delay)函数执行完毕后，才能继续执行，因此程序前后执行总共花费 3s，类似单线程同步的效果。

最后举一个例子，来看下如何创建多个用户线程并运行，代码如下。

```
#coding=utf-8
import threading
import time

def say_after(what, delay):
```

```
    print (what)
    time.sleep(delay)
    print (what)

print (f"程序于 {time.strftime('%X')} 开始执行\n")
threads = []

for i in range(1,6):
    t = threading.Thread(target=say_after, name="线程" + str(i), args=('hello',3))
    print(t.name + '开始执行。')
    t.start()
    threads.append(t)

for i in threads:
    i.join()
print (f"\n 程序于 {time.strftime('%X')} 执行结束")
```

- 我们使用 for 循环配合 range(1,6)创建了 5 个线程，并且将它们以多线程的形式执行，也就是把 say_after(what, delay)函数以多线程的形式执行了 5 次。每个线程都作为元素加入 threads 空列表，然后使用 for 语句遍历已经有 5 个线程的 threads 列表，对其中的每个线程都调用join()方法，确保直到它们都执行结束，才执行内核线程的 print (f"程序于 {time.strftime('%X')} 执行结束")。

运行代码看效果，如下图所示。

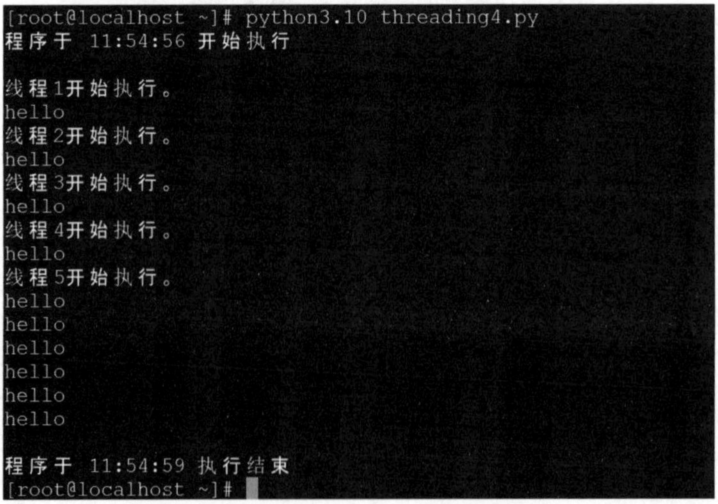

```
[root@localhost ~]# python3.10 threading4.py
程序于 11:54:56 开始执行

线程1开始执行。
hello
线程2开始执行。
hello
线程3开始执行。
hello
线程4开始执行。
hello
线程5开始执行。
hello
hello
hello
hello
hello
hello

程序于 11:54:59 执行结束
[root@localhost ~]#
```

可以看到，我们成功地使用多线程将程序执行，如果以单线程来执行 5 次 say_after (what,delay)函数，则需要花费 $3\times5=15$s 才能运行完整个脚本，而在多线程形式下，整个程序只花费 3s 就运行完毕。

6.6.3　多线程在 Netmiko 中的应用

在掌握了 threading 模块的基本用法后，我们来看如何将它与 Netmiko 结合，实现通过 Netmiko 对网络进行多线程登录和操作。

为了让大家更直观地感受到多线程的速度，首先我们用 Netmiko 配合单线程（Netmiko 默认为单线程运行）对 5 台交换机的 line vty 5 15 下面加上 login local 和 transport input ssh，脚本如下。

```
import time
from netmiko import ConnectHandler

f = open('ip_list.txt')

print (f"程序于 {time.strftime('%X')} 开始执行\n")
for ips in f.readlines():
    commands = ["line vty 5 15", "login local", "transport input ssh"]
    SW = {'device_type': 'cisco_ios', 'ip': ips, 'username': 'python', 'password': '123'}
    ssh_session = ConnectHandler(**SW)
    output = ssh_session.send_config_set(commands)
    print (output)

print (f"\n程序于 {time.strftime('%X')} 执行结束")
```

运行脚本看效果，如下图所示。

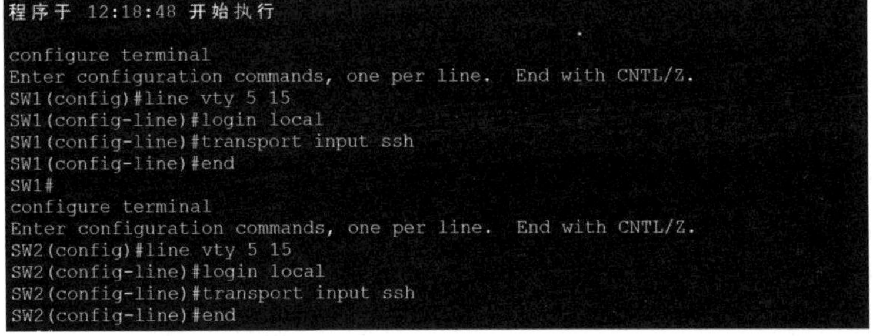

```
[root@localhost ~]# python3.10 threading5.py
程序于 12:18:48 开始执行

configure terminal
Enter configuration commands, one per line.  End with CNTL/Z.
SW1(config)#line vty 5 15
SW1(config-line)#login local
SW1(config-line)#transport input ssh
SW1(config-line)#end
SW1#
configure terminal
Enter configuration commands, one per line.  End with CNTL/Z.
SW2(config)#line vty 5 15
SW2(config-line)#login local
SW2(config-line)#transport input ssh
SW2(config-line)#end
```

```
SW2#
configure terminal
Enter configuration commands, one per line.  End with CNTL/Z.
SW3(config)#line vty 5 15
SW3(config-line)#login local
SW3(config-line)#transport input ssh
SW3(config-line)#end
SW3#
configure terminal
Enter configuration commands, one per line.  End with CNTL/Z.
SW4(config)#line vty 5 15
SW4(config-line)#login local
SW4(config-line)#transport input ssh
SW4(config-line)#end
SW4#
configure terminal
Enter configuration commands, one per line.  End with CNTL/Z.
SW5(config)#line vty 5 15
SW5(config-line)#login local
SW5(config-line)#transport input ssh
SW5(config-line)#end
SW5#

程序于 12:18:53 执行结束
[root@localhost ~]#
```

可以看到，使用单线程时，程序总共耗时 5s 才执行完毕。

接下来我们用 Netmiko 配合多线程再执行一次同样的任务，脚本如下。

```
#coding=utf-8
import threading
from queue import Queue
import time
from netmiko import ConnectHandler

f = open('ip_list.txt')
threads = []

def ssh_session(ip, output_q):
   commands = ["line vty 5 15", "login local", "transport input ssh"]
   SW = {'device_type': 'cisco_ios', 'ip': ip, 'username': 'python', 'password': '123'}
   ssh_session = ConnectHandler(**SW)
   output = ssh_session.send_config_set(commands)
   print (output)

print (f"程序于 {time.strftime('%X')} 开始执行\n")
for ips in f.readlines():
     t = threading.Thread(target=ssh_session, args=(ips.strip(), Queue()))
     t.start()
     threads.append(t)
```

```
for i in threads:
    i.join()
print (f"\n程序于 {time.strftime('%X')} 执行结束")
```

- 在使用 Netmiko 实现多线程时，需要导入内置队列模块 queue。在 Python 中，队列（queue）是线程间最常用的交换数据的形式，这里引用 queue 模块中的 Queue 类，也就是先进先出（First Input First Output，FIFO）队列，并将它作为出队列参数配置给 ssh_session(ip, output_q)函数。有关 queue 模块的具体介绍不在本书的讨论范围内，只需要知道这是使用 Netmiko 实现多线程时必备的步骤即可。
- 其余代码的讲解从略。

运行脚本看效果，如下图所示。

```
[root@localhost ~]# python3.10 threading6.py
程序于 12:27:52 开始执行

configure terminal
Enter configuration commands, one per line.  End with CNTL/Z.
SW1(config)#line vty 5 15
SW1(config-line)#login local
SW1(config-line)#transport input ssh
SW1(config-line)#end
SW1#
configure terminal
Enter configuration commands, one per line.  End with CNTL/Z.
SW5(config)#line vty 5 15
SW5(config-line)#login local
SW5(config-line)#transport input ssh
SW5(config-line)#end
SW5#
configure terminal
Enter configuration commands, one per line.  End with CNTL/Z.
SW4(config)#line vty 5 15
SW4(config-line)#login local
SW4(config-line)#transport input ssh
SW4(config-line)#end
SW4#
configure terminal
Enter configuration commands, one per line.  End with CNTL/Z.
SW3(config)#line vty 5 15
SW3(config-line)#login local
SW3(config-line)#transport input ssh
SW3(config-line)#end
SW3#
configure terminal
Enter configuration commands, one per line.  End with CNTL/Z.
SW2(config)#line vty 5 15
SW2(config-line)#login local
SW2(config-line)#transport input ssh
SW2(config-line)#end
SW2#

程序于 12:27:54 执行结束
[root@localhost ~]#
```

可以看到，使用 Netmiko 多线程总共仅耗时 2s（12:27:52—12:27:54）便完成了对 5 台交换机的配置（在 vty 5 15 下面配置 login local 和 transport input ssh），和单线程相比节省了 3s 的时间。如果设备数越多，那么多线程的优势会越明显，尤其是当设备超过 100 台甚至 1000 台的时候，读者可以自行尝试做实验验证。

6.7　CSV 和 Jinja2

在平时的网络运维工作中，网络工程师少不了要给新设备做批量配置。其中一些配置比如 line vty 下的 transport input ssh 及 AAA、SNMP 之类的属于共有配置，所有此类命令在所有设备上是统一没有差异的，写一个 Python 脚本用 Paramiko、Netmiko 或者 pexpect 等 SSH 模块就能轻松搞定。但是针对那些有差异的配置，比如每个端口下的 description（如用来描述每个端口下面所连接的服务器的名称，服务器的物理端口（网卡）号，以及服务器的用途等）、每个端口的功能（access port 还是 trunk port？如果是 access port，access 哪个 VLAN？因为存在差异，你不能简单地用一个 interface range 来一次性统一完成所有端口的配置），以及在不同交换机上划分和创建不同的 VLAN ID 等，光靠上述 SSH 模块是很难搞定的，就算能搞定，写出来的脚本的代码量也相当庞大，不便于维护。这时就有必要使用 Jinja2 这个模板引擎来帮助我们完成这些量大又有差异的配置。

本节将以一台思科的 Nexus 9300 交换机为例，讲解如何使用 Jinja2 来完成对该交换机 12 个端口的配置。

6.7.1　CSV 配置文件及 csv 模块在 Python 中的使用

在讲解 Jinja2 之前，首先要知道怎么写配置文件。配置文件可以用 YAML，也可以用 CSV 来写，但是鉴于后者的受众更多（会 Excel 的人肯定比会 YAML 的人多，你身边不懂编程的同事和同行更是如此），因此本节的示例配置文件将用 CSV 来写，并且会简单介绍如何使用 Python 内置的 csv 模块来导入我们用 Excel 写的 CSV 文件的内容。

首先来看如下图所示的在 Excel 里写好的 CSV 文件，文件名为“switch ports.csv”。

该 CSV 文件包含了 Nexus 9300 交换机的 12 个端口的端口号，其中 Server、Port、Purpose 下的内容对应每个端口的 description 配置，用来描述交换机每个端口下面所连接的服务器的名称、服务器的物理端口（网卡）号，以及服务器的用途，具体命令的格式为 description Link to 【Server】 port 【port】 for 【purpose】，最后的 VLAN 将用来区别该端口是 trunk port 还是 access port，方便我们配置 switchport mode access|trunk，以及 switchport access vlan xxx。

	A	B	C	D	E
1	Interface	Server	Port	Purpose	VLAN
2	Eth1/1	esxi-01	nic 0	VM Host	Trunk
3	Eth1/2	esxi-01	nic 1	VM Host	Trunk
4	Eth1/3	esxi-01	nic 2	VM Host	Trunk
5	Eth1/4	esxi-01	nic 3	VM Host	Trunk
6	Eth1/5	esxi-02	nic 0	VM Host	Trunk
7	Eth1/6	esxi-02	nic 1	VM Host	Trunk
8	Eth1/7	esxi-02	nic 2	VM Host	Trunk
9	Eth1/8	esxi-02	nic 3	VM Host	Trunk
10	Eth1/9	db-01	nic 0	Database Server	101
11	Eth1/10	db-02	nic 0	Database Server	101
12	Eth1/11	app-01	nic 0	Application Server	102
13	Eth1/12	app-02	nic 0	Application Server	102

接下来看下如何在 Python 中使用 csv 模块将上面的 CSV 文件里的内容导入 Python 中为我们所用，代码如下。

```
import csv

f = open('switch ports.csv')
reader = csv.DictReader(f)

for row in reader:
        print (row)
f.close()
```

代码讲解如下。

- csv 是 Python 内置模块，无须安装即可导入直接使用。
- 用 open()函数打开上面提到的 CSV 文件，对其调用 csv 模块的 DictReader()函数，并将它赋值给变量 reader，该变量是一组特殊的包含“有序的字典”的可迭代对象（字典本身是无序的，但是 csv 模块的 DictReader()函数返回的可迭代对象里的元素是一组有序的字典）。
- 用 for 语句遍历 reader 里的内容，然后将它们按顺序打印出来，最后用 close()关闭刚才打开的 CSV 文件。

执行代码后效果如下图所示。

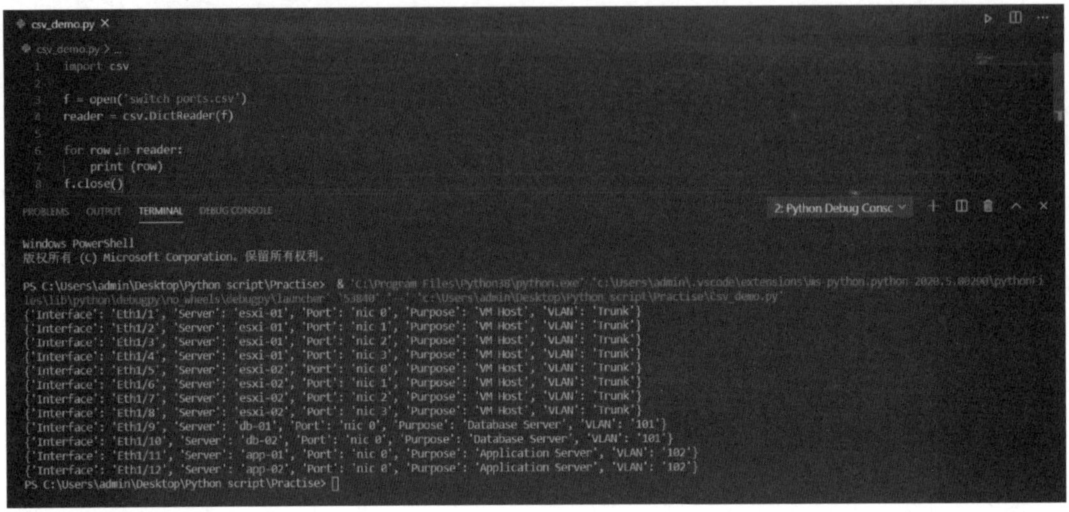

```
csv_demo.py
1  import csv
2
3  f = open('switch ports.csv')
4  reader = csv.DictReader(f)
5
6  for row in reader:
7      print (row)
8  f.close()
```

```
Windows PowerShell
版权所有 (C) Microsoft Corporation。保留所有权利。

PS C:\Users\admin\Desktop\Python script\Practise>
{'Interface': 'Eth1/1', 'Server': 'esxi-01', 'Port': 'nic 0', 'Purpose': 'VM Host', 'VLAN': 'Trunk'}
{'Interface': 'Eth1/2', 'Server': 'esxi-01', 'Port': 'nic 1', 'Purpose': 'VM Host', 'VLAN': 'Trunk'}
{'Interface': 'Eth1/3', 'Server': 'esxi-01', 'Port': 'nic 2', 'Purpose': 'VM Host', 'VLAN': 'Trunk'}
{'Interface': 'Eth1/4', 'Server': 'esxi-01', 'Port': 'nic 3', 'Purpose': 'VM Host', 'VLAN': 'Trunk'}
{'Interface': 'Eth1/5', 'Server': 'esxi-02', 'Port': 'nic 0', 'Purpose': 'VM Host', 'VLAN': 'Trunk'}
{'Interface': 'Eth1/6', 'Server': 'esxi-02', 'Port': 'nic 1', 'Purpose': 'VM Host', 'VLAN': 'Trunk'}
{'Interface': 'Eth1/7', 'Server': 'esxi-02', 'Port': 'nic 2', 'Purpose': 'VM Host', 'VLAN': 'Trunk'}
{'Interface': 'Eth1/8', 'Server': 'esxi-02', 'Port': 'nic 3', 'Purpose': 'VM Host', 'VLAN': 'Trunk'}
{'Interface': 'Eth1/9', 'Server': 'db-01', 'Port': 'nic 0', 'Purpose': 'Database Server', 'VLAN': '101'}
{'Interface': 'Eth1/10', 'Server': 'db-02', 'Port': 'nic 0', 'Purpose': 'Database Server', 'VLAN': '101'}
{'Interface': 'Eth1/11', 'Server': 'app-01', 'Port': 'nic 0', 'Purpose': 'Application Server', 'VLAN': '102'}
{'Interface': 'Eth1/12', 'Server': 'app-02', 'Port': 'nic 0', 'Purpose': 'Application Server', 'VLAN': '102'}
PS C:\Users\admin\Desktop\Python script\Practise>
```

可以看到，我们通过 csv 模块打印出了 CSV 文件下的所有内容，每行的内容格式均为有序的字典，方便我们后续引用。

6.7.2　Jinja2 配置模板

有了 CSV 写好的配置文件后，接下来我们要用 Jinja2 来写配置模板了。首先在 switch ports.csv 文件所在的同一目录下新建一个文本文件，用记事本将它打开，写入下面的配置模板内容后，单击记事本左上角的文件(F)→另存为(A)，将文件名设为 interface-template.j2，将它的“保存类型”改为“所有文件”，最后单击“保存”，这样我们就得到了一个扩展名为.j2 的 Jinja2 模板文件，如下图所示。

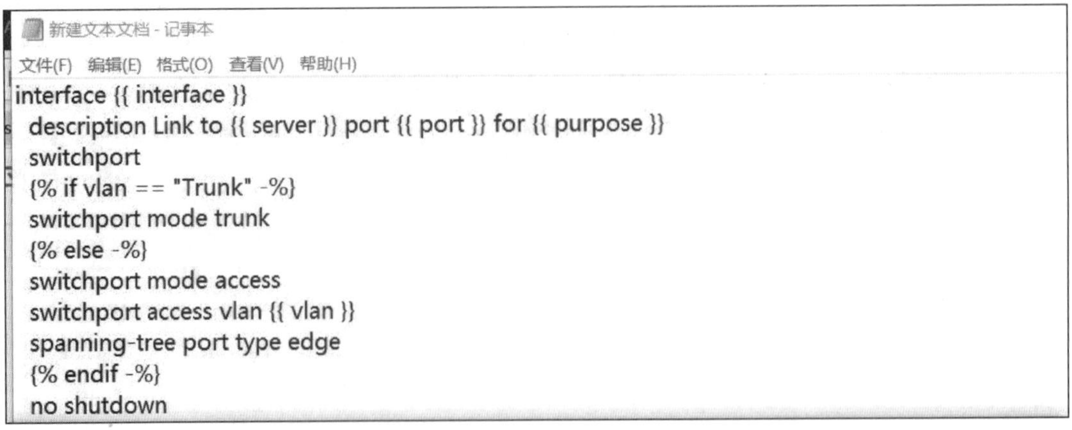

```
新建文本文档 - 记事本
文件(F) 编辑(E) 格式(O) 查看(V) 帮助(H)
interface {{ interface }}
 description Link to {{ server }} port {{ port }} for {{ purpose }}
 switchport
 {% if vlan == "Trunk" -%}
 switchport mode trunk
 {% else -%}
 switchport mode access
 switchport access vlan {{ vlan }}
 spanning-tree port type edge
 {% endif -%}
 no shutdown
```

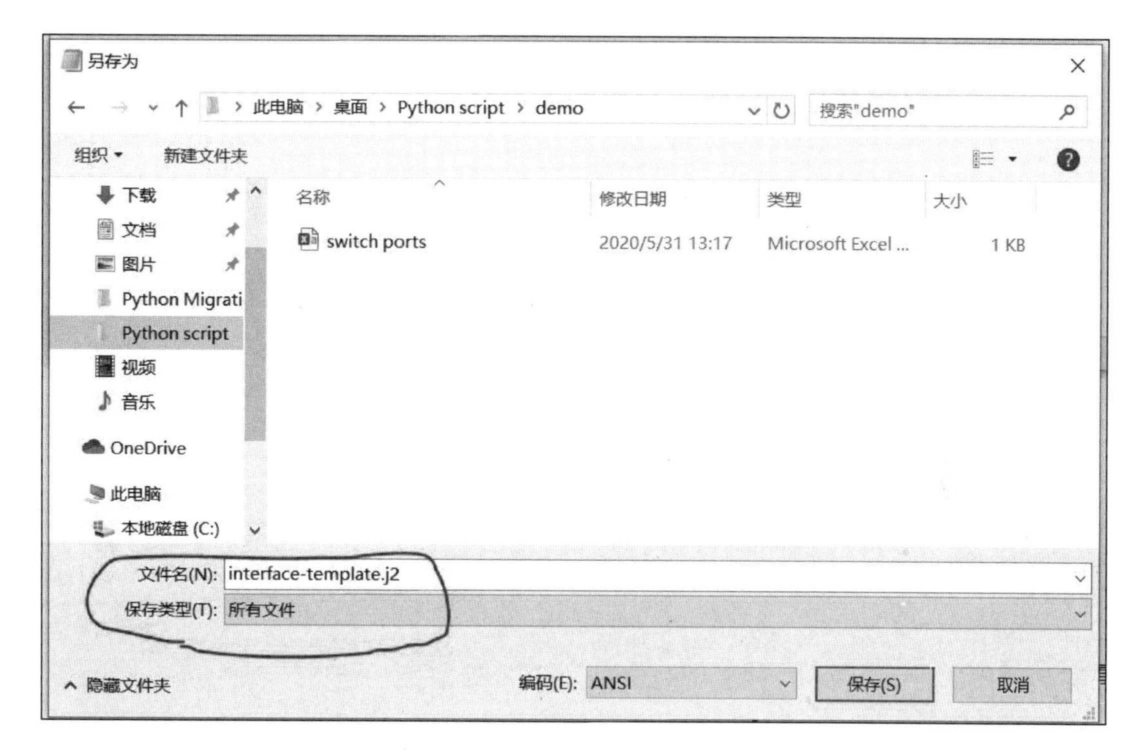

配置模板里的具体内容如下。

```
interface {{ interface }}
 description Link to {{ server }} port {{ port }} for {{ purpose }}
 switchport
 {% if vlan == "Trunk" -%}
 switchport mode trunk
 {% else -%}
 switchport mode access
 switchport access vlan {{ vlan }}
 spanning-tree port type edge
 {% endif -%}
 no shutdown
```

代码分段讲解如下。

- 在 Jinja2 中，我们用{{ 变量名 }}的格式来定义变量（注意变量名和左右两边的大括号之间分别隔了一个空格）。很显然，这里我们定义的{{ interface }}、{{ server }}、{{ port }}、{{ purpose }}、{{ vlan }}分别对应的是 CSV 配置文件里的 A1 栏至 E1 栏里的 Interface、Server、Port、Purpose、VLAN。

```
interface {{ interface }}
 description Link to {{ server }} port {{ port }} for {{ purpose }}
 switchport access vlan {{ vlan }}
```

- 在 Jinja2 中，if、else 语句的格式为{% if 条件 -%}和{% else -%}，这里要表达的意思是：如果 CSV 配置文件里的 VLAN（E 栏）下对应的内容为 Trunk，那么我们将该端口配置为 trunk port，如果端口模式为非 Trunk，那我们将端口配置为 access port，并分配对应的 VLAN 给它，最后将其生成树的端口模式改为 edge。注意 Jinja2 中的 if 语句最后需要用{% endif -%}来收尾。

```
 {% if vlan == "Trunk" -%}
 switchport mode trunk
 {% else -%}
 switchport mode access
 switchport access vlan {{ vlan }}
 spanning-tree port type edge
 {% endif -%}
```

- 最后不管端口类型是 trunk port 还是 access port，我们都用 no shutdown 命令开启它（因为 no shutdown 写在{% endif -%}下面）。

```
 no shutdown
```

这里介绍了如何使用 Jinja2 来创建配置模板及 Jinja2 的部分语法，下面我们来看如何在 Python 中使用 Jinja2。

6.7.3　Jinja2 在 Python 中的使用

前面讲到，我们在 Jinja2 的配置模板里定义了{{ interface }}、{{ server }}、{{ port }}、{{ purpose }}、{{ vlan }}5 个变量，来分别对应 CSV 配置文件里的 A1 ~ E1 栏里的 Interface、Server、Port、Purpose、VLAN。接下来要做的是将两者真正结合，生成最后我们需要的 Nexus 9300 交换机的实际配置命令，这里我们需要借助 Python 来完成，代码如下。

```
from jinja2 import Template
import csv

csv_file = open('switch_ports.csv')
template_file = open('interface-template.j2')
```

```
reader = csv.DictReader(csv_file)
interface_template = Template(template_file.read(), keep_trailing_newline=True)

interface_configs = ''

for row in reader:
    interface_config = interface_template.render(
    interface = row['Interface'],
    vlan = row['VLAN'],
    server = row['Server'],
    port = row['Port'],
    purpose = row['Purpose']
    )
    interface_configs += interface_config

print(interface_configs)
```

代码分段讲解如下。

- Jinja2 从 Python 2.4 起被引入为 Python 的内置模块，无须安装即可 import 导入使用，其中 Template 是 Jinja2 模块下最常用的类，这里我们用 from jinja2 import Template 将其导入。随后我们用 open() 函数打开之前创建好的 switch ports.csv 和 interface-template.j2 两个文件，并分别赋值给 csv_file 和 template_file 两个变量。

```
from jinja2 import Template
import csv

csv_file = open('switch ports.csv')
template_file = open('interface-template.j2')
```

- reader = csv.DictReader(csv_file)的作用在前面已经讲到了，这里不再赘述。我们用 Jinja2 的 Template() 函数将 j2 配置模板文件模板化，并将它赋值给变量 interface_template，因为这个模板要被反复使用 12 次（我们有 12 个交换机端口需要配置），Template()函数里的 keep_trailing_newline=True 这个参数的作用是在每使用完一次模板，自动替我们在配置命令的最后加入一个换行符\n，避免下面这种命令格式问题的出现（在之前创建的 interface-termplate.j2 文件中也需要在最后一行多输入一个回车，让文件最后多一个空白行）。

```
interface Eth1/1
 description Link to esxi-01 port  for VM Host
 switchport
 switchport mode trunk
 no shutdowninterface Eth1/2
 description Link to esxi-01 port  for VM Host
 switchport
 switchport mode trunk
 no shutdowninterface Eth1/3
 description Link to esxi-01 port  for VM Host
 switchport
 switchport mode trunk
 no shutdown
```

```
reader = csv.DictReader(csv_file)
interface_template = Template(template_file.read(), keep_trailing_newline=True)
```

- 这里创建一个空的字符串，将其赋值给变量 interface_configs，它的作用在下面会讲到。

```
interface_configs = ''
```

- 前面讲到了 **csv 的 DictReader()函数返回的是可迭代对象，这个可迭代对象里的元素是一组一组有序的字典**，这里我们使用 for 语句来遍历 reader 这个 DictReader()函数返回的可迭代对象，并调用 Template()函数下的 render()函数对配置模板做**渲染**，这样配置文件和配置模板就被结合起来了。每调用一次配置文件和配置模板，我们就能生成一个接口下的配置命令（从 Eth1/1 开始，直到 Eth1/12），每个接口配置命令的数据格式都为字符串。最后我们用 interface_configs += interface_config 将 12 个接口的配置（字符串）全部汇总进上面创建的空字符串变量 interface_configs，然后通过 print (interface_configs)来验证其内容。

```
interface_configs = ''

for row in reader:
    interface_config = interface_template.render(
    interface = row['Interface'],
    vlan = row['VLAN'],
    server = row['Server'],
    port = row['Port'],
    purpose = row['Purpose']
    )
    interface_configs += interface_config
```

```
print(interface_configs)
```

执行代码后效果如下图所示。

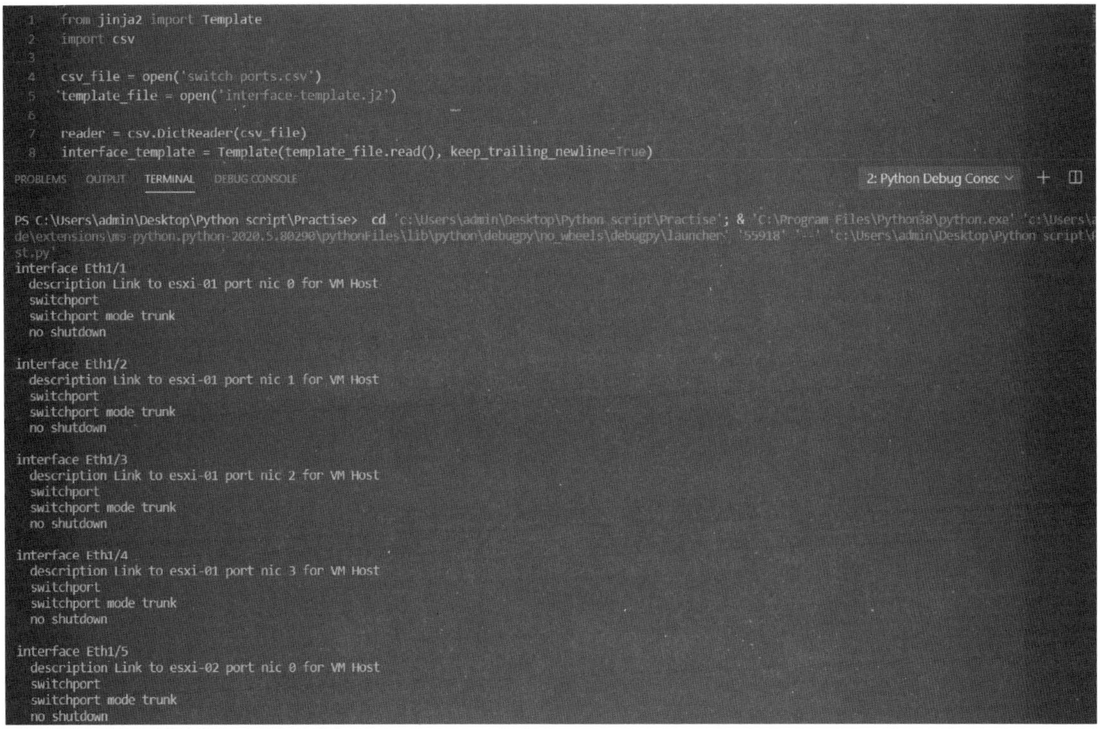

```
1   from jinja2 import Template
2   import csv
3
4   csv_file = open('switch ports.csv')
5   template_file = open('interface-template.j2')
6
7   reader = csv.DictReader(csv_file)
8   interface_template = Template(template_file.read(), keep_trailing_newline=True)
PROBLEMS   OUTPUT   TERMINAL   DEBUG CONSOLE                    2: Python Debug Consc

PS C:\Users\admin\Desktop\Python script\Practise>  cd 'c:\Users\admin\Desktop\Python script\Practise'; & 'C:\Program Files\Python38\python.exe' 'c:\Users\a
de\extensions\ms-python.python-2020.5.80290\pythonFiles\lib\python\debugpy\no_wheels\debugpy\launcher' '55918' '--' 'c:\Users\admin\Desktop\Python script\
st.py'
interface Eth1/1
  description Link to esxi-01 port nic 0 for VM Host
  switchport
  switchport mode trunk
  no shutdown

interface Eth1/2
  description Link to esxi-01 port nic 1 for VM Host
  switchport
  switchport mode trunk
  no shutdown

interface Eth1/3
  description Link to esxi-01 port nic 2 for VM Host
  switchport
  switchport mode trunk
  no shutdown

interface Eth1/4
  description Link to esxi-01 port nic 3 for VM Host
  switchport
  switchport mode trunk
  no shutdown

interface Eth1/5
  description Link to esxi-02 port nic 0 for VM Host
  switchport
  switchport mode trunk
  no shutdown
```

可以看到，在运行脚本后，代码将 Jinja2 配置模板渲染成了我们实际需要的 12 个端口的配置命令。

6.7.4 将生成的配置命令上传到交换机并执行

用 Jinja2 配置模板配合 CSV 配置文件生成我们想要的实际配置命令后，最后一步自然是将该配置命令上传到交换机并执行，这里我们使用 Netmiko 来完成，在上面的代码中添加一些内容，如下。

```
from jinja2 import Template
import csv
from netmiko import ConnectHandler

csv_file = open('switch ports.csv')
```

```
template_file = open('interface-template.j2')

reader = csv.DictReader(csv_file)
interface_template = Template(template_file.read(), keep_trailing_newline=True)

interface_configs = ''

for row in reader:
    interface_config = interface_template.render(
    interface = row['Interface'],
    vlan = row['VLAN'],
    server = row['Server'],
    link = row['Port'],
    purpose = row['Purpose']
    )
    interface_configs += interface_config

config_set = interface_configs.split('\n')

SW = {
 'device_type': 'cisco_nxos',
 'ip': '10.1.1.1',
 'username': 'admin',
 'password': 'pass',
    }

connect = ConnectHandler(**SW)
print ('Connected to switch')

output = connect.send_config_set(config_set, cmd_verify=False)
print(output)
```

代码讲解如下。

- 这里我们用 config_set = interface_configs.split('\n')将字符串的配置命令用 split('\n')隔行拆解成一个列表，方便我们配合 Netmiko 的 send_config_set()函数来对交换机做配置。
- output=connect.send_config_set(config_set, cmd_verify=False) 中之所以要把参数

cmd_verify 设为 False，其原因为：Netmiko 3 和 Netmiko 4 默认要等到输入的命令在屏幕上打印出来才会执行后面的命令（因为 Netmiko 3 和 Netmiko 4 默认将 send_config_set()里的 cmd_verify 参数设为 True)，像我们这种一次性对交换机输入多达 60 条命令的情况（12 个端口要配置），经常会出现网络延迟的问题，导致在执行脚本时 Netmiko 会返回“netmiko.sshexception.NemikoTimeoutException：Time-out reading channel, data not available”异常（我们写的代码本身没有问题，这是 Netmiko 自身的一个“bug“），如下图所示。

```
  File "C:\Users\admin\AppData\Roaming\Python\Python38\site-packages\netmiko\base_connection.py", line 1757, in send_config_set
    new_output = self.read_until_pattern(pattern=re.escape(cmd.strip()))
  File "C:\Users\admin\AppData\Roaming\Python\Python38\site-packages\netmiko\base_connection.py", line 627, in read_until_pattern
    return self._read_channel_expect(*args, **kwargs)
  File "C:\Users\admin\AppData\Roaming\Python\Python38\site-packages\netmiko\base_connection.py", line 560, in _read_channel_expect
    raise NetmikoTimeoutException(
netmiko.ssh_exception.NetmikoTimeoutException: Timed-out reading channel, data not available.
```

笔者使用的是 Netmiko 4.1.2，因此必须将 cmd_verify 参数设为 False，写成 output = connect.send_config_set(config_set, cmd_verify=False)；如果你使用的是 Netmiko 2，则没有这个顾虑（可以通过 pip freeze 来查看你的 Netmiko 版本)。

最后，在执行脚本前，我们先验证 Nexus 9300 交换机上现有的配置，这里我们选取 Eth1/1、Eth1/2、Eth1/9、Eth1/11 进行查看，可以发现，目前这些端口下没有任何配置，并且 show interface description 也可以看到前面 12 个端口没有任何 description 配置（除了 Eth1/5 当前有一个“L3 LINK”的 description，但是不影响我们的配置和验证)，如下图所示。

```
sbx-n9kv-ao# show run int e1/1

!Command: show running-config interface Ethernet1/1
!Running configuration last done at: Sat May 30 14:01:33 2020
!Time: Sat May 30 14:01:43 2020

version 9.3(3) Bios:version

interface Ethernet1/1

sbx-n9kv-ao# show run int e1/2

!Command: show running-config interface Ethernet1/2
!Running configuration last done at: Sat May 30 14:01:33 2020
!Time: Sat May 30 14:01:47 2020

version 9.3(3) Bios:version

interface Ethernet1/2

sbx-n9kv-ao# show run int e1/9
```

```
version 9.3(3) Bios:version

interface Ethernet1/2

sbx-n9kv-ao# show run int e1/9

!Command: show running-config interface Ethernet1/9
!Running configuration last done at: Sat May 30 14:01:33 2020
!Time: Sat May 30 14:01:52 2020

version 9.3(3) Bios:version

interface Ethernet1/9

sbx-n9kv-ao# show run int e1/11

!Command: show running-config interface Ethernet1/11
!Running configuration last done at: Sat May 30 14:01:33 2020
!Time: Sat May 30 14:02:13 2020

version 9.3(3) Bios:version

interface Ethernet1/11

sbx-n9kv-ao# _
sbx-n9kv-ao# show int description

-------------------------------------------------------------------------------
Interface                Description
-------------------------------------------------------------------------------
mgmt0                    DO NOT TOUCH CONFIG ON THIS INTERFACE

-------------------------------------------------------------------------------
Port          Type   Speed   Description
-------------------------------------------------------------------------------
Eth1/1        eth    10G     --
Eth1/2        eth    10G     --
Eth1/3        eth    10G     --
Eth1/4        eth    10G     --
Eth1/5        eth    10G     L3 Link
Eth1/6        eth    10G     --
Eth1/7        eth    10G     --
Eth1/8        eth    10G     --
Eth1/9        eth    10G     --
Eth1/10       eth    10G     --
Eth1/11       eth    10G     --
Eth1/12       eth    10G     --
```

执行脚本看效果，如下图所示。

```
PS C:\Users\admin\Desktop\Python script\Practise>  cd 'c:\Users\admin\Desktop\Python script\Practise'; & 'C:\Program Files\Python38\python.exe' 'c:\Users\admin\.vscode\extensions\ms-python.python-2020.5.80290\pythonFiles\lib\python\debugpy\no_wheels\debugpy\launcher' '64814' '--' 'c:\Users\admin\Desktop\Python script\Practise\jinja2_demo.py'
Connected to switch
config term
Enter configuration commands, one per line. End with CNTL/Z.

sbx-ao(config)# interface Eth1/1

sbx-ao(config-if)#   description Link to esxi-01 port nic 0 for VM Host

sbx-ao(config-if)#   switchport
                               ^
% Incomplete command at '^' marker.

sbx-ao(config-if)#   switchport mode trunk
                                ^
% Invalid command at '^' marker.

sbx-ao(config-if)#   no shutdown

sbx-ao(config-if)#

sbx-ao(config-if)# interface Eth1/2
  description Link to esxi-01 port nic 1 for VM Host

sbx-ao(config-if)#   description Link to esxi-01 port nic 1 for VM Host
```

配置完成后，再次验证交换机的端口配置，并用 show interface description 进一步验证，如下图所示。

```
sbx-n9kv-ao# show run int e1/1

!Command: show running-config interface Ethernet1/1
!Running configuration last done at: Sat May 30 13:59:15 2020
!Time: Sat May 30 13:59:29 2020

version 9.3(3) Bios:version

interface Ethernet1/1
  description Link to esxi-01 port nic 0 for VM Host
  switchport mode trunk

sbx-n9kv-ao# show run int e1/2

!Command: show running-config interface Ethernet1/2
!Running configuration last done at: Sat May 30 13:59:15 2020
!Time: Sat May 30 13:59:32 2020

version 9.3(3) Bios:version

interface Ethernet1/2
  description Link to esxi-01 port nic 1 for VM Host
  switchport mode trunk

sbx-n9kv-ao#
sbx-n9kv-ao# show run int e1/9

!Command: show running-config interface Ethernet1/9
!Running configuration last done at: Sat May 30 13:59:15 2020
!Time: Sat May 30 13:59:45 2020

version 9.3(3) Bios:version

interface Ethernet1/9
```

```
  description Link to db-01 port nic 0 for Database Server
  switchport access vlan 101
  spanning-tree port type edge

sbx-n9kv-ao# show run int e1/11

!Command: show running-config interface Ethernet1/11
!Running configuration last done at: Sat May 30 13:59:15 2020
!Time: Sat May 30 13:59:51 2020

version 9.3(3) Bios:version

interface Ethernet1/11
  description Link to app-01 port nic 0 for Application Server
  switchport access vlan 102
  spanning-tree port type edge
```

第7章

TextFSM详解

在第 1 版出版后，我们收到了很多读者朋友们的积极反馈。其中，一个重要反馈集中在国产设备的联机适配问题上，即在思科设备上的操作，是否可以，以及如何才能迁移适配到国产设备上？于是，我们在第 2 版增加了国产设备适配的内容。这部分虽以华为设备做示例，但实际上也可覆盖至其他国产品牌设备。另一个重要反馈集中在登录设备抓取回显报文后如何能快速有效地解析，诸如 ntc-templates 中涉及国产设备的模板少之又少。于是，我们在第 2 版专门安排一章，介绍一个强大的文本解析工具——TextFSM，旨在通过几个实战案例练习后，读者朋友们可按自己的需求自制模板，从而有效地开展联机解析或离线解析。

TextFSM 是 Google 开发的一个用来处理网络设备输出信息的第三方库，旨在通过以匹配自定义模板的方式，将无规律文本内容打造成自己想要的有序数据格式，方便操控数据。虽然 TextFSM 设计之初旨在解析网络设备输出，但实际上其可应用的范围非常广，可以处理任何文本类信息，比如邮件内容、短信内容。

TextFSM 允许我们制作模板来解析设备配置，“成块”处理指令回显。因此，使用 TextFSM 要比传统的逐行读取解析等方法方便很多。此外，模板对处理套路还有累积沉淀效应，分享起来也很方便，拿到别人写的模板，等于前人给你造了轮子；分享自己制作的模板，等于你给后人造了轮子。这里顺便提一下，我们使用 TextFSM 联动思科等国外厂商设备时，往往还会辅助使用另一个第三方库——ntc-templates。该库是 network To Code 团队用 TextFSM 花费了无数心血开发出来的一套模板集，可以理解成有一个造轮子的厂。但目前在 ntc-templates 中，鲜有国产设备的相关模板，因此本章不重点介绍。正因为 ntc-templates 支持国产品牌的模板很少，本章就以华为设备举例。既然没有现成的轮子，那么我们更有理由自己学习 TextFSM 的基本用法，写出自己想要的模板来。

学习 TextFSM 之前，能知道一点正则表达式基础知识会更好些；当然，没有正则表达式基础则兼顾学习起来即可。从理论上讲，TextFSM 能解决的问题，正则表达式都能解决。但是，TextFSM 使用起来会比正则表达式简化且清晰，可以有效屏蔽过于复杂的正则规则，相应的 Python 代码量会大幅降低。学习某个陌生的 Python 第三方模块，一个有效途径就是看具体案例教程和实战总结分享，另一个有效手段则是直达官网，查阅官网文档。官网手册为原汁原味的第一手材料，能原文阅读则效果更佳。参考案例教程（比如本章）了解实战，并逐步探索官网手册，两者结合起来，这就是我们推荐的学习路径。

为避免行文冗长和表述枯燥，我们不安排专门的 TextFSM 模板基础语法介绍章节，而直接通过实验切入，逐层带入常用语法点，并逐步引导读者能自己在官方手册中进行信息检索，从而掌握该库的使用。好了，我们开始吧！

7.1　TextFSM 的安装及引例

7.1.1　TextFSM 的安装

本章节在 Windows 10 上开展实验。我们先安装 TextFSM 库，之后通过一个小例子作为引入，同时展示一下 TextFSM 的魅力。

```
pip3 install textfsm
```

安装过程比较简单，与其他第三方模块类似，在本书前面章节已多次提及，这里不再赘述。

```
C:\>pip show textfsm
Name: textfsm
Version: 1.1.3
Summary: Python module for parsing semi-structured text into python tables.
Home-page: https://github.com/google/textfsm
Author:
Author-email:
License: Apache License, Version 2.0
Location: c:\program files\python310\lib\site-packages
Requires: future, six
Required-by: netmiko, ntc-templates
```

从上述信息，我们可以获得当前安装的 TextFSM 版本、官网地址、依赖包等信息。在你阅读时，TextFSM 版本或许已有差异，但基本知识通常不会有太大变化。

7.1.2　TextFSM 引例类比

TextFSM 初看复杂，我们先用一个“简单”例子，结合日常生活、生产场景做知识迁移，串讲一下 TextFSM 到底是干吗的、能怎么用，以便快速入门。

我们把整个 Python 环境下 TextFSM 引擎解析的过程，类比成一个生产车间生产产品的过程，来一点点剖析。

1. 生产车间（生产原料）

找一台华为数通设备，CLI 上执行 tracert 跟踪指令，获取回显信息，如下所示。

```
<R1>tracert 192.5.5.5
tracert 192.5.5.5
 traceroute to  192.5.5.5(192.5.5.5), max hops: 30 ,packet length: 40,press CTRL_C to
break
 1 36.1.1.6 60 ms  50 ms  10 ms
 2 67.1.1.7 100 ms  90 ms  80 ms
 3 78.1.1.8 90 ms  80 ms  60 ms
 4 59.1.1.9 100 ms  90 ms  80 ms
 5 59.1.1.5 90 ms  100 ms  60 ms
```

此时，不用关心实际组网与拓扑，聚焦这个交互报文即可。在 Python 中，它就是一个字符串数据（str 类型）。通常，对网络工程师而言，此类信息要么是设备配置数据，要么是指令回显内容。我们把这些信息当成“生产原料”。

2. 生产车间（生产模具）

“生产原料”有了，就要出产品，就得有“生产模具”辅助成型。在 TextFSM 中，“生产模具”也叫“模板”。模板也是文本信息，同样可以放入文本文件中。

```
Value ID (\d+)
Value Hop (\S+)

Start
  ^ ${ID} ${Hop} +\d+ -> Record
```

3. 生产车间（生产机器）

生产原料、生产模具都到齐后，自然得把“生产机器”搬上来。“生产机器”其实就是 Python

代码！具体的代码这里先按下不表，后面再安排它出场。

总结一下，在某生产车间中，一批不锈钢板被送入生产机器，配合着不同的模具，“咔咔咔”变成各种不锈钢零件。这样的类比，你能不能有点感性认识了呢？

- 不锈钢板（“生产原料”）→设置配置、指令回显（待匹配字符串、输入文本）。
- 不同模具（“生产模具”）→TextFSM 模块。
- 生产工具（“生产机器”）→Python 脚本。

7.1.3　TextFSM 引例详解

1. TextFSM 模板

生产原料、生产机器都已在 Python 生态中，如文本文件打开读写、循环分支等内容。现在我们重点关注“生产模具”——TextFSM 模板，通常以文本文件的形式存在。我们重新来逐行探究模板内容，列表如下。

1	Value ID (\d+)
2	Value Hop (\S+)
3	——
4	Start
5	^ ${ID} ${Hop} +\d+ -> Record

（1）第 1 行。

- Value 是关键字，用于声明变量。
- ID 是变量名，用户自定义，不能与关键字重复。
- ()圆括号内为变量定义区域。\d+为正则表达式，用户自定义。

（2）第 2 行与（1）类似，请读者试着自行补充完整。

（3）第 3 行。

- 空白行，为固定结构，隔开变量定义和状态定义区域。

（4）第 4 行。

- 状态名，从这里开启 TextFSM 匹配，为固定格式。

（5）第 5 行。

- 行首 1 ~ 2 个空格（通常为两个），以^符号开始，随后为规则及动作描述。
- ^字符在正则表达式中表行首，此处也可以这么理解。
- $字符在正则表达式中表行末，但在 TextFSM 中$配合变量使用，改用$$表行末。
- 规则描述支持正则表达式，动作描述有若干种。（后续随实验逐步展开。）
- 规则描述可以联动已定义的变量，如${ID}中 ID 是已定义的变量。
- ->用于连接规则和动作，Record 即动作。

TextFSM 直译为"文本有限状态机"，Start 就是一个状态机。整个匹配过程可在多个状态之间跳转。当然，太多的状态跳转容易让初学者不知所措，建议先由 Start 一个状态解决，而后视情况进阶。

初学者往往要被空格"折腾"几下，需要耐心和细心，我们再把一些必要的空格标注出来。对于规则中的空格也可以用正则表达式\s 配合+和*来表示，可能会更清晰一点。

```
Value【空格】ID【空格】(\d+)
Value【空格】Hop【空格】(\S+)
【空行】
Start
【空格】【空格】^ ${ID} ${Hop} +\d+【空格】->【空格】Record
```

2. Python 脚本代码

我们终于要把"生产工具"搬上来了。因为本章重点是学习 TextFSM 模板，参考"控制变量法"思想，对于 Python 代码，我们尽量做到简单统一。换言之，本章涉及的 Python 代码都差不多。

```
import textfsm
traceroute = '''
<R1>tracert 192.5.5.5
tracert 192.5.5.5
 traceroute to  192.5.5.5(192.5.5.5), max hops: 30 ,packet length: 40,press CTRL_C to break
 1 36.1.1.6 60 ms  50 ms  10 ms
 2 67.1.1.7 100 ms  90 ms  80 ms
```

```
 3 78.1.1.8 90 ms  80 ms  60 ms
 4 59.1.1.9 100 ms  90 ms  80 ms
 5 59.1.1.5 90 ms  100 ms  60 ms
'''
with open('traceroute.template') as template:
    fsm = textfsm.TextFSM(template)
    result = fsm.ParseText(traceroute)
    print(fsm.header)
    print(result)
```

在 Python 脚本中，剔除变量 traceroute，代码量很少。

（1）引入 import textfsm 模块。

（2）变量 traceroute 装 CLI 回显内容，如回显过长，可另存文本再读取。

（3）with open('traceroute.template') as template: 以上下文方式打开模板文件。请注意，变量 template 不是模板中的内容，而是模板文件对象。

（4）fsm = textfsm.TextFSM(template)，把模板文件导入 textfsm 模块的 TextFSM()类里，创建一个有限状态机，赋值给变量 fsm。我们可配合 help 函数或到官网检索查阅，如下图所示。

```
>>> import textfsm
>>> help(textfsm.TextFSM)
Help on class TextFSM in module textfsm.parser:

class TextFSM(builtins.object)
 |  TextFSM(template, options_class=<class 'textfsm.parser.TextFSMOptions'>)
 |
 |  Parses template and creates Finite State Machine (FSM).
 |
 |  Attributes:
 |    states: (str), Dictionary of FSMState objects.
 |    values: (str), List of FSMVariables.
 |    value_map: (map), For substituting values for names in the expressions.
 |    header: Ordered list of values.
 |    state_list: Ordered list of valid states.
 |
 |  Methods defined here:
```

（5）result = fsm.ParseText(traceroute)，把待解析的文本，即变量 traceroute，传入刚创建好的有限状态机，即变量 fsm 对象的 ParseText 方法（其实也是一个类），处理后返回一个列表。

（6）print(fsm.header)，打印该有限状态机的 header，其实就是 traceroute.template 文件中模板定义的两个变量。

（7）print(result)，打印结果如下所示。

```
File  Edit  Format  Run  Options  Window  Help
import textfsm
traceroute = '''
<R1>tracert 192.5.5.5
tracert 192.5.5.5
 traceroute to  192.5.5.5(192.5.5.5), max hops: 30 ,packet length: 40,press CTRL_C to break
 1 36.1.1.6 60 ms  50 ms  10 ms
 2 67.1.1.7 100 ms  90 ms  80 ms
 3 78.1.1.8 90 ms  80 ms  60 ms
 4 59.1.1.9 100 ms  90 ms  80 ms
 5 59.1.1.5 90 ms  100 ms  60 ms
'''
with open('traceroute.template') as template:
    fsm = textfsm.TextFSM(template)
    result = fsm.ParseText(traceroute)
    print(fsm.header)
    print(result)
```

关键两句，联动模板和待处理文本

3. Python 脚本执行

下面我们可以直接执行 Python 脚本。Windows 系统可能会有隐藏后缀名的功能，造成代码中的文件名与真实文件名有出入，需要特别留意。对于 TextFSM 模板的后缀名，并无特殊要求，使用.txt 则可以直接用记事本双击打开，如下图所示。

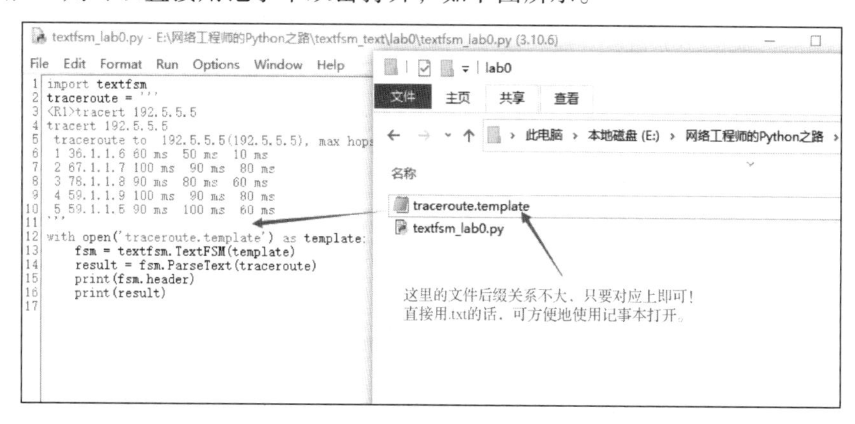

我们看一下代码执行后的结果，如下图所示。

```
IDLE Shell 3.10.6
File  Edit  Shell  Debug  Options  Window  Help
Python 3.10.6 (tags/v3.10.6:9c7b4bd, Aug  1 2022, 21:53:49) [MSC v.1932 64 bit (AMD64)] on win32
Type "help", "copyright", "credits" or "license()" for more information.
>>>
========= RESTART: E:\网络工程师的Python之路\textfsm_text\lab0\textfsm_lab0.py =========
['ID', 'Hop']
[['1', '36.1.1.6'], ['2', '67.1.1.7'], ['3', '78.1.1.8'], ['4', '59.1.1.9'], ['5', '59.1.1.5']]
>>> |
```

通过这个“简单”引例，我们把一个 str 文本数据处理成一个嵌套列表数据。此后，我们就可以通过 Python 相关丰富的轮子，来操作这些数据。

我们归纳一下这个全过程：**基于 Python 的 TextFSM 引擎，在模板（可自定义）的配合下，把半结构文本解析（parse）成结构化数据**，如下图所示。TextFSM 模板基础语法和模板拟写是本章重点，后续小节将从网络工程师运维视角，安排 4 个实战场景化案例，通过实验带出理论知识，并逐步引导读者查阅官方手册文档。

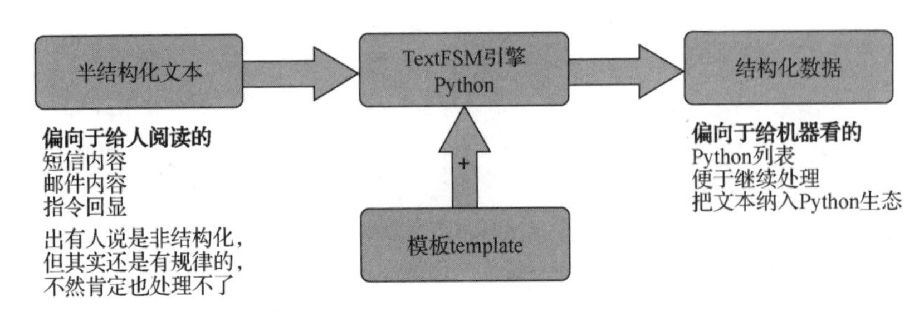

基于Python的TextFSM引擎，在模板（可自定义）的配合下，
把半结构化文本“解析”（parse）成结构化数据

读者们试着做一个小练习，在上图中圈出“生产原料”“生产模具”“生产机器”。

7.2　实验 1　单行回显单行 rule

经过上一节，你对 TextFSM 的应用多少有点实际感受了吧？或许此时你去看模块手册的基础语法，依然会云里雾里的。没关系，从本节起，我们将从头开始，一点一点“拨开迷雾”，把 TextFSM 这个强大工具用起来。

7.2.1　安装 tabulate 模块

模块 tabulate 并非必需，其作用是辅助把结果显示成表格，这样看起来会舒服很多。安装 tabulate，与 textfsm 模块安装过程基本类似。如之前已经安装，则会提示，可使用 show 查看一下版本，如下图所示。

```
pip3 install tabulate
pip3 show tabulate
```

```
C:\>pip3 install tabulate
Defaulting to user installation because normal site-packages is not writeable
Requirement already satisfied: tabulate in c:\program files\python310\lib\site-packages (0.8.10)

C:\>pip3 show tabulate
Name: tabulate
Version: 0.8.10
Summary: Pretty-print tabular data
Home-page: https://github.com/astanin/python-tabulate
Author: Sergey Astanin
Author-email: s.astanin@gmail.com
License: MIT
Location: c:\program files\python310\lib\site-packages
Requires:
Required-by:

C:\>
```

打开 IDLE，尝试通过 import 引入模块，没有报错即可，如下图所示。

```
C:\>python
Python 3.10.6 (tags/v3.10.6:9c7b4bd, Aug  1 2022, 21:53:49) [MSC v.1932 64 bit (AMD64)] on win32
Type "help", "copyright", "credits" or "license" for more information.
>>> import tabulate
>>>
>>>
>>> _
```

7.2.2　创建实验文件夹

创建一个实验文件夹，用来放置实验需要用到的 3 个文件。除了 Python 文件的后缀有要求，其他两个文件与后缀关系不大，在 Windows 上做实验，用 txt 作后缀方便记事本打开。

（1）**output.txt**——放置设备配置或命令回显信息。在现实生产中，我们则可能会结合 netmiko、nornir 等模块，自动提取回显文本再查找 TextFSM 进行联动。现在我们只是要针对性地学习 TextFSM，所以以手工方式保存进文本文件。

（2）**template.txt**——放置 TextFSM 模板信息。在现实生产中，也是这样一个模板，与 netmiko、nornir 模块配合时，仅仅是被集成到 ntc-templates 中用 index 文件进行索引得到而已。当然 TextFSM 也有一个叫作 clitable 的类，可以做映射。本章重点关注“TextFSM 语法”学习，其他暂作屏蔽处理。

（3）**lab1.py**——放置 python 脚本，如下图所示。

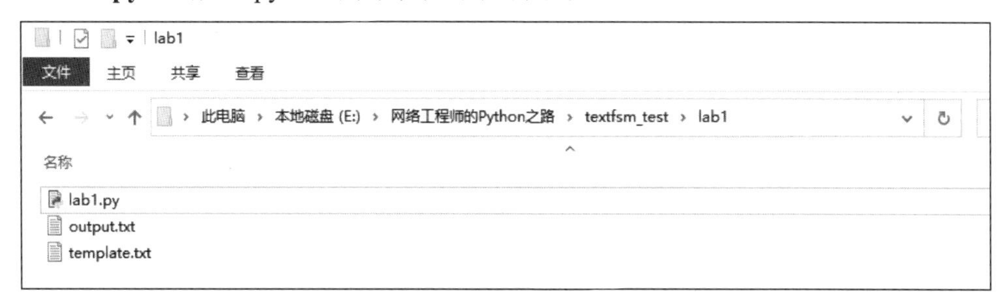

在 Windows 上有一个小细节要注意，一定要把文件扩展名这个复选框勾选，如下图所示。

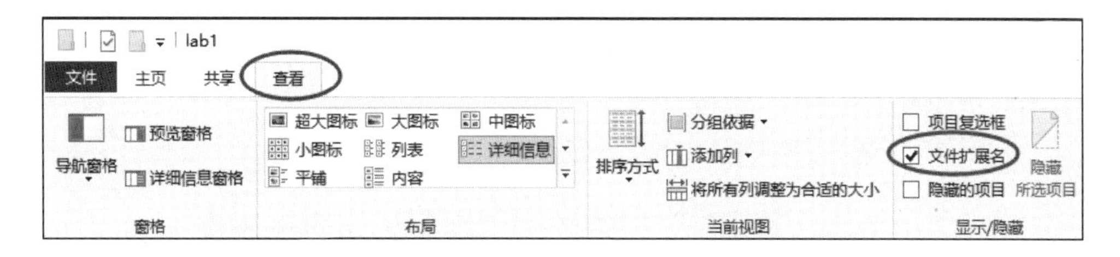

7.2.3　准备 output.txt

使用记事本等文本工具，手工编辑并保存 **output**.txt 文件。打开华为实验拓扑，模拟器或真机都可以，执行如下图所示的命令。

```
<Layer3Switch-1>disp clock
2022-08-09 22:15:35-08:00
Tuesday
Time Zone(China-Standard-Time) : UTC-08:00
<Layer3Switch-1>
<Layer3Switch-1>
<Layer3Switch-1>
<Layer3Switch-1>
<Layer3Switch-1>
```

命令回显总共 3 行。简单起见，我们暂只提第 1 行，放入 output.txt 文件，保存关闭。

```
2022-08-09 22:15:35-08:00
```

这样，output.txt 文件就准备好了。

7.2.4　准备 template.txt

1. 模板语法解释

根据引入示例内容中的模板，我们“依葫芦画瓢”，修改一下。

```
Value Date (....-..-..)
Value Time (..:..:..)

Start
  ^${Date} ${Time} -> Record
```

这里会用到一些较为基础的正则表达式知识，读者也可以在学习 TextFSM 模板的过程中，一并编织学习。我们先看变量定义中的内容。

```
....-..-..    可对应    2022-08-09
..:..:..      可对应    22:15:35
```

再看状态定义中的内容。

```
  ^${Date} ${Time}   变量替换后   ^....-..-.. ..:..:..
```

我们的目标是在 Python 脚本（后文进行介绍）的帮助下，Start 状态启动，“ ^${Date} ${Time} ”被 TextFSM 翻译成了“ ^....-..-.. ..:..:..”，然后开始处理“2021-06-23 22:12:28-08:00”

这段文本，设置的两个 Value 变量如同安插两个“侦察员”，在对应的位置上爬取信息，之后记录（Record）起来，最后把全部记录作为结果输出。

2. 模板拟写补充

变量定义和状态定义中规则（rule）的正则表达式写法因人而异，没有标准答案，通常能实现效果即可。不同模板之间会有粗匹配、精匹配、效率高、效率低的区别，针对上面的 template 模板，我们再多写两个版本，大体效果类似。

版本 1（更细匹配）

```
Value Date (\d\d\d\d-\d\d-\d\d)
Value Time (\d\d:\d\d:\d\d)

Start
  ^${Date} ${Time}-08:00 -> Record
```

其中，\d 表示数字；-08:00 表示后面的也进行匹配，但没有设置提取。

版本 2（更粗匹配）

```
Value Date (\S+)
Value Time (\S+)

Start
  ^${Date} ${Time}- -> Record
```

其中，\S+表示一个或多个非空字符。

7.2.5 准备 Python 脚本

我们把 Python 代码组织一下，基本在引例基础上调整并增加了一点。

```
import textfsm
from tabulate import tabulate

with open('template.txt') as template, open('output.txt') as output:
    re_table = textfsm.TextFSM(template)
    header = re_table.header
    result = re_table.ParseText(output.read())
```

```
    print(result)
    print(tabulate(result, headers=header))
```

代码讲解如下。

（1）引入 textfsm、tabulate 模块。

```
import textfsm
from tabulate import tabulate
```

（2）使用上下文管理器，打开文件。

```
with open('template.txt') as template, open('output.txt') as output:
```

打开当前目录的两个文件 template.txt、output.txt（注意要加单引号，以字符串形式传入参数），并分别给它们两个 Python 脚本范围内的名字（代号、变量），即 template 代表 template.txt 文件，output 代表 output.txt 文件。此时，这两个文件都是打开的状态。

（3）通过模板建立状态机 FSM（生产模具）。

```
    re_table = textfsm.TextFSM(template)
```

模板文件对象 template 作为参数传入 textfsm.TextFSM(template)，初始化一个 “Finite State Machine”（有限状态机）的 Python 类，并且给这个状态机起一个名字（代号、变量）——re_table。即 re_table 代表用 template 文件内容处理后的一个状态机，如下图所示。

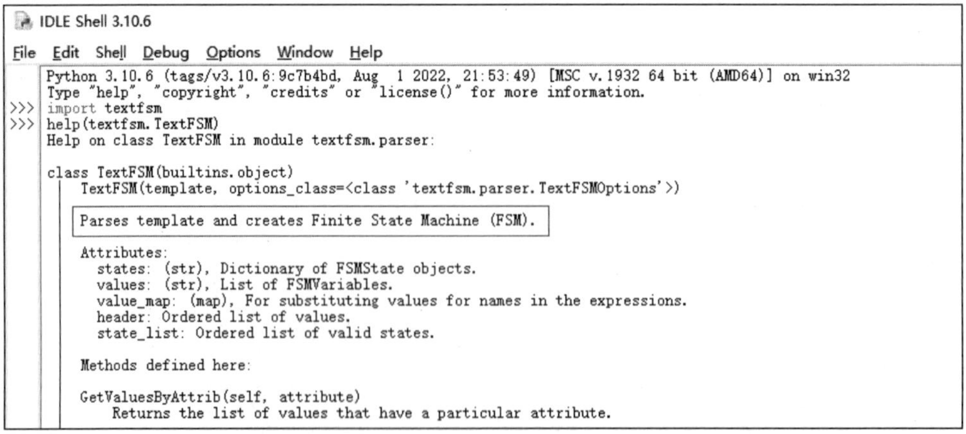

```
IDLE Shell 3.10.6
File Edit Shell Debug Options Window Help
    Python 3.10.6 (tags/v3.10.6:9c7b4bd, Aug  1 2022, 21:53:49) [MSC v.1932 64 bit (AMD64)] on win32
    Type "help", "copyright", "credits" or "license()" for more information.
>>> import textfsm
>>> help(textfsm.TextFSM)
    Help on class TextFSM in module textfsm.parser:

    class TextFSM(builtins.object)
     |  TextFSM(template, options_class=<class 'textfsm.parser.TextFSMOptions'>)
     |
     |  Parses template and creates Finite State Machine (FSM).
     |
     |  Attributes:
     |    states: (str), Dictionary of FSMState objects.
     |    values: (str), List of FSMVariables.
     |    value_map: (map), For substituting values for names in the expressions.
     |    header: Ordered list of values.
     |    state_list: Ordered list of valid states.
     |
     |  Methods defined here:
     |
     |  GetValuesByAttrib(self, attribute)
     |      Returns the list of values that have a particular attribute.
```

换句话说，生产模具已经装到生产机器上，这个装好模具的机器叫 “re_table”。

（4）提取 FSM 关键字段（表头）。

```
    header = re_table.header
```

变量 header 代表这个有限状态机（FSM）的 header 属性，其实就是 template.txt 文件中定义的 Value，即字段，也可以理解成表头。

（5）用状态机 FSM 处理 CLI 回显信息。

```
result = re_table.ParseText(output.read())
```

变量 re_table 实例有 ParseText 方法，可利用内置函数 help 获取帮助，如下图所示。

```
IDLE Shell 3.10.6
File  Edit  Shell  Debug  Options  Window  Help

ParseText(self, text, eof=True)
    Passes CLI output through FSM and returns list of tuples.

    First tuple is the header, every subsequent tuple is a row.

    Args:
      text: (str), Text to parse with embedded newlines.
      eof: (boolean), Set to False if we are parsing only part of the file.
            Suppresses triggering EOF state.

    Raises:
      TextFSMError: An error occurred within the FSM.

    Returns:
      List of Lists.

ParseTextToDicts(self, *args, **kwargs)
    Calls ParseText and turns the result into list of dicts.
```

这方法也常用。

换言之，读取 output（output.txt 文件）内容，即把 CLI 回显（生产原料）放入 re_table 状态机（生产模具），结果（产品）就出来了，如下图所示。

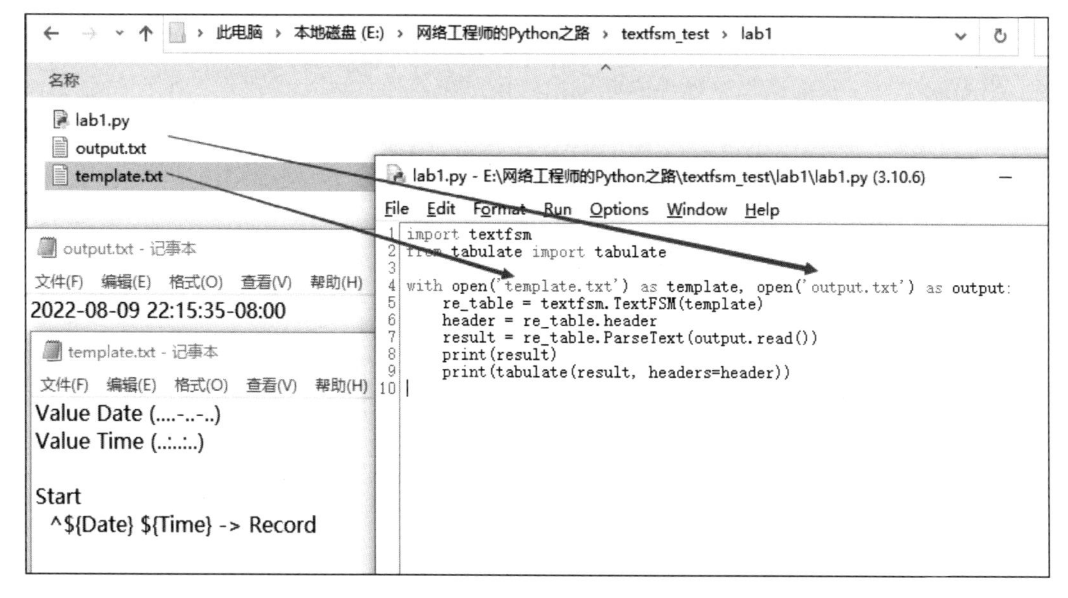

（6）print 函数把匹配后的结果打印出来，在实际生产中，我们可以写入文本或其他操作。

7.2.6　执行 Python 脚本

准备工作似乎比较烦琐，执行起来却比较简单。大家可以根据对正则表达式的理解，自己修改 template.txt 文件中的模板内容，如下图所示。

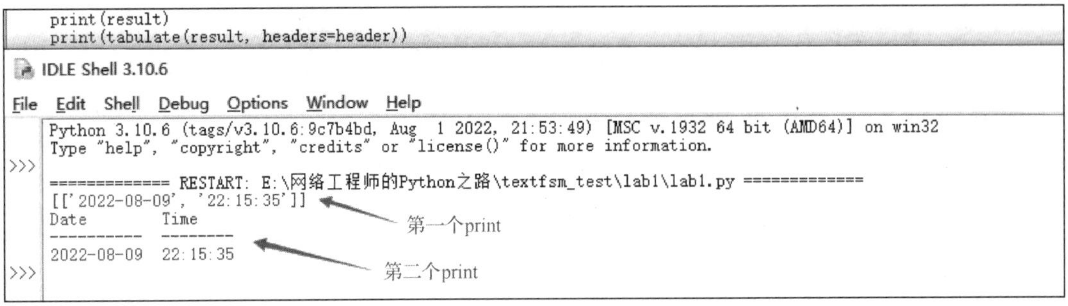

```
    print(result)
    print(tabulate(result, headers=header))
IDLE Shell 3.10.6
File  Edit  Shell  Debug  Options  Window  Help
    Python 3.10.6 (tags/v3.10.6:9c7b4bd, Aug  1 2022, 21:53:49) [MSC v.1932 64 bit (AMD64)] on win32
    Type "help", "copyright", "credits" or "license()" for more information.
>>>
    ============= RESTART: E:\网络工程师的Python之路\textfsm_test\lab1\lab1.py =============
    [['2022-08-09', '22:15:35']]          第一个print
    Date        Time
    ----------  --------
    2022-08-09  22:15:35                  第二个print
>>>
```

7.2.7　模板匹配过程

我们将上面的内容简化一下，如下表所示。

output.txt 中的内容	template.txt 中的内容
line1	rule 1

TextFSM 状态机是这样一个匹配过程：首先从 output.txt 读取 Aaa，尝试匹配 rule 1，如匹配中，则提取变量，执行动作；如果未匹配中，则结束返回。因为现在是“单行回显 vs.单行规则”，我们还很难看出什么门道来。没关系，请继续往下做实验。

7.2.8　实验小结

这个实验只涉及一行回显文本、一行规则（rule）的情况，不过我们梳理起来还是挺冗杂的。这也证明了使用 TextFSM 并不容易，需要花精力和心思。我们运用科学的学习方法，化繁为简，一步一步来。这一节打基础后，实验环境和实验 Python 代码基本就固定下来了，后面的实验就可以聚焦到“模板语法”学习上来了。

7.3　实验 2　多行回显单行 rule

实验 1 指令回显是单行，TextFSM 模板中的 rule 规则也是单行。实际生产中没有这么简单的。我们再进一小步，先来看看指令回显是多行的情况。我们可以把这种匹配过程叫作“跨行提取”。

7.3.1　准备 output.txt

我们在实验环境中，登录一台华为交换机，随意配置一些接口 IP 地址，部分端口关闭（shutdown）。

```
vlan 12
vlan 34
vlan 192
vlan 172
vlan 10

int vlanif 12
ip address 12.1.1.3 29
int vlanif 34
ip address 34.1.1.5 24
int vlanif 192
ip address 192.168.123.5 27
int vlanif 172
ip address 172.18.123.5 26
int vlanif 10
ip address 10.18.123.5 25

int vlanif 172
shut
int vlanif 192
shut
int vlanif 12
shut
```

配置完成后，我们执行如下命令。

```
display ip int bri
```

手工复制命令回显到 output.txt 文件中，随后保存即可。

```
#前面有部分文本省略
Interface                         IP Address/Mask      Physical     Protocol
LoopBack0                         11.11.11.11/32       up           up(s)
MEth0/0/1                         unassigned           down         down
NULL0                             unassigned           up           up(s)
Vlanif1                           192.168.11.11/24     up           up
Vlanif10                          10.18.123.5/25       down         down
Vlanif12                          12.1.1.3/29          *down        down
Vlanif34                          34.1.1.5/24          down         down
Vlanif172                         172.18.123.5/26      *down        down
Vlanif192                          192.168.123.5/27    *down        down
```

7.3.2　准备 template.txt

这种命令回显文本是给人类看的，机器“看不懂”。我们配合模板，通过 TextFSM 引擎，可抓取想要的数据，并将其放入列表等 Python 数据类型中。这样机器就能“看懂”并且能处理，TextFSM 工具的意义也在于此。

莎翁曾说：“There are a thousand Hamlets in a thousand people's eyes.”一千个人眼里有一千个哈姆雷特。同样地，实现同一目标的 TextFSM 模板，有各种各样不同的写法。我们这里把 4 列数据都提取出来，实际使用中按需即可，比如提取 4 列中的某 2 列。

◎　版本 1

```
Value Interface (\S+)
Value Address (\S+)
Value Physical (up|down|\*down)
Value Protocol (up|down)

Start
  ^${Interface}\s+${Address}\s+${Physical}\s+${Protocol} -> Record
```

◎　版本 2

Value Protocol (up|down)变为 Value Protocol (\S+)，属于较粗匹配。

```
Value Interface (\S+)
Value Address (\S+)
Value Physical (up|down|\*down)
Value Protocol (\S+)
```

```
Start
  ^${Interface}\s+${Address}\s+${Physical}\s+${Protocol} -> Record
```

◎　版本 3

Value Physical (up|down|*down)变成 Value Physical (\S+)，属于更粗的匹配。

```
Value Interface (\S+)
Value Address (\S+)
Value Physical (\S+)
Value Protocol (\S+)

Start
  ^${Interface}\s+${Address}\s+${Physical}\s+${Protocol} -> Record
```

如果模板的正则表达式写得太粗，有时候会把干扰信息也提取出来，会影响结果；如果模板的正则表达式写得太细，则增加模板复杂度，降低模板可读性，可能会影响调测。因此，正则表达式的粗细要适当把控为佳。

7.3.3　准备 Python 脚本

Python 脚本与实验 1 基本一致，照搬即可。

```
import textfsm
from tabulate import tabulate

with open('template.txt') as template, open('output.txt') as output:
    re_table = textfsm.TextFSM(template)
    header = re_table.header
    result = re_table.ParseText(output.read())
    #print(result)          #适当做一点调测
    #print('\n\n')          #适当做一点调测
    print(tabulate(result, headers=header))
```

7.3.4　执行 Python 脚本

结合模板版本一，执行 Python 脚本，如下图所示。

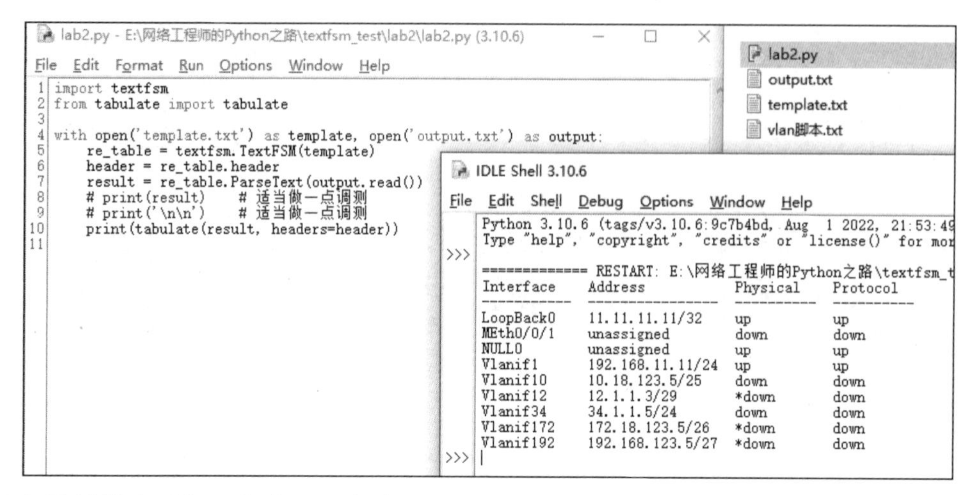

运行结果符合预期。版本二和版本三跑起来效果如何？与版本一有何差异？如果只需取两列（如 Interface、Address），要如何调整模板？请读者自行尝试。

7.3.5　模板匹配过程

现在我们讨论一下，回显文本和模板状态里面的规则是怎么匹配的。我们把整个过程抽象梳理一下，如下表所示。

output.txt 中的内容	template.txt 中的内容
line1	rule1
line2	
line3	

结合当前知识点，TextFSM 引擎的匹配过程如下。

（1）读取 line1，尝试与 rule1 做匹配。

（2）如能匹配中，则提取变量，执行动作。接着读取 line2 重回 rule1，如此反复。

（3）如未匹配中，则当什么都没发生，接着读取 line 重回 rule1，如此反复。

（4）最终，line 一行一行都读取匹配完毕，返回结果。

知识要点如下。

（1）记录动作（Record Action），Record 执行后，变量中的值默认会被清空，如不想被清空，需配合变量关键字，后面再介绍。

（2）行动作（Line Action），Next 默认值，无须指明。因此，Next. Record 和 Record 是等价的。这个默认值可以修改，后面再介绍。

（3）状态跳转（State Transition），后面再介绍。

7.3.6　实验小结

TextFSM 模板用上了正则表达式，是否记录可设置，默认动作也能修改，状态还能迁移，灵活度很高。太灵活则往往对初学者不够友好。因此，在实验编排过程中，我们适当屏蔽一些信息，将部分知识点进行后置处理，并把 Python 脚本稍微固定下来。在这个基础上，我们一点点地从“单行回显 vs.单行规则”迈向“多行回显 vs.单行规则”，最终征服“多行回显 vs.多行规则”，同时慢慢地感受和探测 TextFSM 引擎的匹配逻辑。在有一定实验感知后，我们再配合着看官方手册文档，研读语法讲解，就不再云里雾里了，有时甚至会恍然大悟。此后，我们实验或生产实战，遇到问题也多尝试自行查阅手册，争取独立解决问题。这样 TextFSM 相关知识就逐渐积淀，越来越牢固。

7.4　实验 3　多行回显多行 rule，初识关键字

在生产环境中，我们遇到更多的场景是“多行回显 vs.多行 rule”。这次我们由一个实验 OSPF 邻居报文解析切入，并动态调整 TextFSM 模板，逐步观察。

7.4.1　准备 output.txt

在实验文件夹中，我们取一段 OSPF 邻居查询指令回显，手工复制保存至 output.txt 中。

```
<R1>display ospf peer

     OSPF Process 1 with Router ID 1.1.1.1
         Neighbors

 Area 0.0.0.0 interface 12.1.1.1(Ethernet0/0/0)'s neighbors
 Router ID: 2.2.2.2          Address: 12.1.1.2
   State: Full  Mode:Nbr is  Master  Priority: 1
   DR: 12.1.1.2  BDR: 12.1.1.1  MTU: 1500
   Dead timer due in 37  sec
```

```
  Retrans timer interval: 5
  Neighbor is up for 00:00:50
  Authentication Sequence: [ 0 ]

          Neighbors

 Area 0.0.0.0 interface 14.1.1.1(Ethernet0/0/1)'s neighbors
 Router ID: 4.4.4.4          Address: 14.1.1.4
   State: Full  Mode:Nbr is  Master  Priority: 1
   DR: 14.1.1.4  BDR: None   MTU: 0
   Dead timer due in 40  sec
   Retrans timer interval: 5
   Neighbor is up for 00:00:01
   Authentication Sequence: [ 0 ]

          Neighbors

 Area 0.0.0.0 interface 17.1.1.1(Ethernet0/0/7)'s neighbors
 Router ID: 7.7.7.7          Address: 17.1.1.7
   State: Full  Mode:Nbr is  Master  Priority: 1
   DR: 14.1.1.7  BDR: None   MTU: 0
   Dead timer due in 39  sec
   Retrans timer interval: 5
   Neighbor is up for 00:00:11
   Authentication Sequence: [ 0 ]
```

7.4.2　准备 template.txt

这里 TextFSM 模板定义了 7 个变量，用于抓取指令回显报文中的特定信息。如果前面的实验你已经"消化"了，那么这模板看起来就并不复杂了，只是有点长而已。现网实战的模板往往不会太短。如果你觉得还是有点"复杂"的话，则可以尝试着把 7 个变量先变成 1 个，而后开启测试，再 2 个、3 个……慢慢增加起来。

```
Value LOCAL_RouterID (\S+)
Value PROCESS (\d+)
Value AREA (\S+)
Value LOCAL_INT_IP (\S+)
```

```
Value LOCAL_INT (\S+)
Value DEST_RouterID (\S+)
Value DEST_INT_IP (\S+)

Start
  ^<${LOCAL_RouterID}>
  ^\s+OSPF\sProcess\s${PROCESS}
  ^\s+Area\s${AREA}\sinterface\s${LOCAL_INT_IP}\(${LOCAL_INT}\)
  ^\s+Router\sID:\s${DEST_RouterID}\s+Address:\s${DEST_INT_IP}
```

7.4.3　准备 Python 脚本

Python 脚本与前面实验完全一致。

```
import textfsm
from tabulate import tabulate

with open('template.txt') as template, open('output.txt') as output:
    re_table = textfsm.TextFSM(template)
    header = re_table.header
    result = re_table.ParseText(output.read())
    #print(result)     #适当做一点调测
    #print('\n\n')     #适当做一点调测
    print(tabulate(result, headers=header))
```

实验文件夹如下图所示。

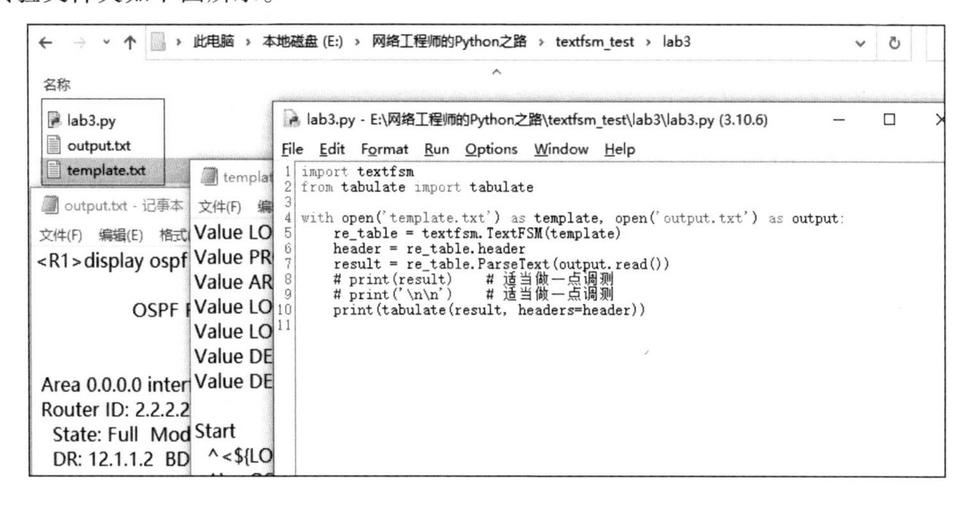

7.4.4　实验调试

1. EOF 状态

直接运行 Python 脚本，如下图所示。

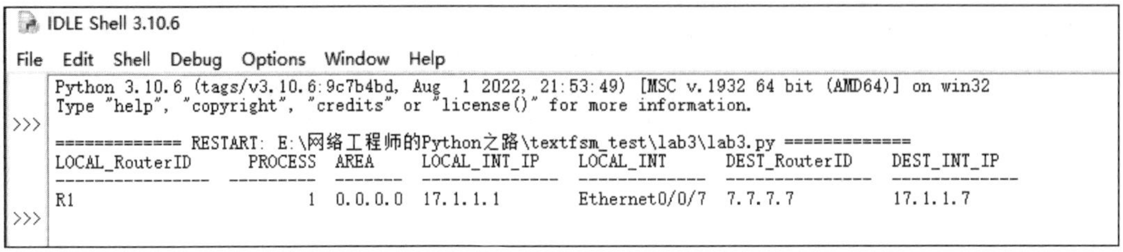

```
IDLE Shell 3.10.6
File  Edit  Shell  Debug  Options  Window  Help
    Python 3.10.6 (tags/v3.10.6:9c7b4bd, Aug  1 2022, 21:53:49) [MSC v.1932 64 bit (AMD64)] on win32
    Type "help", "copyright", "credits" or "license()" for more information.
>>>
    ============= RESTART: E:\网络工程师的Python之路\textfsm_test\lab3\lab3.py =============
    LOCAL_RouterID      PROCESS  AREA     LOCAL_INT_IP    LOCAL_INT       DEST_RouterID     DEST_INT_IP
    ----------------    -------- -------  --------------  --------------  ----------------  -------------
    R1                        1  0.0.0.0  17.1.1.1        Ethernet0/0/7   7.7.7.7           17.1.1.7
>>>
```

TextFSM 模板没有 Record 动作，即 NoRecord，不记录。但似乎出现了不符合逻辑的结果，最后竟然打印出一行记录。此外，原始报文显示路由器 R1 有 3 个 OSPF 邻居，为何 TextFSM 引擎偏偏只提取到 1 个，而且还是最后 1 个？

为了解释这个问题，我们先得补充一个知识点。TextFSM 模板中没指明 Record 等动作，即为 NoRecord，默认动作。匹配全过程并不记录，变量自上而下轮番匹配更新。比如，变量 DEST_RouterID 会从 2.2.2.2 变成 4.4.4.4 再变成 7.7.7.7；变量 LOCAL_RouterID 一直是 R1，变量 PROCESS 一直是 1，它们全过程只匹配了一次；变量 AREA 看似没变，其实有变，只是每次都变成 0.0.0.0。

那全程都是 NoRecord，为什么还有结果被打印出来呢？其实 TextFSM 引擎处理到最后，会从 Start 状态跳转到 EOF 状态。这个 EOF 状态是隐式的，模板没有指明的话，它就默认，如下所示。

```
EOF
  ^.* -> Record
```

这个 EOF 状态有一条规则：匹配任意，做一次 Record 动作。最后一条记录就是这么来的。这里我们也可以体验一下所谓的“有限状态机”的概念，即匹配过程可在不同状态间跳转。

那么能不能重写 EOF 状态的规则呢？答案是可以的。

```
Value LOCAL_RouterID (\S+)
Value PROCESS (\d+)
Value AREA (\S+)
Value LOCAL_INT_IP (\S+)
Value LOCAL_INT (\S+)
```

```
Value DEST_RouterID (\S+)
Value DEST_INT_IP (\S+)

Start
  ^<${LOCAL_RouterID}>
  ^\s+OSPF\sProcess\s${PROCESS}
  ^\s+Area\s${AREA}\sinterface\s${LOCAL_INT_IP}\(${LOCAL_INT}\)
  ^\s+Router\sID:\s${DEST_RouterID}\s+Address:\s${DEST_INT_IP}

EOF
```

此时，EOF 状态被重新定义了，删掉 Record，脚本执行结果就没有记录了，如下图所示。

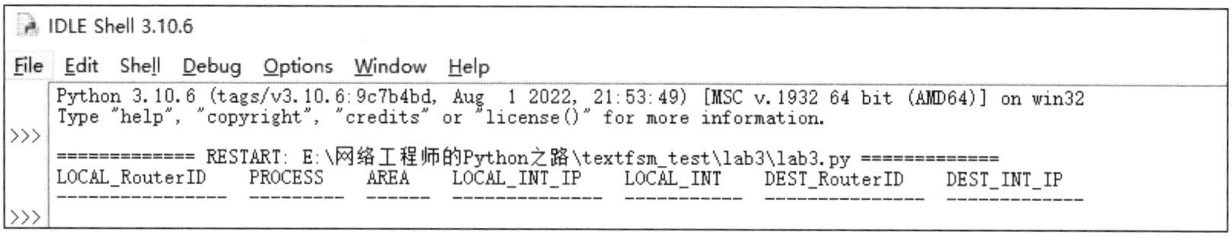

```
IDLE Shell 3.10.6
File Edit Shell Debug Options Window Help
Python 3.10.6 (tags/v3.10.6:9c7b4bd, Aug  1 2022, 21:53:49) [MSC v.1932 64 bit (AMD64)] on win32
Type "help", "copyright", "credits" or "license()" for more information.
>>>
============= RESTART: E:\网络工程师的Python之路\textfsm_test\lab3\lab3.py =============
LOCAL_RouterID    PROCESS    AREA    LOCAL_INT_IP    LOCAL_INT    DEST_RouterID    DEST_INT_IP
----------------  ---------  ------  --------------  -----------  ---------------  -------------
>>>
```

来个小练习，如果此时我们在模板中 EOF 前加一个“#”呢？这就是注释。请读者们自行尝试！

2. Record

修改一下模板，EOF 状态重新保持默认，Start 状态的最后一行规则带上 Record 动作。

```
Value LOCAL_RouterID (\S+)
Value PROCESS (\d+)
Value AREA (\S+)
Value LOCAL_INT_IP (\S+)
Value LOCAL_INT (\S+)
Value DEST_RouterID (\S+)
Value DEST_INT_IP (\S+)

Start
  ^<${LOCAL_RouterID}>
  ^\s+OSPF\sProcess\s${PROCESS}
  ^\s+Area\s${AREA}\sinterface\s${LOCAL_INT_IP}\(${LOCAL_INT}\)
  ^\s+Router\sID:\s${DEST_RouterID}\s+Address:\s${DEST_INT_IP} -> Record
```

继续运行脚本，结果如下图所示。

```
IDLE Shell 3.10.6
File Edit Shell Debug Options Window Help
Python 3.10.6 (tags/v3.10.6:9c7b4bd, Aug  1 2022, 21:53:49) [MSC v.1932 64 bit (AMD64)] on win32
Type "help", "copyright", "credits" or "license()" for more information.
>>>
============= RESTART: E:\网络工程师的Python之路\textfsm_test\lab3\lab3.py =============
LOCAL_RouterID    PROCESS    AREA     LOCAL_INT_IP    LOCAL_INT      DEST_RouterID    DEST_INT_IP
----------------  ---------  -------  --------------  -------------  ---------------  -------------
R1                1          0.0.0.0  12.1.1.1        Ethernet0/0/0  2.2.2.2          12.1.1.2
                             0.0.0.0  14.1.1.1        Ethernet0/0/1  4.4.4.4          14.1.1.4
                             0.0.0.0  17.1.1.1        Ethernet0/0/7  7.7.7.7          17.1.1.7
>>>
```

此时 3 条记录、3 个邻居都有了。我们仔细看看结果，发现依然还有问题。为什么 LOCAL_RouterID 和 Process 这两个变量只在第一个记录有信息，而在其他记录中都是空白的呢？结合 output.txt 的内容，我们可以看到 LOCAL_RouterID 和 Process 在报文中只出现了一次。这时我们要引入一个新知识点，在默认情况下，Record 动作执行后，会把原先的变量清空。此时，我们可以尝试查阅官方手册，找到这一"空白"现象的"理论依据"，如下图所示。

https://github.com/google/textfsm/wiki/TextFSM

Record Actions

After the line action is the optional record action, these are separated by a full stop '.'.

Action	Description
NoRecord	Do nothing. This is the default behavior if no record action is specified.
Record	Record the values collected so far as a row in the return data. Non Filldown values are cleared. Note: No record will be output if there are any 'Required' values that are unassigned.
Clear	Clear non Filldown values.
Clearall	Clear all values.

我们看到这么一句"Non Filldown values are cleared."，那么，是不是在对应的变量中加一个 Filldown 关键字就能"向下填充"呢？

3. Filldown

我们借助关键字 Filldown 来"加持"LOCAL_RouterID、PROCESS 这两个变量。

```
Value Filldown LOCAL_RouterID (\S+)
Value Filldown PROCESS (\d+)
Value AREA (\S+)
Value LOCAL_INT_IP (\S+)
Value LOCAL_INT (\S+)
Value DEST_RouterID (\S+)
Value DEST_INT_IP (\S+)
```

```
Start
  ^<${LOCAL_RouterID}>
  ^\s+OSPF\sProcess\s${PROCESS}
  ^\s+Area\s${AREA}\sinterface\s${LOCAL_INT_IP}\(${LOCAL_INT}\)
  ^\s+Router\sID:\s${DEST_RouterID}\s+Address:\s${DEST_INT_IP} -> Record
```

再次运行 Python 脚本，看看效果，如下图所示。

```
IDLE Shell 3.10.6
File  Edit  Shell  Debug  Options  Window  Help
Python 3.10.6 (tags/v3.10.6:9c7b4bd, Aug  1 2022, 21:53:49) [MSC v.1932 64 bit (AMD64)] on win32
Type "help", "copyright", "credits" or "license()" for more information.
>>>
============= RESTART: E:\网络工程师的Python之路\textfsm_test\lab3\lab3.py =============
LOCAL_RouterID    PROCESS   AREA     LOCAL_INT_IP    LOCAL_INT       DEST_RouterID    DEST_INT_IP
----------------  --------  -------  --------------  --------------  ---------------  -------------
R1                       1  0.0.0.0  12.1.1.1        Ethernet0/0/0   2.2.2.2          12.1.1.2
R1                       1  0.0.0.0  14.1.1.1        Ethernet0/0/1   4.4.4.4          14.1.1.4
R1                       1  0.0.0.0  17.1.1.1        Ethernet0/0/7   7.7.7.7          17.1.1.7
R1                       1
>>>
```

“空白”问题解决了,但新的问题又出现了!怎么又多了一行呢?我们回想一下前面的 EOF 状态相关内容,应该就能够解释这个现象了。TextFSM 引擎最后从 Start 状态跳转至 EOF 状态,此时 LOCAL_RouterID、PROCESS 因 Filldown 关键字而非空,其他变量则已被清空。EOF 状态默认有一个 Record 动作,因此结果多出来一行。如何解决?前面已有答案,改写 EOF 状态,如下图所示。这是一种方法,请自行尝试,并回到官网手册中找“理论依据”。

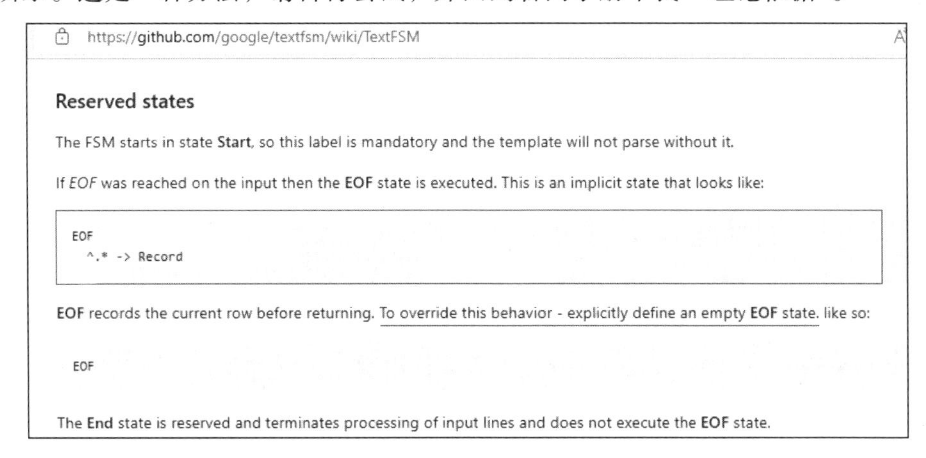
https://github.com/google/textfsm/wiki/TextFSM

Reserved states

The FSM starts in state **Start**, so this label is mandatory and the template will not parse without it.

If *EOF* was reached on the input then the **EOF** state is executed. This is an implicit state that looks like:

```
EOF
  ^.* -> Record
```

EOF records the current row before returning. To override this behavior - explicitly define an empty **EOF** state, like so:

```
EOF
```

The **End** state is reserved and terminates processing of input lines and does not execute the **EOF** state.

除了改写 EOF 状态，还有另一种方法，即配合关键字 Required。

4. Required

LOCAL_RouterID、PROCESS 添加 Filldown 关键字后，导致最后多了一行冗杂记录。我们可以找另一个变量（如 DEST_RouterID，用其他变量也行）设置 Required 关键字，作为必选项，值为空时整行都不记录，从而巧妙解决。

```
Value Filldown LOCAL_RouterID (\S+)
Value Filldown PROCESS (\d+)
Value AREA (\S+)
Value LOCAL_INT_IP (\S+)
Value LOCAL_INT (\S+)
Value Required DEST_RouterID (\S+)
Value DEST_INT_IP (\S+)

Start
  ^<${LOCAL_RouterID}>
  ^\s+OSPF\sProcess\s${PROCESS}
  ^\s+Area\s${AREA}\sinterface\s${LOCAL_INT_IP}\(${LOCAL_INT}\)
  ^\s+Router\sID:\s${DEST_RouterID}\s+Address:\s${DEST_INT_IP} -> Record
```

再重新运行一下 Python，效果如下图所示。

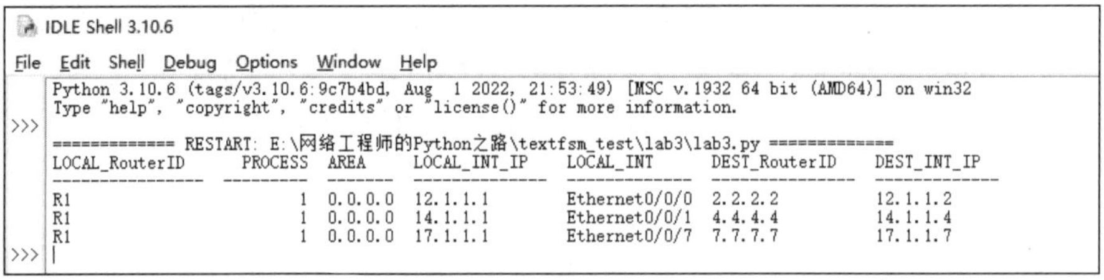

```
IDLE Shell 3.10.6
File Edit Shell Debug Options Window Help
Python 3.10.6 (tags/v3.10.6:9c7b4bd, Aug  1 2022, 21:53:49) [MSC v.1932 64 bit (AMD64)] on win32
Type "help", "copyright", "credits" or "license()" for more information.
>>>
============ RESTART: E:\网络工程师的Python之路\textfsm_test\lab3\lab3.py ============
LOCAL_RouterID      PROCESS  AREA     LOCAL_INT_IP    LOCAL_INT        DEST_RouterID    DEST_INT_IP
----------------  ---------  -------  --------------  ---------------  ---------------  -------------
R1                        1  0.0.0.0  12.1.1.1        Ethernet0/0/0    2.2.2.2          12.1.1.2
R1                        1  0.0.0.0  14.1.1.1        Ethernet0/0/1    4.4.4.4          14.1.1.4
R1                        1  0.0.0.0  17.1.1.1        Ethernet0/0/7    7.7.7.7          17.1.1.7
>>> |
```

这样一张结构“完美”的表格已经形成。我们可以再好好感受一下 TextFSM 引擎把半结构化文本信息转化为结构化数据的过程。

7.4.5　模板匹配过程

我们继续来探究一下 TextFSM 多行文本和多行规则的匹配过程，如下表所示。

output.txt 中的内容	template.txt 中的内容
line1	rule 1
line2	rule 2
line3	rule 3

（1）从 output.txt 读取 line1，尝试匹配 rule1，如匹配中，则提取变量（符合就提取），记录结果（默认 NoRecord，不记录），执行动作（默认动作 Next，结束本行 line 不再匹配后面

的 rule，取下一行 line 重新开始），跳转状态。如果未匹配中，则接着依次尝试匹配 rule2，rule2 未匹配中则匹配 rule3，处理过程与 rule1 一致。

（2）完成 line1 后，接着从 output.txt 读取 line2，其过程和 line1 的匹配过程一致，如此反复。

（3）最后跳转到 EOF 状态，而后结束。

也可以这样理解。

- 在次序方面，这个过程从每个 line 出发去找每个 rule，即 output 队伍中 line1 出来"站稳"，逐条处理 rule1、rule2、rule3。之后 line2 再出来，重复 rule1、rule2、rule3。
- 在动作方面，默认为 Next，不用指明。Next 时，某 rule 一旦匹配中，则后续 rule 不再尝试匹配，line 取下一行重新开始。与默认行动作 Next 相对的还有 Continue，我们在下个实验中介绍。

7.4.6　实验小结

本实验主要就 TextFSM 模板中涉及的 EOF 状态，Record 动作，变量关键字 Filldown、Required，默认行动作 Next 等做了动态调整演示。实验中每个变量最多只加了一个关键字，实际上可以加多个，用逗号隔开。在完成这个实验后，日常生产中一些简单的匹配需求就能够满足了。TextFSM 模板的有些默认值是隐式的，不用指明。我们需要加深理解默认动作，按需灵活调整。

7.5　实验 4　关键字 List 和动作 Continue.Record

Eth-Trunk 链路聚合可实现传输双路由、设备跨板保护及提升带宽的目的，在现实网络运维中应用广泛。我们取一份查询 Eth-Trunk 指令回显，先按前面的实验做法复习一下已学知识，再推进使用关键字 List 和动作 Continue.Record，看看能实现什么新效果。

7.5.1　准备 output.txt

我们准备一份关于 Eth-Trunk 的指令回显，截取其中一部分。

```
display interface brief main
```

回显如下。

```
Interface                       PHY    Protocol  InUti   OutUti   inErrors   outErrors
Eth-Trunk3                      up     down      0.34%   0.49%       0           0
  GigabitEthernet8/1/2(10G)     up     up        0.27%   0.48%       0           0
  GigabitEthernet10/0/4(10G)    up     up        0.41%   0.49%       0           0
Eth-Trunk4                      up     down      2.14%   29.39%      0           0
  GigabitEthernet8/1/0(10G)     up     up        2.33%   29.44%      0           0
  GigabitEthernet10/0/3(10G)    up     up        1.96%   29.33%      0           0
Eth-Trunk5                      up     down      0.13%   0.30%       0           0
  GigabitEthernet6/0/3(10G)     up     up        0.24%   0.33%       0           0
  GigabitEthernet8/1/6(10G)     up     up        0.03%   0.26%       0           0
Eth-Trunk6                      up     down      0.01%   0.01%       0           0
  GigabitEthernet2/1/10         up     up        0.01%   0.01%       0           0
Eth-Trunk8                      *down  down      0%      0%          0           0
Eth-Trunk11                     up     down      3.02%   0.72%       0           0
  GigabitEthernet2/1/20         up     up        4.14%   0.93%       0           0
  GigabitEthernet2/1/21         up     up        3.40%   1.30%       0           0
  GigabitEthernet2/1/22         up     up        1.33%   0.40%       0           0
```

从回显信息可以看出，如果每个 Eth-Trunk 都有物理端口，则会多行显示。我们的目标是把这些回显数据加工成一个 Eth-Trunk 一行的形式。与前序实验一致，我们建一个实验文件夹。

7.5.2　准备 template.txt

相对于实验 3，变量少了，但正则表达式稍微复杂些。这次 rule 从 1 行开始，后面边调测边增加规则。

```
Value Tru_INT (Eth-Trunk\d+)
Value PHY (up|down|\*down)
Value Protocol (up|down)
Value Phy_INT (GigabitEthernet\d+/\d+/\d+)

Start
  ^${Tru_INT}\s+${PHY}\s+${Protocol}\s -> Record
```

7.5.3　准备 Python 脚本

前面已经多次提到过，我们使用几乎相同的 Python 脚本，这对于我们针对性地学习

TextFSM 相关知识是很有好处的。这可以说是"控制变量法"。

```
import textfsm
from tabulate import tabulate

with open('template.txt') as template, open('output.txt') as output:
    re_table = textfsm.TextFSM(template)
    header = re_table.header
    result = re_table.ParseText(output.read())
    #print(result)      #适当做一点调测
    #print('\n\n')      #适当做一点调测
    print(tabulate(result, headers=header))
```

实验文件夹如下图所示。

7.5.4　实验调测

1. 原始模板调测

对于原始模板，直接运行 Python 脚本看效果，如下图所示。

```
lab4.py - E:\网络工程师的Python之路\textfsm_test\lab4\lab4.py (3.10.6)
IDLE Shell 3.10.6
File Edit Shell Debug Options Window Help
Python 3.10.6 (tags/v3.10.6:9c7b4bd, Aug  1 2022, 21:53:49) [MSC v.1932 64 bit (AMD64)] on win32
Type "help", "copyright", "credits" or "license()" for more information.
>>>
============ RESTART: E:\网络工程师的Python之路\textfsm_test\lab4\lab4.py ============
Tru_INT      PHY    Protocol    Phy_INT
-----------  -----  ----------  ---------
Eth-Trunk3   up     down
Eth-Trunk4   up     down
Eth-Trunk5   up     down
Eth-Trunk6   up     down
Eth-Trunk8   *down  down
Eth-Trunk11  up     down
```

Phy_INT 这一列为空是正常的，因为此时还没有把它加入匹配规则中。

2. 修改模板调测（1）

修改模板，最后增加一行，把物理端口提取出来。

```
Value Tru_INT (Eth-Trunk\d+)
Value PHY (up|down|\*down)
Value Protocol (up|down)
Value Phy_INT (GigabitEthernet\d+/\d+/\d+)

Start
  ^${Tru_INT}\s+${PHY}\s+${Protocol}\s -> Record
  ^\s+${Phy_INT} -> Record
```

运行 Python 脚本，如下图所示。

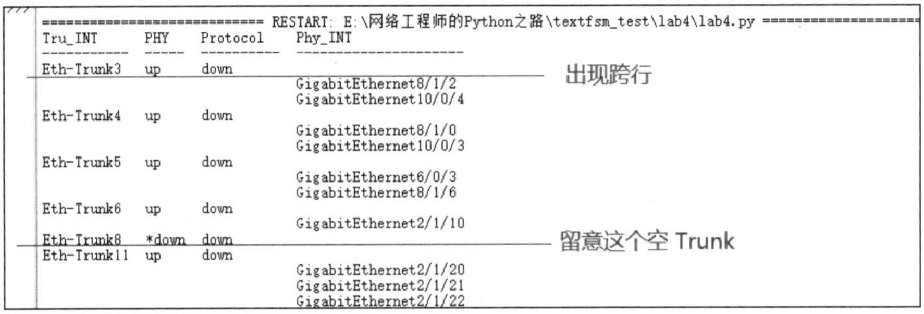

```
============================ RESTART: E:\网络工程师的Python之路\textfsm_test\lab4\lab4.py ====================
Tru_INT       PHY     Protocol    Phy_INT
-----------   -----   ----------  ---------------------
Eth-Trunk3    up      down
                                  GigabitEthernet8/1/2
                                  GigabitEthernet10/0/4
Eth-Trunk4    up      down
                                  GigabitEthernet8/1/0
                                  GigabitEthernet10/0/3
Eth-Trunk5    up      down
                                  GigabitEthernet6/0/3
                                  GigabitEthernet8/1/6
Eth-Trunk6    up      down
                                  GigabitEthernet2/1/10
Eth-Trunk8    *down   down
Eth-Trunk11   up      down
                                  GigabitEthernet2/1/20
                                  GigabitEthernet2/1/21
                                  GigabitEthernet2/1/22
```

出现跨行

留意这个空 Trunk

主 Trunk 口和物理端口的显示跨行了，看起来与 CLI 命令回显的呈现有点相似。此时如果直接打印 result，则数据结构还是很“混乱”，显然不是我们想要的结果，如下图所示。

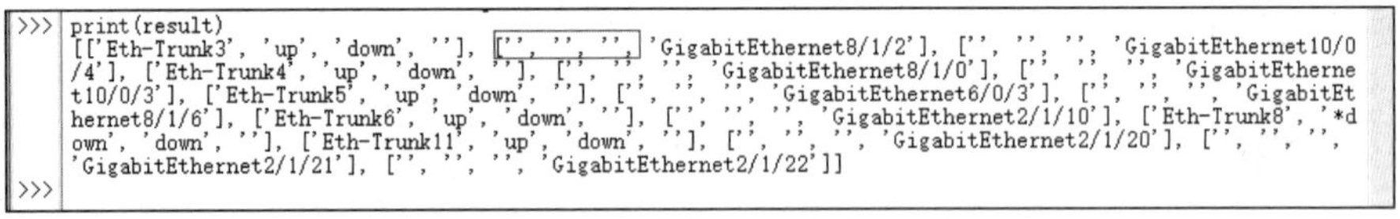

```
>>> print(result)
[['Eth-Trunk3', 'up', 'down', ''], ['', '', '', 'GigabitEthernet8/1/2'], ['', '', '', 'GigabitEthernet10/0/4'], ['Eth-Trunk4', 'up', 'down', ''], ['', '', '', 'GigabitEthernet8/1/0'], ['', '', '', 'GigabitEthernet10/0/3'], ['Eth-Trunk5', 'up', 'down', ''], ['', '', '', 'GigabitEthernet6/0/3'], ['', '', '', 'GigabitEthernet8/1/6'], ['Eth-Trunk6', 'up', 'down', ''], ['', '', '', 'GigabitEthernet2/1/10'], ['Eth-Trunk8', '*down', 'down', ''], ['Eth-Trunk11', 'up', 'down', ''], ['', '', '', 'GigabitEthernet2/1/20'], ['', '', '', 'GigabitEthernet2/1/21'], ['', '', '', 'GigabitEthernet2/1/22']]
>>>
```

如上图所示，在 Python 的 IDLE 脚本模式下运行代码，出现对话模式后，可直接敲代码进行调测。

3. 修改模板调测（2）

这里我们更关注变量 Tru_INT 和 Phy_INT 的实现，而变量 PHY 和 Protocol 经过简化处理，直接取 Eth-Trunk 口信息，而不取物理端口信息，因而直接加 Filldown 关键字即可。模板修改如下。

```
Value Filldown Tru_INT (Eth-Trunk\d+)
```

```
Value Filldown PHY (up|down|\*down)
Value Filldown Protocol (up|down)
Value Phy_INT (GigabitEthernet\d+/\d+/\d+)

Start
  ^${Tru_INT}\s+${PHY}\s+${Protocol}\s -> Record
  ^\s+${Phy_INT} -> Record
```

运行 Python 脚本，如下图所示。

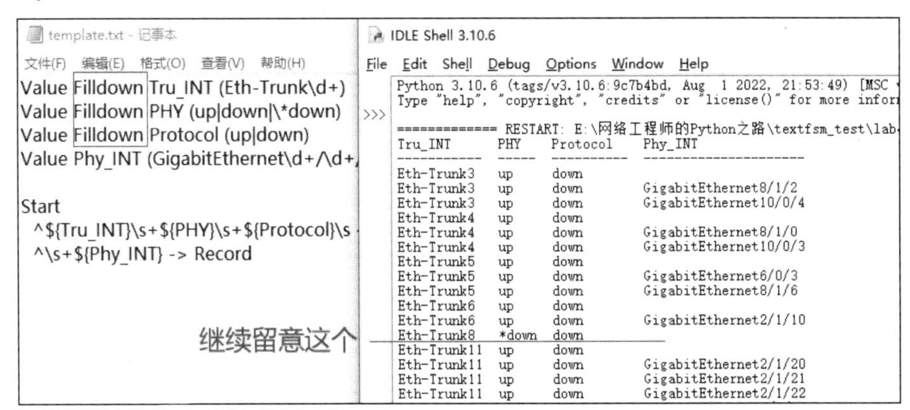

在 3 个变量中加入 Filldown 关键字后，这些变量就会向下填充，也就复习了早前的知识点。

4. 修改模板调测（3）

为了解决最后“又复现了多一行”的情况，我们继续修改模板，在 Phy_INT 变量中加入 Required 关键字，继续复习。

```
Value Filldown Tru_INT (Eth-Trunk\d+)
Value Filldown PHY (up|down|\*down)
Value Filldown Protocol (up|down)
Value Required Phy_INT (GigabitEthernet\d+/\d+/\d+)

Start
  ^${Tru_INT}\s+${PHY}\s+${Protocol}\s -> Record
  ^\s+${Phy_INT} -> Record
```

运行 Python 脚本，如下图所示。

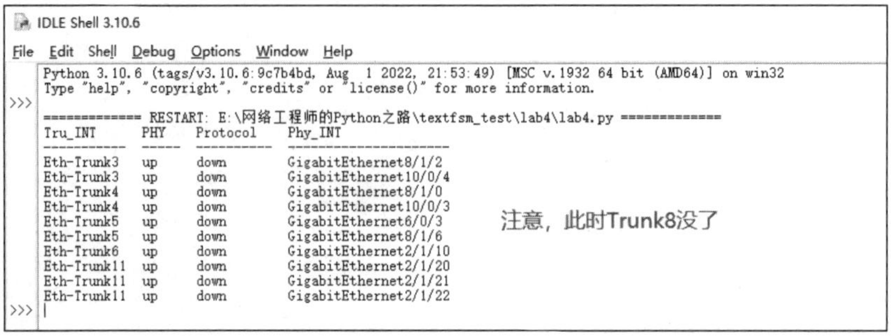

```
IDLE Shell 3.10.6
File Edit Shell Debug Options Window Help
Python 3.10.6 (tags/v3.10.6:9c7b4bd, Aug  1 2022, 21:53:49) [MSC v.1932 64 bit (AMD64)] on win32
Type "help", "copyright", "credits" or "license()" for more information.
>>>
============= RESTART: E:\网络工程师的Python之路\textfsm_test\lab4\lab4.py =============
Tru_INT      PHY    Protocol    Phy_INT
-----------  -----  ----------  ----------------------
Eth-Trunk3   up     down        GigabitEthernet8/1/2
Eth-Trunk3   up     down        GigabitEthernet10/0/4
Eth-Trunk4   up     down        GigabitEthernet8/1/0
Eth-Trunk4   up     down        GigabitEthernet10/0/3
Eth-Trunk5   up     down        GigabitEthernet6/0/3
Eth-Trunk5   up     down        GigabitEthernet8/1/6
Eth-Trunk6   up     down        GigabitEthernet2/1/10
Eth-Trunk11  up     down        GigabitEthernet2/1/20
Eth-Trunk11  up     down        GigabitEthernet2/1/21
Eth-Trunk11  up     down        GigabitEthernet2/1/22
>>> |
```

注意，此时Trunk8没了

此时，这份回显文本已经被整理成常规表格，可以和 csv、excel、数据库表等进行对接。需要留意一个现象，在这种方法中，空 Trunk 会被过滤掉。

5. 修改模板调测（4）

此时，我们可以配合查阅模块官方手册，检索与 Filldown、Required 等关键字类似的内容。从手册上可以看到，还有一个 List 关键字。我们重新修改一下模板，这次我们去掉 Filldown、Required，在变量 Phy_INT 中加入关键字 List。同时，增加 Record 动作，由之前遇到物理端口触发 Record 的规则，调整为无论是物理端口还是 Trunk 口都触发 Record 动作。

```
Value Tru_INT (Eth-Trunk\d+)
Value PHY (up|down|\*down)
Value Protocol (up|down)
Value List Phy_INT (GigabitEthernet\d+/\d+/\d+)

Start
  ^${Tru_INT}\s+${PHY}\s+${Protocol}\s -> Record
  ^\s+${Phy_INT} -> Record
```

运行 Python 脚本，如下图所示。

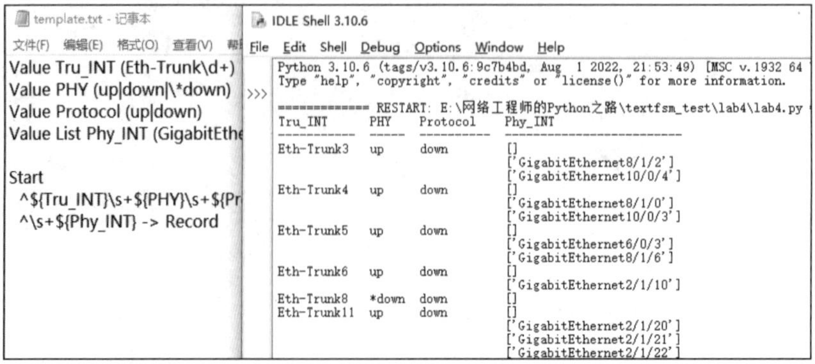

```
template.txt - 记事本
文件(F) 编辑(E) 格式(O) 查看(V) 帮
Value Tru_INT (Eth-Trunk\d+)
Value PHY (up|down|\*down)
Value Protocol (up|down)
Value List Phy_INT (GigabitEthe

Start
  ^${Tru_INT}\s+${PHY}\s+${Pr
  ^\s+${Phy_INT} -> Record
```

```
IDLE Shell 3.10.6
File Edit Shell Debug Options Window Help
Python 3.10.6 (tags/v3.10.6:9c7b4bd, Aug  1 2022, 21:53:49) [MSC v.1932 64
Type "help", "copyright", "credits" or "license()" for more information.
>>>
============= RESTART: E:\网络工程师的Python之路\textfsm_test\lab4\lab4.py
Tru_INT      PHY    Protocol    Phy_INT
-----------  -----  ----------  --------------------------
Eth-Trunk3   up     down        []
                                ['GigabitEthernet8/1/2']
                                ['GigabitEthernet10/0/4']
Eth-Trunk4   up     down        []
                                ['GigabitEthernet8/1/0']
                                ['GigabitEthernet10/0/3']
Eth-Trunk5   up     down        []
                                ['GigabitEthernet6/0/3']
                                ['GigabitEthernet8/1/6']
Eth-Trunk6   up     down        []
                                ['GigabitEthernet2/1/10']
Eth-Trunk8   *down  down        []
Eth-Trunk11  up     down        []
                                ['GigabitEthernet2/1/20']
                                ['GigabitEthernet2/1/21']
                                ['GigabitEthernet2/1/22']
```

初用 List 关键字，其结果看起来比较乱。请不要紧张，我们接着调测。

6. 修改模板调测（5）

我们再调整一下 Record 动作位置，删掉物理端口触发 Record 动作，仅保留 Trunk 口触发。

```
Value Tru_INT (Eth-Trunk\d+)
Value PHY (up|down|\*down)
Value Protocol (up|down)
Value List Phy_INT (GigabitEthernet\d+/\d+/\d+)

Start
  ^${Tru_INT}\s+${PHY}\s+${Protocol}\s -> Record
  ^\s+${Phy_INT}
```

运行 Python 脚本，如下图所示。

```
IDLE Shell 3.10.6
File Edit Shell Debug Options Window Help
Python 3.10.6 (tags/v3.10.6:9c7b4bd, Aug  1 2022, 21:53:49) [MSC v.1932 64 bit (AMD64)] on win32
Type "help", "copyright", "credits" or "license()" for more information.
>>>
============= RESTART: E:\网络工程师的Python之路\textfsm_test\lab4\lab4.py =============
Tru_INT      PHY    Protocol    Phy_INT
-----------  -----  ----------  ------------------------------------------------------------------
Eth-Trunk3   up     down        []
Eth-Trunk4   up     down        ['GigabitEthernet8/1/2', 'GigabitEthernet10/0/4']
Eth-Trunk5   up     down        ['GigabitEthernet8/1/0', 'GigabitEthernet10/0/3']
Eth-Trunk6   up     down        ['GigabitEthernet6/0/3', 'GigabitEthernet8/1/6']
Eth-Trunk8   *down  down        ['GigabitEthernet2/1/10']
Eth-Trunk11  up     down        []
                                ['GigabitEthernet2/1/20', 'GigabitEthernet2/1/21', 'GigabitEthernet2/1/22']
>>> |
```

仔细看运行结果，我们发现错行了。比如 Trunk3 的物理端口放到了 Trunk4 的记录中。于是问题转换成解决这个错行。那么如何攻克呢？我们来看一下这个匹配过程。

原始报文我们截取前 6 行来看，并忽略后面的物理状态和协议状态字段，我们只关注两个字段——Tru_INT 和 PHY_INT，如下表所示。

output.txt 中的内容	template.txt 中的内容
Eth-Trunk3	Value Tru_INT (Eth-Trunk\d+)
GigabitEthernet8/1/2(10G)	Value List Phy_INT (GigabitEthernet\d+/\d+/\d+)
GigabitEthernet10/0/4(10G)	
Eth-Trunk4	Start
GigabitEthernet8/1/0(10G)	^${Tru_INT} -> Record
GigabitEthernet10/0/3(10G)	^\s+${Phy_INT}

（1）TextFSM 引擎读取第 1 行输入文本——Eth-Trunk3。第一条规则命中，执行 Record 动作，形成一条记录，随后所有变量被清空，后面的匹配规则不再进行。

（2）TextFSM 引擎读取第 2 行输入文本——GigabitEthernet8/1/2(10G)。第一条规则未命中，继续匹配第二条规则。第二条规则命中，提取物理端口信息到 Phy_INT 列表中（此时它有 List 参数）。此时，模板中的规则都匹配完成。

（3）TextFSM 引擎读取第 3 行输入文本——GigabitEthernet10/0/4(10G)。类似（2），第一条规则未命中，而第二条规则命中。Phy_INT 列表继续追加匹配到的变量，列表中已有两个元素。

（4）TextFSM 引擎读取第 4 行输入文本——Eth-Trunk4。类似（1），第一条规则命中，执行 Record 动作，形成一条记录，随后所有变量被清空，后面的匹配规则不再进行。于是，错行就这么产生了，如下图所示。

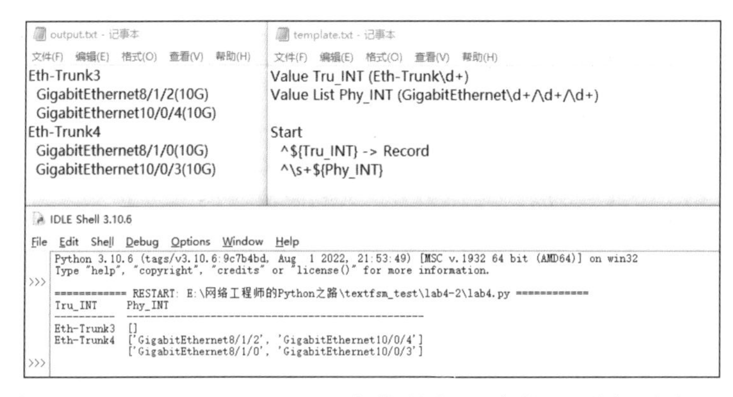

（5）最后一行 Tru_INT 空而 Phy_INT 有值是怎么来的？别忘了在匹配结束时会跳转至 EOF 状态，该状态默认会再做一个 Record 动作，于是就有了最后一行记录。

7. 修改模板调测（6）

查阅手册，我们介绍一下 Next 和 Continue，以及 Record 和 Record。这是原文引用，可以尝试翻译一下。

（1）行动作。

Line Actions	Description
Next	Finish with the input line, read in the next and start matching again from the start of the state. This is the default behavior if no line action is specified.
Continue	Retain the current line and do not resume matching from the first rule of the state. Continue processing rules as if a match did not occur (value assignments still occur).

（2）记录动作。

Record Actions	Description
NoRecord	Do nothing. This is the default behavior if no record action is specified.
Record	Record the values collected so far as a row in the return data. Non Filldown values are cleared. Note: No record will be output if there are any 'Required' values that are unassigned.

从这里可以知道，每一行规则，后面都会有 Line Actions 和 Record Actions，默认值是 Next.NoRecord，并且这个默认值是隐式的，可以不写出来。它的意思是，读取下一行输入文本（即 output.txt 里的 line），然后回头重新匹配规则（即 template.txt 里的 rule）。

我们有了这个概念以后，重新回过头来调整模板，用上 Continue.Record，其作用是 TextFSM 引擎在记录后不读取下一行（默认 Next，即读取下一行），带着原来的文本行，继续进行下一行规则匹配。

```
Value Tru_INT (Eth-Trunk\d+)
Value PHY (up|down|\*down)
Value Protocol (up|down)
Value List Phy_INT (GigabitEthernet\d+/\d+/\d+)

Start
  ^Eth-Trunk\d+ -> Continue.Record
  ^${Tru_INT}\s+${PHY}\s+${Protocol}\s
  ^\s+${Phy_INT}
```

运行 Python 脚本，如下图所示。

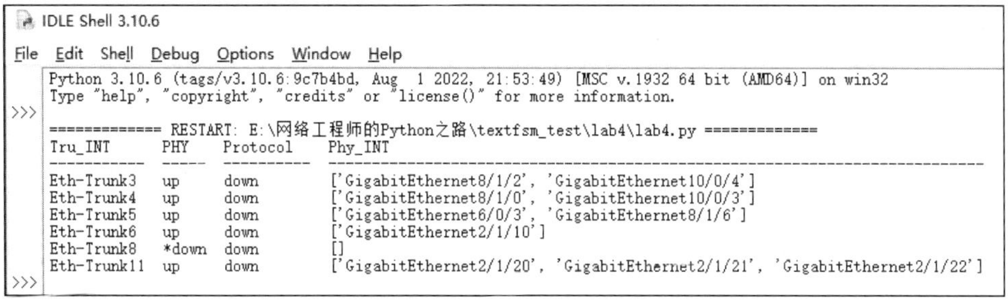

```
IDLE Shell 3.10.6
File  Edit  Shell  Debug  Options  Window  Help
Python 3.10.6 (tags/v3.10.6:9c7b4bd, Aug  1 2022, 21:53:49) [MSC v.1932 64 bit (AMD64)] on win32
Type "help", "copyright", "credits" or "license()" for more information.
>>>
============= RESTART: E:\网络工程师的Python之路\textfsm_test\lab4\lab4.py =============
Tru_INT      PHY    Protocol    Phy_INT
-----------  -----  ----------  ---------------------------------------------------------------------------
Eth-Trunk3   up     down        ['GigabitEthernet8/1/2', 'GigabitEthernet10/0/4']
Eth-Trunk4   up     down        ['GigabitEthernet8/1/0', 'GigabitEthernet10/0/3']
Eth-Trunk5   up     down        ['GigabitEthernet6/0/3', 'GigabitEthernet8/1/6']
Eth-Trunk6   up     down        ['GigabitEthernet2/1/10']
Eth-Trunk8   *down  down        []
Eth-Trunk11  up     down        ['GigabitEthernet2/1/20', 'GigabitEthernet2/1/21', 'GigabitEthernet2/1/22']
>>>
```

现在就符合我们的实验预期了。每一个 Trunk 口都占一行，归属于该 Trunk 的物理端口都被装入一个列表中。Python 脚本运行后，我们还可以在 IDLE 窗口中继续进行调测，如下图所示。

```
>>> print(result[0])
['Eth-Trunk3', 'up', 'down', ['GigabitEthernet8/1/2', 'GigabitEthernet10/0/4']]
>>> print(result[1])
['Eth-Trunk4', 'up', 'down', ['GigabitEthernet8/1/0', 'GigabitEthernet10/0/3']]
>>> print(result[2])
['Eth-Trunk5', 'up', 'down', ['GigabitEthernet6/0/3', 'GigabitEthernet8/1/6']]
>>> print(result[3])
['Eth-Trunk6', 'up', 'down', ['GigabitEthernet2/1/10']]
>>> |
```

7.5.5 模板匹配过程

我们用伪代码的方式，对 TextFSM 引擎的匹配过程再次进行一次归纳梳理。

```
for line in output:
    for rule in template:
        if 匹配中:

            #L - Line Action
            Next（默认）:跳出本层循环结构体，类似于 Python 的 break
            Continue:进入下一轮循环，类似于 Python 的 continue

            #R - Record Action
            NoRecord（默认）:不记录
            Record:记录
            #此外，R 还有 Clear、Clearall，比较少用

            #S - State Transition
            跳转至其他状态（如果有定义的话）

            #Error Action
            停止匹配，清除已收集，返回提示
            ^regex -> Error "some words."
```

7.5.6 实验小结

本实验在复习变量关键字 Filldown、Required 的基础上，继续推进变量关键字 List 的使用。同时，我们还尝试了 Continue.Record 动作进行跨行匹配，提取记录后，继续匹配。大家也可以感悟一下 TextFSM 是“有状态”的。

7.6 TextFSM 场景梳理及拓展

本章通过 1 个引入、4 个实战共计 5 个场景化实验，逐步带出 TextFSM 的基础语法，旨在引导读者们快速入门和有效应用。网络工程师在日常运维中，跨行解析有两个常见的文本解析场景，如下表所示。

场景一，提取标识符在后（知后）	场景二，提取标识符在前（知前）
实验 3，OSPF 邻居	实验 4，Eth-Trunk 下物理端口
AAA bbb ccc 一定符合 ccc 规则时提取 ……	AAA bbb ccc 不一定符合 ccc 规则时提取，可能缺 ccc，也可能 ccc 后还有 ddd ……

TextFSM 纯文本看起来有点辛苦，如果使用 Visual Studio Code 软件，可加载如下图所示的插件，会有一些语法高亮，辅助阅读。它需要将模板名改成.textfsm 后缀方可识别。读者如有兴趣请自行尝试。

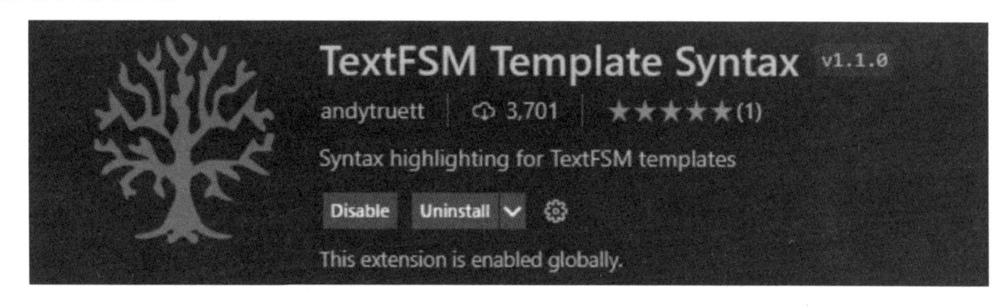

在日常实战应用中，我们不可为了 TextFSM 而 TextFSM，不要试图通过 TextFSM 解决所有解析问题，可以综合使用不同的手段。比如，有些比较复杂的文本解析任务，使用单个 TextFSM 模板往往需要跳转其他状态来实现，难度较大。此时，我们可以结合 Python 基本操作，切割一下文本，用不同的 TextFSM 模板来实现，这样往往会方便和清晰很多。我们的“初心”是高效且高质量地完成我们的工作任务，自动化是手段，不是目的。

最后，TextFSM 的内容远不限于此，日常生产可结合实际场景需求，配合查阅官网手册，逐步实施。当我们熟练使用之后，也可以尝试往状态跳转、TextFSM CLI Table、给 ntc-templates 贡献适配国产设备的模板等方向进行拓展。在本书编写过程中，我们也将 TextFSM 官方 Wiki 手册全文进行翻译并将长期维护，动态与官网保持同步，详见作者专栏博文信息。

第 8 章

Netmiko 详解

作为 Paramiko 最为成功的衍生模块，Netmiko 成了很多学习 Python 网络运维自动化技术的网络工程师日常工作中最常用的模块之一。相较于 Paramiko，Netmiko 将很多细节优化和简化，比如不需要导入 time 模块做休眠，不需要在输入的每条命令后都加换行符\n，不需要执行 config term、exit、end 等命令，提取、打印回显内容更方便，可以配合 Jinja2 模块调用配置模板，以及配合 TextFSM、pyATS、Genie 等模块将回显内容以有序的 JSON 格式输出，方便我们过滤和提取出所需的数据，等等。此外，在 Netmiko 的基础上也诞生出 NAPALM、pyntc、netdev 等第三方模块，甚至 Nornir 这样成功的网络运维自动化框架。

在本书第 1 版里，笔者刻意减少了 Netmiko 的相关内容，重点讲解了 Paramiko，因为道理很简单：Netmiko 将太多功能简化的做法其实并不利于初学者学习。Netmiko 和 Paramiko 的区别就像自动挡汽车和手动挡汽车的区别。会驾车的人都知道，一开始就学手动挡汽车的人 100%会开自动挡汽车，而从一开始就学自动挡汽车的人，除非额外加课，否则是 100%不会开手动挡汽车的。Paramiko 虽然复杂、烦琐一些，但是就像手动挡汽车一样，整体“操控感”更强，运维脚本中的所有细节和各种参数都在我们自己的掌控之中，更利于我们从整体来把握，进而写出自己需要的脚本，并且无须像 Netmiko 那样担心对各家厂商设备及各种 OS 的支持的问题。

随着越来越多网络工程师读者逐渐上手和适应 Paramiko，为了弥补本书第 1 版“重 Paramiko，轻 Netmiko”的遗憾，特此在第 2 版里补上一章“Netmiko 详解”，由简到难涵盖 Netmiko 的各个知识点及使用技巧。

2022 年 3 月 23 日，在 Netmiko 3 的最后一版 Netmiko 3.4.0 发布整整 11 个月后，“Netmiko 之父”Kirk Byers 正式公布并发行了 Netmiko 4 的初版即 Netmiko 4.0.0，与很多模块的重大更新类似，Netmiko 4.0.0 发布后存在很多 Bug，尤其是华为、H3C 等很多国产设备受影响十分严重，很多用户通过 pip 安装 Netmiko 的时候不得不手动指定版本，用回稳定的 3.4.0 版本。

不过 Kirk 行动迅速，在 4 月 27 日、6 月 29 日及 8 月 9 日相继发布了 4.1.0、4.1.1 和 4.1.2 版本，不仅优化了 Netmiko 4 的性能，解决了 4.0.0 版本的很多 Bug，也提供了对更多设备的支持。

在生产环境中，可能基于框架已部署、应用求稳定等因素的考虑，我们会持续使用老版本。这无可厚非，也不失为一个好办法！但是，从把握行业动态、了解领域前沿出发，我们在学习和测试方面，要勇于尝鲜，勇于使用最新版本，遇到问题解决问题。寻求解决问题的一个重要手段，是与作者、维护者良性互动起来。本章安排 9 个实验代码，在 Netmiko 3(3.4.0)、Netmiko 4（4.1.2）均测试成功。此外，本章将独立安排一节，专门介绍 Netmiko 4 的新特性。

本章涉及的实验拓扑、操作系统类型、设备型号承接第 4 章的内容。

◎　实验准备

通过 pip 安装 Netmiko（参考第 4.2.2 节）后在 Python 解释器里用 import 验证，截至 2022 年 8 月，目前 Netmiko 的最新版本为 4.1.2，如下图所示。

```
[root@localhost ~]# pip3.10 freeze | grep netmiko
netmiko==4.1.2
[root@localhost ~]# python3.10
Python 3.10.6 (main, Aug  6 2022, 13:49:15) [GCC 8.5.0 20210514 (Red Hat 8.5.0-1
5)] on linux
Type "help", "copyright", "credits" or "license" for more information.
>>> import netmiko
>>> exit()
[root@localhost ~]#
```

如果依然要使用 Netmiko 3，建议在安装 Netmiko 时指定下载最稳定的 Netmiko 3.4.0，方法是 pip3.10 install netmiko==3.4.0，如下图所示。

```
P:\>pip3.10 install netmiko==3.4.0
Defaulting to user installation because normal site-packages is not writeable
Collecting netmiko==3.4.0
  Using cached netmiko-3.4.0-py3-none-any.whl (178 kB)
Collecting scp>=0.13.2
  Downloading scp-0.14.4-py2.py3-none-any.whl (8.6 kB)
Requirement already satisfied: paramiko>=2.6.0 in c:\users\wangy01\appdata\roaming\python\python310\site-packages (from netmiko==3.4.0) (2.11.0)
Collecting ntc-templates
  Downloading ntc_templates-3.0.0-py3-none-any.whl (303 kB)
     ---------------------------------------- 303.5/303.5 kB 1.3 MB/s eta 0:00:00
Collecting tenacity
  Downloading tenacity-8.0.1-py3-none-any.whl (24 kB)
Requirement already satisfied: setuptools>=38.4.0 in c:\program files\python310\lib\site-packages (from netmiko==3.4.0) (63.2.0)
Collecting pyserial
  Using cached pyserial-3.5-py2.py3-none-any.whl (90 kB)
```

8.1　实验 1　通过 Netmiko 登录一台交换机（思科设备）

要通过 Netmiko 登录一台设备需要用到它的核心对象 ConnectHandler。ConnectHandler() 包含几个必要的参数和可选参数，必要参数包括 device_type、ip(也可以是 host)、username 和

password，可选参数包括 port、secret、use_keys、key_file、conn_timeout 等，可选参数的作用和用法在后面的实验中会陆续讲到。

首先创建第一个脚本 netmiko1.py，然后写入下面的代码。

```
from netmiko import ConnectHandler
sw1 = {
    'device_type': 'cisco_ios',
    'ip': '192.168.2.11',
    'username': 'python',
    'password': '123'
}

with ConnectHandler(**sw1) as connect:
    print ("已经成功登录交换机" + sw1['ip'])
```

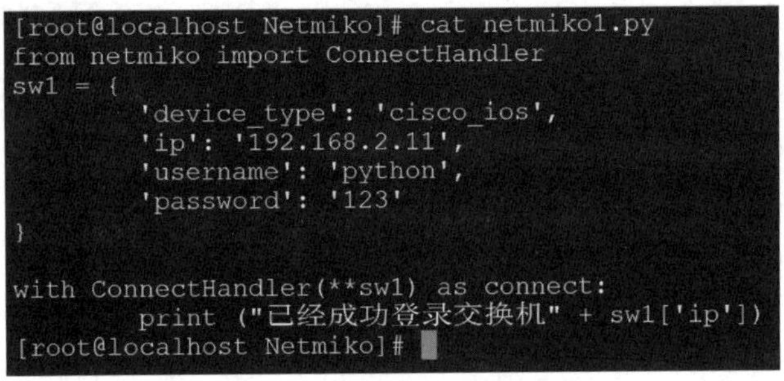

```
[root@localhost Netmiko]# cat netmiko1.py
from netmiko import ConnectHandler
sw1 = {
        'device_type': 'cisco_ios',
        'ip': '192.168.2.11',
        'username': 'python',
        'password': '123'
}

with ConnectHandler(**sw1) as connect:
        print ("已经成功登录交换机" + sw1['ip'])
[root@localhost Netmiko]#
```

代码分段讲解如下。

- 首先通过 import 语句从 Netmiko 导入它的核心对象 ConnectHandler()。

```
from netmiko import ConnectHandler
```

- 然后创建一个名为 sw1 的字典，该字典包含 device_type、ip、username 和 password 4 个必选的键。sw1 字典里的 4 组键值对会以关键字参数的形式（**kwargs）在 ConnectHandler()里打开，被用作前面讲到的 ConnectHandler()的 4 个必选参数（device_type、ip、username 和 password）。其中 ip、username、password 3 个键比较好理解（在实际工作中，username 和 password 建议通过 input()和 getpass 模块来输入，这里因为只是实验演示，所以直接把 username 和 password 明文写进脚本里），这里主要说下 device_type。截至 2022 年 8 月，Netmiko 支持 Arista、Cisco、HP、Juniper、Alcatel、Huawei、H3C、Extreme 和 Palo Alto 等绝大多数主流厂商的设备。除此之外，

Netmiko 同样支持拥有多种不同 OS 类型的厂商的设备，比如针对思科的设备，Netmiko 能同时支持 Cisco ASA、Cisco IOS、Cisco IOS-XE、Cisco IOS-XR、Cisco NX-OS 和 Cisco SG300 共 6 种不同 OS 类型的设备。由于不同厂商的设备登录 SSH 后命令行界面和特性不尽相同，因此我们必须通过 device_type 来指定需要登录的设备的类型。因为实验里我们用到的是思科 IOS 设备，因此 deivce_type 的键值为 cisco_ios。

```
sw1 = {
    'device_type': 'cisco_ios',
    'ip': '192.168.2.11',
    'username': 'python',
    'password': '123'
}
```

- 最后调用 ConnectHandler()，以关键字参数的形式将 sw1 这个字典里的键值对提取出来，作为自己的 ConnectHandler()本身的参数使用，最后打印出"已经成功登录交换机 xx.xx.xx.xx"来提示 SSH 登录成功，注意这里用了 context manager（上下文管理器，也就是 with 语句）来调用 ConnectHandler，它的好处是可以在脚本运行完毕后自动关闭 SSH 会话（如果你不想使用 context manager，而是写成 connect = ConnectHandler (**sw1)也是可以的，不影响 Netmiko 登录设备）。

```
with ConnectHandler(**sw1) as connect:
    print ("已经成功登录交换机" + sw1['ip'])
```

- 这里额外提一点：如果脚本只需要单独登录一台设备，也可以不创建 sw1 字典，而是直接在 ConnectHandler()里面放入 device_type、ip、username 和 password 4 个参数，如下所示。

```
from netmiko import ConnectHandler
with ConnectHandler(device_type='cisco_ios', ip='192.168.2.11', username='python',
password='123') as connect:
   print ("已经成功登录交换机 192.168.2.11")
```

运行脚本前先在 S1 上开启 debug ip ssh，以便验证脚本是否真正 SSH 登录了交换机，如下图所示。

```
S1#debug ip ssh
Incoming SSH debugging is on
S1#
```

运行脚本，结果如下图所示。

```
[root@localhost Netmiko]# python3 netmiko1.py
已经成功登录交换机192.168.2.11
[root@localhost Netmiko]#
```

然后回到 S1 上，如果看到如下图所示的 debug 日志，则说明 Netmiko 登录交换机成功。

```
S1#debug ip ssh
Incoming SSH debugging is on
S1#
*Apr 25 15:20:10.180: SSH0: starting SSH control process
*Apr 25 15:20:10.180: SSH0: sent protocol version id SSH-2.0-Cisco-1.25
*Apr 25 15:20:10.188: SSH0: protocol version id is - SSH-2.0-paramiko_2.7.2
*Apr 25 15:20:10.188: SSH2 0: send:packet of  length 368 (length also includes padlen of 5)
*Apr 25 15:20:10.189: SSH2 0: SSH2_MSG_KEXINIT sent
*Apr 25 15:20:10.189: SSH2 0: ssh_receive: 880 bytes received
*Apr 25 15:20:10.189: SSH2 0: input: total packet length of 880 bytes
*Apr 25 15:20:10.189: SSH2 0: partial packet length(block size)8 bytes,needed 872 bytes,
              maclen 0
*Apr 25 15:20:10.189: SSH2 0: input: padlength 10 bytes
*Apr 25 15:20:10.189: SSH2 0: SSH2_MSG_KEXINIT received
*Apr 25 15:20:10.189: SSH2 0: kex: client->server enc:aes128-ctr mac:hmac-sha1
*Apr 25 15:20:10.189: SSH2 0: kex: server->client enc:aes128-ctr mac:hmac-sha1
*Apr 25 15:20:10.189: SSH2 0: Using kex_algo = diffie-hellman-group-exchange-sha1
*Apr 25 15:20:10.191: SSH2 0: ssh_receive: 24 bytes received
*Apr 25 15:20:10.191: SSH2 0: input: total packet length of 24 bytes
*Apr 25 15:20:10.191: SSH2 0: partial packet length(block size)8 bytes,needed 16 bytes,
              maclen 0
*Apr 25 15:20:10.191: SSH2 0: input: padlength 6 bytes
*Apr 25 15:20:10.191: SSH2 0: SSH2_MSG_KEX_DH_GEX_REQUEST received
*Apr 25 15:20:10.191: SSH2 0: Range sent by client is - 1024 < 2048 < 8192
*Apr 25 15:20:10.191: SSH2 0:  Modulus size established : 2048 bits
```

8.2　实验 1　通过 Netmiko 登录一台交换机（华为设备）

有了第 4 章 Paramiko 实验的经验，我们要把 Netmiko 代码由思科设备迁移适配到华为设备自然就轻车熟路了。这个实验只需要在 netmiko1.py 中修改 device_type 值，将其改为 huawei 即可，如下图所示。

```
from netmiko import ConnectHandler
sw1 = {'device_type': 'huawei',
    'ip': '192.168.2.11',
    'username': 'python',
    'password': '123'}

with ConnectHandler(**sw1) as connect:
    print ("已经成功登录交换机" + sw1['ip'])
```

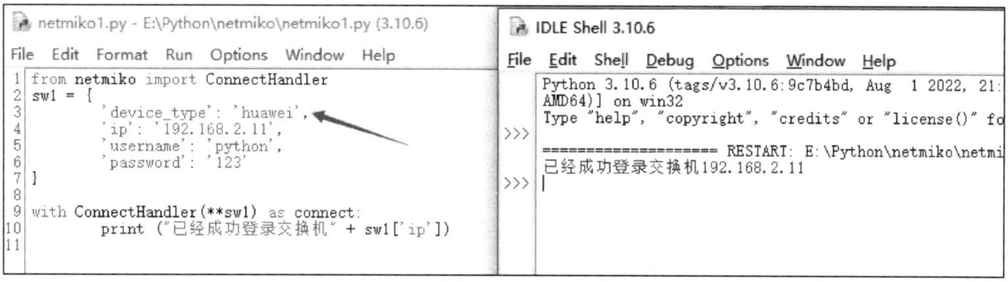

```
netmiko1.py - E:\Python\netmiko\netmiko1.py (3.10.6)
File  Edit  Format  Run  Options  Window  Help
1  from netmiko import ConnectHandler
2  sw1 = {
3          'device_type': 'huawei',
4          'ip': '192.168.2.11',
5          'username': 'python',
6          'password': '123'
7  }
8
9  with ConnectHandler(**sw1) as connect:
10         print ("已经成功登录交换机" + sw1['ip'])
11
```

```
IDLE Shell 3.10.6
File  Edit  Shell  Debug  Options  Window  Help
    Python 3.10.6 (tags/v3.10.6:9c7b4bd, Aug  1 2022, 21:
    AMD64)] on win32
    Type "help", "copyright", "credits" or "license()" fo
>>>
    =================== RESTART: E:\Python\netmiko\netmi
    已经成功登录交换机192.168.2.11
>>> |
```

Netmiko 对华为设备原生支持，关于设备类型也分多种，如本实验中的 V5 用 huawei 适配，到 V8 则改用 huawei_vrp 适配。

再次强调，Python 脚本的文件名称可以是 netmiko1.py，但千万不能是 netmiko.py，这几乎是所有初学者必踩之坑。

细心的读者可能会发现，这里我们把 print 提示信息由英文变成了中文。这与第 4 章实验中提及的尽量用英文的观点不一致。其实，我们在 Python 网络自动化这个领域持续精进，起初紧抓主线大胆前行，适当“后置”一些旁系知识。在这个过程中，我们慢慢地拾起一些旁系知识，则实施过程即可按我们的一些“个性化”需求来尝试。当然，如 Print 内容的中文版对用户友好一些，这是有意义的，但代码中的变量名、文件路径等还是坚持尽量用纯英文字符。

8.3　实验 2　通过 Netmiko 向设备做配置（思科设备）

Netmiko 主要有 4 种函数向设备做配置：send_command()、send_config_set()、send_config_from_file()和 send_command_timing()。其中，send_command_timing()不太常用。它们的用法和区别如下。

- send_command()：只支持向设备发送一条命令，通常是 show/display 之类的查询、排错命令或者 wr mem 这样的保存配置的命令。发出命令后，默认情况下这个函数会一直等待，直到接收到设备的完整回显内容（以收到设备提示符为准，比如要一直等到读到“S1#”），如果在一定时间内依然没读到完整的回显内容，Netmiko 则会返回一个 OSError: Search pattern never detected in send_command: xxxxx 的异常。如果想要指定 Netmiko 在回显内容中读到我们需要的内容，则需要用到 expect_string 参数（expect_string 默认值为 None）。如果 send_command()从回显内容中读到了 expect_string 参数指定的内容，则 send_command()依然返回完整的回显内容；如果没读到 expect_string 参数指定的内容，则 Netmiko 同样会返回一个 OSError: Search pattern never detected in send_command: xxxxx 的异常，关于 expect_string 参数的用法会在稍后的实验中演示。

- send_config_set()：向设备发送一条或多条配置命令，注意是配置命令，不是 show/display 之类的查询命令，因为 send_config_set()本身会自动替我们加上一个 config terminal 命令进入配置模式（以及在命令末尾自动替我们加上一个 end 命令），在 config terminal 下，除非在 show 命令前加上一个 do，如 do show ip int brief，否则 show 命令无效（以上以思科 IOS 设备为例）。send_config_set()一般配合列表使用。
- send_config_from_file()：在配置命令数量较多时，将所有配置命令写入列表显然是比较笨拙的，因为会造成代码过长，不方便阅读，并且在部分厂商的设备上（如华为）还会出现超时报错的情况。我们可以先将所有的配置命令写入一个配置文件，然后使用 send_config_from_file()去读取该文件的内容，帮助我们完成配置。和 send_config_set()一样，send_config_from_file()也会自动帮我们添加 config terminal 和 end 两个命令，所以在配置文件里无须加入这两个命令。
- send_command_timing()：和 send_command()一样，只支持向设备发送一条 show/display 命令。区别是在用 send_command()输入一条 show 命令后，send_command()会一直等待，直到收到设备的完整回显内容（以收到设备提示符为准）。send_command_timing()则会自己去“猜”什么时候停止运行，它的原理是，如果没有从设备收到更多新的回显内容，它会继续等待 2s，然后自动停止运行，并且不会抛出任何异常。send_command_timing()里有一个叫作 delay_factor 的参数，默认为 1，如果将它修改为 2，则 send_command_timing()会等待 4s，修改为 3，则等待 6s，修改为 4，等待 8s，依此类推。因为 send_command_timing()不会抛出 OSError: Search pattern never detected in send_command: xxxxx 异常，所以它是 send_command()的一个良好替代方案。有时候在输入 show tech-support 或者 show log（logging buffer 特别大，日志特别长的那种）这种回显内容巨多的命令时，即使输入了 term len 0 来取消分屏显示，设备在返回回显内容时依然会有停顿，有时停顿时间会大于 2s，这样就会导致截屏不完整，必须手动调整 delay_factor 这个参数。

接下来我们做实验，首先创建一个 config.txt 文件，写入下列配置命令。这里我们将 gi0/0 端口的 description 改为 Netmiko2.py，如下图所示。

```
int gi0/0
description Netmiko2.py
```

```
[root@localhost Netmiko]# cat config.txt
int gi0/0
description Netmiko2.py
[root@localhost Netmiko]#
```

然后创建第二个实验脚本 netmiko.py2（确保实验脚本和 config.txt 配置命令位于同一个文件夹），在脚本中放入下列代码。

```
from netmiko import ConnectHandler
sw1 = {
        'device_type': 'cisco_ios',
        'ip': '192.168.2.11',
        'username': 'python',
        'password': '123'
}

commands = ['interface gi0/1', 'description Nornir2.py']
with ConnectHandler(**sw1) as connect:
        print ("已经成功登录交换机" + sw1['ip'])
        output = connect.send_command('show interface description')
        print(output)
        output = connect.send_config_set(commands)
        print(output)
        output = connect.send_config_from_file('config.txt')
        print(output)
        output = connect.send_command('show interface description')
        print(output)
        output = connect.send_command('wr mem')
        print(output)
```

```
[root@localhost Netmiko]# cat netmiko2.py
from netmiko import ConnectHandler
sw1 = {
        'device_type': 'cisco_ios',
        'ip': '192.168.2.11',
        'username': 'python',
        'password': '123'
}

commands = ['interface gi0/1', 'description Nornir2.py']
with ConnectHandler(**sw1) as connect:
        print ("已经成功登录交换机" + sw1['ip'])
        output = connect.send_command('show interface description')
        print(output)
        output = connect.send_config_set(commands)
        print(output)
        output = connect.send_config_from_file('config.txt')
        print(output)
        output = connect.send_command('show interface description')
        print(output)
[root@localhost Netmiko]#
```

代码分段讲解如下。

- 首先通过 import 语句从 Netmiko 导入它的核心对象 ConnectHandler()，创建字典 sw1，用作 ConnectHandler()登录交换机的参数。

```
from netmiko import ConnectHandler
sw1 = {
        'device_type': 'cisco_ios',
        'ip': '192.168.2.11',
        'username': 'python',
        'password': '123'
}
```

- 然后创建一个列表 commands，放入要对 Gi0/1 端口修改 description 的配置，该 commands 列表会被 send_config_set()函数调用。

```
commands = ['interface gi0/1', 'description Nornir2.py']
```

- 通过 ConnectHandler()连入交换机 S1 后，首先使用 send_command('show interface description')查询配置前 Gi0/0 和 Gi0/1 两个端口当前的 description，然后通过 send_config_set(commands)调用 commands 列表对 Gi0/1 做配置，然后使用 send_config_from_file('config.txt')读取之前我们创建的配置文件里的命令对 Gi0/0 做配置，最后用 send_command('show interface description')查询配置完毕后两个端口的 description，并 send_command('wr mem')保存配置。

```
with ConnectHandler(**sw1) as connect:
        print ("已经成功登录交换机" + sw1['ip'])
        output = connect.send_command('show interface description')
        print(output)
        output = connect.send_config_set(commands)
        print(output)
        output = connect.send_config_from_file('config.txt')
        print(output)
        output = connect.send_command('show interface description')
        print(output)
        output = connect.send_command('wr mem')
        print(output)
```

运行脚本看效果，如下图所示。

```
[root@localhost Netmiko]# python3 netmiko2.py
已经成功登陆交换机192.168.2.11
Interface                      Status         Protocol Description
Gi0/0                          up             up
Gi0/1                          up             up
Gi0/2                          up             up       Access Port to VLAN 999 (Nornir)
Lo0                            up             up
Lo1                            up             up
Vl1                            up             up
configure terminal
Enter configuration commands, one per line.  End with CNTL/Z.
S1(config)#interface gi0/1
S1(config-if)#description Nornir2.py
S1(config-if)#end
S1#
configure terminal
Enter configuration commands, one per line.  End with CNTL/Z.
S1(config)#int gi0/0
S1(config-if)#description Netmiko2.py
S1(config-if)#end
S1#

Interface                      Status         Protocol Description
Gi0/0                          up             up       Netmiko2.py
Gi0/1                          up             up       Nornir2.py
Gi0/2                          up             up       Access Port to VLAN 999 (Nornir)
Lo0                            up             up
Lo1                            up             up
Vl1                            up             up
Building configuration...
Compressed configuration from 6901 bytes to 2951 bytes[OK]
[root@localhost Netmiko]#
```

8.4　实验 2　通过 Netmiko 向设备做配置（华为设备）

华为设备的实验代码适配如下。

```
from netmiko import ConnectHandler
sw1 = {'device_type': 'huawei',
       'ip': '192.168.2.11',
       'username': 'python',
       'password': '123'}

commands = ['interface gi 0/0/1', 'desc descby_send_config_set()']

with ConnectHandler(**sw1) as connect:
      print ("已经成功登录交换机" + sw1['ip'])

      print('\n1.交互形式推送一条指令:')
      output = connect.send_command('dis int desc | inc GE0/0/[12][^0-9]')
      print(output)

      print('\n2.列表形式推送多条指令:')
```

```
output = connect.send_config_set(commands)
print(output)

print('\n3.文件形式推送多条指令')
output = connect.send_config_from_file('config.txt')
print(output)

print('\n4.最后检查配置:')
output = connect.send_command('dis int desc | inc GE0/0/[12][^0-9]')
print(output)
```

与思科设备实验稍微有点差别，本实验华为设备在保存配置 save 后需要输入 y 进行确认，后面的实验会再演示，如下图所示。

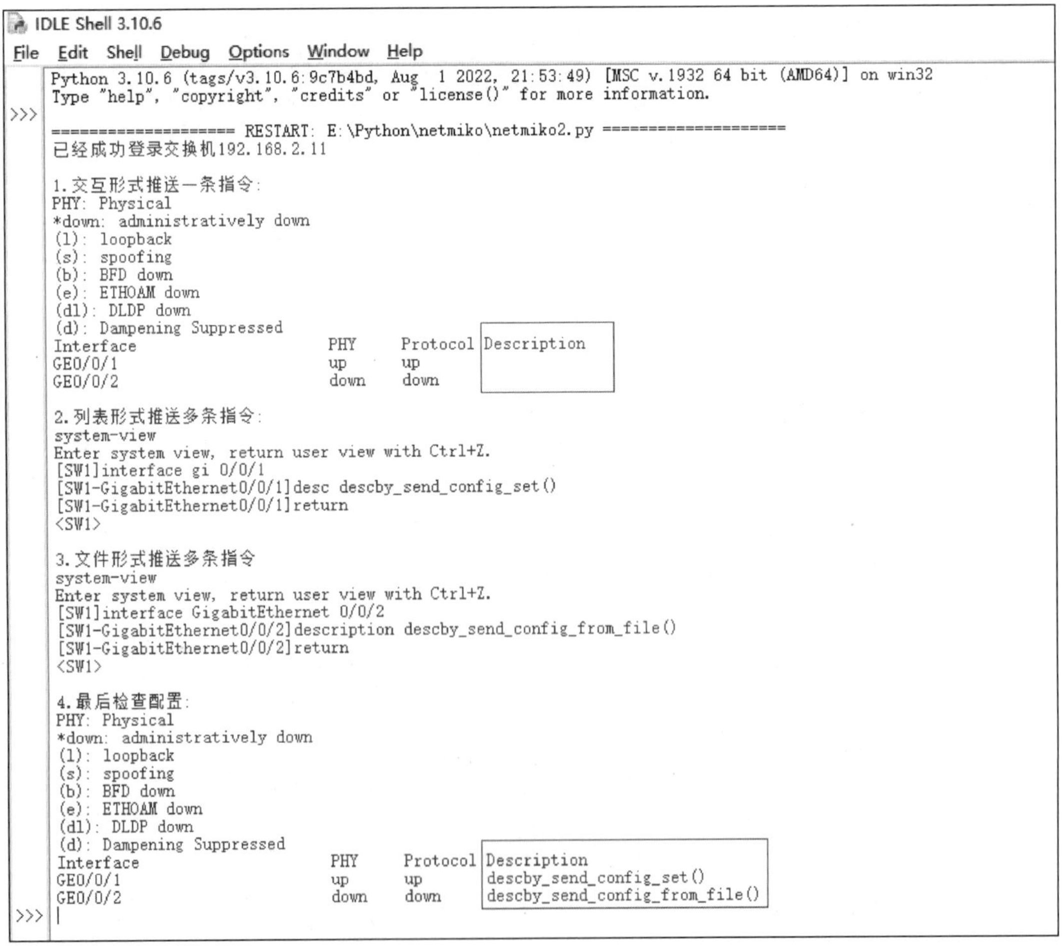

```
IDLE Shell 3.10.6
File Edit Shell Debug Options Window Help
Python 3.10.6 (tags/v3.10.6:9c7b4bd, Aug  1 2022, 21:53:49) [MSC v.1932 64 bit (AMD64)] on win32
Type "help", "copyright", "credits" or "license()" for more information.
>>>
==================== RESTART: E:\Python\netmiko\netmiko2.py ====================
已经成功登录交换机192.168.2.11

1.交互形式推送一条指令:
PHY: Physical
*down: administratively down
(l): loopback
(s): spoofing
(b): BFD down
(e): ETHOAM down
(dl): DLDP down
(d): Dampening Suppressed
Interface                     PHY     Protocol Description
GE0/0/1                       up      up
GE0/0/2                       down    down

2.列表形式推送多条指令:
system-view
Enter system view, return user view with Ctrl+Z.
[SW1]interface gi 0/0/1
[SW1-GigabitEthernet0/0/1]desc descby_send_config_set()
[SW1-GigabitEthernet0/0/1]return
<SW1>

3.文件形式推送多条指令
system-view
Enter system view, return user view with Ctrl+Z.
[SW1]interface GigabitEthernet 0/0/2
[SW1-GigabitEthernet0/0/2]description descby_send_config_from_file()
[SW1-GigabitEthernet0/0/2]return
<SW1>

4.最后检查配置:
PHY: Physical
*down: administratively down
(l): loopback
(s): spoofing
(b): BFD down
(e): ETHOAM down
(dl): DLDP down
(d): Dampening Suppressed
Interface                     PHY     Protocol Description
GE0/0/1                       up      up       descby_send_config_set()
GE0/0/2                       down    down     descby_send_config_from_file()
>>> |
```

正则表达式不仅能运用在脚本代码中，在 CLI 指令交互过程中也常用于对回显信息进行规则过滤。网络工程师日常可积累正则表达式的常用语法。在 Python 脚本调测中，print 函数除显示提示外，如果我们觉察到脚本哪里可能异常，就可以在哪里“安插”print 函数，配合换行符、数字编号、type 函数、id 函数等能更清晰地进行定位。

8.5　实验 3　用 Netmiko 配合 TextFSM 或 Genie 将回显格式化（思科设备）

除 Textfsm 外，思科也出过类似的工具 pyATS 和 Genie 来帮助我们将网络设备无序的回显内容以有序的 JSON 格式输出（不过只限于思科设备）。在 Netmiko 的 send_command()和 send_command_timing()中已经内置了可以直接调用 TextFSM 和 Genie 的参数，非常方便。现在就来分别介绍一下它们的用法。

首先通过 pip3.10 install 来下载 TextFSM、pyATS 和 Genie，这里要注意的是 pyATS 和 Genie 只支持 Linux，不支持 Windows（Windows 下运行 pip install pyats genie 会报错），如果你是 Windows 用户，则需要使用 WLS2 或者通过虚拟机运行 Linux，再来安装 pyATS 和 Genie。强烈建议先通过 pip install --upgrade pip 将 pip 升级到最新版本，否则下载的 pyATS 和 Genie 会碰到各种奇怪的问题。

```
Pip3.10 install --upgrade pip
Pip3.10 install textfsm
Pip3.10 install pyats
Pip3.10 install genie
```

我们以 show interfaces 命令为例，首先创建一个 test.py 来看下不使用 TextFSM 或 Genie 时输入该命令后的回显内容，test.py 的代码如下。

```
from netmiko import ConnectHandler
import pprint

connection_info = {
        'device_type': 'cisco_ios',
        'host': '192.168.2.11',
        'username': 'python',
        'password': '123'
}
```

```
with ConnectHandler(**connection_info) as conn:
    out = conn.send_command("show interfaces")
    print(out)
```

运行脚本看效果，如下图所示。

```
[root@localhost Netmiko]# python3 test.py

GigabitEthernet0/0 is up, line protocol is up (connected)
  Hardware is iGbE, address is 00e2.d5cc.2e00 (bia 00e2.d5cc.2e00)
  Description: Netmiko2.py
  MTU 1500 bytes, BW 1000000 Kbit/sec, DLY 10 usec,
     reliability 255/255, txload 1/255, rxload 1/255
  Encapsulation ARPA, loopback not set
  Keepalive set (10 sec)
  Unknown, Unknown, link type is auto, media type is unknown media type
  output flow-control is unsupported, input flow-control is unsupported
  Auto-duplex, Auto-speed, link type is auto, media type is unknown
  input flow-control is off, output flow-control is unsupported
  ARP type: ARPA, ARP Timeout 04:00:00
  Last input 00:00:00, output 00:00:00, output hang never
  Last clearing of "show interface" counters never
  Input queue: 0/75/0/0 (size/max/drops/flushes); Total output drops: 0
  Queueing strategy: fifo
  Output queue: 0/0 (size/max)
  5 minute input rate 0 bits/sec, 0 packets/sec
  5 minute output rate 0 bits/sec, 0 packets/sec
     436021 packets input, 41753878 bytes, 0 no buffer
     Received 0 broadcasts (0 multicasts)
     0 runts, 0 giants, 0 throttles
     0 input errors, 0 CRC, 0 frame, 0 overrun, 0 ignored
     0 watchdog, 0 multicast, 0 pause input
     122230 packets output, 12813325 bytes, 0 underruns
     0 output errors, 0 collisions, 2 interface resets
     0 unknown protocol drops
     0 babbles, 0 late collision, 0 deferred
     0 lost carrier, 0 no carrier, 0 pause output
     0 output buffer failures, 0 output buffers swapped out
GigabitEthernet0/1 is up, line protocol is up (connected)
  Hardware is iGbE, address is 00e2.d5cc.2e01 (bia 00e2.d5cc.2e01)
  Description: Nornir2.py
  MTU 1500 bytes, BW 1000000 Kbit/sec, DLY 10 usec,
     reliability 255/255, txload 1/255, rxload 1/255
  Encapsulation ARPA, loopback not set
  Keepalive set (10 sec)
  Unknown, Unknown, link type is force-up, media type is unknown media type
  output flow-control is unsupported, input flow-control is unsupported
  Auto-duplex, Auto-speed, link type is force-up, media type is unknown
  input flow-control is off, output flow-control is unsupported
  ARP type: ARPA, ARP Timeout 04:00:00
  Last input never, output 00:00:01, output hang never
  Last clearing of "show interface" counters never
  Input queue: 0/75/0/0 (size/max/drops/flushes); Total output drops: 0
  Queueing strategy: fifo
```

接下来创建脚本 netmiko3_1.py，写入下面的代码。

```
from netmiko import ConnectHandler

connection_info = {
        'device_type': 'cisco_ios',
        'host': '192.168.2.11',
```

```
        'username': 'python',
        'password': '123'
}

with ConnectHandler(**connection_info) as conn:
    out = conn.send_command("show interfaces", use_textfsm=True)
    print(out)
```

可以看到和 Nornir 一样，想要在 Netmiko 里调用 TextFSM 很简单，只需要在 send_command()或者 send_command_timing()里直接添加参数 use_textfsm=True 即可（默认为 False）。接下来运行脚本看效果，如下图所示。

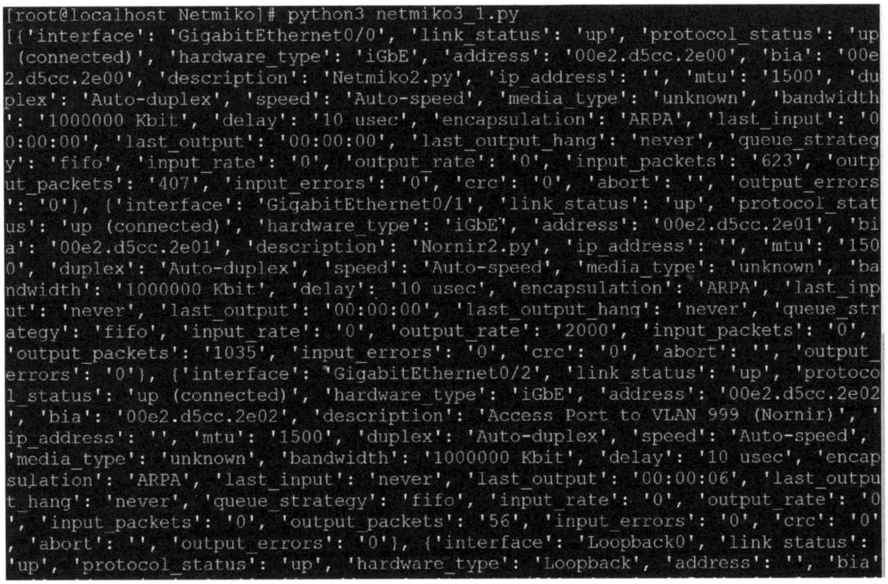

```
[root@localhost Netmiko]# python3 netmiko3_1.py
[{'interface': 'GigabitEthernet0/0', 'link_status': 'up', 'protocol_status': 'up
 (connected)', 'hardware_type': 'iGbE', 'address': '00e2.d5cc.2e00', 'bia': '00e
2.d5cc.2e00', 'description': 'Netmiko2.py', 'ip_address': '', 'mtu': '1500', 'du
plex': 'Auto-duplex', 'speed': 'Auto-speed', 'media_type': 'unknown', 'bandwidth
': '1000000 Kbit', 'delay': '10 usec', 'encapsulation': 'ARPA', 'last_input': '0
0:00:00', 'last_output': '00:00:00', 'last_output_hang': 'never', 'queue_strateg
y': 'fifo', 'input_rate': '0', 'output_rate': '0', 'input_packets': '623', 'outp
ut_packets': '407', 'input_errors': '0', 'crc': '0', 'abort': '', 'output_errors
': '0'}, {'interface': 'GigabitEthernet0/1', 'link_status': 'up', 'protocol_stat
us': 'up (connected)', 'hardware_type': 'iGbE', 'address': '00e2.d5cc.2e01', 'bi
a': '00e2.d5cc.2e01', 'description': 'Nornir2.py', 'ip_address': '', 'mtu': '150
0', 'duplex': 'Auto-duplex', 'speed': 'Auto-speed', 'media_type': 'unknown', 'ba
ndwidth': '1000000 Kbit', 'delay': '10 usec', 'encapsulation': 'ARPA', 'last_inp
ut': 'never', 'last_output': '00:00:00', 'last_output_hang': 'never', 'queue_str
ategy': 'fifo', 'input_rate': '0', 'output_rate': '2000', 'input_packets': '0',
'output_packets': '1035', 'input_errors': '0', 'crc': '0', 'abort': '', 'output_
errors': '0'}, {'interface': 'GigabitEthernet0/2', 'link_status': 'up', 'protoco
l_status': 'up (connected)', 'hardware_type': 'iGbE', 'address': '00e2.d5cc.2e02
', 'bia': '00e2.d5cc.2e02', 'description': 'Access Port to VLAN 999 (Nornir)', '
ip_address': '', 'mtu': '1500', 'duplex': 'Auto-duplex', 'speed': 'Auto-speed',
'media_type': 'unknown', 'bandwidth': '1000000 Kbit', 'delay': '10 usec', 'encap
sulation': 'ARPA', 'last_input': 'never', 'last_output': '00:00:06', 'last_outpu
t_hang': 'never', 'queue_strategy': 'fifo', 'input_rate': '0', 'output_rate': '0
', 'input_packets': '0', 'output_packets': '56', 'input_errors': '0', 'crc': '0'
, 'abort': '', 'output_errors': '0'}, {'interface': 'Loopback0', 'link_status':
'up', 'protocol_status': 'up', 'hardware_type': 'Loopback', 'address': '', 'bia'
```

这里可以看到虽然 TextFSM 输出的内容为 JSON 格式，但是因为没有换行，因此依然不具备可读性。为了将内容更美观地打印出来，我们可以用 Python 另外一个很强大的内置库——pprint。pprint 的全称是 pretty printer，它的功能是将各种数据结构更美观地输出。我们将 netmiko3_1.py 的代码稍作修改，在第二行加上 from pprint import pprint，在最后一行将 print(out)改为 pprint(out)即可。

```
from netmiko import ConnectHandler
from pprint import pprint

connection_info = {
        'device_type': 'cisco_ios',
```

```
        'host': '192.168.2.11',
        'username': 'python',
        'password': '123'
}

with ConnectHandler(**connection_info) as conn:
    out = conn.send_command("show interfaces", use_textfsm=True)
    pprint(out)
```

运行代码看效果，如下图所示，输出内容是不是更美观、可读性更强了？

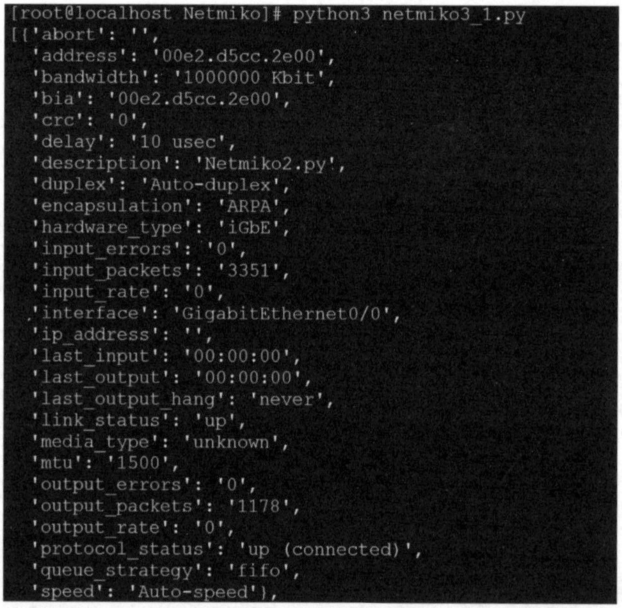

```
[root@localhost Netmiko]# python3 netmiko3_1.py
[{'abort': '',
  'address': '00e2.d5cc.2e00',
  'bandwidth': '1000000 Kbit',
  'bia': '00e2.d5cc.2e00',
  'crc': '0',
  'delay': '10 usec',
  'description': 'Netmiko2.py',
  'duplex': 'Auto-duplex',
  'encapsulation': 'ARPA',
  'hardware_type': 'iGbE',
  'input_errors': '0',
  'input_packets': '3351',
  'input_rate': '0',
  'interface': 'GigabitEthernet0/0',
  'ip_address': '',
  'last_input': '00:00:00',
  'last_output': '00:00:00',
  'last_output_hang': 'never',
  'link_status': 'up',
  'media_type': 'unknown',
  'mtu': '1500',
  'output_errors': '0',
  'output_packets': '1178',
  'output_rate': '0',
  'protocol_status': 'up (connected)',
  'queue_strategy': 'fifo',
  'speed': 'Auto-speed'},
```

和 TextFSM 一样，在 Netmiko 中使用 Genie 也很简单，只需要在 send_command()或者 send_command_timing()里直接添加参数 use_genie=True 即可（默认为 False），当然前提是你之前通过 pip 安装了 pyATS 和 Genie（虽然参数只是 use_genie，但是 pyATS 也必须下载安装，缺一不可）。接下来我们创建 netmiko3_2.py，看一下用 Genie 配合 pprint 将数据内容输出后是什么效果，代码如下。

```
from netmiko import ConnectHandler
from pprint import pprint

connection_info = {
        'device_type': 'cisco_ios',
```

```
        'host': '192.168.2.11',
        'username': 'python',
        'password': '123'
}

with ConnectHandler(**connection_info) as conn:
    out = conn.send_command("show interfaces", use_genie=True)
    pprint(out)
```

运行代码看效果，如下图所示，可以看到 Netmiko 使用 Genie 配合 pprint 后输出的内容比 Textfsm+pprint 更美观，层次更分明。

```
[root@localhost Netmiko]# python3 netmiko3_2.py
{'GigabitEthernet0/0': {'arp_timeout': '04:00:00',
                        'arp_type': 'arpa',
                        'auto_negotiate': True,
                        'bandwidth': 1000000,
                        'connected': True,
                        'counters': {'in_broadcast_pkts': 0,
                                     'in_crc_errors': 0,
                                     'in_errors': 0,
                                     'in_frame': 0,
                                     'in_giants': 0,
                                     'in_ignored': 0,
                                     'in_mac_pause_frames': 0,
                                     'in_multicast_pkts': 0,
                                     'in_no_buffer': 0,
                                     'in_octets': 422745,
                                     'in_overrun': 0,
                                     'in_pkts': 4370,
                                     'in_runts': 0,
                                     'in_throttles': 0,
                                     'in_watchdog': 0,
                                     'last_clear': 'never',
                                     'out_babble': 0,
                                     'out_buffer_failure': 0,
                                     'out_buffers_swapped': 0,
                                     'out_collision': 0,
                                     'out_deferred': 0,
                                     'out_errors': 0,
                                     'out_interface_resets': 2,
                                     'out_late_collision': 0,
                                     'out_lost_carrier': 0,
                                     'out_mac_pause_frames': 0,
                                     'out_no_carrier': 0,
                                     'out_octets': 187509,
                                     'out_pkts': 1517,
                                     'out_underruns': 0,
                                     'out_unknown_protocl_drops': 0,
                                     'rate': {'in_rate': 1000,
                                              'in_rate_pkts': 1,
                                              'load_interval': 300,
                                              'out_rate': 0,
                                              'out_rate_pkts': 0}},
                        'delay': 10,
                        'description': 'Netmiko2.py',
                        'duplex_mode': 'auto',
                        'enabled': True,
                        'encapsulations': {'encapsulation':
                        'flow_control': {'receive': False, 'send': False},
```

注：虽然 TextFSM 和 Genie 输出的内容同为 JSON 格式，但是它们返回的值的数据类型不同。TextFSM 返回的是一个列表（列表里的元素是字典），Genie 则是直接返回字典。从这一点来说，显然 Genie 更优，因为在过滤数据时可以少写一个索引号。比如下面这个脚本就是笔者平时使用的，可以用 Genie 配合 show interfaces 命令过滤出的思科设备的每个端口的状态、MAC 地址及 Duplex 模式，供大家参考。

```
from netmiko import ConnectHandler
import pprint

connection_info = {
        'device_type': 'cisco_ios',
        'host': '192.168.2.11',
        'username': 'python',
        'password': '123'
}

with ConnectHandler(**connection_info) as conn:
    out = conn.send_command("show interfaces", use_genie=True)
    for name, details in out.items():
        print(f"{name}")
        print(f"- Status: {details.get('enabled', None)}")
        print(f"- Physical address: {details.get('phys_address', None)}")
        print(f"- Duplex mode: {details.get('duplex_mode', None)}")
    for counter, count in details.get('counters', {}).items():
            if isinstance(count, int):
                if count > 0:
                    print(f"- {counter}: {count}")
            elif isinstance(count, dict):
                for sub_counter, sub_count in count.items():
                    if sub_count > 0:
                        print(f"- {counter}::{sub_counter}: {sub_count}")
```

脚本运行后的效果如下图所示。

```
[root@localhost Netmiko]# python3 test.py
GigabitEthernet0/0
- Status: True
- Physical address: 00e2.d5cc.2e00
- Duplex mode: auto
GigabitEthernet0/1
- Status: True
- Physical address: 00e2.d5cc.2e01
- Duplex mode: auto
GigabitEthernet0/2
- Status: True
- Physical address: 00e2.d5cc.2e02
- Duplex mode: auto
Loopback0
- Status: True
- Physical address: None
- Duplex mode: None
Loopback1
- Status: True
- Physical address: None
- Duplex mode: None
Vlan1
- Status: True
- Physical address: 00e2.d5cc.8001
- Duplex mode: None
- rate::load_interval: 300
- in_pkts: 1816
- in_octets: 177280
- out_pkts: 970
- out_octets: 143870
```

8.6　实验 3　用 Netmiko 配合 TextFSM 或 Genie 将回显格式化（华为设备）

Genie 是思科设备特有的，华为设备则可以使用 TextFSM。TextFSM 是一个功能强大的模板解析工具。Netmiko 的安装过程会自动加载 TextFSM 和 Ntc-templates，Netmiko 中直接通过 use_textfsm=True 实际调用了 ntc-templates 库中的模板，如下图所示。目前，支持华为设备的模板少之又少。因此，本书推荐读者自学 TextFSM 的基础语法，自己根据实际需求拟写模板。本书也在第 7 章专门安排讲解了 Textfsm 的基础语法。

此电脑 › 本地磁盘 (C:) › Program Files › Python310 › Lib › site-packages › ntc_templates › templates

名称	修改日期	类型	大小
hp_procurve_show_vlans.textfsm	2022/8/7 9:36	TEXTFSM 文件	1 KB
huawei_vrp_display_interface_brief.textfsm	2022/8/7 9:36	TEXTFSM 文件	1 KB
huawei_vrp_display_interface_description.textfsm	2022/8/7 9:36	TEXTFSM 文件	1 KB
huawei_vrp_display_lldp_neighbor.textfsm	2022/8/7 9:36	TEXTFSM 文件	2 KB
huawei_vrp_display_port_vlan.textfsm	2022/8/7 9:36	TEXTFSM 文件	8 KB
huawei_vrp_display_temperature.textfsm	2022/8/7 9:36	TEXTFSM 文件	1 KB
huawei_vrp_display_version.textfsm	2022/8/7 9:36	TEXTFSM 文件	1 KB
index ← 索引文件，很重要	2022/8/7 9:36	文件	41 KB
juniper_junos_show_arp_no-resolve.textfsm	2022/8/7 9:36	TEXTFSM 文件	1 KB

自己拟写的模板可以放入 ntc-templates 模板库（各系统因安装 Python 的路径不同可能有稍微差别），这样由 Netmiko 通过 use_textfsm=True 即可调取使用。下面进行演示。

华为命令：display interface brief

模板文件名：huawei_display_interface_brief.textfsm

模板内容如下，并将该文件放入 ntc_templates\templates 文件夹中。

```
Value INTERFACE (\S+)
Value PHY (down|\*down|#down|up\(s\)|up)
Value PROTOCOL (down|\*down|#down|up\(s\)|up)
Value INUTI (\d*\.?\d*%|\-\-)
Value OUTUTI (\d*\.?\d*%|\-\-)
Value INERRORS (\d+)
Value OUTERRORS (\d+)

Start

^\s*${INTERFACE}\s+${PHY}\s+${PROTOCOL}\s+${INUTI}\s+${OUTUTI}\s+${INERRORS}\s+${OU
TERRORS} -> Record
  ^PHY:\s+Physical
  ^(?:\*|#)down:
  ^\(\w+\):\s+\S+
  ^InUti/OutUti:
  ^Interface\s+PHY\s+Protocol\s+InUti\s+OutUti\s+inErrors\s+outErrors\s*$$
  ^\s*$$
  ^. -> Error
```

修改 ntc_templates\templates 文件夹中的 index 文件。huawei_vrp 的版本是 V8，本次实验拓扑的版本为 V5，使用 huawei，如下图所示。

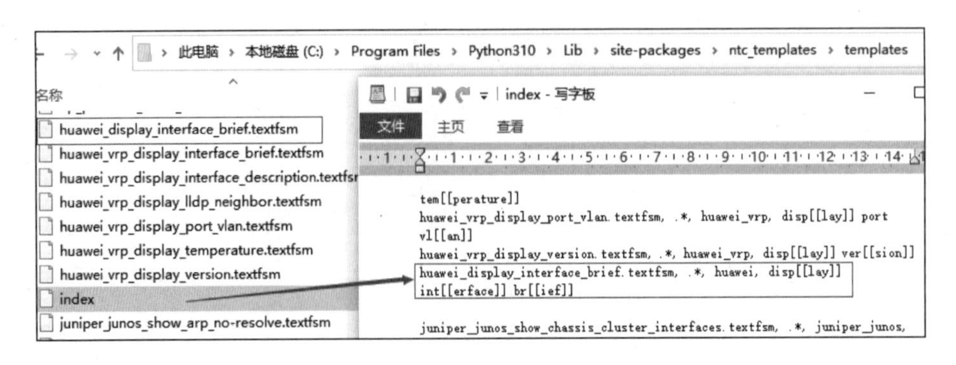

```
huawei_display_interface_brief.textfsm, .*, huawei, disp[[lay]] int[[erface]]
br[[ief]]
```

此时，有可能会提示用户权限问题不予保存，可以使用管理员身份打开编辑或把 index 文件复制出来后修改完再用管理员身份复制回去替换掉。

接着，在系统上需要改一下环境变量，如下图所示。

```
变量名：NET_TEXTFSM
变量值：【按你系统的路径】......\Lib\site-packages\ntc_templates\templates
```

此处我们继续把思科设备实验的 Python 代码迁移到华为设备上来。在 netmik03.py 脚本中，我们调用 send_command 方法时，将参数 use_textfsm 设为 True，即表示使用 net_templates 仓库中的模板。

```
from netmiko import ConnectHandler
from pprint import pprint

connection_info = {'device_type': 'huawei',
        'host': '192.168.2.11',
        'username': 'python',
        'password': '123'}
```

```
with ConnectHandler(**connection_info) as conn:
    out = conn.send_command("disp int bri", use_textfsm=True)
    pprint(out)
```

运行结果如下图所示。

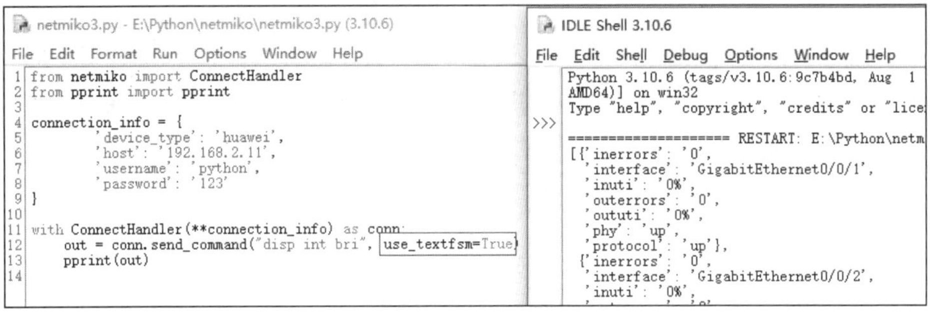

```
netmiko3.py - E:\Python\netmiko\netmiko3.py (3.10.6)
File Edit Format Run Options Window Help
1  from netmiko import ConnectHandler
2  from pprint import pprint
3
4  connection_info = {
5          'device_type': 'huawei',
6          'host': '192.168.2.11',
7          'username': 'python',
8          'password': '123'
9  }
10
11 with ConnectHandler(**connection_info) as conn:
12     out = conn.send_command("disp int bri", use_textfsm=True)
13     pprint(out)
14
```

```
IDLE Shell 3.10.6
File Edit Shell Debug Options Window Help
Python 3.10.6 (tags/v3.10.6:9c7b4bd, Aug  1
AMD64)] on win32
Type "help", "copyright", "credits" or "lice
>>>
==================== RESTART: E:\Python\netm
[{'inerrors': '0',
  'interface': 'GigabitEthernet0/0/1',
  'inuti': '0%',
  'outerrors': '0',
  'oututi': '0%',
  'phy': 'up',
  'protocol': 'up'},
 {'inerrors': '0',
  'interface': 'GigabitEthernet0/0/2',
  'inuti': '0%',
```

如果把 use_textfsm=True 改成 use_textfsm=Flase 或直接删掉这个参数，脚本运行结果会有怎样的差异？请读者自行尝试！另外，如果不手工设置系统环境变量，可否通过 Python 的 os 模块直接在运行过程中临时带上一个 NET_TEXTFSM 值呢？也请读者自行尝试。

```
import os
#路径因安装操作等，有差别，具体以实际为准
os.environ['NET_TEXTFSM']=r'C:\Program
Files\Python310\Lib\site-packages\ntc_templates\templates'
```

8.7　实验 4　通过 Netmiko 连接多台交换机（思科设备）

前面 3 个实验我们都是通过 Netmiko 登录一台设备，在实验 4 里我们来看下怎么通过 Netmiko 登录多台设备，根据情况不同主要有两种方法，分两个脚本讲解。

1. 生产网络里所有设备的 username、password、port 等参数都一样

我们可以参照 Paramiko 的方法，创建一个 ip_list.txt 文件，将所有设备的管理 IP 地址都写进去，这里将 SW1 ~ SW5 总共 5 个交换机的 IP 地址都写入 ip_list.txt 文件里。如下图所示。

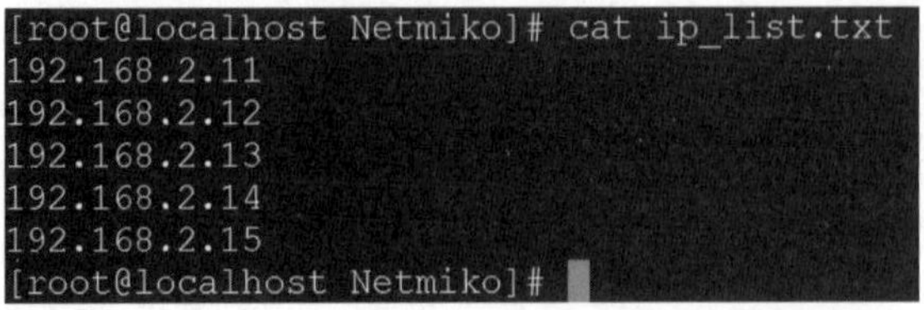

```
[root@localhost Netmiko]# cat ip_list.txt
192.168.2.11
192.168.2.12
192.168.2.13
192.168.2.14
192.168.2.15
[root@localhost Netmiko]#
```

然后创建脚本 netmiko4_1.py，写入下列代码。

```
from netmiko import ConnectHandler

with open('ip_list.txt') as f:
    for ips in f.readlines():
        ip = ips.strip()
        connection_info = {
                'device_type': 'cisco_ios',
                'ip': ip,
                'username': 'python',
                'password': '123',
        }
        with ConnectHandler(**connection_info) as conn:
            print (f'已经成功登录交换机{ip}')
            output = conn.send_command('show run | i hostname')
            print(output)
```

```
[root@localhost Netmiko]# cat netmiko4_1.py
from netmiko import ConnectHandler

with open('ip_list.txt') as f:
    for ips in f.readlines():
        ip = ips.strip()
        connection_info = {
                'device_type': 'cisco_ios',
                'ip': ip,
                'username': 'python',
                'password': '123',
        }
        with ConnectHandler(**connection_info) as conn:
            print (f'已经成功登录交换机{ip}')
            output = conn.send_command('show run | i hostname')
            print(output)
[root@localhost Netmiko]#
```

代码分段讲解如下。

- 我们用 context manager（with 语句）打开保存 5 台交换机管理 IP 地址的 ip_list.txt 文件，然后用 for 循环配合 readlines()遍历里面的每一个 IP 地址，因为 readlines()返回的列表里的每个元素后面都会接一个换行符\n，所以我们用 strip()函数将其拿掉然后赋值给变量 ip，这个变量 ip 则作为字典 connection_info 里 “ip” 这个键的值放入字典，因为所有的交换机用的 SSH 端口号都是默认的 22，所以没有必要在字典里明文给出来。

```
from netmiko import ConnectHandler
```

```
with open('ip_list.txt') as f:
    for ips in f.readlines():
        ip = ips.strip()
        connection_info = {
                'device_type': 'cisco_ios',
                'ip': ip,
                'username': 'python',
                'password': '123',
        }
```

- 然后通过 ConnetHandler(**connection_info)依次登录每台交换机，这里除了打印“已经成功登录交换机 xxx.xxx.xxx.xxx”，还额外向每台交换机发送了一个“show run | i hostname”，并将回显内容（即交换机各自的 hostname）打印出来，目的是验证我们确实通过 Netmiko 登录了每台交换机。

```
        with ConnectHandler(**connection_info) as conn:
            print (f'已经成功登录交换机{ip}')
            output = conn.send_command('show run | i hostname')
            print(output)
```

运行脚本看效果，如下图所示。

[root@localhost Netmiko]# python3 netmiko4_1.py
已经成功登录交换机192.168.2.11
hostname S1
已经成功登录交换机192.168.2.12
hostname S2
已经成功登录交换机192.168.2.13
hostname S3
已经成功登录交换机192.168.2.14
hostname S4
已经成功登录交换机192.168.2.15
hostname S5
[root@localhost Netmiko]#

2. 生产网络里设备的 username、password、port 等参数不尽相同

这种情况就不能再用 ip_list.txt 的方法来做了，必须给每一个参数不同的交换机创建一个字典，因为每个字典一般都要占据 5、6 排代码的空间，设备一多，脚本代码量会非常恐怖。我们可以将这些字典写入一个额外的文件类型为 json 的文件，将其取名为 switches.json，在 Python 脚本里可以导入 Python 内置的 JSON 模块，利用 json.load()函数来加载 switches.json 文件里的内容，因为 json.load()会返回一个列表类型的 JSON 数据，我们可以使用 for 循环来遍历该列表里的 JSON 数据，达到依次登录多台设备的目的。具体方法如下。

首先创建 switches.json 文件，将包含交换机 S1 和 S2 的数据以字典形式写入，如下图所示。

```
[
  {
    "name": "SW1",
    "connection": {
      "device_type": "cisco_ios",
      "host": "192.168.2.11",
      "username": "python",
      "password": "123"
    }
  },
  {
    "name": "SW2",
    "connection": {
      "device_type": "cisco_ios",
      "host": "192.168.2.12",
      "username": "python",
      "password": "123"
    }
  }
]
```

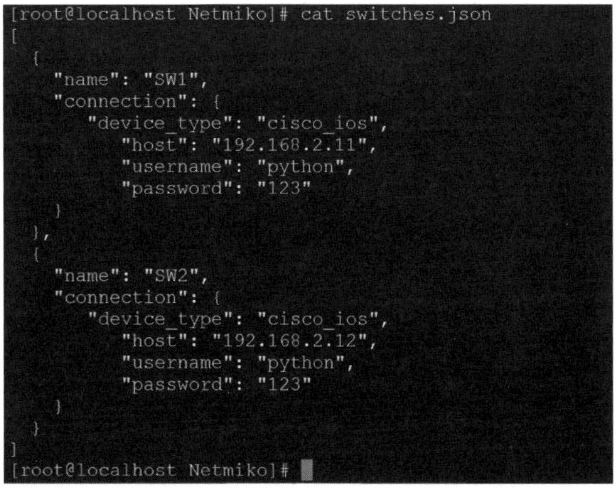

```
[root@localhost Netmiko]# cat switches.json
[
  {
    "name": "SW1",
    "connection": {
      "device_type": "cisco_ios",
        "host": "192.168.2.11",
        "username": "python",
        "password": "123"
    }
  },
  {
    "name": "SW2",
    "connection": {
      "device_type": "cisco_ios",
        "host": "192.168.2.12",
        "username": "python",
        "password": "123"
    }
  }
]
[root@localhost Netmiko]#
```

然后创建实验 4 的第二个脚本 netmiko4_2.py，如下图所示。

```
import json
from netmiko import ConnectHandler
```

```
with open("switches.json") as f:
    devices = json.load(f)
for device in devices:
    with ConnectHandler(**device['connection']) as conn:
        hostname = device['name']
        print (f'已经成功登录交换机{hostname}')
        output = conn.send_command('show run | i hostname')
        print(output)
```

```
[root@localhost Netmiko]# cat netmiko4_2.py
import json
from netmiko import ConnectHandler

with open("switches.json") as f:
    devices = json.load(f)
for device in devices:
    with ConnectHandler(**device['connection']) as conn:
        hostname = device['name']
        print (f'已经成功登录交换机{hostname}')
        output = conn.send_command('show run | i hostname')
        print(output)
[root@localhost Netmiko]#
```

代码分段讲解如下。

- 导入 Python 内置的 JSON 模块，然后调用它的 load()函数将我们之前创建的 switches.json 里面的 SW1 和 SW2 的登录信息导出并赋值给变量 devices，注意 json.load() 返回值的数据类型为列表。

```
import json
from netmiko import ConnectHandler

with open("switches.json") as f:
    devices = json.load(f)
```

- 随后用 for 循环遍历 devices，并对每台交换机执行命令 show run | i hostname，打印出它们的 hostname。

```
for device in devices:
    with ConnectHandler(**device['connection']) as conn:
        hostname = device['name']
        print (f'已经成功登录交换机{hostname}')
        output = conn.send_command('show run | i hostname')
```

```
    print(output)
```

运行脚本看效果，如下图所示。

```
[root@localhost Netmiko]# python3 netmiko4_2.py
已经成功登录交换机SW1

hostname S1
已经成功登录交换机SW2

hostname S2
```

8.8　实验 4　通过 Netmiko 连接多台交换机（华为设备）

第一种情况，这种场景属于理想情况，即每台设备的登录账号和密码是一致的。我们简单适配一下即可，直接运行，如下图所示。

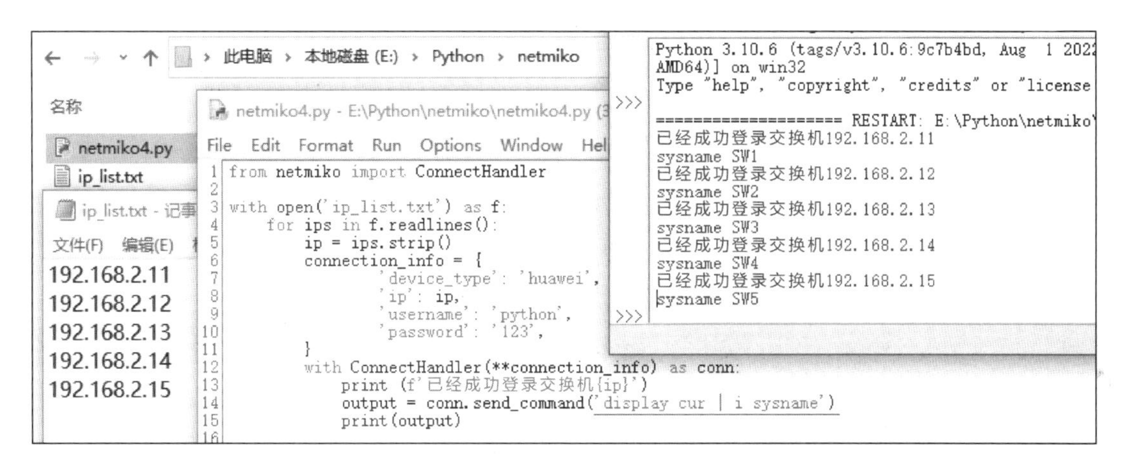

在实际网络生产中，往往出于安全考虑或者历史原因，导致全部或部分设备的用户名和密码并不一致。此时我们可以通过 json 的方式来传递所需参数，打通 SSH 链接，如下图所示。

```
import json
from netmiko import ConnectHandler

with open("switches.json") as f:
    devices = json.load(f)

for device in devices:
    with ConnectHandler(**device['connection']) as conn:
        hostname = device['name']
        print (f'已经成功登录交换机{hostname}')
```

```
    output = conn.send_command('display cur | i sysname')
    print(output)
```

有了 json 这个轻量级文本实现信息传输，Python 脚本就可以与 csv、excel、数据库等信息存储介质进行交互对接了。在日常运维中，经过多次调优打磨，我们最终把 Python 脚本固定下来，而后通过维护 excel 表、数据库表等手段来管理网络设备的 IP 地址、端口、账号密码等信息。这是把资源数据和脚本代码做分离的一种思想。

8.9　实验 5　Netmiko 配合 Jinja2 配置模板为设备做配置（思科设备）

Jinja2 是 Python 的内置模块，无须通过 pip 下载安装。关于 Jinja2 的使用方法在 6.7 节中介绍过。

这里我们将使用 Netmiko 配合 Jinja2 为 SW1 的 loopback 1 端口配置一个 inbound 方向的 ACL 1，该 ACL 1 会分别允许和拒绝一些 IP 地址。首先创建一个文件夹，取名 templates，进入该文件夹，创建我们的 Jinja2 模板文件，将其取名为 acl.conf.tpl，然后放入下面的 Jinja2 语句，如下图所示。

```
interface {{ interface }}
ip access-group 1 in
{% for host in disallow_ip %}
access-list 1 deny host {{ host }}
{% endfor %}
{% for host in allow_ip %}
```

```
access-list 1 permit host {{ host }}
{% endfor %}
```

```
[root@localhost templates]# cat acl.conf.tpl
interface {{ interface }}
ip access-group 1 in
{% for host in disallow_ip %}
access-list 1 deny host {{ host }}
{% endfor %}
{% for host in allow_ip %}
access-list 1 permit host {{ host }}
{% endfor %}
[root@localhost templates]#
```

然后从 templates 文件里退出来，返回之前所在的文件夹，创建 netmiko5.py，并放入下列代码。

```
from netmiko import ConnectHandler
from jinja2 import Environment, FileSystemLoader

sw1 = {
        'device_type': 'cisco_ios',
        'ip': '192.168.2.11',
        'username': 'python',
        'password': '123'
}

loader = FileSystemLoader('templates')
environment = Environment(loader=loader)
tpl = environment.get_template('acl.conf.tpl')

allow_ip = ['10.1.1.1', '10.1.1.2']
disallow_ip = ['10.1.1.3', '10.1.1.4']
out = tpl.render(allow_ip=allow_ip, disallow_ip=disallow_ip, interface='loopback 1')

with open("configuration.conf", "w") as f:
      f.write(out)

with ConnectHandler(**sw1) as conn:
        print ("已经成功登录交换机" + sw1['ip'])
        output = conn.send_config_from_file("configuration.conf")
        print (output)
```

代码分段讲解如下。

- 我们导入 Jinja2 的 Enviroment 和 FileSystemLoader 两个子类，用来读取 templates 文件夹下的 Jinja2 模板文件 acl.conf.tpl，然后赋值给 tpl 这个函数。

```
from netmiko import ConnectHandler
from jinja2 import Environment, FileSystemLoader

sw1 = {
        'device_type': 'cisco_ios',
        'ip': '192.168.2.11',
        'username': 'python',
        'password': '123'
}

loader = FileSystemLoader('templates')
environment = Environment(loader=loader)
tpl = environment.get_template('acl.conf.tpl')
```

- 然后我们将允许和拒绝的 IP 地址写入两个列表，分别取名 allow_ip 和 disallow_ip，然后通过 render()函数对模板文件做渲染（注意这里 render()里的三个参数 allow_ip、disallow_ip 和 interface 就是模板文件里的变量）,将渲染后的模板文件赋值给变量 out。

```
allow_ip = ['10.1.1.1', '10.1.1.2']
disallow_ip = ['10.1.1.3', '10.1.1.4']
out = tpl.render(allow_ip=allow_ip, disallow_ip=disallow_ip, interface='loopback 1')
```

- 有了渲染后的模板文件，我们需要将它写入一个扩展名为 conf 的文件。我们创建一个 configuration.conf 文件，将模板文件内容全部写入其中。

```
with open("configuration.conf", "w") as f:
     f.write(out)
```

- 最后通过 ConnectHandler()登录 SW1，然后使用 send_config_from_file()读取 configuration.conf 文件里的内容对 SW1 做配置。

```
with ConnectHandler(**sw1) as conn:
      out = conn.send_config_from_file("configuration.conf")
```

在运行脚本前，首先确认 SW1 里没有 ACL 1，并且 loopback 1 下也没有放入任何 ACL，如下图所示。

```
S1#show ip access-lists
Extended IP access list preauth_ipv4_acl (per-user)
    10 permit udp any any eq domain
    20 permit tcp any any eq domain
    30 permit udp any eq bootps any
    40 permit udp any any eq bootpc
    50 permit udp any eq bootpc any
    60 deny ip any any
S1#show run int loopback 1
Building configuration...

Current configuration : 67 bytes
!
interface Loopback1
 ip address 11.11.11.11 255.255.255.255
end

S1#
```

```
[root@localhost Netmiko]# python3 netmiko5.py
已经成功登录交换机192.168.2.11
configure terminal
Enter configuration commands, one per line.  End with CNTL/Z.
S1(config)#interface loopback 1
S1(config-if)#ip access-group 1 in
S1(config-if)#
S1(config-if)#access-list 1 deny host 10.1.1.3
S1(config)#
S1(config)#access-list 1 deny host 10.1.1.4
S1(config)#
S1(config)#
S1(config)#access-list 1 permit host 10.1.1.1
S1(config)#
S1(config)#access-list 1 permit host 10.1.1.2
S1(config)#end
S1#
[root@localhost Netmiko]#
```

脚本运行完毕，再次返回 SW1 验证，如下图所示。

```
S1#show ip access-lists
Standard IP access list 1
    40 permit 10.1.1.2
    10 deny   10.1.1.3
    30 permit 10.1.1.1
    20 deny   10.1.1.4
Extended IP access list preauth_ipv4_acl (per-user)
    10 permit udp any any eq domain
    20 permit tcp any any eq domain
    30 permit udp any eq bootps any
    40 permit udp any any eq bootpc
    50 permit udp any eq bootpc any
    60 deny ip any any
S1#show run int loop1
Building configuration...

Current configuration : 89 bytes
!
interface Loopback1
 ip address 11.11.11.11 255.255.255.255
 ip access-group 1 in
end
```

8.10　实验 5　Netmiko 配合 Jinja2 配置模板为设备做配置（华为设备）

在现网设备配置文件中，有些配置模块的行间配置非常相似，比如 ACL、IP 地址前缀列表、用户列表配置等。如果我们在核查配置或者制作执行脚本时，能够配合 Jinja2 模板，则会有事半功倍的效果。相对于思科设备的实验，这次华为实验做一个调整，修改实际网络中可能更常用的 vty 数据。

实验背景及需求：

（1）利用 Jinja2 模板，制作 SW1 的 vty 登录限制脚本，用 ACL 方式。

（2）用 Netmiko 登录 SW1，执行（1）脚本并回显。

（3）实验目的及效果预期：PC、SW2 可登录 SW1；SW3、SW4 无法登录 SW1；SW5 无相关配置，留作观察比对。

Jinja2 模板 acl.conf.tpl 制作代码如下。

```
user-interface {{ interface }}

acl 2000 inbound

acl 2000

{% for host in disallow_ip %}
rule deny source {{ host }} 0
{% endfor %}

{% for host in allow_ip %}
rule permit source {{ host }} 0
{% endfor %}
```

Python 脚本 netmiko4.py 代码如下。

```
import netmiko
from netmiko import ConnectHandler
from jinja2 import Environment, FileSystemLoader
```

```
allow_ip = ['192.168.2.2', '192.168.2.12']
disallow_ip = ['192.168.2.13', '192.168.2.14']

sw1 = {'device_type':'huawei',
     'ip':'192.168.2.11',
     'username':'python',
     'password':'123'}

loader = FileSystemLoader('templates')
environment = Environment(loader=loader)
tpl = environment.get_template('acl.conf.tpl')
out = tpl.render(allow_ip=allow_ip, disallow_ip=disallow_ip, interface='vty 0 4')

with open("configuration.conf", "w") as f:
      f.write(out)

with ConnectHandler(**sw1) as conn:
       print ("已经成功登录交换机" + sw1['ip'])
       output = conn.send_config_from_file("configuration.conf")
       print (output)
```

华为设备实验虽然实验背景和需求做了一点调整，脚本逻辑和思科实验如出一辙。这里留意一下实验的目录结构，如下图所示。

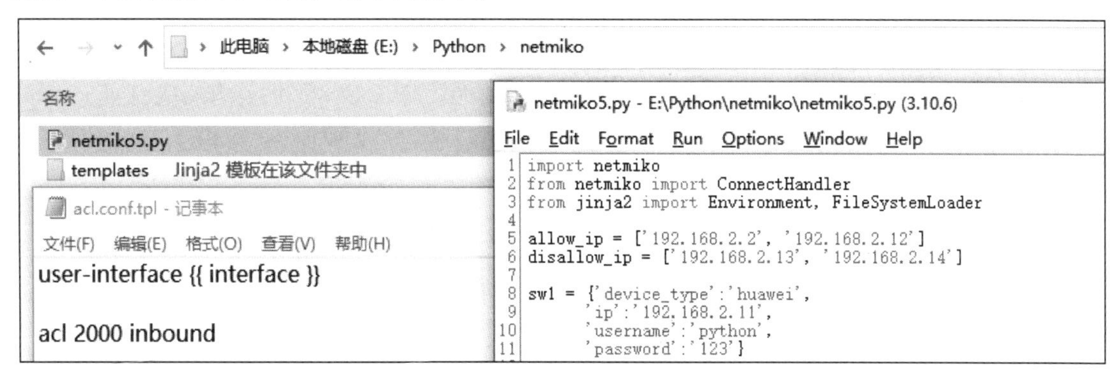

运行一下脚本！此时，实验目录还会生成一个文件 configuration.conf，也就是配置脚本，如下图所示。

```
==================== RESTART: E:\Python\netmiko\netmiko5.py ====================
已经成功登录交换机192.168.2.11
system-view
Enter system view, return user view with Ctrl+Z.
[SW1]user-interface vty 0 4
[SW1-ui-vty0-4]
[SW1-ui-vty0-4]acl 2000 inbound
[SW1-ui-vty0-4]
[SW1-ui-vty0-4]acl 2000
[SW1-acl-basic-2000]
[SW1-acl-basic-2000]
[SW1-acl-basic-2000]rule deny source 192.168.2.13 0
[SW1-acl-basic-2000]
[SW1-acl-basic-2000]rule deny source 192.168.2.14 0
[SW1-acl-basic-2000]
[SW1-acl-basic-2000]
[SW1-acl-basic-2000]
[SW1-acl-basic-2000]rule permit source 192.168.2.2 0
[SW1-acl-basic-2000]
[SW1-acl-basic-2000]rule permit source 192.168.2.12 0
[SW1-acl-basic-2000]return
<SW1>
```

```
SW1
[SW1]user-interface vty 0 4
[SW1-ui-vty0-4]dis this
#
user-interface con 0
user-interface vty 0 4
 acl 2000 inbound
 authentication-mode aaa
 protocol inbound all
#
return
[SW1-ui-vty0-4]acl 2000
[SW1-acl-basic-2000]dis this
#
acl number 2000
 rule 5 deny source 192.168.2.13 0
 rule 10 deny source 192.168.2.14 0
 rule 15 permit source 192.168.2.2 0
 rule 20 permit source 192.168.2.12 0
#
return
[SW1-acl-basic-2000]
```

从其他 SW（如 SW2）进行 SSH 登录到 SW1 的验证过程如下图所示，采用命令 stelnet。

```
[SW2]stelnet 192.168.2.11
Please input the username:python
Trying 192.168.2.11 ...
Press CTRL+K to abort
Connected to 192.168.2.11 ...
Error: Failed to verify the server's public key.
Please run the command "ssh client first-time enable"to enable the first-time ac
cess function and try again.
[SW2]ssh client first-time enable
[SW2]stelnet 192.168.2.11
Please input the username:python
Trying 192.168.2.11 ...
Press CTRL+K to abort
Connected to 192.168.2.11 ...
The server is not authenticated. Continue to access it? [Y/N] :y
Save the server's public key? [Y/N] :y
The server's public key will be saved with the name 192.168.2.11. Please wait...


Enter password:
Info: The max number of VTY users is 5, and the number
      of current VTY users on line is 1.
      The current login time is 2022-08-29 11:54:48.
<SW1>
```

实验结果 SW2 可正常登录 SW1，而 SW3 和 SW4 无法登录 SW1，均符合实验预期。另外，作为参照对比的 SW5（SW1 上没有其相关配合）同样登录不了 SW1，说明华为设备这种基本 ACL 是默认 deny 的。

8.11　实验 6　在 Netmiko 中使用 enable 密码进入设备特权模式（思科设备）

在之前的章节中，我们都是将 SSH 用户名的特权级别设为 15，跳过了输入 enable 密码这一步。鉴于部分读者的生产网络中仍然要求使用 enable 密码来进入设备的特权模式，本节将介绍如何在 Netmiko 中使用 enable 密码。

开始实验之前，首先将 SW1 上的用户 python 的特权级别从之前的 15 改为 1，如下图所示。

```
S1(config)#username python privilege 1 password 123
S1(config)#end
S1#
*May  2 07:07:29.106: %SYS-5-CONFIG_I: Configured from console by console
S1#
```

然后使用用户名 python 手动 SSH 登录 SW1，可以发现输入 SSH 密码后，来到了用户模式（>提示符），必须再次输入 enable 密码才能进入特权模式，如下图所示。

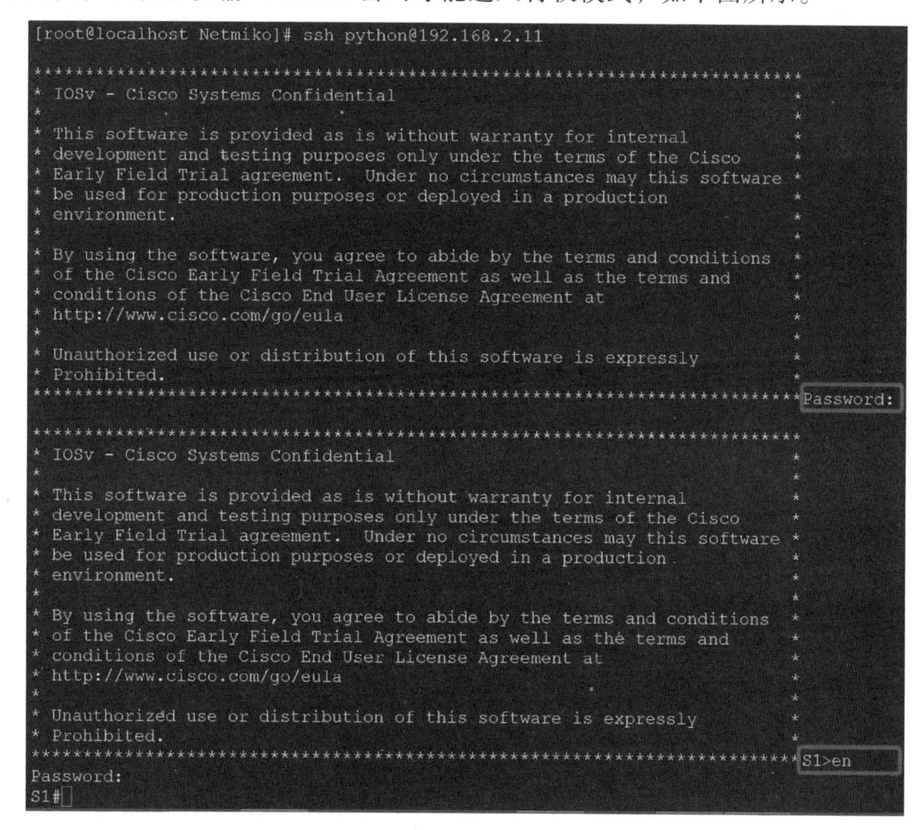

验证完毕后，开始创建实验 6 的脚本 netmiko6.py，写入下列代码。

```
from netmiko import ConnectHandler
sw1 = {
        'device_type': 'cisco_ios',
        'ip': '192.168.2.11',
        'username': 'python',
        'password': '123',
```

```
        'secret': '123'
}

with ConnectHandler(**sw1) as connect:
    connect.enable()
    print ("已经成功登录交换机" + sw1['ip'])
    output = connect.send_command('show run')
    print (output)
```

```
[root@localhost Netmiko]# cat netmiko6.py
from netmiko import ConnectHandler
sw1 = {
        'device_type': 'cisco_ios',
        'ip': '192.168.2.11',
        'username': 'python',
        'password': '123',
        'secret': '123'
}

with ConnectHandler(**sw1) as connect:
    connect.enable()
    print ("已经成功登录交换机" + sw1['ip'])
    output = connect.send_command('show run')
    print (output)
[root@localhost Netmiko]#
```

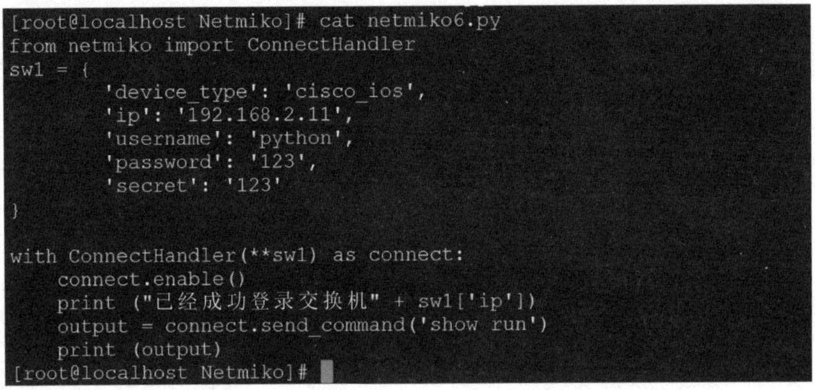

代码分段讲解如下。

- 要在 Netmiko 中使用 enable 密码，需要在设备登录信息的字典里加入 secret 键，该键对应的值就是设备的 enable 密码，这里笔者用的 enable 密码是 123。

```
from netmiko import ConnectHandler
sw1 = {
        'device_type': 'cisco_ios',
        'ip': '192.168.2.11',
        'username': 'python',
        'password': '123',
        'secret': '123'
}
```

- 之后在使用 ConnectHandler()登录交换机时，这里要多一个步骤：调用 enable()方法，这样 Netmiko 才能配合之前的 secret 键，将它的值作为 enable 的密码输入，登录设备的特权模式。注意这里我们特意用 send_command()输入了 show run 这个只能在特权模式下使用而用户模式下不支持的命令来验证 Netmiko 确实帮我们输入了 enable 密码、进入了特权模式。

```
with ConnectHandler(**sw1) as connect:
    connect.enable()
    print ("已经成功登录交换机" + sw1['ip'])
    output = connect.send_command('show run')
    print (output)
```

运行脚本看效果，如下图所示。

```
[root@localhost Netmiko]# python3 netmiko6.py
已经成功登录交换机192.168.2.11

Building configuration...

Current configuration : 8113 bytes
!
! Last configuration change at 08:26:34 UTC Sun May 2 2021
!
version 15.0
service timestamps debug datetime msec
service timestamps log datetime msec
no service password-encryption
service compress-config
!
hostname S1
!
boot-start-marker
boot-end-marker
!
!
enable password 123
!
username python privilege 15 password 0 123
no aaa new-model
!
!
!
!
!
vtp domain CISCO-vIOS
vtp mode transparent
!
!
!
no ip domain-lookup
ip domain-name gns3.com
ip name-server 8.8.6.6
ip name-server 8.8.8.8
ip cef
no ipv6 cef
!
crypto pki trustpoint TP-self-signed-0
 enrollment selfsigned
 subject-name cn=IOS-Self-Signed-Certificate-0
 revocation-check none
 rsakeypair TP-self-signed-0
```

8.12　实验 6　在 Netmiko 中使用 enable 密码进入设备特权模式（华为设备）

对于华为交换机提权操作，Netmiko 实际不支持直接使用 enable()方法。源码中有这些方法，但实际没有效果。目前随着版本升级，华为的提权操作也不是主流了。但是，作为 Paramiko 到 Netmiko 的过渡，这是一个典型例子，可以探讨下怎么使用 Netmiko 操作一些带交互的指令操作场景，也可以看看怎么把 Netmiko 当成 Paramiko 来用。Netmiko 是基于 Paramiko 二次开发的，从稳扎稳打的角度出发，建议学习 Netmiko 之前稍微了解一些 Paramiko 的基础操作，再逐步过渡。

我们先调整一下 SW1 的权限，设置完成后可手工测试一下看是否符合预期，如下图所示。

```
system-view
aaa
local-user python privilege level 1

quit
super password level 3 cipher 123
```

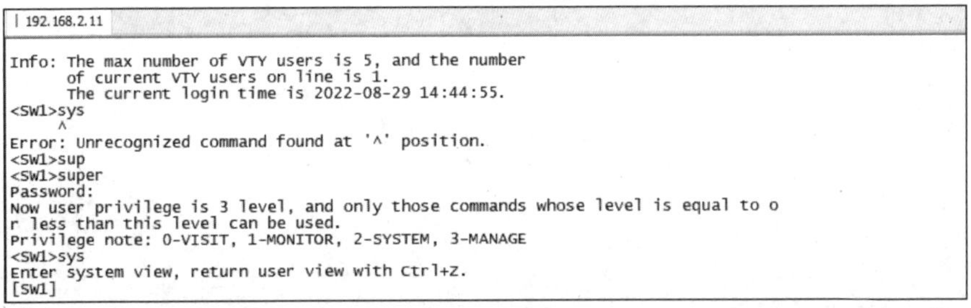

```
| 192.168.2.11

Info: The max number of VTY users is 5, and the number
      of current VTY users on line is 1.
      The current login time is 2022-08-29 14:44:55.
<SW1>sys
      ^
Error: Unrecognized command found at '^' position.
<SW1>sup
<SW1>super
Password:
Now user privilege is 3 level, and only those commands whose level is equal to o
r less than this level can be used.
Privilege note: 0-VISIT, 1-MONITOR, 2-SYSTEM, 3-MANAGE
<SW1>sys
Enter system view, return user view with Ctrl+Z.
[SW1]
```

```
import netmiko
from netmiko import ConnectHandler

sw1 = {'device_type': 'huawei',
      'ip': '192.168.2.11',
      'username': 'python',
      'password': '123',
      'secret': r'123', #提权密码放在这，也在 command_string 写入
      'session_log': 'netmiko_log.log'}

with ConnectHandler(**sw1) as connect:
```

```
    print("已经成功登录交换机" + sw1['ip'])
    output = connect.send_command(command_string='su',expect_string = 'Password:')
    output += connect.send_command(command_string=sw1['secret'],expect_string =
'>',cmd_verify = False)
    output += connect.send_command(command_string='sys',expect_string = ']')
    #这里可根据实际需要做一些配置操作
    output += connect.send_command(command_string='quit',expect_string = '>')
    print(output)
```

这份脚本的逻辑即发送指令，等待特定回显，再发送新指令，等待新特定回显，如下图所示。这么一个过程是不是像极了 Paramiko 的相关操作，只是在 Paramiko 实现中常使用 time 模块的 sleep 函数进行休眠等待，而 Netmiko 的这个操作可以精准地设定到底等什么回显，而且这个字符串还可以是正则表达式形式的。

- command_string：发送特定指令。
- expect_string：等待特定回显（支持正则表达式）。
- cmd_verify：屏幕校验，即 command_string 送指令后，默认情况下 Netmiko 会检测这个命令是不是真的被“敲”到“屏幕”上去了。因此，在密码输入这种场景下，必须把 cmd_verify 置成 Flase。

```
Python 3.10.6 (tags/v3.10.6:9c7b4bd, Aug  1 2022, 21:53:49) [MSC v.1932 64 bit (AMD64)] on win32
Type "help", "copyright", "credits" or "license()" for more information.
>>>
==================== RESTART: E:\Python\netmiko\netmiko6.py ====================
已经成功登录交换机192.168.2.11
Password:
Now user privilege is 3 level, and only those commands whose level is equal to or less than this level can be
used.
Privilege note: 0-VISIT, 1-MONITOR, 2-SYSTEM, 3-MANAGEEnter system view, return user view with Ctrl+Z.
>>>
```

践行 Python 网络自动化犹如航海行船，在 Netmiko 的应用中，send_command 可以说是一个重点锚点，值得在此“抛锚作业”。学习到这里，我们需要学会通过 help、dir 等函数查阅，或者检索 Netmiko 的官方 API 手册，从而获取解决问题的资料。下图为 Netmiko 的官方 API 网站的截图。

```
def send_command(self, command_string, expect_string=None, delay_factor=1.0, max_loops=500,
                 auto_find_prompt=True, strip_prompt=True, strip_command=True, normalize=True,
                 use_textfsm=False, textfsm_template=None, use_ttp=False, ttp_template=None,
                 use_genie=False, cmd_verify=True)
```

下面有个参数的解释。

Execute command_string on the SSH channel using a pattern-based mechanism. Generally used for show commands. By default this method will keep waiting to receive data until the network device prompt is detected. The current network device prompt will be determined automatically.

:param command_string: The command to be executed on the remote device. :type command_string: str

完成了前面几个实验，你是否发现有些 Netmiko 函数在回显中会带有<SW1>、[SW1]之类的提示符，有些则没有，这是为什么呢？可以通过手册说明解释一下吗？我们要如何通过调整参数控制这些内容呢？代码中还在变量 sw1 字典中加入了如下信息。

```
'session_log': ' netmiko_log.log'
```

这样设置又能起到什么作用呢？请读者自行尝试！践行 Python 网络自动化的过程，基本是一个反复“折腾”的过程，注重动手实践，结合文档手册反复测试。

8.13　实验 7　使用 Netmiko 向设备传送文件（思科设备）

开始实验 7 之前不要忘记先把用户 python 的特权级别从 1 改回 15。

平常的网络运维工作中少不了要向设备传送文件，比如新的 OS 的镜像文件。实验 7 将演示如何使用 Netmiko 配合 scp 协议向设备（SW1）传送文件。

首先在 SW1 上开启 scp，如下图所示。

```
S1(config)#ip scp server enable
S1(config)#end
S1#
*May  2 09:07:02.677: %SYS-5-CONFIG_I: Configured from console by console
S1#show run | i scp
ip scp server enable
S1#
```

然后创建一个名为 test.txt 的文本文件，在其中写入 test 这个单词，如下图所示。

```
[root@localhost Netmiko]# vi test.txt
[root@localhost Netmiko]# cat test.txt
test
[root@localhost Netmiko]#
```

接着创建脚本 netmiko7.py，写入下列代码。

```
from netmiko import ConnectHandler, file_transfer

sw1 = {
      'device_type': 'cisco_ios',
      'ip': '192.168.2.11',
      'username': 'python',
      'password': '123'
}
```

```
with ConnectHandler(**sw1) as connect:
    print ("已经成功登录交换机" + sw1['ip'])
    output = file_transfer(connect,
               source_file="test.txt",
               dest_file="test.txt",
               file_system="flash:",
               direction="put")
    print (output)
```

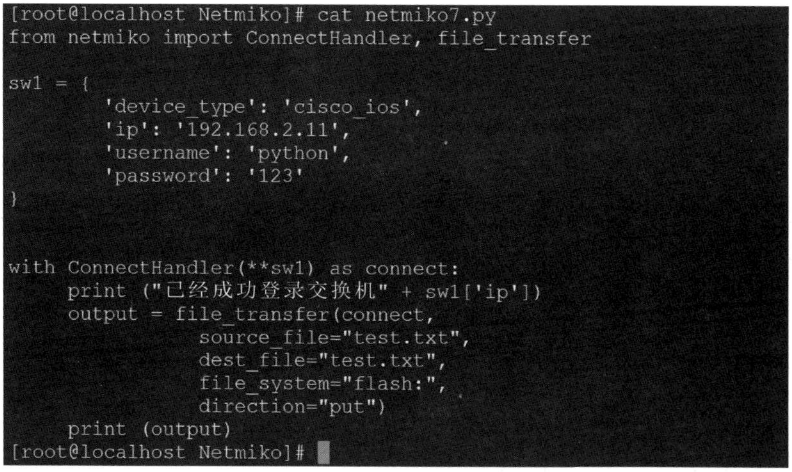

```
[root@localhost Netmiko]# cat netmiko7.py
from netmiko import ConnectHandler, file_transfer

sw1 = {
        'device_type': 'cisco_ios',
        'ip': '192.168.2.11',
        'username': 'python',
        'password': '123'
}


with ConnectHandler(**sw1) as connect:
    print ("已经成功登录交换机" + sw1['ip'])
    output = file_transfer(connect,
               source_file="test.txt",
               dest_file="test.txt",
               file_system="flash:",
               direction="put")
    print (output)
[root@localhost Netmiko]#
```

代码分段讲解如下。

- 这里除 ConnectHandler 外，我们额外从 Netmiko 引入 file_transfer 这个子类，用来向设备传送文件。

```
from netmiko import ConnectHandler, file_transfer

sw1 = {
        'device_type': 'cisco_ios',
        'ip': '192.168.2.11',
        'username': 'python',
        'password': '123'
}
```

- 通过 ConnectHandler()登录 SW1 后，调用 file_transfer()，里面的各个参数不难理解。因为笔者用的 SW1 是思科的 IOS 设备，所以 file_system 参数后面放的是 flash:；如果是思科 NX-OS，则需要放 bootflash:；如果是 Arista 设备，则需要放/mnt/flash；如果是 Junos 设备，则要放/var/tmp。读者可以根据自己的情况来改变。最后的 direction 参数，根据 Netmiko 官网的 API 资料只能放 put 这一个值。

```
with ConnectHandler(**sw1) as connect:
    print ("已经成功登录交换机" + sw1['ip'])
    output = file_transfer(connect,
              source_file="test.txt",
              dest_file="test.txt",
              file_system="flash:",
              direction="put")
    print (output)
```

运行脚本前，首先查看 SW1 的文件系统 flash0，确认里面没有 test.txt 文件，如下图所示。

```
S1#dir
Directory of flash0:/

    1  drw-           0  Jan 30 2013 00:00:00 +00:00  boot
  264  drw-           0  Oct 14 2013 00:00:00 +00:00  config
  267  -rw-   103134052   Jun 5 2014 00:00:00 +00:00  vios_l2-adventerprisek9-m
  268  -rw-      524288  Oct 11 2014 10:27:10 +00:00  nvram
  269  -rw-         439   May 2 2021 08:32:50 +00:00  e1000_bia.txt
  270  -rw-         924  Jan 20 2021 06:58:30 +00:00  vlan.dat
  271  -rw-       35380  Oct 17 2020 13:34:48 +00:00  crashinfo_20201017-133448-UTC
  272  -rw-           0  Jan 24 2021 18:08:44 +00:00  merge_config.txt
  273  -rw-        7931  Nov 15 2020 07:24:02 +00:00  rollback_config.txt
  274  -rw-       36183  Nov 26 2020 09:55:44 +00:00  crashinfo_20201126-095545-UTC
  275  -rw-        8095  Jan 24 2021 18:08:44 +00:00  candidate_config.txt
  276  -rw-           5   May 2 2021 09:14:32 +00:00  test.txt

2142715904 bytes total (2034348032 bytes free)
S1#
```

运行脚本，如下图所示。

```
[root@localhost Netmiko]# python3 netmiko7.py
已经成功登录交换机192.168.2.11
{'file_exists': True, 'file_transferred': True, 'file_verified': True}
[root@localhost Netmiko]#
```

注意，这里 print(output)返回的值为一个字典，如果文件传输失败，则该字典 file_transferred 键的值将为 False。

最后回到 SW1 上验证，可以看到 test.txt 文件已经成功上传到 SW1 的 flash0:，如下图所示。

```
S1#dir
Directory of flash0:/

    1  drw-           0  Jan 30 2013 00:00:00 +00:00  boot
  264  drw-           0  Oct 14 2013 00:00:00 +00:00  config
  267  -rw-   103134052   Jun 5 2014 00:00:00 +00:00  vios_l2-adventerprisek9-m
  268  -rw-      524288  Oct 11 2014 10:27:10 +00:00  nvram
  269  -rw-         439   May 2 2021 08:32:50 +00:00  e1000_bia.txt
  270  -rw-         924  Jan 20 2021 06:58:30 +00:00  vlan.dat
  271  -rw-       35380  Oct 17 2020 13:34:48 +00:00  crashinfo_20201017-133448-UTC
  272  -rw-           0  Jan 24 2021 18:08:44 +00:00  merge_config.txt
  273  -rw-        7931  Nov 15 2020 07:24:02 +00:00  rollback_config.txt
  274  -rw-       36183  Nov 26 2020 09:55:44 +00:00  crashinfo_20201126-095545-UTC
  275  -rw-        8095  Jan 24 2021 18:08:44 +00:00  candidate_config.txt
  276  -rw-           5   May 2 2021 09:29:26 +00:00  test.txt

2142715904 bytes total (2034348032 bytes free)
S1#
```

8.14　实验 7　使用 Netmiko 向设备传送文件（华为设备）

按照思科设备的实验脚本，简单修改一下试图进行迁移适配。结果很遗憾，从报错信息来看，Netmiko 并不支持华为设备的这个操作（后续新版本可能会支持），如下图所示。

此时通过报错信息中的文件路径（路径因 Windows 系统和具体账号可能有差别）和行号，可以定位到相应的源文件中去，如下图所示。

```
 File
"C:\Users\zhuji\AppData\Roaming\Python\Python310\site-packages\netmiko\ssh_dispatch
er.py", line 473, in FileTransfer
```

```
  raise ValueError(
ValueError: Unsupported SCP device_type: currently supported platforms are:
```

```
466 def FileTransfer(*args: Any, **kwargs: Any) -> "BaseFileTransfer":
467     """Factory function selects the proper SCP class and creates object based on device_type."""
468     if len(args) >= 1:
469         device_type = args[0].device_type
470     else:
471         device_type = kwargs["ssh_conn"].device_type
472     if device_type not in scp_platforms:
473         raise ValueError(
474             "Unsupported SCP device_type: "
475             "currently supported platforms are: {}".format(scp_platforms_str)
476         )
```

如果你的 Python 基础扎实，且对 Netmiko 模块已经很熟悉，可以重写这些类和方法，实现目标，自我"参与"到 Netmiko 的开发中。当然，我们并不建议直接修改源码。那么，要如何把文件从 PC 送到设备上去呢？下面提供一种做法，我们先在 SW1 上开启 FTP 服务。

```
sys
ftp server enable
aaa
 local-user python ftp-directory flash:
 local-user python service-type telnet ssh ftp
```

开启 FTP 后，可以适当做手工测试，并且把 PC 上的 test.txt 文件成功上传到 SW1 上，如下图所示。

```
C:\Users\zhuji>ftp 192.168.2.11
连接到 192.168.2.11。
220 FTP service ready.
530 Please login with USER and PASS.
用户(192.168.2.11:(none)): python
331 Password required for python.
密码:
230 User logged in.
ftp> lcd E:\Python\netmiko
目前的本地目录 E:\Python\netmiko。
ftp> bin
200 Type set to I.
ftp> put test.txt
200 Port command okay.
150 Opening BINARY mode data connection for test.txt.
226 Transfer complete.
ftp>
```

```
Idx  Attr     Size(Byte)  Date        Time       FileName
  0  drw-              -  Aug 06 2015 21:26:42   src
  1  drw-              -  Mar 17 2021 15:06:12   compatible
  2  drw-              -  Aug 26 2022 00:00:03   resetinfo
  3  -rw-            771  Aug 26 2022 23:58:53   vrpcfg.zip
  4  -rw-              0  Aug 29 2022 16:32:50   test.txt

32,004 KB total (31,936 KB free)

<SW1>
```

此时，稍微调整一下需求，用 Netmiko 登录 SW2，而后从 SW1 复制 test.txt 文件到 SW2 上。使用 send_command 配合 command_string、expect_string 等参数可以完成，这里我们尝试另一个 Netmiko 的常用方法——send_command_timing。

```
from netmiko import ConnectHandler

sw2 = {'device_type':'huawei',
```

```
    'ip':'192.168.2.12',
    'username':'python',
    'password':'123'}

with ConnectHandler(**sw2) as connect:
    print ("已经成功登录交换机" + sw2['ip'])

    output = connect.send_command_timing(command_string="dir",)
    output += connect.send_command_timing(command_string="ftp 192.168.2.11")
    output += connect.send_command_timing(command_string="python")
    output += connect.send_command_timing(command_string="123")
    output += connect.send_command_timing(command_string="bin")
    output += connect.send_command_timing(command_string="get test.txt")
    output += connect.send_command_timing(command_string="quit")
    output += connect.send_command_timing(command_string="dir")
    print(output)
```

运行结果，SW2 已经通过 FTP 成功登录到 SW1 并复制了 test.txt 文件，如下图所示。

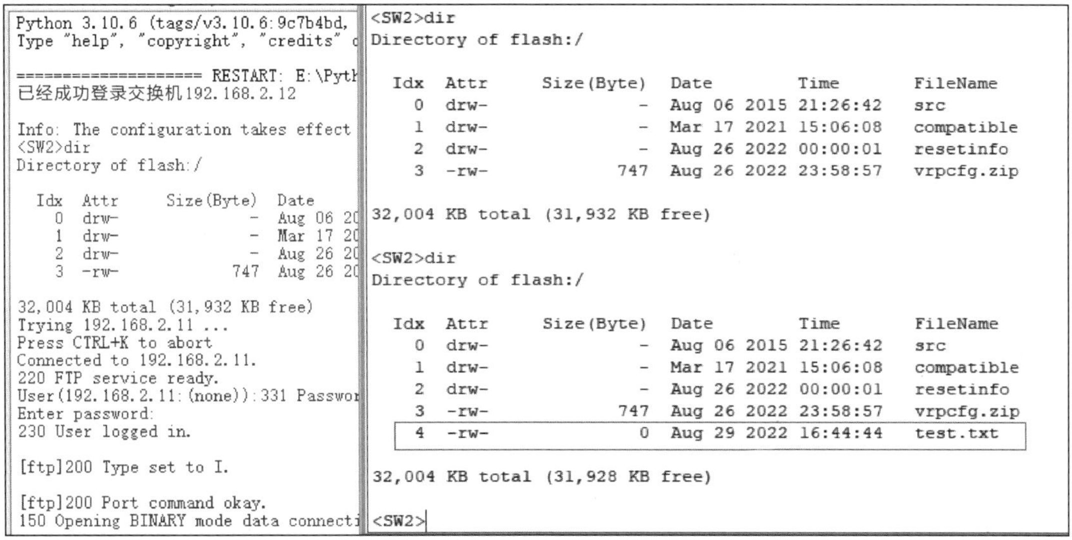

```
Python 3.10.6 (tags/v3.10.6:9c7b4bd,
Type "help", "copyright", "credits" c

==================== RESTART: E:\Pyth
已经成功登录交换机192.168.2.12

Info: The configuration takes effect
<SW2>dir
Directory of flash:/

  Idx  Attr     Size(Byte)  Date
    0  drw-              -  Aug 06 20
    1  drw-              -  Mar 17 20
    2  drw-              -  Aug 26 20
    3  -rw-            747  Aug 26 20

32,004 KB total (31,932 KB free)
Trying 192.168.2.11 ...
Press CTRL+K to abort
Connected to 192.168.2.11.
220 FTP service ready.
User(192.168.2.11:(none)):331 Passwor
Enter password:
230 User logged in.

[ftp]200 Type set to I.

[ftp]200 Port command okay.
150 Opening BINARY mode data connecti
```

```
<SW2>dir
Directory of flash:/

  Idx  Attr     Size(Byte)  Date         Time       FileName
    0  drw-              -  Aug 06 2015  21:26:42   src
    1  drw-              -  Mar 17 2021  15:06:08   compatible
    2  drw-              -  Aug 26 2022  00:00:01   resetinfo
    3  -rw-            747  Aug 26 2022  23:58:57   vrpcfg.zip

32,004 KB total (31,932 KB free)

<SW2>dir
Directory of flash:/

  Idx  Attr     Size(Byte)  Date         Time       FileName
    0  drw-              -  Aug 06 2015  21:26:42   src
    1  drw-              -  Mar 17 2021  15:06:08   compatible
    2  drw-              -  Aug 26 2022  00:00:01   resetinfo
    3  -rw-            747  Aug 26 2022  23:58:57   vrpcfg.zip
    4  -rw-              0  Aug 29 2022  16:44:44   test.txt

32,004 KB total (31,928 KB free)

<SW2>
```

在开始掌握 Python 网络自动化的逻辑后，很多情形开始变得灵活起来，就需要发挥运维人员的主观能动性。Netmiko 暂时不支持这种功能，但我们可以通过逻辑推演改用另一种解决方法。Netmiko 暂时解决不了的，我们甚至也可以退阶用回朴素的 Paramiko。运维自动化的目的是帮助网络工程师有效提升生产效能，按质按量完成工作任务。至于使用何种方式完成，并没有优劣之分。

8.15　实验 8　使用 Netmiko 处理设备提示命令（思科设备）

在网络设备中输入某些命令后，系统会返回一个提示命令，询问你是继续执行命令还是撤销命令。比如我们在实验 7 中向 SW1 传入了 test.txt 这个文件，如果这时我们想将它删除，则需要输入命令 del flash0:/test.txt。输入该命令后，系统会询问你是否 confirm，如下图所示。

```
S1#del flash0:/te
S1#del flash0:/test.txt
Delete flash0:/test.txt? [confirm]n
Delete of flash0:/test.txt aborted!
```

这时如果输入字母 y 或者按下回车键，则系统会执行 del 命令将该文件删除；如果输入字母 n，则系统会中断该命令，该文件不会被删除。

在 Netmiko 3 中，我们可以调用 send_command()函数中的 command_string、expect_string、strip_prompt 和 strip_command 4 个参数来应对这种情况。

首先创建实验 8 的脚本 netmiko8.py，写入下列代码。

```
from netmiko import ConnectHandler

sw1 = {
        'device_type': 'cisco_ios',
        'ip': '192.168.2.11',
        'username': 'python',
        'password': '123'
}

with ConnectHandler(**sw1) as connect:
    print ("已经成功登录交换机" + sw1['ip'])
    output = connect.send_command(command_string="del flash0:/test.txt",
                  expect_string=r"Delete flash0:/test.txt?",
                  strip_prompt=False,
                  strip_command=False)
    output += connect.send_command(command_string="y",
                  expect_string=r"#",
                  strip_prompt=False,
                  strip_command=False)

print(output)
```

```
[root@localhost Netmiko]# cat netmiko8.py
from netmiko import ConnectHandler
sw1 = {
        'device_type': 'cisco_ios',
        'ip': '192.168.2.11',
        'username': 'python',
        'password': '123'
}

with ConnectHandler(**sw1) as connect:
    print ("已经成功登录交换机" + sw1['ip'])
    output = connect.send_command(command_string="del flash0:/test.txt",
                  expect_string=r"Delete flash0:/test.txt?",
                  strip_prompt=False,
                  strip_command=False)
    output += connect.send_command(command_string="y",
                  expect_string=r"#",
                  strip_prompt=False,
                  strip_command=False)

print(output)
```

代码分段讲解如下。

- 首先创建字典用以登录交换机 SW1。

```
from netmiko import ConnectHandler

sw1 ={
     'device_type':'cisco_ios',
     'ip':'192.168.2.11',
     'username':'python',
     'password':'123'
     }
```

- 通过 ConnectHandler()登录 SW1 后，我们前后两次调用 send_command()函数。我们在第一个 send_command()函数中通过 command_string 参数向 SW1 输入命令 del flash0:/test.txt，然后在 expect_string 参数里告知 Netmiko 去 SW1 的回显内容里查找 Delete flash0:/test.txt? 这段系统返回的提示命令，如果查到了，则继续输入命令 y（第二个 send_command()），让脚本删除 test.txt 文件；之后如果收到命令提示符#，则继续执行脚本后面的代码。strip_prompt 和 strip_command 两个参数在这里放 Fasle 就行，目的是让代码最后的 print(output)输出的内容格式更好看一点。

```
with ConnectHandler(**sw1) as connect:
  print ("已经成功登录交换机" + sw1['ip'])
  output = connect.send_command(command_string="del flash0:/test.txt",
```

```
                expect_string=r"Delete flash0:/test.txt?",
                strip_prompt=False,
                strip_command=False)
    output += connect.send_command(command_string="y",
                expect_string=r"#",
                strip_prompt=False,
                strip_command=False)

print(output)
```

运行脚本前，首先确认目前 SW1 的 flash0 下还有 test.txt 文件，如下图所示。

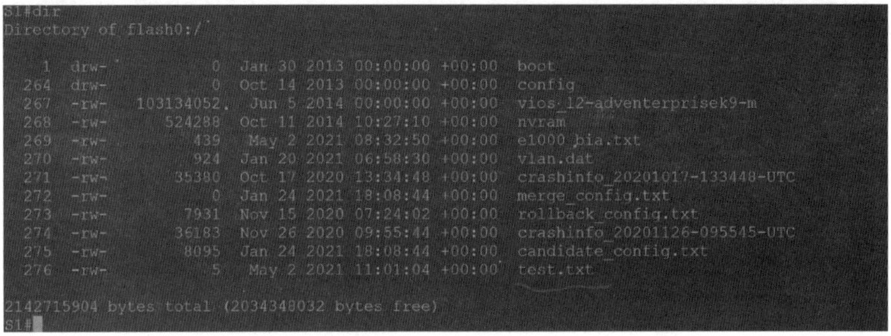

```
S1#dir
Directory of flash0:/

    1  drw-           0  Jan 30 2013 00:00:00 +00:00  boot
  264  drw-           0  Oct 14 2013 00:00:00 +00:00  config
  267  -rw-   103134052   Jun 5 2014 00:00:00 +00:00  vios_l2-adventerprisek9-m
  268  -rw-      524288  Oct 11 2014 10:27:10 +00:00  nvram
  269  -rw-         439   May 2 2021 08:32:50 +00:00  e1000_bia.txt
  270  -rw-         924  Jan 20 2021 06:58:30 +00:00  vlan.dat
  271  -rw-       35380  Oct 17 2020 13:34:48 +00:00  crashinfo_20201017-133448-UTC
  272  -rw-           0  Jan 24 2021 18:08:44 +00:00  merge_config.txt
  273  -rw-        7931  Nov 15 2020 07:24:02 +00:00  rollback_config.txt
  274  -rw-       36183  Nov 26 2020 09:55:44 +00:00  crashinfo_20201126-095545-UTC
  275  -rw-        8095  Jan 24 2021 18:08:44 +00:00  candidate_config.txt
  276  -rw-           5   May 2 2021 11:01:04 +00:00  test.txt

2142715904 bytes total (2034348032 bytes free)
S1#
```

然后运行脚本 netmiko8.py，如下图所示。

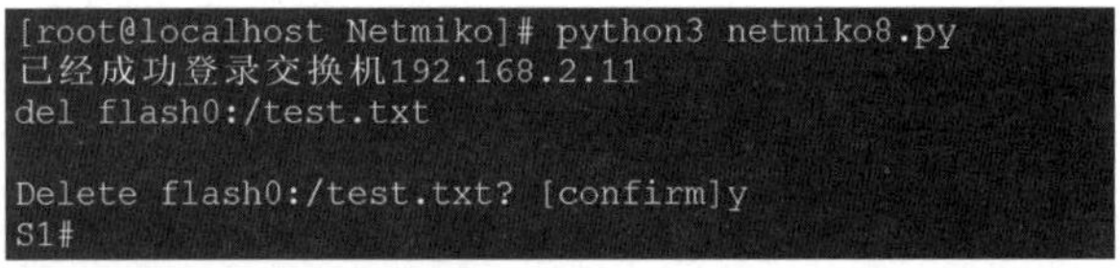

```
[root@localhost Netmiko]# python3 netmiko8.py
已经成功登录交换机192.168.2.11
del flash0:/test.txt

Delete flash0:/test.txt? [confirm]y
S1#
```

可以看到，Netmiko 在遇到提示命令 Delete flash0:/test.txt? 后帮我们自动输入了命令 y，删除了 test.txt 文件。

回到 SW1 上验证，test.txt 已被删除，如下图所示。

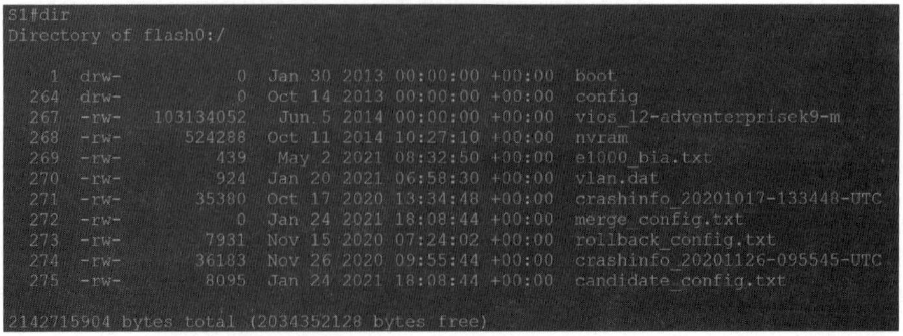

```
S1#dir
Directory of flash0:/

    1  drw-           0  Jan 30 2013 00:00:00 +00:00  boot
  264  drw-           0  Oct 14 2013 00:00:00 +00:00  config
  267  -rw-   103134052   Jun 5 2014 00:00:00 +00:00  vios_l2-adventerprisek9-m
  268  -rw-      524288  Oct 11 2014 10:27:10 +00:00  nvram
  269  -rw-         439   May 2 2021 08:32:50 +00:00  e1000_bia.txt
  270  -rw-         924  Jan 20 2021 06:58:30 +00:00  vlan.dat
  271  -rw-       35380  Oct 17 2020 13:34:48 +00:00  crashinfo_20201017-133448-UTC
  272  -rw-           0  Jan 24 2021 18:08:44 +00:00  merge_config.txt
  273  -rw-        7931  Nov 15 2020 07:24:02 +00:00  rollback_config.txt
  274  -rw-       36183  Nov 26 2020 09:55:44 +00:00  crashinfo_20201126-095545-UTC
  275  -rw-        8095  Jan 24 2021 18:08:44 +00:00  candidate_config.txt

2142715904 bytes total (2034352128 bytes free)
```

前面提到，strip_prompt 和 strip_command 两个参数在这里放 Fasle 的目的是让代码最后的 print(output)输出的内容格式更好看一点，如果把它们改成 True，结果会怎么样呢？来看效果，如下图所示。

```
[root@localhost Netmiko]# python3 netmiko8.py
已经成功登录交换机192.168.2.11

Delete flash0:/test.txt? [confirm]S1#

[root@localhost Netmiko]#
```

效果显然不如放 False 的时候更美观，并且看不到 Netmiko 帮我们输入的 y 命令。

8.16　实验 8　使用 Netmiko 处理设备提示命令（华为设备）

其实在设备提权实验中我们已经体验了设备指令交互。现在我们通过处理提示再来复习一下指令交互过程，如下图所示。

```
from netmiko import ConnectHandler

sw1 = {'device_type':'huawei',
     'ip':'192.168.2.11',
     'username':'python',
     'password':'123'
      }

with ConnectHandler(**sw1) as connect:
   print ("已经成功登录交换机" + sw1['ip'])

   output = connect.send_command(command_string="delete flash:/test.txt",
              expect_string=r"Delete flash:/test.txt?",
              strip_prompt=False,
              strip_command=False)
   output += connect.send_command(command_string="y",
              expect_string=r">",
              strip_prompt=False,
              strip_command=False)

   print(output)
```

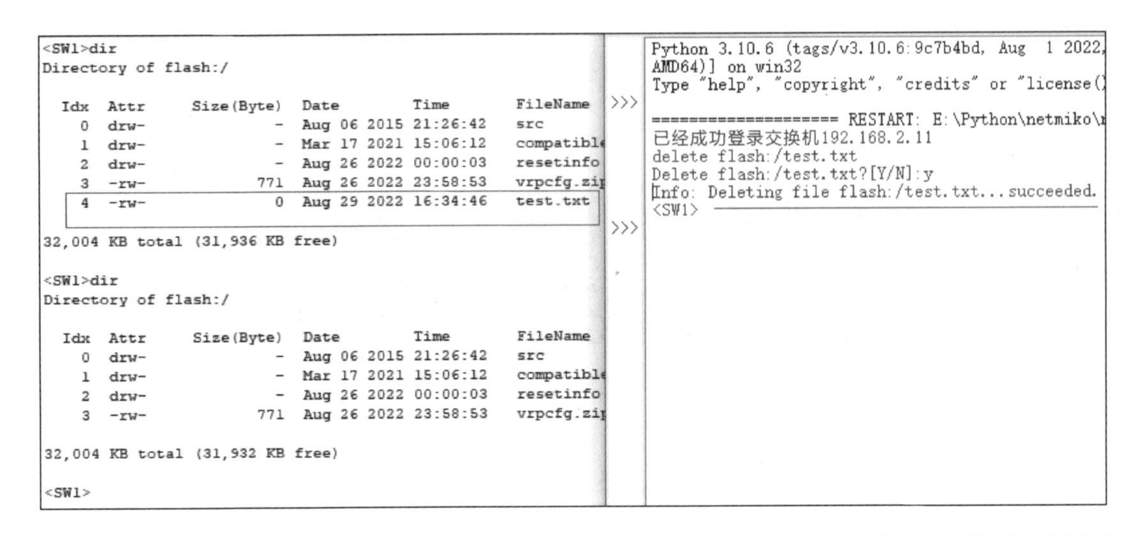

沿着主线学习，最开始后置部分旁系知识。现在我们已经比较有把握了，就可以深入探讨一些看起来有点“无关紧要”（偶尔可能是“事关重要”）的知识内容了，比如 strip_prompt 和 strip_command 两个参数。学习这些知识的时候，最好还能配合手册检索与查阅。掌握快速学习的方法，有时候可能比学多几个知识点更重要。

8.17　实验 9　使用 Netmiko 获取设备主机名（思科设备）

在平常的网络运维工作中，我们有时需要通过 Python 脚本获取设备的主机名，虽然这种需求可以用正则表达式实现，但是因为每家公司对主机名格式的标准各不相同，正则表达式没有统一的写法。而在那些没有对设备主机名格式制定标准的公司里，主机名杂乱无章，毫无规律可循，这样会大大提高通过使用正则表达式获取主机名的难度。

在 6.4 节中介绍的 NAPALM 可以通过 get_facts()方法来简单地解决这个需求，代码及输出结果如下。

```
from napalm import get_network_driver
import json

driver = get_network_driver('ios')
SW1 = driver('192.168.2.11','python','123')
SW1.open()

output = SW1.get_facts()
print (json.dumps(output, indent=2))
```

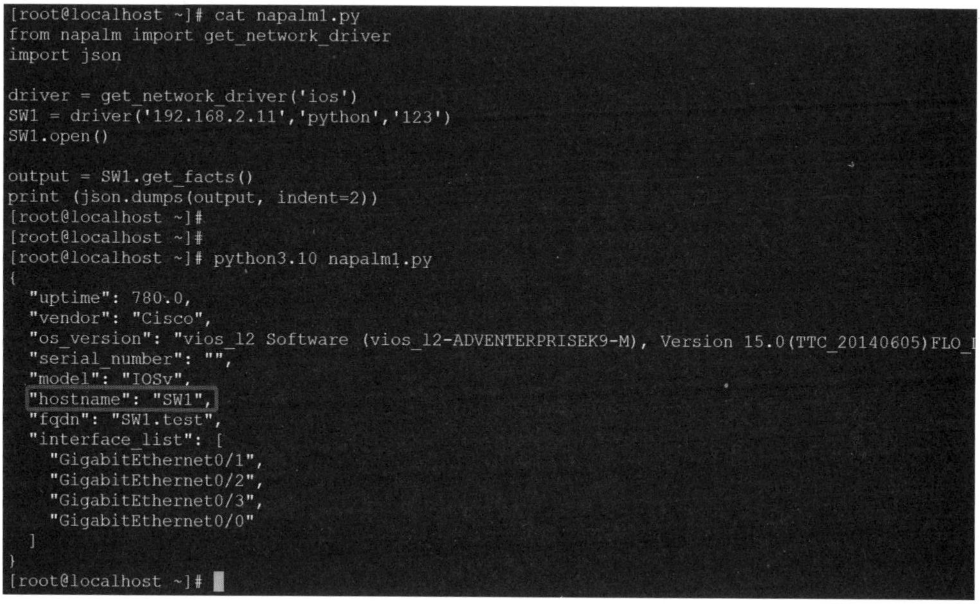

```
[root@localhost ~]# cat napalm1.py
from napalm import get_network_driver
import json

driver = get_network_driver('ios')
SW1 = driver('192.168.2.11','python','123')
SW1.open()

output = SW1.get_facts()
print (json.dumps(output, indent=2))
[root@localhost ~]#
[root@localhost ~]#
[root@localhost ~]# python3.10 napalm1.py
{
  "uptime": 780.0,
  "vendor": "Cisco",
  "os_version": "vios_l2 Software (vios_l2-ADVENTERPRISEK9-M), Version 15.0(TTC_20140605)FLO_
  "serial_number": "",
  "model": "IOSv",
  "hostname": "SW1",
  "fqdn": "SW1.test",
  "interface_list": [
    "GigabitEthernet0/1",
    "GigabitEthernet0/2",
    "GigabitEthernet0/3",
    "GigabitEthernet0/0"
  ]
}
[root@localhost ~]#
```

从上图中可以看到，NAPALM 的 get_infacts()方法返回字典中的“hostname”这个键对应的值即设备的主机名，我们可以将倒数第二行的代码 output = SW1.get_facts()改成 output = SW1.get_facts()['hostname']即可直接打印出设备的主机名。

除 NAPALM 的 get_facts()方法外，Netmiko 也提供了 find_prompt()方法来帮助我们获取设备的主机名，代码如下。

```
from netmiko import ConnectHandler

sw1 = {
        'device_type': 'cisco_ios',
        'ip': '192.168.2.11',
        'username': 'python',
        'password': '123'
}

with ConnectHandler(**sw1) as connect:
    print ("已经成功登录交换机" + sw1['ip'])
    print (connect.find_prompt())
```

这段代码十分简单，在通过 ConnectHandler 连接交换机 SW1 后，我们随即通过 print (connect.find_prompt())来获取设备的主机名，运行脚本后的效果如下图所示。

```
[root@localhost ~]# python3.10 netmiko9.py
已经成功登录交换机 192.168.2.11
SW1#
[root@localhost ~]#
```

可以看到，find_prompt()返回的主机名包含了设备提示符（这里是#），如果想去除提示符，可以用切片的方法将 print (connect.find_prompt())写成 print (connect.find_prompt()[:-1])，这里留给读者自行去实验。

8.18　实验 9　使用 Netmiko 获取设备主机名（华为设备）

到这里，读者们应该已经轻车熟路了，直接把思科实验的脚本修改适配即可，如下图所示。

```
from netmiko import ConnectHandler

sw1 = {'device_type': 'huawei',
       'ip': '192.168.2.11',
       'username': 'python',
       'password': '123'}

with ConnectHandler(**sw1) as connect:
   print ("已经成功登录交换机" + sw1['ip'])
   print (connect.find_prompt()[:-1])
```

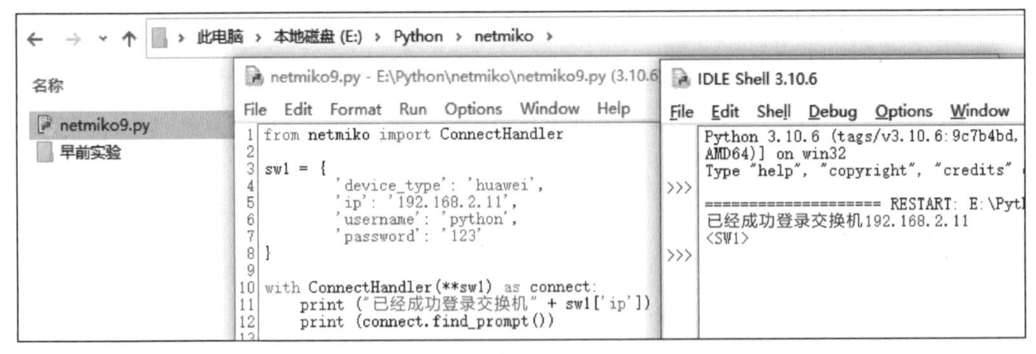

Netmiko 的相关实验，最后我们用一个“简单”的需求来暂时告一段落。下一节，我们将对 Netmiko 4 相对于 Netmiko 3 的改进特性，做一次新功能介绍。本章 Netmiko 的 9 个实验到此结束，通过一些生产场景逻辑抽象，层层引出知识点，但是，相对于若干知识点，我们更注重传授学习方法和实践思路。

8.19　Netmiko 4 的新功能介绍

鉴于 Netmiko 4 性能逐渐稳定，笔者认为有必要在本章加入一节来介绍它的一些新功能。

1. read_timeout 和 read_timeout_override

Netmiko 的作者 Kirk Byers 曾坦诚，Netmiko 4 诞生之前的所有版本中最困扰他的就是截屏（screen-scraping）问题。所谓截屏问题指在用户向设备输入一条 show 或者 display 命令后，Netmiko 无法判断回显内容是否已经完整返回。其实不仅 Netmiko，任何需要用到截屏功能的模块都会遇到类似的问题，比如 Paramiko 需要我们用 time.sleep()手动指定休眠时间，完全把问题抛回给用户自己解决。

众所周知，Netmiko 是 Paramiko 的衍生模块，它帮我们省去了手动指定休眠时间的步骤。在 Netmiko 4 之前，Netmiko 会用 delay_factor 参数（在 send_command()中使用，只对在 send_command()里使用的命令有效）或 global_delay_factor 参数（在设定登录设备参数的字典中使用，对在该设备上输入的所有命令都有效）配合 expect_string 参数来处理这个问题。比如通过 send_command()在思科交换机里输入 show ip int brief 命令，我们可以通过 expect_string=r'#'告诉 Netmiko 去回显内容中抓取最后一个字符，即思科设备的命令提示符#（如果没有指定 expect_string 并且 device_type 为 cisco_ios，则 expect_string 的默认值为#），如下图所示。

```
device = {
    "device_type": "cisco_ios",
    "ip": "192.168.2.11",
    "username": "python",
    "password": "123",
    "fast_cli": True, #device_type为cisco_ios时，默认值为True
    "global_delay_factor": 0.1 #默认值为1, fast_cli为True时，该值为0.1
}
```

```
with ConnectHandler(**device) as conn:
    output = conn.send_command("show ip int brief",
                               expect_string=r'#', #device_type为cisco_ios时，默认值为#
                               delay_factor=0.1) #默认值为1, fast_cli为True时，该值为0.1
```

在向设备输入一条命令后，如果 fast_cli=False，那么 Netmiko 默认最多等待 100s 从回显内容中抓取 expect_string 指定的字符，如果 fast_cli=True，则 Netmiko 最多只等待 10s，这 10s 是怎么来的？也就是由 100（默认等待时间）乘以 0.1（fast_cli=True 时，delay_factor 和 global_delay_factor 的值）得来的。如果抓取到了 expect_string 指定的回显字符，则 Netmiko

会立即返回回显内容（注意是立即返回，而不是必须等到第 10s 或第 100s 才返回），如果过了 10s 或 100s 都还没抓取到#，则 Netmiko 会返回一个异常。

一般来说，show ip int brief 的回显内容肯定能在 10s 内顺利返回，但是 show run 则不一定，show tech-support 就更不可能，这个时候我们必须手动修改 fast_cli、delay_factor 或者 global_delay_factor 几个参数来调整 Netmiko 等待回显内容的时间。

为了简化这个步骤，Netmiko 4 特意在 send_command()中引入了 read_timeout 参数，read_timeout 参数可以让我们直接设定 Netmiko 最多等待多少秒从回显内容中抓取 expect_string 指定的字符，比如说在生产网络里一台思科 6800 核心三层交换机上输入 show tech-support 后要等 45 ~ 50s 以上才能返回完整的回显内容，那么我们可以把 read_timeout 参数设为 60（read_timeout 默认值为 10s），如下图所示。

```
with ConnectHandler(**device) as conn:
    output = conn.send_command("show tech-support",
                               read_timeout=60) #让Netmiko最多等待60秒
```

可以看到，read_timeout 最大的好处是它大大节省了新手的学习成本，不用再去学习 fast_cli、delay_factor 和 global_delay_factor 等参数，并且也免去了额外做乘法运算的麻烦。另外，和 global_delay_factor 类似，我们也可以在设定登录设备参数的字典中使用 read_timeout_override 来**全局修改** read_timeout 的值，如下图所示。

```
device = {
    "device_type": "cisco_ios",
    "ip": "192.168.2.11",
    "username": "python",
    "password": "123",
    "read_timeout_override": 90 #全局有效
}
```

另外，出于兼容性考虑，还可以在字典里将 delay_factor_compact 设为 True，这样 Netmiko 4 会按照 Netmiko 3 的模式继续使用 fast_cli、delay_factor 和 global_delay_factor 来计算等待时间，如下图所示。

```
device = {
    "device_type": "cisco_ios",
    "ip": "192.168.2.11",
    "username": "python",
    "password": "123",
    "read_timeout_override": 90 #全局有效
    "delay_factor_compact": True #默认为False，不兼容Netmiko4之前的版本
}
```

2. send_multiline()和 send_multiline_timing()

实验 8 介绍了如何处理设备交互命令的场景，比如在思科交换机上输入 del flash0:/test.txt 这个删除 flash:下文件的命令后，系统会询问是否 confirm，如下图所示。

```
SW1#del flash0:/test.txt
Delete flash0:/test.txt? [confirm]n
Delete of flash0:/test.txt aborted!
SW1#
```

另外，如果在思科设备里使用 ping 命令的 extended 模式，系统也会让用户输入一系列与 ping 相关的参数，如下图所示。

```
SW1#ping
Protocol [ip]:
Target IP address: 8.8.8.8
Repeat count [5]:
Datagram size [100]:
Timeout in seconds [2]:
Extended commands [n]:
Sweep range of sizes [n]:
Type escape sequence to abort.
Sending 5, 100-byte ICMP Echos to 8.8.8.8, timeout is 2 seconds:
.....
Success rate is 0 percent (0/5)
SW1#
```

在实验 8 中，我们通过 Netmiko 3 中的 send_command()配合 expect_string 来应对这个问题。但是这种做法非常复杂并且要写大量代码，比如说应对第一个 del flash0:/test.txt 的场景时，因为该交互场景只需要我们输入一个参数（是否 confirm），所以代码相对还比较简洁，如下图所示。

```
with ConnectHandler(**sw1) as connect:
    print ("已经成功登录交换机" + sw1['ip'])
    output = connect.send_command(command_string="del flash0:/test.txt",
                 expect_string=r"Delete flash0:/test.txt?",
                 strip_prompt=False,
                 strip_command=False)
    output += connect.send_command(command_string="y",
                 expect_string=r"#",
                 strip_prompt=False,
                 strip_command=False)

print(output)
```

但是遇到 extended ping 这种需要用户输入多个参数的交互场景时，代码量就非常多了，如下图所示。

```
with ConnectHandler(**device) as net_connect:

    cmd = "ping"
    target_ip = "8.8.8.8"
    count = "30"

    output = net_connect.send_command_timing(
        cmd, strip_prompt=False, strip_command=False
    )
    output += net_connect.send_command_timing(
        "\n", strip_prompt=False, strip_command=False
    )
    output += net_connect.send_command_timing(
        target_ip, strip_prompt=False, strip_command=False
    )
    output += net_connect.send_command_timing(
        count, strip_prompt=False, strip_command=False
    )
    output += net_connect.send_command_timing(
        "\n", strip_prompt=False, strip_command=False
    )
    output += net_connect.send_command_timing(
        "\n", strip_prompt=False, strip_command=False
    )
    output += net_connect.send_command_timing(
        "\n", strip_prompt=False, strip_command=False
    )
    output += net_connect.send_command_timing(
        "\n", strip_prompt=False, strip_command=False
    )
    print(output)
```

究其原因，就是 send_command()函数是基于内容的（pattern-based），它必须要等到用户告诉它等到什么回显内容后才会执行后面的代码。同样，在应对 extended ping 的交互命令场景时，如果我们用基于时间（time-based）的 send_command_timing()函数来处理，代码量会相对小很多，如下图所示。

```
with ConnectHandler(**device) as conn:
    data = ""
    commands = [
        "ping",
        "\n",
        "8.8.8.8",
        "\n",
        "\n",
        "\n",
        "\n",
        "\n"
    ]
    for cmd in commands:
        data += conn.send_command_timing(
            cmd,
            strip_command=False,
            strip_prompt=False
        )
    print(data)
```

而在 Netmiko 4 中加入的 send_multiline()和 send_multiline_timing()则将类似的需求变得更简单，其中前者为 pattern-based，后者为 time-based。首先来看怎么用 send_multiline()应对第一个 del flash0:/test.txt 的场景，如下图所示。

```
with ConnectHandler(**sw1) as conn:
    cmd_list =[
        ["del flash0:/test.txt", r"Delete flash0:/test.txt?"],
        ["n", r"confirm"]
    ]

output = conn.send_multiline(cmd_list)
print (output)
```

从上图中可以看到，我们在 cmd_list 列表里额外添加了两组子列表，每组子列表的元素为我们输入的命令，以及执行该命令后我们想要 Netmiko 在回显内容中抓取的字符（类似 send_command()的 expect_string）。比如在第一组子列表中，我们输入命令 del flash0:/test.txt，希望抓取的回显内容为"Delete flash0:/test.txt?"]；在第二组子列表中，我们输入命令 n，希望抓取的回显内容改为 confirm，依此类推。

如果用 time-based 的 send_multiline_timing()来做，上述代码还可以更简洁，如下图所示。

```
with ConnectHandler(**sw1) as conn:
    cmd_list =[
        "del flash0:/test.txt",
        "n"
    ]

output = conn.send_multiline_timing(cmd_list)
print (output)
```

而在 extended ping 场景中，如果用 send_multiline_timing()来做，代码会比 send_command_timing()更加简洁，如下图所示。

```
with ConnectHandler(**device) as net_connect:
    target_ip = "8.8.8.8"
    count = "30"
    cmd_list = [
        "ping",
        "\n",
        target_ip,
        count,
        "\n",
        "\n",
        "\n",
        "\n",
    ]

    output = net_connect.send_multiline_timing(cmd_list)
    print(output)
```

很显然在处理多交互命令的场景时，在 Netmiko 4 中加入的 send_multline()和 send_multiline_timing()将 Netmiko 3 的 send_command()和 send_command_timing()的代码大大简化了。另外，我们注意到 send_multiline_timing()的代码比 send_multiline()更简单易懂，不过相较于 send_multiline()，使用 send_multiline_timing()有一个劣势，那就是每输入一条命令，Netmiko 会默认固定等待 2s 才会执行下一条命令（因为 send_multiline_timing()是 time-based 的），而 pattern-based 的 send_multiline()则会在读取到指定的回显内容后立即执行后面的代码。鱼和熊掌不可兼得，一个代码简单但脚本运行速度慢，一个代码稍微复杂但脚本运行速度快，如何取舍完全看用户自己的决定。

3. ConnLogOnly

在使用 Netmiko 3 或之前的版本时，用户需要写很多 try/except 异常处理来应对各种各样会导致脚本停止工作的错误或异常，比如最常见的因为 SSH 用户名/密码验证不通过导致的 NetmikoAuthenticationException 和设备链接超时无响应导致的 NetmikoTimeoutException，类似这样的异常处理在设备数量众多的大型网络里基本是标配（设备数量越多，发生问题的概率越大），如下图所示。

```
try:
    conn = ConnectHandler(**device)
except NetmikoAuthenticationException:
    return
except NetmikoTimeoutException:
    return
```

在 Netmiko 4 中，我们可以用 ConnLogOnly 替代 ConnectHandler 来统一处理这个问题，如下图所示。

```
from netmiko import ConnLogOnly

conn = ConnLogOnly(**device)
if conn is None:
    print ("登录设备失败！")
```

使用 ConnLogOnly 时，如果其返回值为 None，则 Netmiko 会直接判定登录设备失败（如果登录成功，则和 ConnectHandler 一样返回一个 Netmiko 连接对象——Netmiko Connection Object）。如果要查看具体登录失败的原因，则可以在运行脚本后 Netmiko 生成的 netmiko.log 文件中查看（netmiko.log 也是 Netmiko 4 新引入而 Netmiko 3 之前没有的功能），netmiko.log 文件和脚本文件在同一文件夹下，如下图所示。

```
[root@localhost Netmiko]# ls -lt
total 64
-rw-r--r-- 1 root root 315 Aug 16 12:36 netmiko.log
-rw-r--r-- 1 root root 629 May  2  2021 netmiko8.py
-rw-r--r-- 1 root root 457 May  2  2021 netmiko7.py
-rw-r--r-- 1 root root   5 May  2  2021 test.txt
-rw-r--r-- 1 root root 345 May  2  2021 netmiko6.py
-rw-r--r-- 1 root root 183 May  2  2021 configuration.con
drwxr-xr-x 2 root root  26 May  2  2021 templates
-rw-r--r-- 1 root root 754 May  2  2021 netmiko5.py
-rw-r--r-- 1 root root 362 May  2  2021 netmiko4_2.py
-rw-r--r-- 1 root root 369 Apr 29  2021 switches.json
-rw-r--r-- 1 root root 504 Apr 29  2021 netmiko4_1.py
-rw-r--r-- 1 root root  65 Apr 29  2021 ip_list.txt
-rw-r--r-- 1 root root 336 Apr 29  2021 netmiko3_2.py
-rw-r--r-- 1 root root 338 Apr 29  2021 netmiko3_1.py
-rw-r--r-- 1 root root 624 Apr 28  2021 netmiko2.py
-rw-r--r-- 1 root root  34 Apr 28  2021 config.txt
-rw-r--r-- 1 root root 230 Apr 27  2021 netmiko1.py
[root@localhost Netmiko]#
```

```
[root@localhost Netmiko]# cat netmiko.log
2022-08-16 11:02:13,203 ERROR netmiko.ssh_dispatcher Authentication failure to: 192.168.2.11:22 (cisco_ios)

Authentication to device failed.

Common causes of this problem are:
1. Invalid username and password
2. Incorrect SSH-key file
3. Connecting to the wrong device

Device settings: cisco_ios 192.168.2.11:22
[root@localhost Netmiko]#
```

第9章

Nornir详解

2018年，笔者在自己的知乎专栏里写了《网络工程师的Python之路——Ansible篇》一文，以实验的方式手把手向读者教授了Ansible的使用方法，受到读者的热捧。本来笔者是想单独出一本Ansible的书，但是没曾想，就在2018年，对标Ansible的Nornir横空出世，并在NetDevOps圈子里迅速抢占Ansible的地盘，短短不到三年的时间，Nornir已经从1.0版迅速更迭到了2020年9月16日发布的3.0.0版。Nornir的发展如此迅猛，主要有如下几个原因。

（1）阵容强大：Nornir的作者是NAPALM的作者David Barroso（GitHub账号：dbarrosop）和Netmiko的作者Kirk Byers，两位联手完成了Nornir初代的框架，因为Nornir对比Ansible确实强大，后来有很多业界大佬，比如思科的Dmitry Figol，在2019年于巴塞罗那举办的Cisco Live上为Nornir站台，力推Nornir，让更多业界开发者加入Nornir的开发中，所以Nornir能迅速从1.0版升级到3.0版。

（2）下面两张图是NetDevOps界知名人物瑞典人Patrick Ogenstad放出的Nornir和Ansible的性能对比，可以看到在设备数量多于1000台的大型网络里，Nornir的效率大大优于Ansible（虽然两者都支持多线程并发，但是处理1万台设备，Ansible需要2300s，Nornir只需要17s；处理5000台设备Ansible需要900s，Nornir只需要9s，差距之大，让很多人怀疑这两个根本不是同一个量级的产品。）

（3）除性能上的差距外，目前业界普遍公认的Nornir对比Ansible的优势还有如下几点。

- Ansible完全基于YAML，和其他编程语言相比，缺乏IDE、Linting、静态类型检查等工具的支持，虽然Nornir也要用到YAML，但是它的runbook是用Python写的，有了Python的助力，我们可以通过Nornir做很多事情。

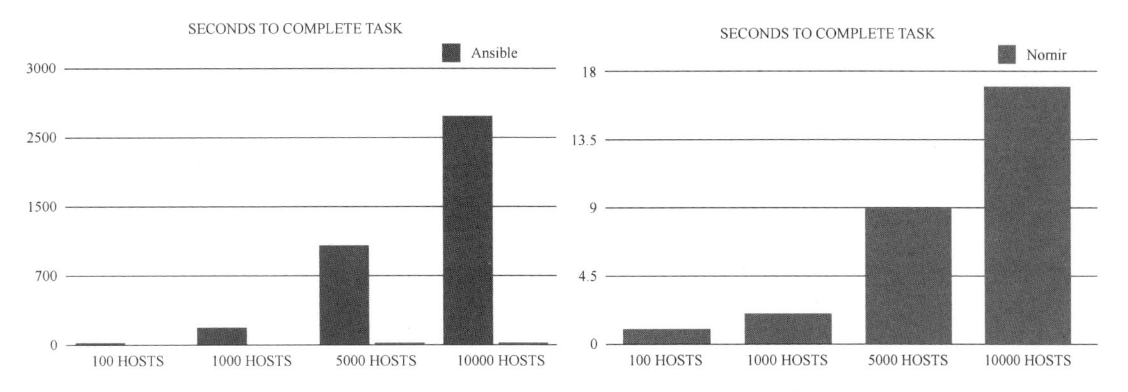

- Ansible 的剧本（Playbook）排错、debug、维护难度相当大，而基于 Python 的 Nornir 在这方面做得比 Ansible 好得多（虽然 Nornir 1.0 到 2.0、2.0 到 3.0 的跨度也比较大，但是总体排错难度比 Ansilbe 简单得多）。
- Ansible 完全基于 YAML，让使用者根本体会不到编程的乐趣，如果太过依赖 Ansible，则使用者的 Python 水平会在原地踏步甚至完全荒废，而使用 Nornir 的用户则不会有这种顾虑。

本章将提供 15 个 Nornir 的实验，全方位地向读者讲解 Nornir 的入门知识，保证读者在读完本章并动手做完 15 个实验后，能熟练地掌握 Nornir 的基本用法，并兼顾到平时网络运维工作中一些比较常见的任务，比如读取设备信息、配置设备、保存设备配置等。尤其是让那些在设备数量成千上万的大型或超大型公司里工作，对 Python 程序运行速度有强烈需求的读者们能够体会到 Nornir 的强大。

9.1　Nornir 实验准备（思科设备，CentOS 系统）

与 Paramiko、Netmiko、NAPALM 等 Python 第三方模块不同，Nornir 是一个 Python 框架，由多种插件组成（这些插件里包含 Netmiko 和 NAPALM）。Nornir 最大的优势在于它默认运行在多线程模式下，不需要写太多额外的与多线程相关的代码，开箱即用，大大节省了入门 NetDevOps 的网络工程师们的学习成本。

Nornir 3.0.0 发布于 2020 年 9 月 16 日，它相比 Nornir 2.X 时代（Nornir 2 最后一个版本为 2.5）最大的区别是将所有插件分拆了出去，这也就意味着，在 2.X 时代光靠 pip3 install nornir 来安装一个 Nornir 模块的做法已经行不通了。在实验开始前，请务必按照下面的方法通过 pip 安装好所有实验需要用到的 Python 模块及 Nornir 插件。

```
pip3.10 install netmiko
pip3.10 install napalm
pip3.10 install nornir
pip3.10 install nornir_utils
pip3.10 install nornir_napalm
pip3.10 install nornir_netmiko
```

安装完毕后，输入命令 pip3.10 freeze | grep nornir 做检查，确保 nornir、nornir_utils、nornir_napalm、nornir_netmiko 等模块都已安装完毕（截至 2022 年 8 月，最新的 Nornir 版本为 3.3.0），如下图所示。

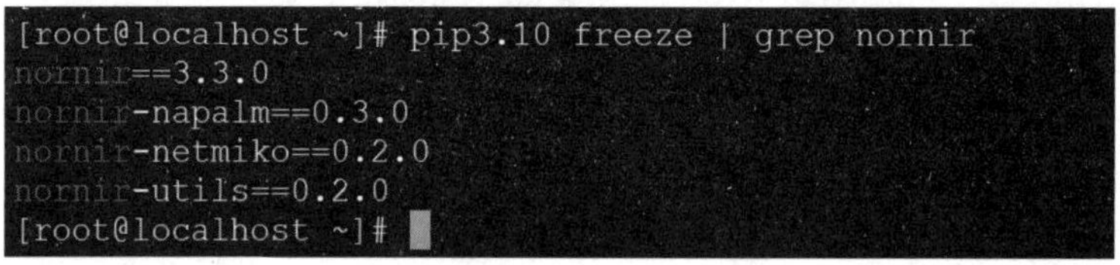

```
[root@localhost ~]# pip3.10 freeze | grep nornir
nornir==3.3.0
nornir-napalm==0.3.0
nornir-netmiko==0.2.0
nornir-utils==0.2.0
[root@localhost ~]#
```

确认 Nornir 版本为 3.3.0 后，我们首先创建一个 YAML 文件，叫作 config.yaml，其内容如下。

```
---
inventory:
  plugin: SimpleInventory
  options:
    host_file: "hosts.yaml"
    group_file: "groups.yaml"
    defaults_file: "defaults.yaml"
runner:
  plugin: threaded
  options:
    num_workers: 100
```

```
[root@localhost Nornir]# cat config.yaml
---
inventory:
    plugin: SimpleInventory
    options:
        host_file: "hosts.yaml"
        group_file: "groups.yaml"
        defaults_file: "defaults.yaml"
runner:
    plugin: threaded
    options:
        num_workers: 100
[root@localhost Nornir]#
```

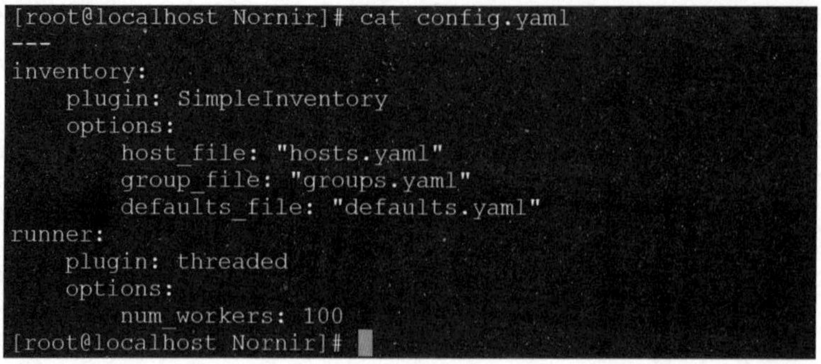

Config.yaml 是 Nornir 最基本的配置文件，其中包含 Nornir 自带的 Inventory 插件 SimpleInventory，用来管理设备，在 options 里可以看到，我们还需要另外创建 3 个 YAML 文件：hosts.yaml、groups.yaml 及 defautls.yaml，分别用来存放设备（hosts）信息、分组（groups）信息及缺省（defaults）信息。

runner 下的 threaded 插件表示我们将使用多线程来让 Nornir 执行任务，num_workers: 100 表示我们最多启用 100 个线程（默认为 20 个）。

接下来创建 defaults.yaml、groups.yaml 及 hosts.yaml 3 个文件，内容如下。

◎　defaults.yaml

```
---
username: python
password: '123'
```

```
[root@localhost Nornir]# cat defaults.yaml
---
username: python
password: '123'
[root@localhost Nornir]#
```

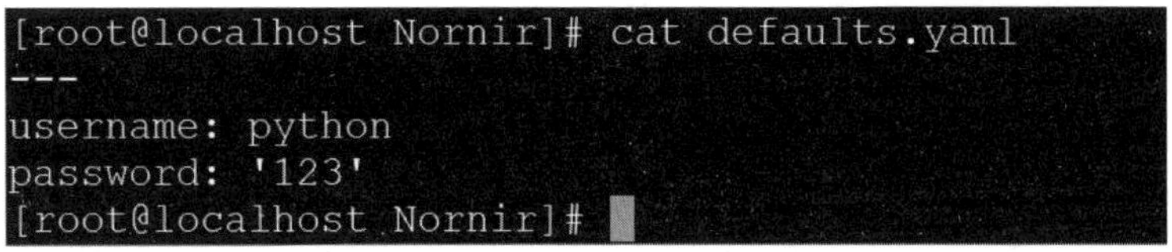

defaults.yaml 的作用是用来填补 hosts.yaml 里遗漏的参数（讲 hosts.yaml 时会讲到它的作用），这里我们在 defaults.yaml 中定义了用来供 Nornir 通过 SSH 登录实验中要用到的 5 台交换机的用户名和密码。

注意：如果密码里含有数字，则必须加引号，比如 password: '123'；如果不含数字或者是字母+数字（字母开头或者数字开头都无所谓）的组合，则无须加引号，比如密码是 python，可以写成 password: python，密码是 python123 或者 123python，可以写成 password: python123 和 password: 123python 均没问题。

◎　groups.yaml

```
---
cisco:
  platform: ios
```

```
[root@localhost Nornir]# cat groups.yaml
---
cisco:
    platform: ios
[root@localhost Nornir]#
```

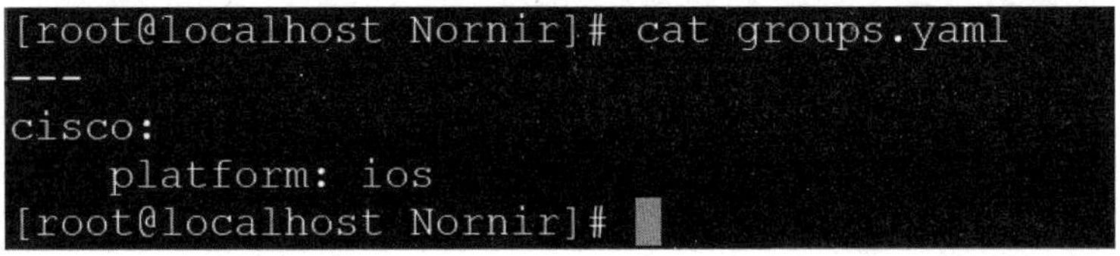

我们可以将 Nornir 要登录管理的设备（hosts）按厂商划为不同的分组，因为 GNS3 模拟器所使用的是虚拟的思科交换机，所以我们在 groups.yaml 里创建了一个叫作 cisco 的分组（组名可以自定义，一般取直观容易理解的组名）。另外，Nornir 是基于 NAPALM 登录设备的，其对应的就是 NAPALM 里的 Driver Name，也就是下图里的 ios。

Supported Devices

General support matrix

_	EOS	Junos	IOS-XR	NX-OS	NX-OS SSH	IOS
Driver Name	eos	junos	iosxr	nxos	nxos_ssh	ios
Structured data	Yes	Yes	No	Yes	No	No
Minimum version	4.15.0F	12.1	5.1.0	6.1 [1]		12.4(20)
Backend library	pyeapi	junos-eznc	pyIOSXR	pynxos	netmiko	netmiko
Caveats	EOS			NXOS	NXOS	IOS

◎　hosts.yaml

```
---
sw1:
   hostname: 192.168.2.11
   username: python
   password: '123'
   groups:
      - cisco
sw2:
   hostname: 192.168.2.12
   groups:
      - cisco
```

```
[root@localhost Nornir]# cat hosts.yaml
---
sw1:
    hostname: 192.168.2.11
    username: python
    password: '123'
    groups:
        - cisco
sw2:
    hostname: 192.168.2.12
    groups:
        - cisco
[root@localhost Nornir]#
```

hosts.yaml 用来存放具体要登录的设备的信息，比如 hostname（也就是 IP 地址）、username、password 及该设备所属的设备组。可以看到，我们给 sw1 设置了 username 和 password，但是对 sw2 却故意漏过了。这里不用担心，前面在讲 defaults.yaml 的时候提到 defaults.yaml 的作用就是填补 hosts.yaml 里遗漏的参数，也就是说，在运行 Nornir 脚本后，如果 Nornir 在登录 sw2 时发现 hosts.yaml 里遗漏了 sw2 的 username 和 password，那么它会直接去看 defaults.yaml 是否有对应的参数并使用。

从这点我们可以看出 defaults.yaml 很好用。举个例子，假设我们要登录 10 台交换机，其中 8 台的用户名和密码都一样，只有剩下两台不一样，那么我们可以把 8 台交换机共同的用户名和密码写入 defaults.yaml 里，然后在 hosts.yaml 里写入这 8 台交换机时故意漏掉它们的 username 和 password（让 Nornir 在 defaults.yaml 里找），只针对剩下的两台交换机添加上它们的 username 和 password 即可。

注：在只有单一类型设备的情况下，groups.yaml 是可以省去的，比如实验里只使用了思科的 IOS 设备，那么我们可以将 groups.yaml 删除，将 hosts.yaml 的内容改为下面这样即可（这个技巧了解即可，实验中我们还是要用到 groups.yaml）。

```
---
sw1:
   hostname: 192.168.2.11
   username: python
   password: '123'
   platform: ios
sw2:
   hostname: 192.168.2.12
   platform: ios
```

最后检查一遍，看是否 config.yaml、defaults.yaml、groups.yaml 和 hosts.yaml 4 个 YAML 文件都已准备就绪，如下图所示。

```
[root@localhost Nornir]# ls -lt
total 16
-rw-r--r--. 1 root root 177 Aug 14 10:20 hosts.yaml
-rw-r--r--. 1 root root  29 Aug 14 10:15 groups.yaml
-rw-r--r--. 1 root root  37 Aug 14 10:06 defaults.yaml
-rw-r--r--. 1 root root 228 Aug 14 09:55 config.yaml
[root@localhost Nornir]#
```

在一切准备妥当后，下面正式进入实验环节。

9.2 Nornir 实验准备（华为设备，Windows 系统）

上一节演示了 CentOS 在 Python 主环境中进行实验准备，这一节将演示 Windows 在 Python 虚拟环境中进行实验准备。本书推荐网络工程师初学 Python 时在原生的 IDLE 上多加练习，充分理解对话模式和脚本模式的运行逻辑后，再自行把握，适当切入第三方 IDE。此次实验采用的 Pycharm 为社区免费版本（PyCharm Community Edition 2022.2.1）。

Nornir 需要安装的配套模块较多，有些模块还要指定版本，这样有可能与早前安装的其他模块冲突，造成不稳定。这里演示采用 Pycharm 基于虚拟环境搭建实验环境，可有效解决模块冲突问题。创建虚拟环境的好处就是这个工程是独立自主的，只服务于此次 Nornir 实验。在这个虚拟环境中，各类操作均不会影响到系统主环境。

1. 创建虚拟环境及安装 Nornir 套件

找一个心仪的目录，单击右键新增一个文件夹，命名为 nornir_test，可以随意命名，但建议整个文件的路径为纯英文。

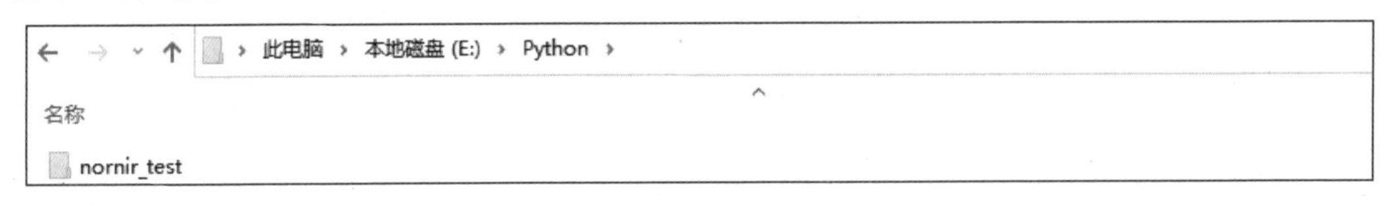

以管理员身份进入 CMD，进入我们建立的目录，用命令创建虚拟环境，如下图所示。

```
python -m venv nornir-venv
```

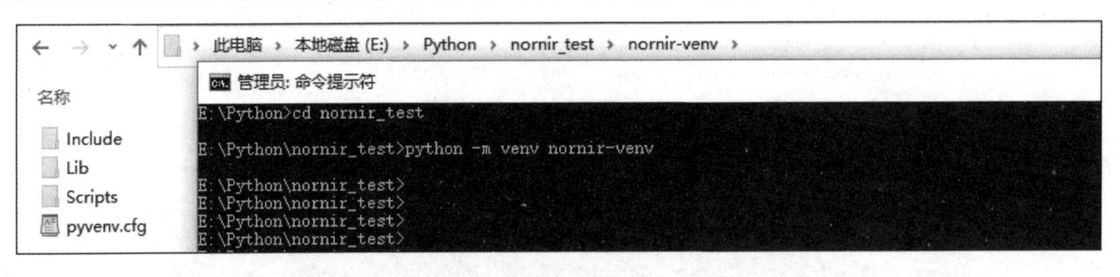

虚拟环境完成创建后，激活虚拟环境，如下图所示。

```
nornir-venv\Scripts\activate
```

```
管理员: 命令提示符

(nornir-venv) E:\Python\nornir_test>pip freeze

(nornir-venv) E:\Python\nornir_test>_
```

进入虚拟环境（路径前会出现标识）后，执行 pip 查看 Python 相关第三方库。因为是全新环境，所以查不到信息。

如果 pip 源要求改成国内，可以执行如下指令。如果不修改，大概率也没有问题，可能就是下载慢一点而已。

```
#临时修改，阿里源
pip3 install xxx[模块名] -i https://mirrors.aliyun.com/pypi/simple/
#永久修改，阿里源
pip3 config set global.index-url https://mirrors.aliyun.com/pypi/simple/
```

安装各种 Nornir 套件，以下未带国内镜像源。

```
#这些模块不是每个实验都用得到，具体什么实验用到什么模块，看实验代码的 import 内容
pip3 install netmiko
pip3 install napalm
pip3 install nornir
pip3 install nornir_utils
pip3 install nornir_napalm
pip3 install nornir_netmiko
pip3 install napalm-huawei-vrp #华为实验特有，将在实验 1 中详细阐述
pip3 install nornir_jinja2 #实验 8 使用
```

这里会有各种跑码，不必细究，留意提示 Successfully。初学者在这里可能会遇到各种报错，有可能是网络问题，需要修改一下镜像源等。虚拟环境的好处就在于，如果安装有异常或误操作，可以直接把文件夹删掉，然后重新来过，完全不影响系统的 Python 主环境。

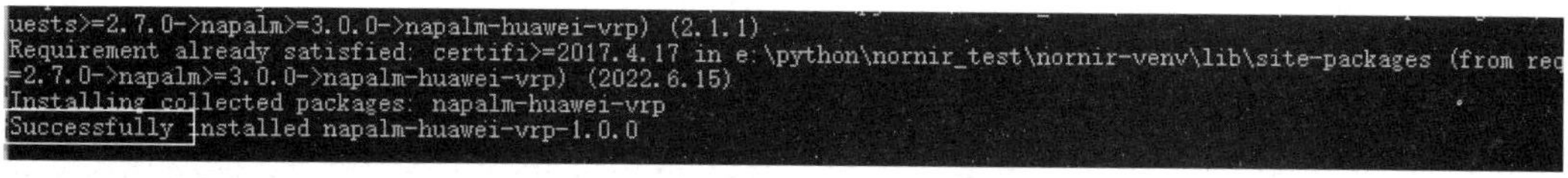

```
uests>=2.7.0->napalm>=3.0.0->napalm-huawei-vrp) (2.1.1)
Requirement already satisfied: certifi>=2017.4.17 in e:\python\nornir_test\nornir-venv\lib\site-packages (from req
=2.7.0->napalm>=3.0.0->napalm-huawei-vrp) (2022.6.15)
Installing collected packages: napalm-huawei-vrp
Successfully installed napalm-huawei-vrp-1.0.0
```

使用 CMD 命令查询一下，留意 Nornir 的版本在 3.0.0 以上。

```
pip freeze | find /I "nornir"
pip3 freeze | find /I "napalm"
```

```
(nornir-venv) E:\Python\nornir_test>
(nornir-venv) E:\Python\nornir_test>pip3 freeze | find /I "nornir"
nornir==3.3.0
nornir-napalm==0.3.0
nornir-netmiko==0.2.0
nornir-utils==0.2.0

(nornir-venv) E:\Python\nornir_test>pip3 freeze | find /I "napalm"
napalm==4.0.0
napalm-huawei-vrp==1.0.0
nornir-napalm==0.3.0

(nornir-venv) E:\Python\nornir_test>_
```

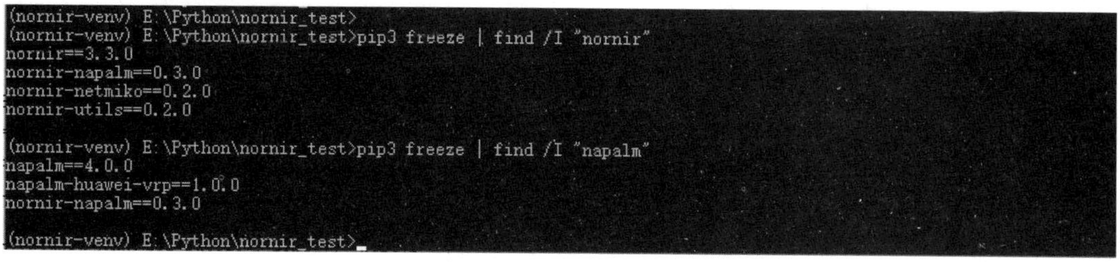

如果要退出虚拟环境，则执行如下命令。

```
Deactivate
```

```
(nornir-venv) E:\Python\nornir_test>deactivate
E:\Python\nornir_test>
```

2. 联动 Pycharm

Pycharm 的具体版本没有特别讲究，本实验从 2019 版尝试到 2022 版都是成功的。找到 Open File or Project 选项即可（通常在 File 菜单的 Open 里），如下图所示。

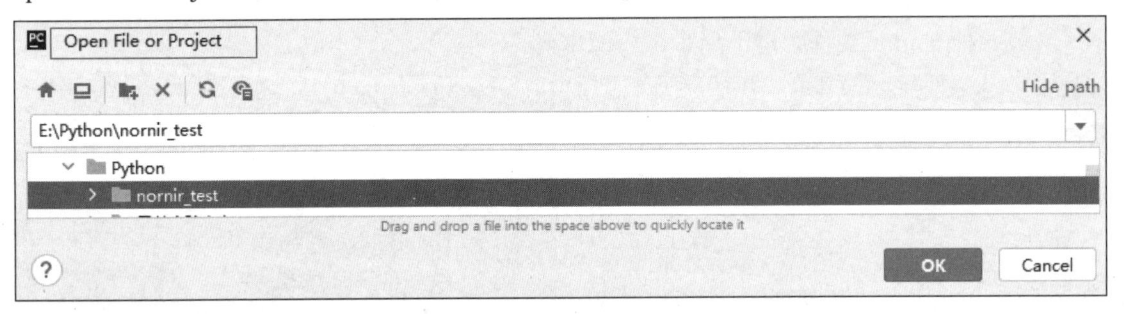

单击“OK”后一番加载，Pycharm 即会切换到这个工程，如下图所示。

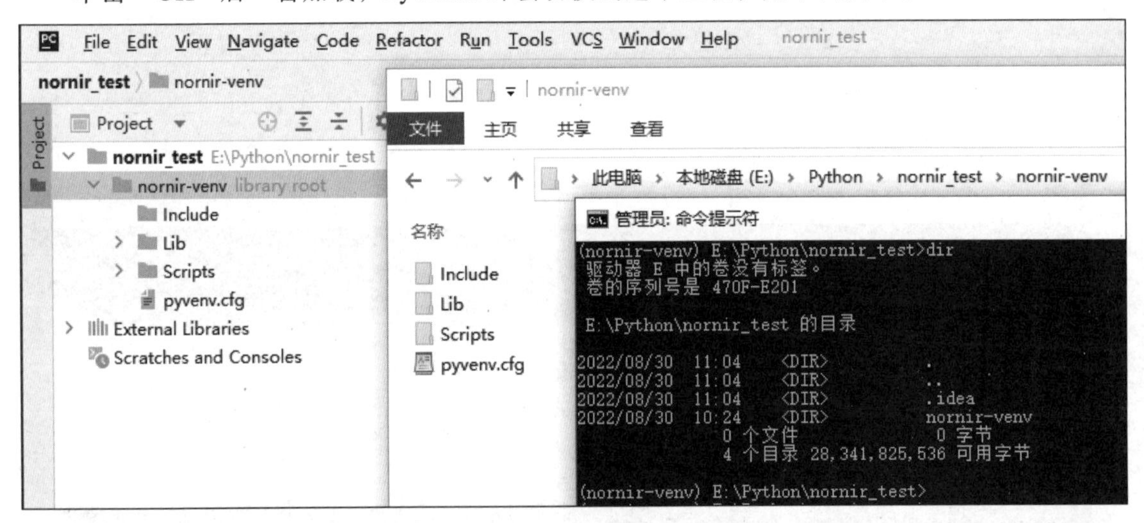

此时，已成功地把 nornir-venv 虚拟环境与 Pycharm 进行联动。检查一下 Python 解释器，看看是不是对应，如下图所示。

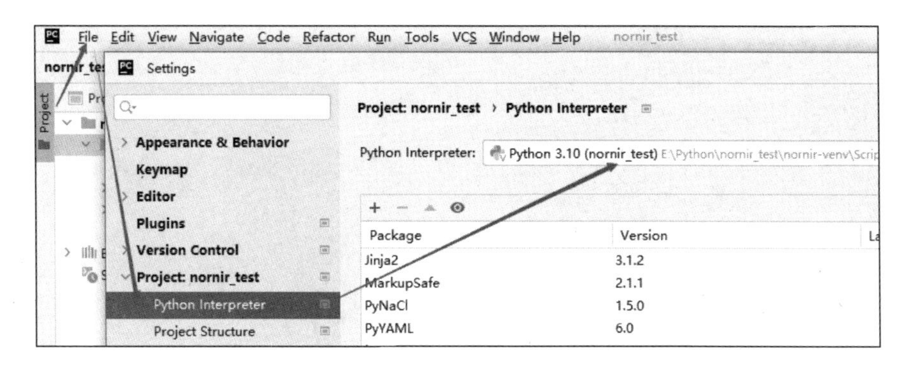

每次实验可按文件夹进行归类。除 Python 脚本外，可能涉及 config.yaml、defaults.yaml、groups.yaml、hosts.yaml 等文件。

◎　config.yaml

```
---
inventory:
  plugin: SimpleInventory
  options:
    host_file: "hosts.yaml"
    group_file: "groups.yaml"
    defaults_file: "defaults.yaml"
runner:
  plugin: threaded
  options:
    num_workers: 100
```

◎　groups.yaml

```
---
username: python
password: '123'
```

◎　groups.yaml

```
---
huawei:
  platform: huawei_vrp
```

◎　hosts.yaml

```
---
sw1:
```

```
    hostname: 192.168.2.11
    username: python
    password: '123'
    groups:
        - huawei
sw2:
    hostname: 192.168.2.12
    groups:
        - Huawei
```

文件内容随每个实验进行相应的调整，其具体含义与思科设备实验准备中的介绍一致，这里不再赘述。

9.3　实验 1　调用 nornir_napalm 获取设备的 facts 和 interfaces 信息（思科设备）

首先创建一个名为 nornir1.py 的 Python 脚本，然后写入如下代码。

```
from nornir import InitNornir
from nornir_napalm.plugins.tasks import napalm_get
from nornir_utils.plugins.functions import print_result

nr = InitNornir(config_file="config.yaml", dry_run=True)
results = nr.run(task=napalm_get, getters=["facts", "interfaces"])

print_result(results)
```

```
[root@localhost Nornir]# cat nornir1.py
from nornir import InitNornir
from nornir_napalm.plugins.tasks import napalm_get
from nornir_utils.plugins.functions import print_result

nr = InitNornir(config_file="config.yaml", dry_run=True)
results = nr.run(task=napalm_get, getters=["facts", "interfaces"])

print_result(results)
[root@localhost Nornir]#
```

代码分段讲解如下。

- 首先从 Nornir 中导入 InitNornir 子类，InitNornir 的作用是初始化 Nornir，是 Nornir 最重要的一个子类。

```
from nornir import InitNornir
```

- 然后从 nornir_napalm.plugins.tasks 中调用 napalm_get。napalm_get 子类允许我们调用 NAPALM 的 getter 类的 API 来获取设备的信息（有关 NAPALM 的 getter 类 API 的内容在 6.5 节中已经详细介绍过）。

```
from nornir_napalm.plugins.tasks import napalm_get
```

- 最后调用 nornir_utils.plugins.functions 中的 print_result，将我们通过 nornir_napalm 的 getter 类 API 得到的设备信息打印出来。

```
from nornir_utils.plugins.functions import print_result
```

- 使用 InitNornir()函数来初始化设备并将它赋值给变量 nr。InitNonir()函数中需要放入 config_file 参数。针对这个参数，我们使用的是前面提到的 config.yaml 文件，后面的 dry_run 参数在调用 nornir_napalm 时必须设为 True。

```
nr = InitNornir(config_file="config.yaml", dry_run=True)
```

- 我们调用 nr.run()函数来正式使用 NAPALM 下的两个 getter 类 API，"facts"和"interfaces"分别对应的是 get_facts 和 get_interfaces 两个 getter 类 API，如下图所示。

```
results = nr.run(task=napalm_get, getters=["facts", "interfaces"])
```

Getters support matrix

Note

The following table is built automatically. Every time there is a release of a supported driver a built is triggered. The result of the tests are aggregated on the following table.

	EOS	IOS	IOSXR	JUNOS	NXOS	NXOS_SSH
get_arp_table	✓	✓	✗	✗	✗	✓
get_bgp_config	✓	✓	✓	✓	✗	✗
get_bgp_neighbors	✓	✓	✓	✓	✓	✓
get_bgp_neighbors_detail	✓	✓	✓	✓	✗	✗
get_config	✓	✓	✓	✓	✓	✓
get_environment	✓	✓	✓	✓	✓	✓
get_facts	✓	✓	✓	✓	✓	✓
get_firewall_policies	✗	✗	✗	✗	✗	✗
get_interfaces	✓	✓	✓	✓	✓	✓
get_interfaces_counters	✓	✓	✓	✓	✗	✓
get_interfaces_ip	✓	✓	✓	✓	✓	✓

- 用 print_result()将结果打印出来（注意这里 nr.run()返回的值的类型为 nornir.core.task.AggregatedResult，必须用 nornir.utils 里的 print_result()才能将它完整打印出来，我们常用的 print()对它是无效的）。

```
results = nr.run(task=napalm_get, getters=["facts", "interfaces"])
```

运行脚本看效果，如下图所示。

```
[root@localhost Nornir]# python3.10 nornir1.py
napalm_get***********************************************************************
* sw1 ** changed : False ********************************************************
vvvv napalm_get ** changed : False vvvvvvvvvvvvvvvvvvvvvvvvvvvvvvvvvvvvvvvvvvvvvvvv INFO
{ 'facts': { 'fqdn': 'SW1.test',
             'hostname': 'SW1',
             'interface_list': [ 'GigabitEthernet0/1',
                                 'GigabitEthernet0/2',
                                 'GigabitEthernet0/3',
                                 'GigabitEthernet0/0'],
             'model': 'IOSv',
             'os_version': 'vios_l2 Software (vios_l2-ADVENTERPRISEK9-M), '
                           'Version 15.0(TTC_20140605)FLO_DSGS7, EARLY '
                           'DEPLOYMENT DEVELOPMENT BUILD, synced to  '
                           'V152_3_0_88_PI4',
             'serial_number': '',
             'uptime': 360.0,
             'vendor': 'Cisco'},
  'interfaces': { 'GigabitEthernet0/0': { 'description': '',
                                          'is_enabled': True,
                                          'is_up': True,
                                          'last_flapped': -1.0,
                                          'mac_address': '0C:1A:91:B4:48:00',
                                          'mtu': 1500,
                                          'speed': 1000.0},
                  'GigabitEthernet0/1': { 'description': '',
                                          'is_enabled': True,
                                          'is_up': False,
                                          'last_flapped': -1.0,
                                          'mac_address': '0C:1A:91:B4:48:01',
                                          'mtu': 1500,
                                          'speed': 1000.0},
                  'GigabitEthernet0/2': { 'description': '',
                                          'is_enabled': True,
                                          'is_up': False,
                                          'last_flapped': -1.0,
                                          'mac_address': '0C:1A:91:B4:48:02',
                                          'mtu': 1500,
                                          'speed': 1000.0},
                  'GigabitEthernet0/3': { 'description': '',
                                          'is_enabled': True,
                                          'is_up': False,
                                          'last_flapped': -1.0,
                                          'mac_address': '0C:1A:91:B4:48:03',
                                          'mtu': 1500,
                                          'speed': 1000.0}}}
^^^^ END napalm_get ^^^^^^^^^^^^^^^^^^^^^^^^^^^^^^^^^^^^^^^^^^^^^^^^^^^^^^^^^^^^^
* sw2 ** changed : False ********************************************************
vvvv napalm_get ** changed : False vvvvvvvvvvvvvvvvvvvvvvvvvvvvvvvvvvvvvvvvvvvvvvvv INFO
{ 'facts': { 'fqdn': 'SW2.test',
             'hostname': 'SW2',
             'interface_list': [ 'GigabitEthernet0/1',
                                 'GigabitEthernet0/2',
                                 'GigabitEthernet0/3',
                                 'GigabitEthernet0/0'],
             'model': 'IOSv',
             'os_version': 'vios_l2 Software (vios_l2-ADVENTERPRISEK9-M), '
                           'Version 15.0(TTC_20140605)FLO_DSGS7, EARLY '
                           'DEPLOYMENT DEVELOPMENT BUILD, synced to  '
```

可以看到，用 print_result()函数打印出来的内容是 json 格式的，便于我们使用键值对来筛选需要的设备信息。

9.4　实验 1　调用 nornir_napalm 获取设备的 facts 和 interfaces 信息（华为设备）

思科设备实验使用了 NAPALM 模块，其原生并不支持华为设备，但已有热心人士基于它开发出了 napalm-huawei-vrp 模块，目前支持华为园区网交换机设备，如下图所示。

napalm-huawei-vrp

NAPALM driver for HUAWEI Campus Network Switch, support the S5700,S6700 etc.

This repository is reference NAPALM-CE and Cisco IOS code, thanks for thdDaniel's contribution

把文件复制到对应的目录中，可根据自己设置的虚拟环境文件结构进行灵活调整，如下图所示。

```
E:\Python\nornir_test\nornir-venv\lab1
```

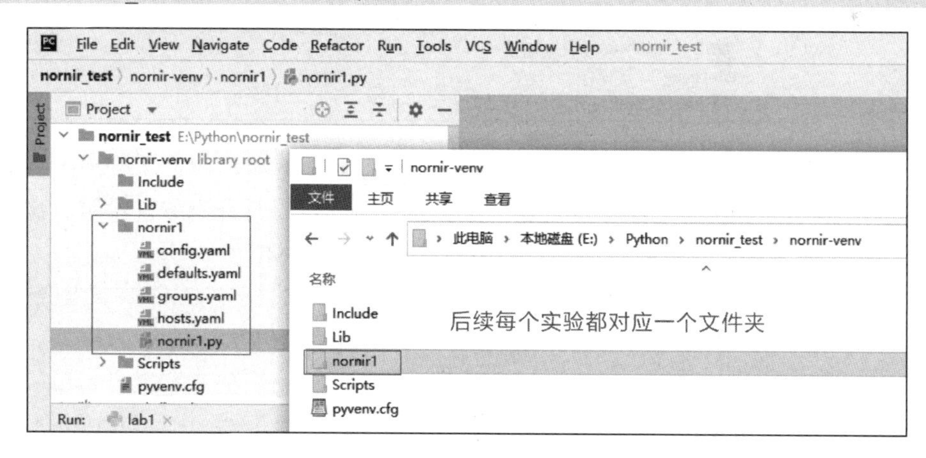

Python 代码如下。

```
from nornir import InitNornir
from nornir_napalm.plugins.tasks import napalm_get
from nornir_utils.plugins.functions import print_result

nr = InitNornir(config_file="config.yaml", dry_run=True)
results = nr.run(task=napalm_get, getters=["facts", "interfaces"])
```

```
print_result(results)
```

如果不用 Pycharm，也可以在虚拟环境中用 CMD 直接运行，如下图所示。

命令提示符

```
(nornir-venv) E:\Python\nornir_test\nornir-venv\nornir1>
(nornir-venv) E:\Python\nornir_test\nornir-venv\nornir1>nornir1.py
napalm_get*************************************************************************
* sw1 ** changed : False ************************************************************
vvvv napalm_get ** changed : False vvvvvvvvvvvvvvvvvvvvvvvvvvvvvvvvvvvvvvvvvvvvvvvvvv INFO
{ 'facts': { 'fqdn': 'Unknown',
             'hostname': 'SW1',
             'interface_list': [ 'GigabitEthernet0/0/1',
                                 'GigabitEthernet0/0/2',
                                 'GigabitEthernet0/0/3',
                                 'GigabitEthernet0/0/4',
```

下图是 Pycharm 运行结果。

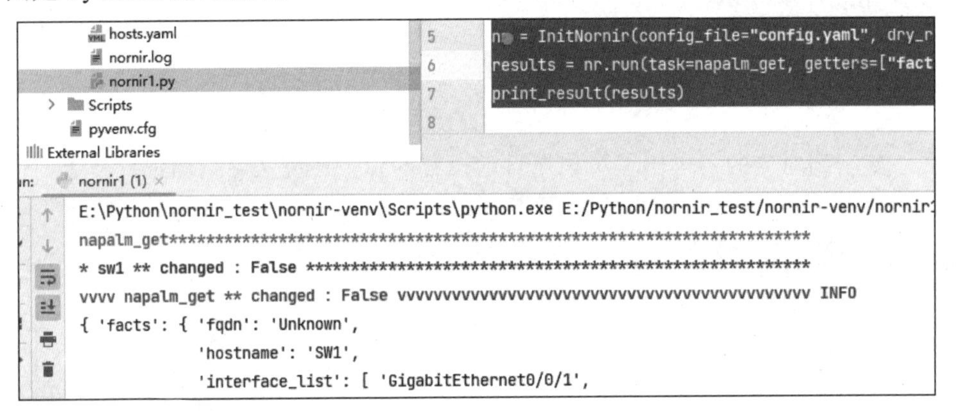

脚本的运行结果符合预期，限于篇幅，仅展示一小部分截图，如下图所示。运行后在文件夹中会生成一个 nornir.log 文件，主要是一些日志记录，可以打开看看。没错，这个实验的 Python 代码与思科设备实验完全一致，差异内容仅在几个 YAML 文件上。

API部分截图

API	Description
load_merge_candidate()	Load config
compare_config()	A string showing the difference between the running configuration and the candidate configuration
discard_config()	Discards the configuration loaded into the candidate

关于 napalm-huawei-vrp 的其他方法，可以查阅其 pypi 网页信息。

napalm-huawei-vrp 算是国产轮子，很感谢有这么些人持续贡献，他们在努力缩小这个领域国内外的差距。期待在不久的未来，我们会有更多国产轮子，甚至是国产原生轮子！

9.5　实验 2　调用 nornir_netmiko 来获取设备信息（思科设备）

除 nornir_napalm 外，我们还可以用 nornir_netmiko 来向设备输入各种 show 命令以获取设备信息，这里我们创建第 2 个实验脚本 nornir2.py 并放入下面的代码。

```
from nornir import InitNornir
from nornir_netmiko import netmiko_send_command
from nornir_utils.plugins.functions import print_result

nr = InitNornir(config_file="config.yaml")
results = nr.run(netmiko_send_command, command_string='sh clock')

print_result(results)
```

```
[root@localhost Nornir]# cat nornir2.py
from nornir import InitNornir
from nornir_netmiko import netmiko_send_command
from nornir_utils.plugins.functions import print_result

nr = InitNornir(config_file="config.yaml")
results = nr.run(netmiko_send_command, command_string='sh clock')

print_result(results)

[root@localhost Nornir]#
```

代码分段讲解如下。

- 因为实验 2 要使用 nornir_netmiko 插件来获取设备信息，所以这里我们只需要把实验 1 中的代码稍作修改，将实验 1 中的 from nornir_napalm.plugins.tasks import napalm_get 换成 from nornir_netmiko import netmiko_send_command 即可。

```
from nornir import InitNornir
from nornir_netmiko import netmiko_send_command
from nornir_utils.plugins.functions import print_result

nr = InitNornir(config_file="config.yaml")
```

- 然后使用 Netmiko 的 send_command 函数向 SW1 和 SW2 执行 show clock 命令，并将回显内容用 print_result()打印出来。

```
results = nr.run(netmiko_send_command, command_string='sh clock')

print_result(results)
```

运行脚本，效果如下图所示。

```
(Nornir) [root@localhost Nornir]# python3 nornir2.py
netmiko_send_command*****************************************************************
* sw1 ** changed : False ************************************************************
vvvv netmiko_send_command ** changed : False vvvvvvvvvvvvvvvvvvvvvvvvvvvvvvvvvvvvvvv INFO

*10:15:19.475 UTC Mon Jan 18 2021
^^^^ END netmiko_send_command ^^^^^^^^^^^^^^^^^^^^^^^^^^^^^^^^^^^^^^^^^^^^^^^^^^^^^^^^^
* sw2 ** changed : False ************************************************************
vvvv netmiko_send_command ** changed : False vvvvvvvvvvvvvvvvvvvvvvvvvvvvvvvvvvvvvvv INFO

*10:15:19.618 UTC Mon Jan 18 2021
^^^^ END netmiko_send_command ^^^^^^^^^^^^^^^^^^^^^^^^^^^^^^^^^^^^^^^^^^^^^^^^^^^^^^^^^
(Nornir) [root@localhost Nornir]#
```

运行代码后，可以看到两台交换机当前的时间，注意：SW1 和 SW2 当前时间只间隔了 0.143s (10:15:19.475—10:15:19.618)，说明 Nornir 是并发进行的（如果想进一步验证，还可以在 hosts 里面加上剩下的 SW3、SW4、SW5，甚至更多的交换机来一起 show clock）。

另外，除了 netmiko_send_command，我们还可以在 Nornir 里使用 netmiko_send_config、netmiko_file_transfer、netmiko_save_config 等函数，如下图所示。关于它们的用法已经在第 8 章里介绍过，这里留给读者自行去尝试。

```
[root@localhost Nornir]# python3.10
Python 3.10.6 (main, Aug  6 2022, 13:49:15) [GCC 8.5.0 20210514 (Red Hat 8.5.0-1
5)] on linux
Type "help", "copyright", "credits" or "license" for more information.
>>> import nornir_netmiko
>>> dir(nornir_netmiko)
['CONNECTION_NAME', 'Netmiko', '__all__', '__builtins__', '__cached__', '__doc__
', '__file__', '__loader__', '__name__', '__package__', '__path__', '__spec__',
'connections', 'netmiko_commit', 'netmiko_file_transfer', 'netmiko_multiline', '
netmiko_save_config', 'netmiko_send_command', 'netmiko_send_config', 'tasks']
>>>
```

9.6　实验 2　调用 nornir_netmiko 获取设备信息（华为设备）

Netmiko 官方支持华为设备，如果考虑设备范围更广泛，更推荐使用 Netmiko。把实验 1 的 nornir1 文件夹直接赋值成 nornir2，需要修改一下 groups.yaml，将 huawei_vrp 修改为 huawei，符合 Netmiko 要求。

```
---
huawei:
  platform: Huawei
```

参照思科设备实验 2 的 Python 脚本，我们将其移植到华为设备上，查询设备时间的指令修改，命名为 nornir2.py。

```
from nornir import InitNornir
from nornir_netmiko import netmiko_send_command
from nornir_utils.plugins.functions import print_result

nr = InitNornir(config_file="config.yaml")
results = nr.run(netmiko_send_command, command_string='disp clock')

print_result(results)
```

此时的文件结构和运行结果如下图所示。

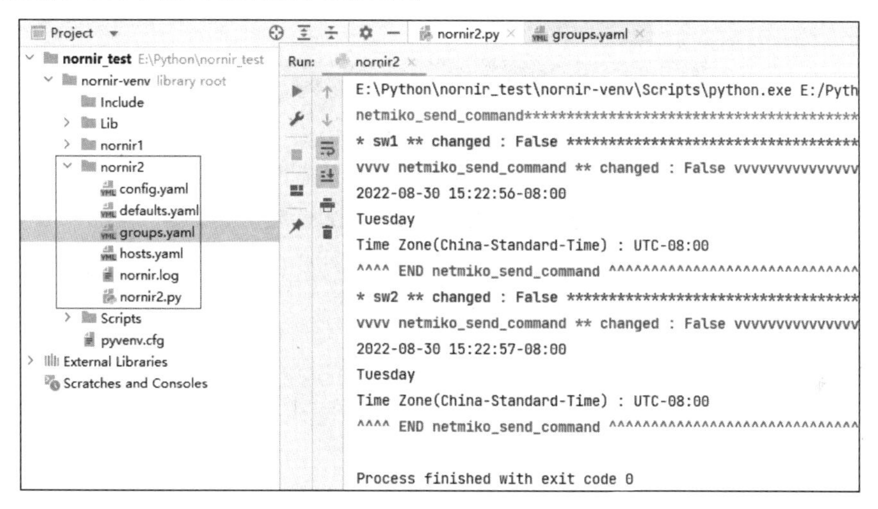

Netmiko 的详细使用方法见本书 Netmiko 详解章节。在 Nornir 中，需要感受 config.yaml 和其他几个配置文件的关系，并且它们是如何与 Netmiko 联动的，比如本实验修改的 huawei 字眼，是会牵动到 Netmiko 的 device_type 的。

另外，还记得如果不在 Pycharm 内运行而用 CMD 运行脚本吗？请读者自行尝试。

9.7　实验 3　使用 filter()配合 F()做高级过滤（思科设备）

在 Nornir 中，可以使用 filter()及在 filter()里配合 F()函数来做过滤，前者叫作简单过滤（simple filtering），后者叫作高级过滤（advanced filtering）。针对过滤的内容，将分成 3 个实验来讲，实验 3 将介绍 filter 配合 F()函数做过滤的方法，实验 4 将介绍使用 filter()过滤的方法，实验 5 将介绍 filter()配合 lambda 做过滤的方法。先讲高级过滤后讲简单过滤的原因是后者涉及一些扩展内容，需要放在后面来讲。

在开始实验 3 前，首先修改 groups.yaml 和 hosts.yaml 两个文件。

在 groups.yaml 里将之前的 cisco 这个 group 拿掉，然后重新划分两个 group，分别取名为 cisco_group1 和 cisco_group2。

```
---
cisco_group1:
   platform: ios

cisco_group2:
   platform: ios
```

```
[root@localhost Nornir]# cat groups.yaml
---
cisco_group1:
    platform: ios

cisco_group2:
    platform: ios
[root@localhost Nornir]#
```

在 hosts.yaml 里添加 sw3 和 sw4，将之前的 sw1 和 sw2 划入 cisco_group1，将 sw3 和 sw4 划入 cisco_group2。

```
---
sw1:
   hostname: 192.168.2.11
   username: python
   password: '123'
   platform: ios
   groups:
      - cisco_group1
sw2:
   hostname: 192.168.2.12
   platform: ios
   groups:
      - cisco_group1
sw3:
   hostname: 192.168.2.13
   platform: ios
   groups:
      - cisco_group2
sw4:
   hostname: 192.168.2.14
```

```
    platform: ios
    groups:
        - cisco_group2
```

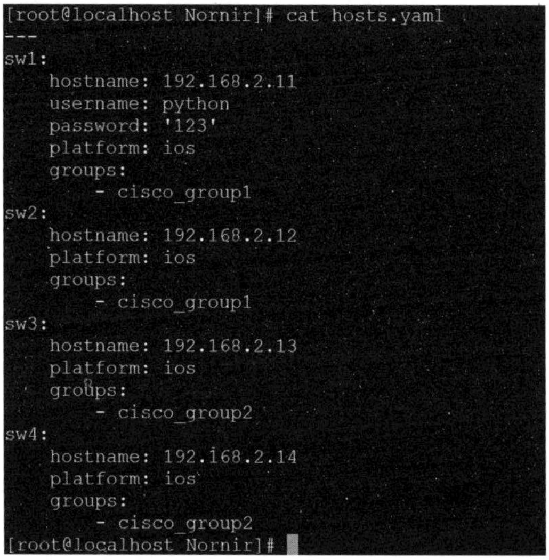

```
[root@localhost Nornir]# cat hosts.yaml
---
sw1:
    hostname: 192.168.2.11
    username: python
    password: '123'
    platform: ios
    groups:
        - cisco_group1
sw2:
    hostname: 192.168.2.12
    platform: ios
    groups:
        - cisco_group1
sw3:
    hostname: 192.168.2.13
    platform: ios
    groups:
        - cisco_group2
sw4:
    hostname: 192.168.2.14
    platform: ios
    groups:
        - cisco_group2
[root@localhost Nornir]#
```

然后创建第 3 个实验脚本 nornir3.py，脚本代码内容如下。

```
from nornir import InitNornir
from nornir_netmiko import netmiko_send_command
from nornir_utils.plugins.functions import print_result
from nornir.core.filter import F

nr = InitNornir(config_file="config.yaml")
group1 = nr.filter(F(groups__contains="cisco_group1"))
group2 = nr.filter(~F(groups__contains="cisco_group1"))
results = group1.run(netmiko_send_command, command_string='sh ip int brief')

print_result(results)
```

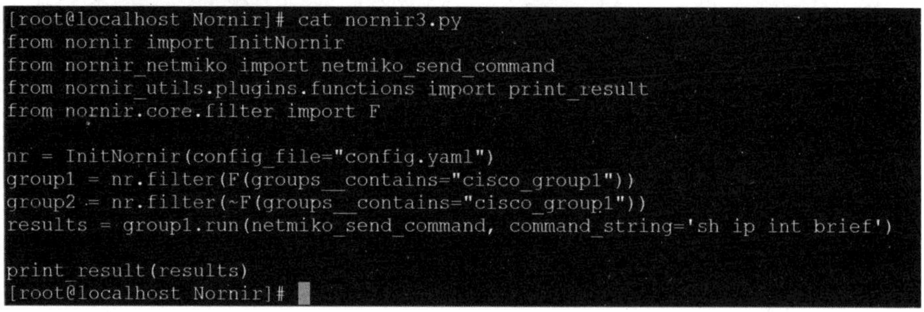

```
[root@localhost Nornir]# cat nornir3.py
from nornir import InitNornir
from nornir_netmiko import netmiko_send_command
from nornir_utils.plugins.functions import print_result
from nornir.core.filter import F

nr = InitNornir(config_file="config.yaml")
group1 = nr.filter(F(groups__contains="cisco_group1"))
group2 = nr.filter(~F(groups__contains="cisco_group1"))
results = group1.run(netmiko_send_command, command_string='sh ip int brief')

print_result(results)
[root@localhost Nornir]#
```

代码分段讲解如下。

- 在 Nornir 中，我们可以使用 nornir.core.filter 的 F()函数来做过滤。

```
from nornir.core.filter import F
```

- Nornir 的过滤支持取非操作，首先调用 nr.filter(F(groups__contains="cisco_group1")来过滤 cisco_group1 下的 SW1 和 SW2，将其赋值给变量 group1，注意 F()函数里的 groups__contains 参数，中间的那个__是双下划线，不是单下划线。

```
group1 = nr.filter(F(groups__contains="cisco_group1"))
```

- 然后在 F 的前面加上一个波浪号~对其进行取非来过滤 cisco_group2 下的 SW3 和 SW4，并将结果赋值给变量 group2（因为这里总共只有 cisco_group1 和 cisco_group2 两个 group，因此对 cisco_group1 取非后过滤出的就是 cisco_group2）。

```
group2 = nr.filter(~F(groups__contains="cisco_group1"))
```

- 接着打印出 group1 的内容，可以看到，它只返回了 SW1 和 SW2 的 show ip int brief 的回显内容。

```
results = group1.run(netmiko_send_command, command_string='sh ip int brief')

print_result(results)
```

```
[root@localhost Nornir]# python3.10 nornir3.py
netmiko_send_command*************************************************************
* sw1 ** changed : False *********************************************************
vvvv netmiko_send_command ** changed : False vvvvvvvvvvvvvvvvvvvvvvvvvvvvvvvvvvvvvv
 INFO
Interface              IP-Address      OK? Method Status                Protocol
GigabitEthernet0/1     unassigned      YES unset  down                  down
GigabitEthernet0/2     unassigned      YES unset  down                  down
GigabitEthernet0/3     unassigned      YES unset  down                  down
GigabitEthernet0/0     192.168.2.11    YES manual up                    up
^^^^ END netmiko_send_command ^^^^^^^^^^^^^^^^^^^^^^^^^^^^^^^^^^^^^^^^^^^^^^^^^^^^^
* sw2 ** changed : False *********************************************************
vvvv netmiko_send_command ** changed : False vvvvvvvvvvvvvvvvvvvvvvvvvvvvvvvvvvvvvv
 INFO
Interface              IP-Address      OK? Method Status                Protocol
GigabitEthernet0/1     unassigned      YES unset  down                  down
GigabitEthernet0/2     unassigned      YES unset  down                  down
GigabitEthernet0/3     unassigned      YES unset  down                  down
GigabitEthernet0/0     192.168.2.12    YES manual up                    up
^^^^ END netmiko_send_command ^^^^^^^^^^^^^^^^^^^^^^^^^^^^^^^^^^^^^^^^^^^^^^^^^^^^^
[root@localhost Nornir]#
```

- 然后在脚本里将 results = group1.run(netmiko_send_command, command_string='sh ip int brief')改为 results = group2.run(netmiko_send_command, command_string='sh ip int brief')，再次执行脚本，打印出 group2 也就是 SW3 和 SW4 的 show ip int brief 的回显内容，如下图所示。

这时你也许会问：怎么取消过滤，一次性将 4 台交换机的 show ip int brief 全部打印出来呢？答案很简单，只需要将 group1.run()或 group2.run()替换成 nr.run()就行了。

```
results = nr.run(netmiko_send_command, command_string='sh ip int brief')

print_result(results)
```

```
[root@localhost Nornir]# python3.10 nornir3.py
netmiko_send_command*****************************************************************
* sw3 ** changed : False ************************************************************
vvvv netmiko_send_command ** changed : False vvvvvvvvvvvvvvvvvvvvvvvvvvvvvvvvvvv INFO
Interface              IP-Address      OK? Method Status                Protocol
GigabitEthernet0/1     unassigned      YES unset  down                  down
GigabitEthernet0/2     unassigned      YES unset  down                  down
GigabitEthernet0/3     unassigned      YES unset  down                  down
GigabitEthernet0/0     192.168.2.13    YES manual up                    up
^^^^ END netmiko_send_command ^^^^^^^^^^^^^^^^^^^^^^^^^^^^^^^^^^^^^^^^^^^^^^^^^^^^
* sw4 ** changed : False ************************************************************
vvvv netmiko_send_command ** changed : False vvvvvvvvvvvvvvvvvvvvvvvvvvvvvvvvvvv INFO
Interface              IP-Address      OK? Method Status                Protocol
GigabitEthernet0/1     unassigned      YES unset  down                  down
GigabitEthernet0/2     unassigned      YES unset  down                  down
GigabitEthernet0/3     unassigned      YES unset  down                  down
GigabitEthernet0/0     192.168.2.14    YES manual up                    up
^^^^ END netmiko_send_command ^^^^^^^^^^^^^^^^^^^^^^^^^^^^^^^^^^^^^^^^^^^^^^^^^^^^
[root@localhost Nornir]#
```

再次运行脚本后，效果如下图所示。

```
[root@localhost Nornir]# cat nornir3.py
from nornir import InitNornir
from nornir_netmiko import netmiko_send_command
from nornir_utils.plugins.functions import print_result
from nornir.core.filter import F

nr = InitNornir(config_file="config.yaml")
group1 = nr.filter(F(groups__contains="cisco_group1"))
group2 = nr.filter(~F(groups__contains="cisco_group1"))
results = nr.run(netmiko_send_command, command_string='sh ip int brief')

print_result(results)
[root@localhost Nornir]# python3.10 nornir3.py
netmiko_send_command*****************************************************************
* sw1 ** changed : False ************************************************************
vvvv netmiko_send_command ** changed : False vvvvvvvvvvvvvvvvvvvvvvvvvvvvvvvvvvv INFO
Interface              IP-Address      OK? Method Status                Protocol
GigabitEthernet0/1     unassigned      YES unset  down                  down
GigabitEthernet0/2     unassigned      YES unset  down                  down
GigabitEthernet0/3     unassigned      YES unset  down                  down
GigabitEthernet0/0     192.168.2.11    YES manual up                    up
^^^^ END netmiko_send_command ^^^^^^^^^^^^^^^^^^^^^^^^^^^^^^^^^^^^^^^^^^^^^^^^^^^^
* sw2 ** changed : False ************************************************************
vvvv netmiko_send_command ** changed : False vvvvvvvvvvvvvvvvvvvvvvvvvvvvvvvvvvv INFO
Interface              IP-Address      OK? Method Status                Protocol
GigabitEthernet0/1     unassigned      YES unset  down                  down
GigabitEthernet0/2     unassigned      YES unset  down                  down
GigabitEthernet0/3     unassigned      YES unset  down                  down
GigabitEthernet0/0     192.168.2.12    YES manual up                    up
^^^^ END netmiko_send_command ^^^^^^^^^^^^^^^^^^^^^^^^^^^^^^^^^^^^^^^^^^^^^^^^^^^^
* sw3 ** changed : False ************************************************************
vvvv netmiko_send_command ** changed : False vvvvvvvvvvvvvvvvvvvvvvvvvvvvvvvvvvv INFO
Interface              IP-Address      OK? Method Status                Protocol
GigabitEthernet0/1     unassigned      YES unset  down                  down
GigabitEthernet0/2     unassigned      YES unset  down                  down
GigabitEthernet0/3     unassigned      YES unset  down                  down
GigabitEthernet0/0     192.168.2.13    YES manual up                    up
^^^^ END netmiko_send_command ^^^^^^^^^^^^^^^^^^^^^^^^^^^^^^^^^^^^^^^^^^^^^^^^^^^^
* sw4 ** changed : False ************************************************************
vvvv netmiko_send_command ** changed : False vvvvvvvvvvvvvvvvvvvvvvvvvvvvvvvvvvv INFO
Interface              IP-Address      OK? Method Status                Protocol
GigabitEthernet0/1     unassigned      YES unset  down                  down
GigabitEthernet0/2     unassigned      YES unset  down                  down
GigabitEthernet0/3     unassigned      YES unset  down                  down
GigabitEthernet0/0     192.168.2.14    YES manual up                    up
^^^^ END netmiko_send_command ^^^^^^^^^^^^^^^^^^^^^^^^^^^^^^^^^^^^^^^^^^^^^^^^^^^^
[root@localhost Nornir]#
```

9.8　实验 3　使用 filter()配合 F()做高级过滤（华为设备）

完整复制实验 2 的文件夹为 nornir3，修改 groups.yaml 和 hosts.yaml 两个文件。

◎　groups.yaml

```
---
huawei_group1:
  platform: huawei

huawei_group2:
  platform: huawei
```

◎　hosts.yaml（请仔细观察，这里比思科实验少了一个信息——platform，做了一些灵活调整，这也是可以的）

```
---
sw1:
  hostname: 192.168.2.11
  username: python
  password: '123'
  groups:
    - huawei_group1
sw2:
  hostname: 192.168.2.12
  groups:
    - huawei_group1
sw3:
  hostname: 192.168.2.13
  groups:
    - huawei_group2
sw4:
  hostname: 192.168.2.14
  groups:
    - huawei_group2
```

此时的目录结构如下图所示。

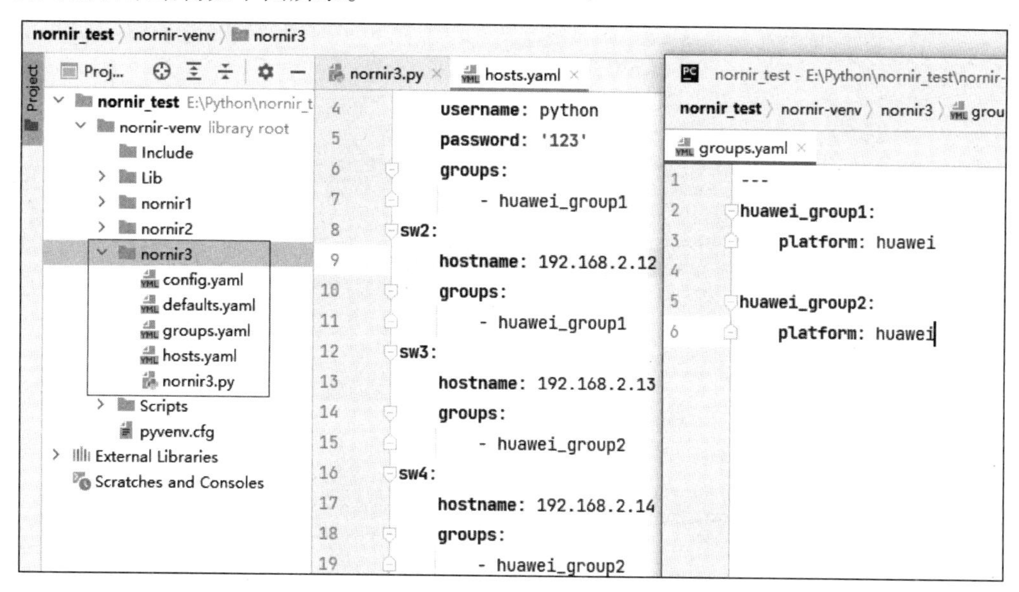

Nornir 通过 host.yaml 获得了用户名、密码及 groups 分组信息；通过 groups.yaml 获得了 platform 信息。如果需要用到的信息都齐全，则不用 defaults.yaml 也可以，即 default 中的信息是用来“兜底”的。在本实验中，我们把 SW1 和 SW2 划入 huawei_group1，把 SW3 和 SW4 划入 huawei_group2。Python 脚本调整如下。

```
from nornir import InitNornir
from nornir_netmiko import netmiko_send_command
from nornir_utils.plugins.functions import print_result
from nornir.core.filter import F

nr = InitNornir(config_file="config.yaml")
group1 = nr.filter(F(groups__contains="huawei_group1"))
#group1 由 huawei_group1 确定，即 SW1 和 SW2
group2 = nr.filter(~F(groups__contains="huawei_group1"))
#group2 由 huawei_group1 取非，即不是 SW1 或 SW2，那就是 SW3 和 SW4
results = group1.run(netmiko_send_command, command_string='disp ip int bri')

print_result(results)
```

运行 Python 脚本，此时 results 是 group1 的执行结果，即 SW1 和 SW2，结果符合预期，限于篇幅，只展示部分截图，如下图所示。

```
* sw1 ** changed : False *************************************************************
vvvv netmiko_send_command ** changed : False vvvvvvvvvvvvvvvvvvvvvvvvvvvvvvvvvvvvvv INFO
*down: administratively down
^down: standby
(l): loopback
(s): spoofing
The number of interface that is UP in Physical is 3
The number of interface that is DOWN in Physical is 1
The number of interface that is UP in Protocol is 3
The number of interface that is DOWN in Protocol is 1

Interface                         IP Address/Mask      Physical   Protocol
LoopBack0                         1.1.1.1/32           up         up(s)
NULL0                             unassigned           up         up(s)
Vlanif1                           192.168.2.11/24      up         up
^^^^ END netmiko_send_command ^^^^^^^^^^^^^^^^^^^^^^^^^^^^^^^^^^^^^^^^^^^^^^^^^^^^^^^^
* sw2 ** changed : False *************************************************************
```

如果把

```
results = group1.run(netmiko_send_command, command_string='disp ip int bri')
```

修改为

```
results = group2.run(netmiko_send_command, command_string='disp ip int bri')
```

则运行的结果会只出来 SW3 和 SW4，请读者自行尝试。

YAML 文件的结构灵活，能在适当的时候做一些分支，Nornir 通过与这些 YAML 文件配合，可以非常方便地进行设备分组和筛选，这对于不同地点、不同客户、不同厂商设备、不同型号、不同软件版本等分组筛选的过滤很有意义。

9.9　实验 4　使用 filter()做简单过滤（思科设备）

实验 4 是本章的重点内容，涉及的知识点会较多。在实验 4 开始讲 filter()函数之前，先来介绍下 inventory.hosts 和 inventory.groups。首先来回顾一下目前为止我们的 hosts.yaml 和 groups.yaml 文件中的内容。

◎　hosts.yaml

```
[root@localhost Nornir]# cat hosts.yaml
---
sw1:
    hostname: 192.168.2.11
    username: python
    password: '123'
    platform: ios
    groups:
        - cisco_group1
```

```
sw2:
    hostname: 192.168.2.12
    platform: ios
    groups:
        - cisco_group1
sw3:
    hostname: 192.168.2.13
    platform: ios
    groups:
        - cisco_group2
sw4:
    hostname: 192.168.2.14
    platform: ios
    groups:
        - cisco_group2
[root@localhost Nornir]#
```

◎　groups.yaml

```
[root@localhost Nornir]# cat groups.yaml
---
cisco_group1:
    platform: ios

cisco_group2:
    platform: ios
[root@localhost Nornir]#
```

本章开篇讲到，Inventory 是 Nornir 中最重要的组成部分，它包含 inventory.hosts 和 inventory.groups，分别对应前面讲到的 hosts.yaml 和 groups.yaml 两个文件。在 Nornir 中，我们可以通过调用 nr.inventory.hosts 和 nr.inventory.groups 来查看 Inventory 中有哪些 hosts 和 groups。下面用 Python 的解释器来做演示。

```
[root@localhost Nornir]#python3.10
Python 3.10.6 (main, Aug  6 2022, 13:49:15) [GCC 8.5.0 20210514 (Red Hat 8.5.0-15)] on linux
Type "help", "copyright", "credits" or "license" for more information.
>>> from nornir import InitNornir
>>> nr = InitNornir(config_file='config.yaml')
>>> nr.inventory.hosts
{'sw1': Host: sw1, 'sw2': Host: sw2, 'sw3': Host: sw3, 'sw4': Host: sw4}
>>> nr.inventory.groups
{'cisco_group1': Group: cisco_group1, 'cisco_group2': Group: cisco_group2}
>>>
```


```
[root@localhost Nornir]# python3.10
Python 3.10.6 (main, Aug  6 2022, 13:49:15) [GCC 8.5.0 20210514 (Red Hat 8.5.0-1
5)] on linux
Type "help", "copyright", "credits" or "license" for more information.
>>> from nornir import InitNornir
>>> nr = InitNornir(config_file='config.yaml')
>>> nr.inventory.hosts
{'sw1': Host: sw1, 'sw2': Host: sw2, 'sw3': Host: sw3, 'sw4': Host: sw4}
>>> nr.inventory.groups
{'cisco_group1': Group: cisco_group1, 'cisco_group2': Group: cisco_group2}
>>>
```

在此基础上，我们还可以继续指定交换机名称，并获取它的某一参数，如下图所示。

```
>>> nr.inventory.hosts['sw1'].platform
'ios'
>>> nr.inventory.hosts['sw1'].username
'python'
>>> nr.inventory.hosts['sw1'].password
'123'
>>> nr.inventory.hosts['sw2'].hostname
'192.168.2.12'
>>> nr.inventory.hosts['sw3'].hostname
'192.168.2.13'
>>> nr.inventory.hosts['sw4'].hostname
'192.168.2.14'
>>>
```


```
>>> nr.inventory.hosts['sw1'].platform
'ios'
>>> nr.inventory.hosts['sw1'].username
'python'
>>> nr.inventory.hosts['sw1'].password
'123'
>>> nr.inventory.hosts['sw2'].hostname
'192.168.2.12'
>>> nr.inventory.hosts['sw3'].hostname
'192.168.2.13'
>>> nr.inventory.hosts['sw4'].hostname
'192.168.2.14'
>>>
```

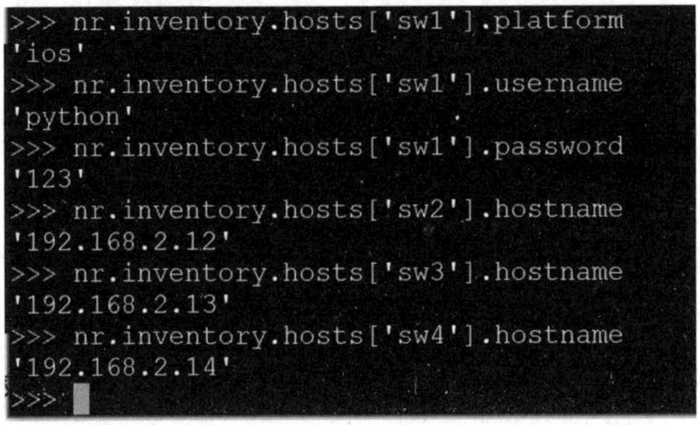

同样的道理，针对 groups，也可以做同样的操作，如下图所示。

```
>>> nr.inventory.groups['cisco_group1'].platform
'ios'
>>> nr.inventory.groups['cisco_group2'].platform
'ios'
>>>
```


```
>>> nr.inventory.groups['cisco_group1'].platform
'ios'
>>> nr.inventory.groups['cisco_group2'].platform
'ios'
>>>
```

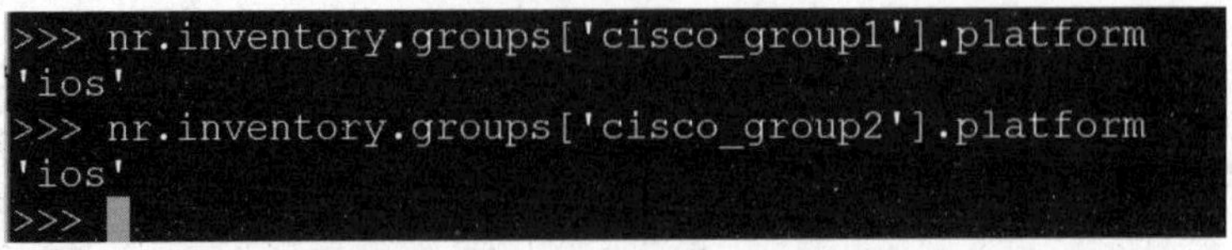

可以看到，只要我们在 hosts.yaml 和 groups.yaml 中放入的键值对越多（也就是信息越多），那么能够筛选的参数也就越多。

讲完 inventory.hosts 和 inventory.groups 后，我们来看下实验 4 的内容，首先将 hosts.yaml 文件里的内容修改如下。

```
---
sw1:
  hostname: 192.168.2.11
  username: python
  password: '123'
  platform: ios
  groups:
    - cisco_group1
  data:
    building: '1'
    level: '1'
sw2:
  hostname: 192.168.2.12
  platform: ios
  groups:
    - cisco_group1
  data:
    building: '1'
    level: '2'
sw3:
  hostname: 192.168.2.13
  platform: ios
  groups:
    - cisco_group2
  data:
    builiding: '2'
    level: '1'
sw4:
  hostname: 192.168.2.14
  platform: ios
  groups:
    - cisco_group2
  data:
    building: '2'
    level: '2'
```

```
[root@localhost Nornir]# cat hosts.yaml
---
sw1:
    hostname: 192.168.2.11
    username: python
    password: '123'
    platform: ios
    groups:
        - cisco_group1
    data:
        building: '1'
        level: '1'
sw2:
    hostname: 192.168.2.12
    platform: ios
    groups:
        - cisco_group1
    data:
        building: '1'
        level: '2'
sw3:
    hostname: 192.168.2.13
    platform: ios
    groups:
        - cisco_group2
    data:
        builiding: '2'
        level: '1'
sw4:
    hostname: 192.168.2.14
    platform: ios
    groups:
        - cisco_group2
    data:
        building: '2'
        level: '2'
[root@localhost Nornir]#
```

这里我们为每台交换机都添加了额外的信息（data），标注了它们所在的位置。比如 SW1 在 building 1 level 1，SW2 在 building 1 level 2 等，和前面提到的密码一样，如果 value 中只使用数字，那么必须对其加上引号才行，所以这里写成 building: '1'、level: '1'；如果 value 中有字母，或者字母、数字混用，则不用加引号，比如 building: A、building: A1、building: 1A 等。

注意：data 下的键值对起的作用是备注，键名（key）可以任意命名，不一定必须是 building、level，也可以使用其他一些键名，比如 country、region、ntp、dhcp 等，就算使用 abc、xyz 做键名也是可以的，并不影响后面的过滤，只是用这些键名没有意义罢了。

接下来创建第 4 个实验脚本 nornir4.py，脚本代码内容如下。

```
from nornir import InitNornir
from nornir_netmiko import netmiko_send_command
```

```
from nornir_utils.plugins.functions import print_result

nr = InitNornir(config_file="config.yaml")
targets = nr.filter(building='1')
results = targets.run(netmiko_send_command, command_string='sh ip arp ')

print_result(results)
```

```
[root@localhost Nornir]# cat nornir4.py
from nornir import InitNornir
from nornir_netmiko import netmiko_send_command
from nornir_utils.plugins.functions import print_result

nr = InitNornir(config_file="config.yaml")
targets = nr.filter(building='1')
results = targets.run(netmiko_send_command, command_string='sh ip arp ')

print_result(results)

[root@localhost Nornir]#
```

代码讲解如下。

- filter()的用法比 F()更简单，其中的参数即 hosts.yaml 里 data 下的键名。这里我们过滤出 4 个交换机中位置在 building 1 里的交换机，也就是 SW1 和 SW2，然后打印出它们输入 show ip arp 命令后的回显内容。

```
targets = nr.filter(building='1')
```

运行脚本看效果，如下图所示。

```
[root@localhost Nornir]# python3.10 nornir4.py
netmiko_send_command*****************************************************************
* sw1 ** changed : False ************************************************************
vvvv netmiko_send_command ** changed : False vvvvvvvvvvvvvvvvvvvvvvvvvvvvvvvvvvvvvv INFO
Protocol  Address          Age (min)  Hardware Addr   Type   Interface
Internet  192.168.2.1             15   000c.29f4.a5ee  ARPA   GigabitEthernet0/0
Internet  192.168.2.11             -   0c1a.91b4.4800  ARPA   GigabitEthernet0/0
^^^^ END netmiko_send_command ^^^^^^^^^^^^^^^^^^^^^^^^^^^^^^^^^^^^^^^^^^^^^^^^^^^^^^
* sw2 ** changed : False ************************************************************
vvvv netmiko_send_command ** changed : False vvvvvvvvvvvvvvvvvvvvvvvvvvvvvvvvvvvvvv INFO
Protocol  Address          Age (min)  Hardware Addr   Type   Interface
Internet  192.168.2.1             15   000c.29f4.a5ee  ARPA   GigabitEthernet0/0
Internet  192.168.2.12             -   0c1a.91e4.8300  ARPA   GigabitEthernet0/0
^^^^ END netmiko_send_command ^^^^^^^^^^^^^^^^^^^^^^^^^^^^^^^^^^^^^^^^^^^^^^^^^^^^^^
[root@localhost Nornir]#
```

也可以用 level 来过滤，写成 targets = nr.filter(level='1')或者 targets = nr.filter(level='2')，这个就留给读者自行去尝试。

9.10　实验 4　使用 filter()做简单过滤（华为设备）

在 nornir3 文件夹的基础上，复制成 nornir4，继续开展实验。对于在 Python 交互模式下调测 inventory.hosts 和 inventory.groups 的对应信息与思科实验基本一致，不再赘述。本实验将 host.yaml 进行如下修改。

```
---
sw1:
   hostname: 192.168.2.11
   username: python
   password: '123'
   groups:
      - huawei_group1
   data:
      building: '1'
      level: '1'
sw2:
   hostname: 192.168.2.12
   groups:
      - huawei_group1
   data:
      building: '1'
      level: '2'
sw3:
   hostname: 192.168.2.13
   groups:
      - huawei_group2
   data:
      building: '2'
      level: '1'
sw4:
   hostname: 192.168.2.14
   groups:
      - huawei_group2
   data:
      building: '2'
      level: '2'
```

通过 data 部署 building、level 后，就可以通过这些信息在不同维度上进行过滤，如下表所示。

SW	building	level
SW1	1	1
SW2	1	2
SW3	2	1
SW4	2	2

Python 代码如下。

```
from nornir import InitNornir
from nornir_netmiko import netmiko_send_command
from nornir_utils.plugins.functions import print_result

nr = InitNornir(config_file="config.yaml")
targets = nr.filter(building='2',level='1')
results = targets.run(netmiko_send_command, command_string='disp arp ')

print_result(results)
```

相对思科设备实验，这里做了点小调整，我们通过 building、level 组合使用，可以更精准地进行定位，比如这段代码只筛选出 SW3 的 ARP 表，如下图所示。

```
Run:  nornir4
E:\Python\nornir_test\nornir-venv\Scripts\python.exe E:/Python/nornir_test/nornir-venv/norn
netmiko_send_command*****************************************************************
* sw3 ** changed : False ***********************************************************
vvvv netmiko_send_command ** changed : False vvvvvvvvvvvvvvvvvvvvvvvvvvvvvvvvvvvvv INFO
IP ADDRESS      MAC ADDRESS     EXPIRE(M) TYPE INTERFACE      VPN-INSTANCE
                                          VLAN
------------------------------------------------------------------------------
192.168.2.13    4c1f-cc37-79d4            I -  Vlanif1
192.168.2.1     0200-4c4f-4f50  20        D-0  GE0/0/1
                                          1
192.168.2.15    4c1f-ccd0-2d35  20        D-0  GE0/0/1
```

这样的话，由这 4 台 SW 编排的所有组合，都可以自行设置筛选，其他组合请读者自行尝试体验。

9.11　实验 5　在 filter()中使用 lambda 过滤单个或多个设备（思科设备）

在 filter()中，我们可以使用参数 filter_func 配合 lambda 来做过滤，可以过滤单个设备，也可以过滤多个设备。首先来看如何使用 lambda 过滤单个设备。

1. 在 filter()中使用 lambda 过滤单个设备

首先创建实验 5 的脚本 nornir5.py，然后放入下列代码。

```
from nornir import InitNornir
from nornir_netmiko import netmiko_send_command
from nornir_utils.plugins.functions import print_result

nr = InitNornir(config_file="config.yaml")
sw4 = nr.filter(filter_func=lambda host: host.name== 'sw4')
results = sw4.run(netmiko_send_command, command_string='sh ip arp')

print_result(results)
```

```
[root@localhost Nornir]# cat nornir5.py
from nornir import InitNornir
from nornir_netmiko import netmiko_send_command
from nornir_utils.plugins.functions import print_result

nr = InitNornir(config_file="config.yaml")
sw4 = nr.filter(filter_func=lambda host: host.name== 'sw4')
results = sw4.run(netmiko_send_command, command_string='sh ip arp')

print_result(results)
[root@localhost Nornir]#
```

代码讲解如下。

- 这里我们调用 filter()里的 filter_func 参数，将 lambda host: host.name=='sw4'赋值给它，即可过滤出 sw4 这台交换机（注意，host.name 对应的 sw4 是我们在 hosts.yaml 文件中为 192.168.2.14 这台交换机取的名称，这个 host.name 对应的设备名称只在本地有效，和交换机真实的 hostname 没有任何关系，如果你在 hosts.yaml 中给这台交换机取名为 abc，那么这里 lambda 后面就要写成 lambda host: host.name=='abc'）。

```
sw4 = nr.filter(filter_func=lambda host: host.name== 'sw4')
```

运行脚本后，效果如下图所示。

```
[root@localhost Nornir]# python3.10 nornir5.py
netmiko_send_command*****************************************************************
* sw4 ** changed : False ************************************************************
vvvv netmiko_send_command ** changed : False vvvvvvvvvvvvvvvvvvvvvvvvvvvvvvvvvvvvvvv INFO
Protocol  Address          Age (min)  Hardware Addr   Type   Interface
Internet  192.168.2.1            19   000c.29f4.a5ee  ARPA   GigabitEthernet0/0
Internet  192.168.2.14            -   0c1a.910e.2c00  ARPA   GigabitEthernet0/0
^^^^ END netmiko_send_command ^^^^^^^^^^^^^^^^^^^^^^^^^^^^^^^^^^^^^^^^^^^^^^^^^^^^^^^
[root@localhost Nornir]#
```

如上图所示，我们通过 lambda 过滤出了 sw4 并打印出了它的 arp 表。如果在生产网络环境下，你记不住设备在 host.yaml 中对应的设备名称，但是你知道该设备的 IP 地址，那么你也可以用 sw4 = nr.filter(filter_func=lambda host: host.hostname=='192.168.2.14')，以 IP 地址来过滤该设备，这里就不演示了。

2. 在 filter()中使用 lambda 过滤多个设备

除了过滤单个设备，在 filter()中还可以用 lambda 配合列表来过滤多个设备，将 nornir5.py 的代码稍作修改，如下所示。

```
from nornir import InitNornir
from nornir_netmiko import netmiko_send_command
from nornir_utils.plugins.functions import print_result

nr = InitNornir(config_file="config.yaml")
#sw4 = nr.filter(filter_func=lambda host: host.name== 'sw4')
switches = ['sw1','sw2','sw3']
sw1_sw2_sw3 = nr.filter(filter_func=lambda host: host.name in switches)

results = sw1_sw2_sw3.run(netmiko_send_command, command_string='sh ip arp')

print_result(results)
```

```
[root@localhost Nornir]# cat nornir5.py
from nornir import InitNornir
from nornir_netmiko import netmiko_send_command
from nornir_utils.plugins.functions import print_result

nr = InitNornir(config_file="config.yaml")
#sw4 = nr.filter(filter_func=lambda host: host.name== 'sw4')
switches = ['sw1','sw2','sw3']
sw1_sw2_sw3 = nr.filter(filter_func=lambda host: host.name in switches)

results = sw1_sw2_sw3.run(netmiko_send_command, command_string='sh ip arp')

print_result(results)
[root@localhost Nornir]#
```

代码分段讲解如下。

- 首先使用 switches = ['sw1', 'sw2', 'sw3']创建一个列表，该列表的元素为 sw1、sw2、sw3 这 3 个将被用 lambda 进行过滤的设备名称。

```
switches = ['sw1','sw2','sw3']
```

- 然后将之前的 sw4 = nr.filter(filter_func=lambda host: host.name== 'sw4')稍作替代，将 host.name=='sw4'这部分改为 host.name in switches 即可，它表示遍历列表 switches 里的所有元素，也就帮我们过滤了 sw1、sw2、sw3 3 台交换机。

```
sw1_sw2_sw3 = nr.filter(filter_func=lambda host: host.name in switches)
```

运行脚本后，效果如下图所示。

```
[root@localhost Nornir]# python3.10 nornir5.py
netmiko_send_command*************************************************************
* sw1 ** changed : False *********************************************************
vvvv netmiko_send_command ** changed : False vvvvvvvvvvvvvvvvvvvvvvvvvvvvvvvvvvvvv
 INFO
Protocol  Address          Age (min)  Hardware Addr   Type   Interface
Internet  192.168.2.1            2    000c.29f4.a5ee  ARPA   GigabitEthernet0/0
Internet  192.168.2.11           -    0c1a.91b4.4800  ARPA   GigabitEthernet0/0
^^^^ END netmiko_send_command ^^^^^^^^^^^^^^^^^^^^^^^^^^^^^^^^^^^^^^^^^^^^^^^^^^^
* sw2 ** changed : False *********************************************************
vvvv netmiko_send_command ** changed : False vvvvvvvvvvvvvvvvvvvvvvvvvvvvvvvvvvvvv
 INFO
Protocol  Address          Age (min)  Hardware Addr   Type   Interface
Internet  192.168.2.1            4    000c.29f4.a5ee  ARPA   GigabitEthernet0/0
Internet  192.168.2.11           2    0c1a.91b4.4800  ARPA   GigabitEthernet0/0
Internet  192.168.2.12           -    0c1a.91e4.8300  ARPA   GigabitEthernet0/0
^^^^ END netmiko_send_command ^^^^^^^^^^^^^^^^^^^^^^^^^^^^^^^^^^^^^^^^^^^^^^^^^^^
* sw3 ** changed : False *********************************************************
vvvv netmiko_send_command ** changed : False vvvvvvvvvvvvvvvvvvvvvvvvvvvvvvvvvvvvv
 INFO
Protocol  Address          Age (min)  Hardware Addr   Type   Interface
Internet  192.168.2.1            4    000c.29f4.a5ee  ARPA   GigabitEthernet0/0
Internet  192.168.2.11           3    0c1a.91b4.4800  ARPA   GigabitEthernet0/0
Internet  192.168.2.13           -    0c1a.912f.2300  ARPA   GigabitEthernet0/0
^^^^ END netmiko_send_command ^^^^^^^^^^^^^^^^^^^^^^^^^^^^^^^^^^^^^^^^^^^^^^^^^^^
[root@localhost Nornir]#
```

同样的原理，如果你想使用 IP 地址来过滤，则只需要将代码稍作修改即可。

```
switches = ['192.168.2.11', '192.168.2.12', '192.168.2.13']
sw1_sw2_sw3 = nr.filter(filter_func=lambda host: host.hostname in switches)
```

验证部分省略，留给读者自行尝试。

9.12　实验 5　在 filter()中使用 lambda 过滤单个或多个设备（华为设备）

lambda 函数也叫匿名函数，在特定场景（如自定义函数需调用的次数很少）下有妙用。借助 lambda 函数，我们就可以不用因为给自定义函数起函数名而烦恼了。这次实验也来体验一下 lambda。

在 nornir4 文件夹的基础上，复制成 nornir5，继续开展实验。这次不用修改 YAML 文件。

1. 在 filter()中使用 lambda 过滤单个设备

```
from nornir import InitNornir
from nornir_netmiko import netmiko_send_command
from nornir_utils.plugins.functions import print_result

nr = InitNornir(config_file="config.yaml")
sw4 = nr.filter(filter_func=lambda host: host.name== 'sw4')
results = sw4.run(netmiko_send_command, command_string='disp arp')

print_result(results)
```

运行代码，结果如下图所示。

```
Run:   nornir5
E:\Python\nornir_test\nornir-venv\Scripts\python.exe E:/Python/nornir_test/nornir-venv/norn
netmiko_send_command*****************************************************************
* sw4 ** changed : False *************************************************************
vvvv netmiko_send_command ** changed : False vvvvvvvvvvvvvvvvvvvvvvvvvvvvvvvvvvv INFO
IP ADDRESS      MAC ADDRESS     EXPIRE(M) TYPE INTERFACE      VPN-INSTANCE
                                          VLAN
------------------------------------------------------------------------------
192.168.2.14    4c1f-ccb9-3de8            I -  Vlanif1
192.168.2.13    4c1f-cc37-79d4  19        D-0  GE0/0/1
```

2. 在 filter()中使用 lambda 过滤多个设备

将上述代码注释一行，补充两行，即可过滤筛选出 sw1、sw2、sw3 3 台设备。

```
#sw4 = nr.filter(filter_func=lambda host: host.name== 'sw4')
switches = ['sw1','sw2','sw3']
sw1_sw2_sw3 = nr.filter(filter_func=lambda host: host.name in switches)
```

运行脚本，限于篇幅，部分如下图所示。

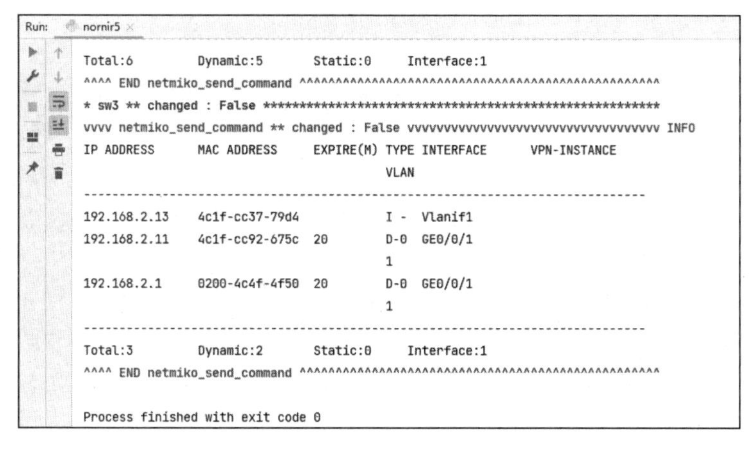

3 个设备（或者说执行对象）过滤，对不同设备来说，差异并不大，关键是 YAML 文件内容要控制好，剩下的操作基本是一模一样的。

9.13　实验 6　用 Nornir 为设备做配置（思科设备）

作为 Nornir 的重要插件，我们可以使用 nornir_netmiko 下的 netmiko_send_config 为设备做配置。首先创建一个 commands.cfg 文件，用来存放需要配置的命令，这里我们为 4 台交换机配置一个 VLAN999，并取名为 nornir_vlan，如下图所示。

```
[root@localhost Nornir]# cat commands.cfg
vlan 999
name nornir_vlan
[root@localhost Nornir]#
```

然后创建脚本 nornir6.py，写入下列代码。

```
from nornir import InitNornir
from nornir_netmiko import netmiko_send_command, netmiko_send_config
from nornir_utils.plugins.functions import print_result, print_title

nr = InitNornir(config_file="config.yaml")

def config(cisco):
  cisco.run(task=netmiko_send_config, config_file='commands.cfg')
  cisco.run(task=netmiko_send_command, command_string='show vlan brief')

print_title('正在配置 VLAN999')
results = nr.run(task=config)

print_result(results)
```

```
[root@localhost Nornir]# cat nornir6.py
from nornir import InitNornir
from nornir_netmiko import netmiko_send_command, netmiko_send_config
from nornir_utils.plugins.functions import print_result, print_title

nr = InitNornir(config_file="config.yaml")

def config(cisco):
    cisco.run(task=netmiko_send_config, config_file='commands.cfg')
    cisco.run(task=netmiko_send_command, command_string='show vlan brief')

print_title('正在配置VLAN999')
results = nr.run(task=config)

print_result(results)

[root@localhost Nornir]#
```

代码分段讲解如下。

- 除 netmiko_send_command 外，我们也从 nornir_netmiko 中另外导入了 netmiko_send_config，用来为交换机配置 VLAN999。

```
from nornir_netmiko import netmiko_send_command, netmiko_send_config
```

- 除 print_result 外，我们还从 nornir_utils.plugins.functions 中导入了 print_title，顾名思义，它的作用是在 Nornir 返回的结果中打印标题，提示这个 Nornir 脚本在做什么。

```
from nornir_utils.plugins.functions import print_result, print_title
```

- 定义一个函数 config(cisco)，用来为 4 台交换机做配置，注意参数名为自定义，这里笔者用的是 cisco。另外，在 run()函数里使用 netmiko_send_config 时，需要配置参数 config_file 来调用实验开始前我们准备的 commands.cfg 这个配置命令文件，配置完成后，我们继续使用 netmiko_send_command 来输入命令 show vlan brief，以验证 VLAN999 是否配置成功。

```
def config(cisco):
  cisco.run(task=netmiko_send_config, config_file='commands.cfg')
  cisco.run(task=netmiko_send_command, command_string='show vlan brief')
```

- 用 print_title()打印出标题，最后使用 nr.run()来运行函数 config()。

```
print_title('正在配置 VLAN999')
results = nr.run(task=config)
```

运行脚本后，效果如下图所示（注意“正在配置 VLAN999"这个标题所在的位置）。

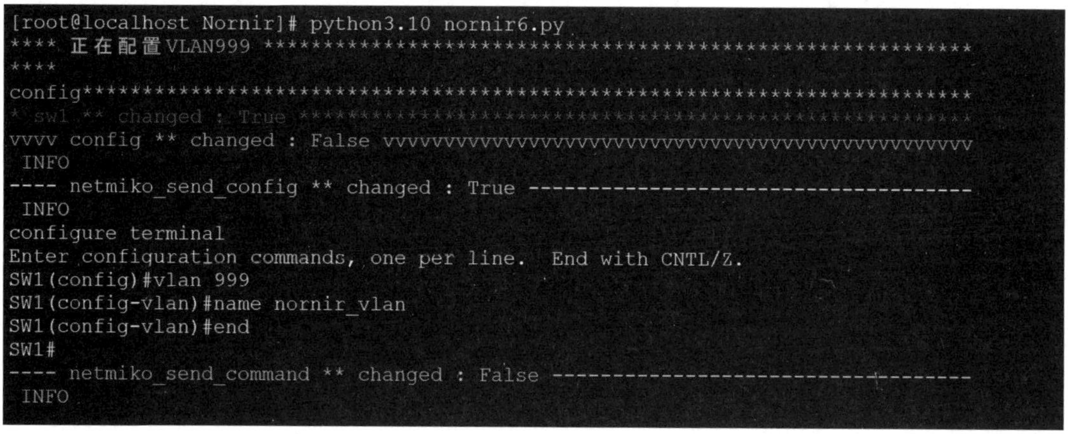

```
[root@localhost Nornir]# python3.10 nornir6.py
**** 正在配置 VLAN999 ***********************************************************
****
config***************************************************************************
* sw1 ** changed : True *********************************************************
vvvv config ** changed : False vvvvvvvvvvvvvvvvvvvvvvvvvvvvvvvvvvvvvvvvvvvvvvvvvvv
 INFO
---- netmiko_send_config ** changed : True --------------------------------------
 INFO
configure terminal
Enter configuration commands, one per line.  End with CNTL/Z.
SW1(config)#vlan 999
SW1(config-vlan)#name nornir_vlan
SW1(config-vlan)#end
SW1#
---- netmiko_send_command ** changed : False ------------------------------------
 INFO
```

```
VLAN Name                             Status    Ports
---- -------------------------------- --------- -------------------------------
1    default                          active    Gi0/1, Gi0/2, Gi0/3
100  VLAN100                          active
200  VLAN0200                         active
300  VLAN0300                         active
999  nornir_vlan                      active
1002 fddi-default                     act/unsup
1003 trcrf-default                    act/unsup
1004 fddinet-default                  act/unsup
1005 trbrf-default                    act/unsup
^^^^ END config ^^^^^^^^^^^^^^^^^^^^^^^^^^^^^^^^^^^^^^^^^^^^^^^^^^^^^^^^^^^^^^^^^
* sw2 ** changed : True *********************************************************
vvvv config ** changed : False vvvvvvvvvvvvvvvvvvvvvvvvvvvvvvvvvvvvvvvvvvvvvvvvvv
 INFO
---- netmiko_send_config ** changed : True ------------------------------------
 INFO
configure terminal
Enter configuration commands, one per line.  End with CNTL/Z.
SW2(config)#vlan 999
SW2(config-vlan)#name nornir_vlan
SW2(config-vlan)#end
SW2#
---- netmiko_send_command ** changed : False ----------------------------------
 INFO
```

9.14　实验 6　用 Nornir 为设备做配置（华为设备）

在 nornir5 文件夹的基础上，复制成 nornir6，继续开展实验。这次不用修改 YAML 文件。实验目录下增加了文件 commands.cfg，内容如下。

```
vlan 999
 description create_by_nornir
```

随着实验的深入开展，实验文件夹会逐渐饱满起来，这就是日常点滴积累的效应。Python 代码如下。

```
from nornir import InitNornir
from nornir_netmiko import netmiko_send_command, netmiko_send_config
from nornir_utils.plugins.functions import print_result, print_title

nr = InitNornir(config_file="config.yaml")

def config(huawei):
   huawei.run(task=netmiko_send_config, config_file='commands.cfg')
   huawei.run(task=netmiko_send_command, command_string='display vlan summary')
```

```
print_title('正在配置 VLAN999')
results = nr.run(task=config)

print_result(results)
```

执行代码，目录结构和执行结果如下图所示。

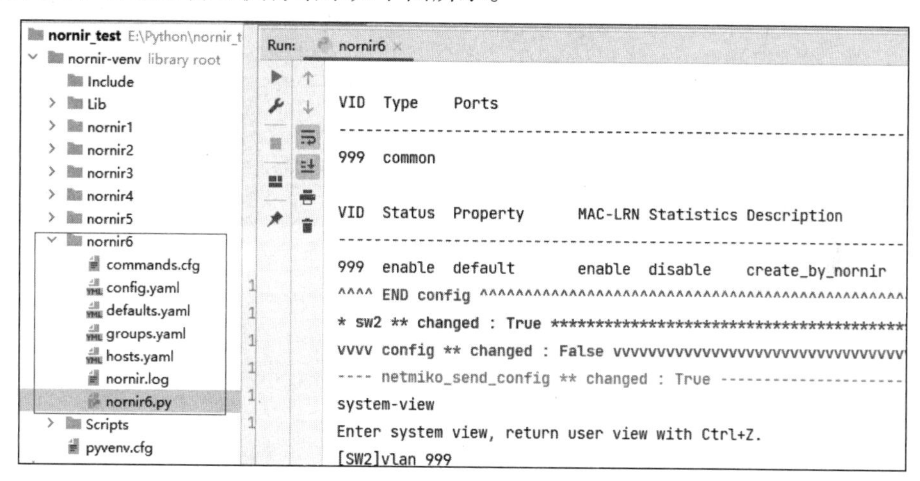

9.15　实验 7　用 Nornir 保存、备份设备配置（思科设备）

在 Nornir 中为设备保存配置的思路很简单，首先用 nornir_naplam 中的 get-config 这个 getter 类 API 获取设备的 show run，然后从 nornir_utils.plugins.taks.files 中调取 write_file 插件，该插件支持创建文本文件来保存设备的配置。

首先创建脚本 nornir7.py，将下列代码写入该脚本。

```
from nornir import InitNornir
from nornir_napalm.plugins.tasks import napalm_get
from nornir_utils.plugins.functions import print_result
from nornir_utils.plugins.tasks.files import write_file
from datetime import date

def backup_configurations(task):
  r = task.run(task=napalm_get, getters=["config"])
  task.run(task=write_file, content=r.result["config"]["running"],
filename=str(task.host.name) + "-" + str(date.today()) + ".txt")
```

```
nr = InitNornir(config_file="config.yaml")
result = nr.run(name="正在备份交换机配置", task=backup_configurations)

print_result(result)
```

```
[root@localhost Nornir]# cat nornir7.py
from nornir import InitNornir
from nornir_napalm.plugins.tasks import napalm_get
from nornir_utils.plugins.functions import print_result
from nornir_utils.plugins.tasks.files import write_file
from datetime import date

def backup_configurations(task):
    r = task.run(task=napalm_get, getters=["config"])
    task.run(task=write_file, content=r.result["config"]["running"], filename=str(task.host.name) + "-" + str(date.today()) + ".txt")

nr = InitNornir(config_file="config.yaml")
result = nr.run(name="正在备份交换机配置", task=backup_configurations)

print_result(result)
[root@localhost Nornir]#
```

代码分段讲解如下。

- 首先从 nornir_utils.plugins.tasks.files 里调用 write_file，用来保存文本文件。

```
from nornir_utils.plugins.tasks.files import write_file
```

- 创建函数 backup_configuration(task)，在该函数中，我们使用 NAPALM 中的 get-config 这个 getter 类 API 来获取设备的 show run 配置，然后调用 write_file 将 show_run 内容写进文本文件，该文本文件名称的格式为 swx-yyyy-mm-dd.txt。

```
def backup_configurations(task):
   r = task.run(task=napalm_get, getters=["config"])
   task.run(task=write_file, content=r.result["config"]["running"],
filename=str(task.host.name) + "-" + str(date.today()) + ".txt")
```

- 除实验 6 中提到的 print_title()外，我们还可以在 nr.run()里加入参数 name，为 Nornir 的输出内容加上标题和脚注。

```
result = nr.run(name="正在备份交换机配置", task=backup_configurations)
```

运行脚本，效果如下图所示。

```
^^^^ END 正在备份交换机配置 ^^^^^^^^^^^^^^^^^^^^^^^^^^^^^^^^^^^^^^^^^^^^^^^^^^^^^^^^^^^^
* sw2 ** changed : True ****************************************************************
vvvv 正在备份交换机配置 ** changed : False vvvvvvvvvvvvvvvvvvvvvvvvvvvvvvvvvvvvvvvvvvvvvvvv INFO
---- napalm_get ** changed : False ------------------------------------------------------ INFO
{ 'config': { 'candidate': '',
              'running': '!\n'
                         '\n'
                         '!\n'
                         'version 15.0\n'
                         'service timestamps debug datetime msec\n'
                         'service timestamps log datetime msec\n'
                         'no service password-encryption\n'
```

```
'service compress-config\n'
'!\n'
'hostname SW2\n'
'!\n'
'boot-start-marker\n'
'boot-end-marker\n'
'!\n'
'!\n'
'!\n'
'username python privilege 15 password 0 123\n'
'no aaa new-model\n'
'!\n'
'!\n'
'!\n'
'!\n'
'!\n'
'vtp domain CISCO-vIOS\n'
'vtp mode transparent\n'
'!\n'
'!\n'
'!\n'
'ip domain-name test\n'
'ip cef\n'
'no ipv6 cef\n'
'!\n'
'!\n'
'spanning-tree mode pvst\n'
'spanning-tree extend system-id\n'
'!\n'
'vlan internal allocation policy ascending\n'
'!\n'
'vlan 100\n'
' name VLAN100\n'
'!\n'
'vlan 200,300 \n'
'!\n'
'vlan 999\n'
' name nornir_vlan\n'
'!\n'
```

脚本执行结束后，可以看到当前目录下多出了 4 台交换机的配置文件，如下图所示。

```
^^^^ END 正在备份交换机配置 ^^^^^^^^^^^^^^^^^^^^^^^^^^^^^^^^^^^^^^^^^^^^^^^^^^^^^^^^
[root@localhost Nornir]#
[root@localhost Nornir]# ls -lt
total 88
-rw-r--r--. 1 root root 5047 Aug 15 00:01 sw1-2022-08-15.txt
-rw-r--r--. 1 root root 5027 Aug 15 00:01 sw4-2022-08-15.txt
-rw-r--r--. 1 root root 5027 Aug 15 00:01 sw2-2022-08-15.txt
-rw-r--r--. 1 root root 5027 Aug 15 00:01 sw3-2022-08-15.txt
-rw-r--r--. 1 root root 7422 Aug 15 00:01 nornir.log
-rw-r--r--. 1 root root  590 Aug 14 23:55 nornir7.py
-rw-r--r--. 1 root root  469 Aug 14 23:28 nornir6.py
-rw-r--r--. 1 root root   26 Aug 14 23:25 commands.cfg
-rw-r--r--. 1 root root  445 Aug 14 23:14 nornir5.py
-rw-r--r--. 1 root root  311 Aug 14 12:46 nornir4.py
-rw-r--r--. 1 root root  590 Aug 14 12:44 hosts.yaml
-rw-r--r--. 1 root root  420 Aug 14 12:31 nornir3.py
-rw-r--r--. 1 root root   70 Aug 14 12:14 groups.yaml
-rw-r--r--. 1 root root  271 Aug 14 11:47 nornir2.py
-rw-r--r--. 1 root root  287 Aug 14 10:32 nornir1.py
-rw-r--r--. 1 root root   37 Aug 14 10:06 defaults.yaml
-rw-r--r--. 1 root root  228 Aug 14 09:55 config.yaml
[root@localhost Nornir]#
```

9.16　实验 7　用 Nornir 保存设备配置（华为设备）

在 nornir6 文件夹的基础上，复制成 nornir7，继续开展实验。这次实验，我们用 nornir_napalm，非 nornir_netmiko，所以要把 groups.yaml 中的 platfrom 从 huawei 修改成 huawei_vrp。实验文件夹中的 commands.cfg 文件可以删掉。

```
---
huawei_group1:
   platform: huawei_vrp

huawei_group2:
   platform: huawei_vrp
```

这里比思科设备实验做了一点拓展，通过对一个个 task 进行封装，把一些功能模块固定下来，以后执行这个 task 即可，清晰明了！在本例子中，task1 取配置信息，task2 取端口信息。

```
from nornir import InitNornir
from nornir_napalm.plugins.tasks import napalm_get
from nornir_utils.plugins.functions import print_result
from nornir_utils.plugins.tasks.files import write_file
from datetime import date

def backup_configurations(task1):
    '''功能 1：取设备备份信息'''
    r = task1.run(task=napalm_get, getters=["config"])
    #print(type(r.result["config"]["running"]))
    task1.run(task=write_file, content=r.result["config"]["running"],
            filename= str(task1.host.name) + "-confs-" + str(date.today()) + ".txt")

def backup_getports(task2):
    '''功能 2：取设备端口信息'''
    r = task2.run(task=napalm_get, getters=["facts"])
    #print(type(r.result["facts"]["interface_list"]))
    task2.run(task=write_file, content='\n'.join(r.result["facts"]["interface_list"]),
            filename=str(task2.host.name) + "-ports-" + str(date.today()) + ".txt")

def other_funtion(taskn):
```

```
    '''功能 n: XXX'''
    pass

nr = InitNornir(config_file="config.yaml")
result1 = nr.run(name="正在备份交换机配置", task=backup_configurations)
result2 = nr.run(name="正在采集交换机端口", task=backup_getports)
```

代码运行结果如下图所示。

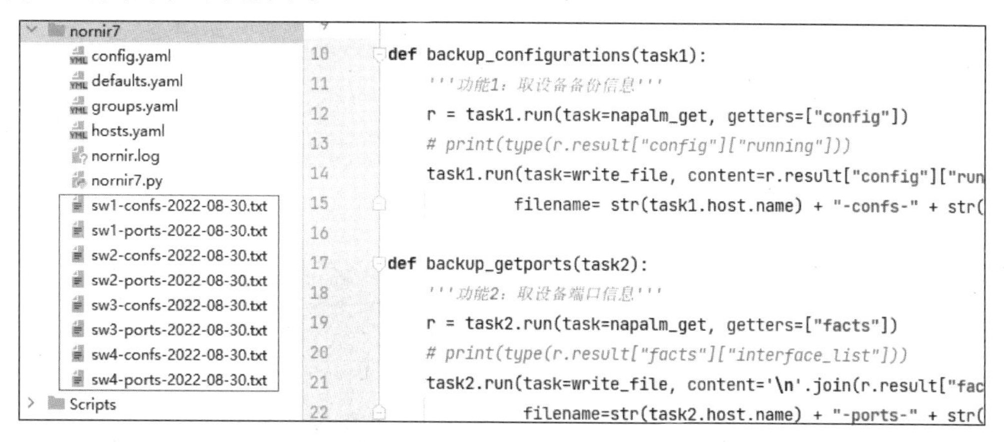

相对于保存配置、获取端口等具体功能，本实验更多的是想传递 Nornir 的一些实现思想。几行代码下去，Nornir 帮我们做了一堆事情。如果封装成函数，则脚本显得更加清晰规整。当然，如果初步接触 Nornir 这些脚本写法，理解起来还是有一定难度的，可能因为脚本高度集成和抽象吧。我们倒大可不必太担心，日常可以用一些示例代码，根据自己需要实现的目标功能，依葫芦画瓢，一点点地迭代和调测，直到成功。

9.17　实验 8　用 Nornir 配合 Jinja2 为设备做配置（思科设备）

6.7 节介绍了 Jinja2 可以用来为有配置差异化的设备做配置，作为一个强大的 Python 框架，Nornir 当然也将 Jinja2 作为插件之一，纳入了自身的阵营，本节就来讲解 Jinja2 在 Nornir 中的用法。

在本节实验中，我们将使用 Nornir 3.0.0 配合 Jinja2 的配置模板为 SW1（192.168.2.11）和 SW2 (192.168.2.12)配置 BGP，让它们建立 eBGP 邻居关系，并各自宣告自己 Loopback0 (SW1: 1.1.1.1, SW2: 2.2.2.2)及 Loopback1(SW1: 11.11.11.11, SW2: 22.22.22.22)的 IP 地址。

SW1 和 SW2 上的 BGP 配置命令分别如下。

◎　SW1 的 BGP 配置命令

```
router bgp 65100
neighbor 192.168.2.12 remote-as 65200
network 1.1.1.1 255.255.255.255
network 11.11.11.11 255.255.255.255
```

◎　SW2 的 BGP 配置命令

```
router bgp 65200
neighbor 192.168.2.11 remote-as 65100
network 2.2.2.2 255.255.255.255
network 22.22.22.22 255.255.255.255
```

在实验开始前，首先通过 pip 安装本节将要用到的插件 nornir_jinja2，如下图所示。

```
[root@localhost Nornir]# pip3.10 install nornir_jinja2
Collecting nornir_jinja2
  Downloading nornir_jinja2-0.2.0-py3-none-any.whl (7.2 kB)
Requirement already satisfied: jinja2<4,>=2.11.2 in /usr/local/lib/python3.10/site-pac
Requirement already satisfied: nornir<4,>=3 in /usr/local/lib/python3.10/site-packages
Requirement already satisfied: MarkupSafe>=2.0 in /usr/local/lib/python3.10/site-packa
Requirement already satisfied: typing_extensions<5.0,>=4.1 in /usr/local/lib/python3.1
Requirement already satisfied: ruamel.yaml>=0.17 in /usr/local/lib/python3.10/site-pac
Requirement already satisfied: mypy_extensions<0.5.0,>=0.4.1 in /usr/local/lib/python3
Requirement already satisfied: ruamel.yaml.clib>=0.2.6 in /usr/local/lib/python3.10/si
Installing collected packages: nornir_jinja2
Successfully installed nornir_jinja2-0.2.0
WARNING: Running pip as the 'root' user can result in broken permissions and conflicti
ttps://pip.pypa.io/warnings/venv

[      ] A new release of pip available:        -> 22.2.2
[      ] To update, run: pip install --upgrade pip
[root@localhost Nornir]#
```

在实验开始前，验证 SW1 的现有配置，如下图所示。

```
S1#show ip int b
Interface              IP-Address      OK? Method Status                Protocol
GigabitEthernet0/0     unassigned      YES unset  up                    up
GigabitEthernet0/1     unassigned      YES unset  up                    up
GigabitEthernet0/2     unassigned      YES unset  up                    up
Loopback0              1.1.1.1         YES NVRAM  up                    up
Loopback1              11.11.11.11     YES NVRAM  up                    up
Vlan1                  192.168.2.11    YES NVRAM  up                    up
S1#
S1#show ip bgp summary
% BGP not active

S1#
```

在实验开始前，验证 SW2 的现有配置，如下图所示。

```
S2#show ip int b
Interface              IP-Address      OK? Method Status                Protocol
GigabitEthernet0/0     unassigned      YES unset  up                    up
GigabitEthernet0/1     unassigned      YES unset  up                    up
GigabitEthernet0/2     unassigned      YES unset  up                    up
Loopback0              2.2.2.2         YES NVRAM  up                    up
Loopback1              22.22.22.22     YES NVRAM  up                    up
Vlan1                  192.168.2.12    YES NVRAM  up                    up
S2#
S2#show ip bgp summary
% BGP not active

S2#
```

首先创建一个名为 nornir8.py 的 runbook 脚本文件，将下列代码写入进去。

```
from nornir import InitNornir
from nornir_utils.plugins.tasks.data import load_yaml
from nornir_jinja2.plugins.tasks import template_file
from nornir_utils.plugins.functions import print_result
from nornir.core.filter import F
from nornir_netmiko import netmiko_send_command, netmiko_send_config

def load_data(task):
    data = task.run(task=load_yaml,file=f'{task.host}.yaml')
    task.host["asn"] = data.result["asn"]
    task.host["neighbor"] = data.result["neighbor"]
    task.host["remoteas"] = data.result["remote-as"]
    task.host["networks"] = data.result["networks"]
    rendering = task.run(task=template_file, template="BGP.j2", path="")
    task.run(task=netmiko_send_config, config_commands=rendering.result.split('\n'))

nr = InitNornir(config_file="config.yaml")
group1 = nr.filter(F(groups__contains="cisco_group1"))
r = group1.run(task=load_data)
print_result(r)
```

```
[root@localhost Nornir]# cat nornir8.py
from nornir import InitNornir
from nornir_utils.plugins.tasks.data import load_yaml
from nornir_jinja2.plugins.tasks import template_file
from nornir_utils.plugins.functions import print_result
from nornir.core.filter import F
from nornir_netmiko import netmiko_send_command, netmiko_send_config
```

```
def load_data(task):
    data = task.run(task=load_yaml,file=f'{task.host}.yaml')
    task.host["asn"] = data.result["asn"]
    task.host["neighbor"] = data.result["neighbor"]
    task.host["remoteas"] = data.result["remote-as"]
    task.host["networks"] = data.result["networks"]
    rendering = task.run(task=template_file, template="BGP.j2", path="")
    task.run(task=netmiko_send_config, config_commands=rendering.result.split('\
n'))

nr = InitNornir(config_file="config.yaml")
group1 = nr.filter(F(groups__contains="cisco_group1"))
r = group1.run(task=load_data)
print_result(r)
[root@localhost Nornir]#
```

然后创建两个 YAML 文件，分别取名为 sw1.yaml 和 sw2.yaml，它们的作用是渲染 Jinja2 的模板配置文件。

◎　sw1.yaml

```
asn: 65100
neighbor: 192.168.2.12
remote-as: 65200
networks:
  - net: 1.1.1.1
    mask: 255.255.255.255
  - net: 11.11.11.11
    mask: 255.255.255.255
```

```
[root@localhost Nornir]# cat sw1.yaml
asn: 65100
neighbor: 192.168.2.12
remote-as: 65200
networks:
  - net: 1.1.1.1
    mask: 255.255.255.255
  - net: 11.11.11.11
    mask: 255.255.255.255
[root@localhost Nornir]#
```

◎　sw2.yaml

```
asn: 65200
neighbor: 192.168.2.11
remote-as: 65100
networks:
  - net: 2.2.2.2
    mask: 255.255.255.255
```

```
  - net: 22.22.22.22
    mask: 255.255.255.255
```

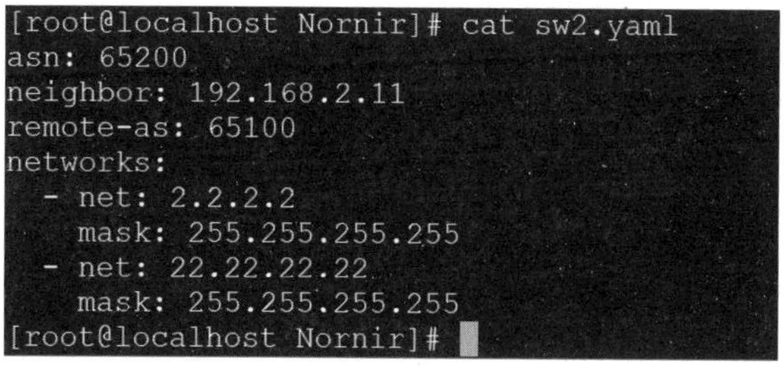

```
[root@localhost Nornir]# cat sw2.yaml
asn: 65200
neighbor: 192.168.2.11
remote-as: 65100
networks:
  - net: 2.2.2.2
    mask: 255.255.255.255
  - net: 22.22.22.22
    mask: 255.255.255.255
[root@localhost Nornir]#
```

创建 Jinja2 的配置模板文件，因为本节实验的目的是为 sw1 和 sw2 配置 BGP，因此我们将该配置模板文件取名为 BGP.j2，其内容如下。

```
router bgp {{ host.asn }}
neighbor {{ host.neighbor }} remote-as {{ host.remoteas }}
{% for n in host.networks %}
network {{ n.net }} mask {{ n.mask }}
{% endfor %}
```

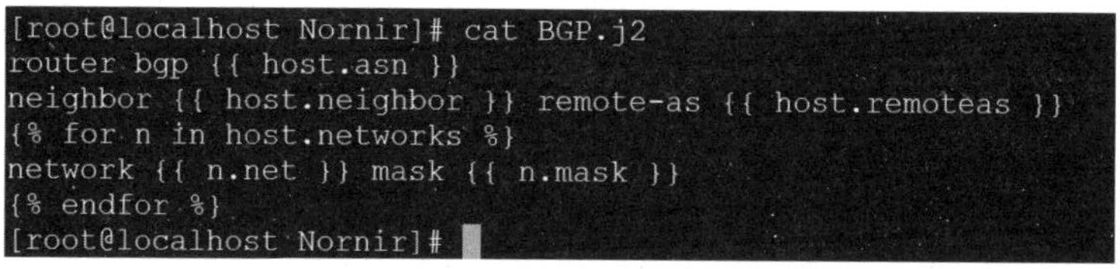

```
[root@localhost Nornir]# cat BGP.j2
router bgp {{ host.asn }}
neighbor {{ host.neighbor }} remote-as {{ host.remoteas }}
{% for n in host.networks %}
network {{ n.net }} mask {{ n.mask }}
{% endfor %}
[root@localhost Nornir]#
```

加上 Nornir 必备的 4 个 YAML 文件，至此我们总共创建了 8 个文件，其中有 6 个 YAML 文件，一个 py 文件，一个 j2 文件，如下图所示。

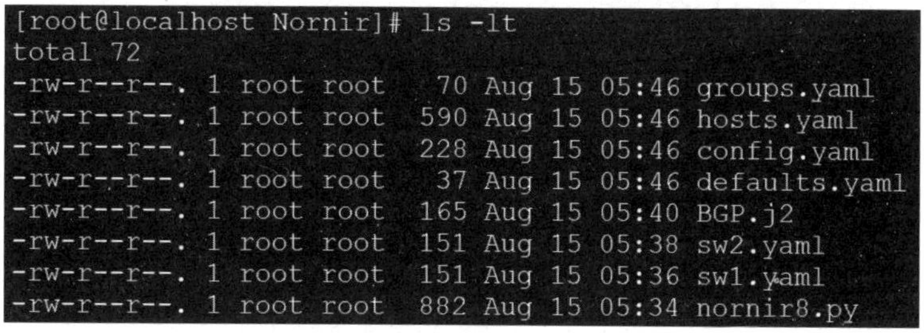

```
[root@localhost Nornir]# ls -lt
total 72
-rw-r--r--. 1 root root  70 Aug 15 05:46 groups.yaml
-rw-r--r--. 1 root root 590 Aug 15 05:46 hosts.yaml
-rw-r--r--. 1 root root 228 Aug 15 05:46 config.yaml
-rw-r--r--. 1 root root  37 Aug 15 05:46 defaults.yaml
-rw-r--r--. 1 root root 165 Aug 15 05:40 BGP.j2
-rw-r--r--. 1 root root 151 Aug 15 05:38 sw2.yaml
-rw-r--r--. 1 root root 151 Aug 15 05:36 sw1.yaml
-rw-r--r--. 1 root root 882 Aug 15 05:34 nornir8.py
```

在确认没有任何遗漏后，接下来分段讲解实验 8 的代码（前面 7 个实验已经涉及并讲解过的知识点将略过不讲）。

- 为了读取 Jinja2 渲染文件（sw1.yaml，sw2.yaml）里的内容，需要调用 nornir_utils 里的 load_yaml 插件。

```
from nornir_utils.plugins.tasks.data import load_yaml
```

- 为了调用 Jinja2 模板文件（BGP.j2）来做配置渲染，需要调用 nornir_jinja2 里的 template_file 插件。

```
from nornir_jinja2.plugins.tasks import template_file
```

- 接下来定义一个 load_data(task)函数来执行任务，首先我们调用插件 load_yaml 读取 sw1.yaml 和 sw2.yaml 里的内容，并将它赋值给 data 变量。注意，这里的 file=f'{task.host}.yaml'，{task.host}的作用是读取 hosts.yaml 里的 4 台交换机的设备名（sw1、sw2、sw3、sw4），配合后面的.yaml 即为读取 sw1.yaml、sw2.yaml、sw3.yaml、sw4.yaml 4 个文件，其中 sw1.yaml 和 sw2.yaml 两个 Jinja2 的渲染文件我们已经有了，不存在的 sw3.yaml 和 sw4.yaml 我们会通过高级过滤将其过滤掉。

```
def load_data(task):
    data = task.run(task=load_yaml,file=f'{task.host}.yaml')
```

再来回顾一下 sw1.yaml 和 sw2.yaml 里的内容。

◎ sw1.yaml

```
[root@localhost Nornir]# cat sw1.yaml
asn: 65100
neighbor: 192.168.2.12
remote-as: 65200
networks:
  - net: 1.1.1.1
    mask: 255.255.255.255
  - net: 11.11.11.11
    mask: 255.255.255.255
[root@localhost Nornir]#
```

◎ sw2.yaml

```
[root@localhost Nornir]# cat sw2.yaml
asn: 65200
neighbor: 192.168.2.11
remote-as: 65100
networks:
  - net: 2.2.2.2
    mask: 255.255.255.255
  - net: 22.22.22.22
    mask: 255.255.255.255
[root@localhost Nornir]#
```

- 接下来我们通过 data.result["asn"]、data.result["neighbor"]、data.result["remote-as"]、data.result["networks"]分别读取 sw1.yaml 和 sw2.yaml 里的 asn、neighbor、remote-as、networks 的内容，然后将得到的赋值依次赋值给 task.host["asn"]，task.host ["neighbor"]、task.host["remoteas"]、task.host["networks"]，以便留给 BGP.j2 这个模板配置文件来做渲染。**注意：BGP.j2 不是直接读取 sw1.yaml 和 sw2.yaml 里的内容来做渲染，而是需要首先通过 load_yaml 插件将 sw1.yaml 和 sw2.yaml 里的内容先读取出来，并赋值给 task.host，再来调用 template_file 插件做配置渲染。**另外，你肯定也会问为什么 task.host["remoteas"]不写成 task.host["remote-as"]，这是因为"-"这个符号在 j2 文件里有另外的含义，如果写成 task.host["remote-as"]，则我们用 template_file 做配置渲染的时候，Jinja 只会读 "remote" 部分，不会读后面的 "-as"，导致 Python 因找不到相应的键名而报错。

```
task.host["asn"] = data.result["asn"]
task.host["neighbor"] = data.result["neighbor"]
task.host["remoteas"] = data.result["remote-as"]
task.host["networks"] = data.result["networks"]
```

- 启动另外一个任务来调用 template_file 插件做渲染（task=template_file），并将结果赋值给 rendering 变量。这里的 template 参数即 BGP.j2 配置模板文件（template="BGP.j2"）。因为 BGP.j2 和 nornir8.py 处于同一个文件夹下，因此这里的 path 参数设为空（path=""）。

```
rendering = task.run(task=template_file, template="BGP.j2", path="")
```

- 注意，使用 template_file 插件配合 BGP.j2 做配置渲染后，其返回的值为字符串（也就是说，下一行代码中的 rendering.result 的数据类型为字符串），该字符串里的内容就是我们最终要配置给 SW1 和 SW2 的 BGP 命令，也就是 "router bgp 65100\nneighbor 192.168.2.12 remote-as 65200\nnetwork 1.1.1.1 mask 255.255.255.255\nnetwork 11.11.11.11 mask 255.255.255.255" 和 "router bgp 65200\nneighbor 192.168.2.11 remote-as 65100\nnetwork 2.2.2.2 mask 255.255.255.255\nnetwork 22.22.22.22 mask 255.255.255.255"。
- 最后我们调用实验 6 讲到的 netmiko_send_config 插件，配合 config_commands 参数（该参数要求数据类型必须为列表，因此我们要对 rendering.result 调用 split()参数，将其转化为列表）将渲染好的 BGP 配置命令配置给 sw1 和 sw2。

```
task.run(task=netmiko_send_config, config_commands=rendering.result.split('\n'))
```

运行脚本看效果，并登录 sw1 和 sw2 做验证，如下图所示。

```
[root@localhost Nornir]# python3.10 nornir8.py
load_data***********************************************************************
* sw1 ** changed : True *********************************************************
vvvv load_data ** changed : False vvvvvvvvvvvvvvvvvvvvvvvvvvvvvvvvvvvvvvvvvvvvvv INFO
---- load_yaml ** changed : False ---------------------------------------------- INFO
{ 'asn': 65100,
  'neighbor': '192.168.2.12',
  'networks': [ {'mask': '255.255.255.255', 'net': '1.1.1.1'},
                {'mask': '255.255.255.255', 'net': '11.11.11.11'}],
  'remote-as': 65200}
---- template_file ** changed : False ------------------------------------------ INFO
router bgp 65100
neighbor 192.168.2.12 remote-as 65200
network 1.1.1.1 mask 255.255.255.255
network 11.11.11.11 mask 255.255.255.255

---- netmiko_send_config ** changed : True ------------------------------------- INFO

SW1#configure terminal
Enter configuration commands, one per line.  End with CNTL/Z.
SW1(config)#router bgp 65100
SW1(config-router)#neighbor 192.168.2.12 remote-as 65200
SW1(config-router)#network 1.1.1.1 mask 255.255.255.255
SW1(config-router)#network 11.11.11.11 mask 255.255.255.255
SW1(config-router)#
SW1(config-router)#end
SW1#
^^^^ END load_data ^^^^^^^^^^^^^^^^^^^^^^^^^^^^^^^^^^^^^^^^^^^^^^^^^^^^^^^^^^^^^^
* sw2 ** changed : True *********************************************************
vvvv load_data ** changed : False vvvvvvvvvvvvvvvvvvvvvvvvvvvvvvvvvvvvvvvvvvvvvv INFO
---- load_yaml ** changed : False ---------------------------------------------- INFO
{ 'asn': 65200,
  'neighbor': '192.168.2.11',
  'networks': [ {'mask': '255.255.255.255', 'net': '2.2.2.2'},
                {'mask': '255.255.255.255', 'net': '22.22.22.22'}],
  'remote-as': 65100}
---- template_file ** changed : False ------------------------------------------ INFO
router bgp 65200
neighbor 192.168.2.11 remote-as 65100
network 2.2.2.2 mask 255.255.255.255
network 22.22.22.22 mask 255.255.255.255

---- netmiko_send_config ** changed : True ------------------------------------- INFO
configure terminal
Enter configuration commands, one per line.  End with CNTL/Z.
SW2(config)#router bgp 65200
SW2(config-router)#neighbor 192.168.2.11 remote-as 65100
SW2(config-router)#network 2.2.2.2 mask 255.255.255.255
SW2(config-router)#network 22.22.22.22 mask 255.255.255.255
SW2(config-router)#
SW2(config-router)#end
SW2#
^^^^ END load_data ^^^^^^^^^^^^^^^^^^^^^^^^^^^^^^^^^^^^^^^^^^^^^^^^^^^^^^^^^^^^^^
[root@localhost Nornir]#
```

```
S1#show run | s r b
router bgp 65100
 bgp log-neighbor-changes
 network 1.1.1.1 mask 255.255.255.255
 network 11.11.11.11 mask 255.255.255.255
 neighbor 192.168.2.12 remote-as 65200
S1#show ip bgp sum
BGP router identifier 11.11.11.11, local AS number 65100
BGP table version is 5, main routing table version 5
4 network entries using 576 bytes of memory
4 path entries using 320 bytes of memory
2/2 BGP path/bestpath attribute entries using 304 bytes of memory
1 BGP AS-PATH entries using 24 bytes of memory
0 BGP route-map cache entries using 0 bytes of memory
0 BGP filter-list cache entries using 0 bytes of memory
BGP using 1224 total bytes of memory
BGP activity 4/0 prefixes, 4/0 paths, scan interval 60 secs

Neighbor        V           AS MsgRcvd MsgSent   TblVer  InQ OutQ Up/Down  State/PfxRcd
192.168.2.12    4        65200       5       5        3    0    0 00:01:16        2
S1#
```

```
S2#show run | s r b
router bgp 65200
 bgp log-neighbor-changes
 network 2.2.2.2 mask 255.255.255.255
 network 22.22.22.22 mask 255.255.255.255
 neighbor 192.168.2.11 remote-as 65100
S2#show ip bgp sum
BGP router identifier 22.22.22.22, local AS number 65200
BGP table version is 5, main routing table version 5
4 network entries using 576 bytes of memory
4 path entries using 320 bytes of memory
2/2 BGP path/bestpath attribute entries using 304 bytes of memory
1 BGP AS-PATH entries using 24 bytes of memory
0 BGP route-map cache entries using 0 bytes of memory
0 BGP filter-list cache entries using 0 bytes of memory
BGP using 1224 total bytes of memory
BGP activity 4/0 prefixes, 4/0 paths, scan interval 60 secs

Neighbor        V           AS MsgRcvd MsgSent   TblVer  InQ OutQ Up/Down  State/PfxRcd
192.168.2.11    4        65100       5       5        5    0    0 00:01:36        2
S2#
```

9.18　实验 8　用 Nornir 配合 Jinja2 给设备做配置（华为设备）

Netmiko 章节中已经有 Jinja2 的相关例子，现在可以把这一块内容一起迁移到 Nornir 上来。在 nornir7 文件夹的基础上，复制成 nornir8，继续开展实验。请别忘了，上个实验是 NAPALM，现在改 Netmiko，需要修改一下 platform，即 groups.yaml。

```
---
huawei_group1:
   platform: huawei
```

```
huawei_group2:
  platform: Huawei
```

在实验文件夹中创建两个 YAML 文件：sw1.yaml 和 sw2.yaml。这两个文件描述了各自的 loopback 端口、BGP 等信息，为 Jinja2 模板渲染（rendering）提供参数。

◎　sw1.yaml

```
loopbacks:
  - loopback: 0
    ipaddress: 1.1.1.1
    mask: 32
  - loopback: 11
    ipaddress: 11.11.11.11
    mask: 32
asn: 65100
neighbor: 192.168.2.12
remote-as: 65200
networks:
  - net: 1.1.1.1
    mask: 255.255.255.255
  - net: 11.11.11.11
    mask: 255.255.255.255
```

◎　sw2.yaml

```
loopbacks:
  - loopback: 0
    ipaddress: 2.2.2.2
    mask: 32
  - loopback: 11
    ipaddress: 22.22.22.22
    mask: 32
asn: 65200
neighbor: 192.168.2.11
remote-as: 65100
networks:
  - net: 2.2.2.2
    mask: 255.255.255.255
  - net: 22.22.22.22
```

```
  mask: 255.255.255.255
```

准备 Jinja2 的配置模板文件 BGP.j2 文件。

```
{% for i in host.loopbacks %}
interface LoopBack {{ i.loopback }}
ip address {{ i.ipaddress }} {{ i.mask }}
{% endfor %}

bgp {{ host.asn }}
peer {{ host.neighbor }} as-number {{ host.remoteas }}
{% for n in host.networks %}
network {{ n.net }} {{ n.mask }}
{% endfor %}
```

两个 YAML 文件中的参数，通过 Python 脚本调用 BGP.j2 中的 Jinja2 模板，就可以按部就班地填入变量中，生成（术语叫“渲染”）配置脚本，如下图所示，如下图所示。

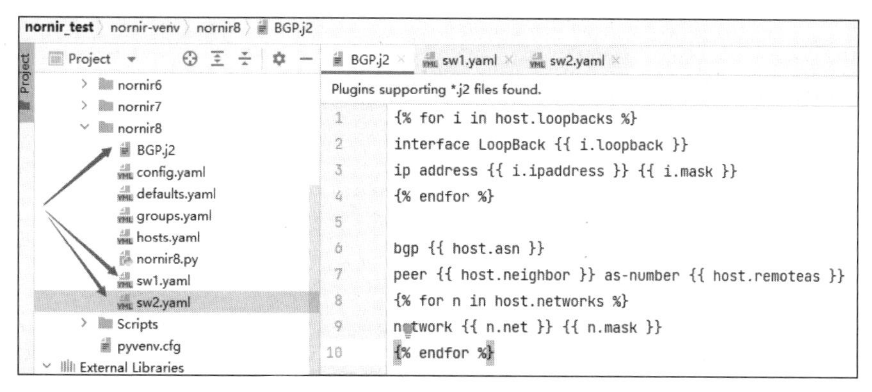

相对于思科实验，华为实验增加了 loopback0 端口的 IP 配置，其他都紧扣思科实验的内容做迁移。下面准备 Python 脚本 nornir8.py。

```
from nornir import InitNornir
from nornir_utils.plugins.tasks.data import load_yaml
from nornir_jinja2.plugins.tasks import template_file
from nornir_utils.plugins.functions import print_result
from nornir.core.filter import F
from nornir_netmiko import netmiko_send_command, netmiko_send_config

def load_data(task):
    data = task.run(task=load_yaml,file=f'{task.host}.yaml')
```

```
    task.host["asn"] = data.result["asn"]
    task.host["neighbor"] = data.result["neighbor"]
    task.host["remoteas"] = data.result["remote-as"]
    task.host["loopbacks"] = data.result["loopbacks"]
    task.host["networks"] = data.result["networks"]
    rendering = task.run(task=template_file, template="BGP.j2", path="")
    task.run(task=netmiko_send_config, config_commands=rendering.result.split('\n'))

nr = InitNornir(config_file="config.yaml")
group1 = nr.filter(F(groups__contains="huawei_group1"))
r = group1.run(task=load_data)
print_result(r)
```

运行脚本，结果如下图所示。

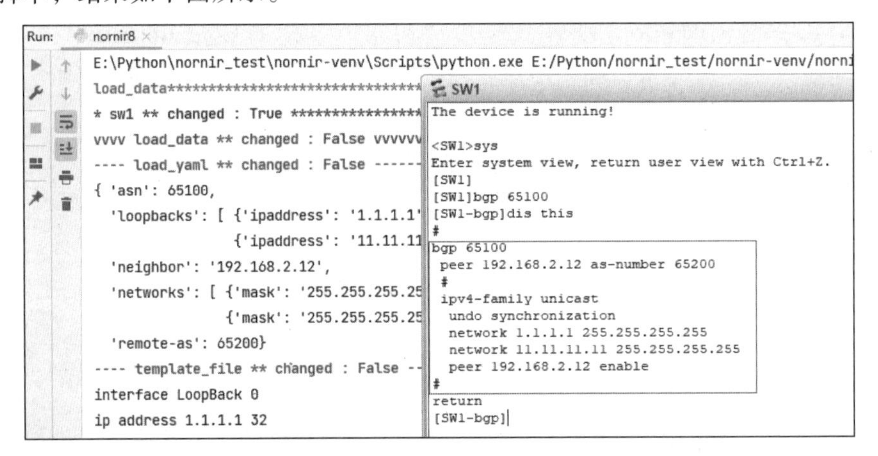

目前网络自动化运维演进的一个方向就是，网络工程师从关注配置制作脚本，人工登录设备下发配置，慢慢地演化为网络工程师关注和确定设备配置的某些重要控制参数，而把制作脚本任务交给 Jinja2 等去做，把下发脚本工作交由 Nornir 等去完成。

9.19　实验 9　Nornir 3 + Scrapli（思科设备）

Nornir 3.0.0 发布 4 个多月以来，相较于 2.X 版本引入了数量众多的新插件，其热度可见一斑。作为 Nornir 3.0.0 众多插件中的一员，听说过 Scrapli 的人不多，有鉴于此，本节将首先系统地介绍 Scrapli 的背景和功能，并举例说明它的独立使用方法，随后介绍 Srcapli 作为插件在 Nornir 3.0.0 中的使用方法。

◎ 什么是 Scrapli

Scrapli 发布于 2020 年 3 月，非常年轻，作者是目前供职于 Twitch 的 Carl Montarari。

根据作者自己的介绍，Scrapli 是一个**专注于向网络设备（不支持主机）**提供 SSH、Telnet 协议应用的 Python 第三方库。是的，这是一个继 Paramiko、Netmiko、Telnetlib、Pexpect 等模块之后 NetDevOps 圈里又一个与 SSH、Telnet 相关的模块，既然是重复造的轮子，又能在短时间内受到青睐，并被 Nornir 3 纳入其插件库，Scrapli 必有其过人之处。

◎ Scrapli 的由来及优势

Scrapli 相较于 Paramiko 的最大的优势，就是 SSH 连接的速度和对 OpenSSH 的支持。Scrapli 基于 ssh2-python 库，ssh2-python 是 libssh（libssh 是 C 语言下的一个库，用来在 C 语言中实现 SSHv2 协议）的一个轻量级的封装（wrapper），libssh 在 Cython 中运行（Cython 是结合了 C 和 Python 的一种编程语言，可以简单地认为，就是给 Python 加上静态类型后的语法，并且由于 Cython 开销小，会直接将代码编译为二进制程序，所以性能较 Python 有很大提升），而作为 libssh 的一个轻量级封装（轻量级封装指 ssh2-python 100%保留了 libssh 的所有功能和特性），ssh2-python 很好地将 libssh 速度快的优势从 Cython 中嫁接到了 Python 里。

而反观 Paramiko，除了其本身是纯 Python 库（基于 Paramiko 二次开发的 Netmiko 是同样的道理），速度上不如 scrapli，另外，Paramiko 也并不直接支持 OpenSSH，而支持 OpenSSH 所有配置选项的 Scrapli 在这一点上则做得很好。读到这里，你也许会问，既然在 Scrapli 诞生之前的 ssh2-python 相较于 Paramiko 有速度上的优势，那为什么今天 Python 上主流的 SSH 库依然是 Paramiko 而不是 ssh2-python 呢？这是因为 ssh2-python 相较于 Paramiko 功能有限，并且对其的开发和支持已经停止，而 Carl Montarari 正是看中了 ssh2-python 的速度优势，才对其二次开发后做出了 Scrapli。

同 Nornir 一样，Scrapli 本身也有插件的概念，除了前面提到的 ssh2-python，Scrapli 还支持 telnet（基于 telnetlib）。除此之外，Scrapli 也用到了 asyncssh 和 asynctelnet 两个插件，也就是说，针对 SSH 和 Telnet 两种协议，Scrapli 对它们都支持同步和异步两种模式。另外，Scrapli 还支持 Netconf、TextFSM、Genie/pyATS 等扩展模块，这些都是 Paramiko、Netmiko、Telnetlib 等所不及的。正因为 Scrapli 同时具备了 ssh2-python 的速度优势，并且还拥有如此丰富的功能，因此很快在 NetDevOps 界受到青睐，并作为插件被引入了 Nornir 3。

1. 安装 Scrapli 和 nornir_scrapli

Scrapli 不支持 Python 2，需要至少 Python 3.6 以上的版本才能使用，如下图所示。

```
pip3.10 install scrapli
pip3.10 install nornir_scrapli
```

```
[root@localhost ~]# pip3.10 install scrapli
Collecting scrapli
  Using cached scrapli-2022.7.30-py3-none-any.whl (138 kB)
Installing collected packages: scrapli
Successfully installed scrapli-2022.7.30
WARNING: Running pip as the 'root' user can result in broken permissions and conflicting behavio
ttps://pip.pypa.io/warnings/venv

[      ] A new release of pip available:        -> 22.2.2
[      ] To update, run: pip install --upgrade pip
[root@localhost ~]# pip3.10 install nornir_scrapli
Collecting nornir_scrapli
  Using cached nornir_scrapli-2022.7.30-py3-none-any.whl (32 kB)
Requirement already satisfied: scrapli-community>=2021.01.30 in /usr/local/lib/python3.10/site-pa
Requirement already satisfied: scrapli-netconf>=2022.01.30 in /usr/local/lib/python3.10/site-pack
Requirement already satisfied: scrapli-cfg>=2022.01.30 in /usr/local/lib/python3.10/site-packages
Requirement already satisfied: ntc-templates>=1.1.0 in /usr/local/lib/python3.10/site-packages (
Requirement already satisfied: scrapli>=2022.01.30 in /usr/local/lib/python3.10/site-packages (f
Requirement already satisfied: nornir>=3.0.0 in /usr/local/lib/python3.10/site-packages (from no
Requirement already satisfied: nornir-utils>=0.1.0 in /usr/local/lib/python3.10/site-packages (f
Requirement already satisfied: textfsm>=1.1.0 in /usr/local/lib/python3.10/site-packages (from n
Requirement already satisfied: mypy_extensions<0.5.0,>=0.4.1 in /usr/local/lib/python3.10/site-pa
Requirement already satisfied: ruamel.yaml>=0.17 in /usr/local/lib/python3.10/site-packages (fro
Requirement already satisfied: typing_extensions<5.0,>=4.1 in /usr/local/lib/python3.10/site-pac
Requirement already satisfied: colorama<0.5.0,>=0.4.3 in /usr/local/lib/python3.10/site-packages
Requirement already satisfied: lxml<5.0.0,>=4.5.1 in /usr/local/lib/python3.10/site-packages (fr
Requirement already satisfied: future in /usr/local/lib/python3.10/site-packages (from textfsm>=
Requirement already satisfied: six in /usr/local/lib/python3.10/site-packages (from textfsm>=1.1
Requirement already satisfied: ruamel.yaml.clib>=0.2.6 in /usr/local/lib/python3.10/site-packages
Installing collected packages: nornir_scrapli
Successfully installed nornir_scrapli-2022.7.30
WARNING: Running pip as the 'root' user can result in broken permissions and conflicting behavio
ttps://pip.pypa.io/warnings/venv

[      ] A new release of pip available:        -> 22.2.2
[      ] To update, run: pip install --upgrade pip
[root@localhost ~]# python3.10
Python 3.10.6 (main, Aug  6 2022, 13:49:15) [GCC 8.5.0 20210514 (Red Hat 8.5.0-15)] on linux
Type "help", "copyright", "credits" or "license" for more information.
>>> import scrapli
>>> import nornir_scrapli
>>>
```

和 NetDevOps 界各类主流的第三方库不同，Scrapli 和 nornir_scrapli 均以日期作为版本号（直观、明了，这是一个很好的尝试和改进），截至 2022 年 8 月 16 日，Scrapli 及 nornir_scrapli 目前最新的版本号为 2022.7.30，如下图所示。

```
[root@localhost ~]# pip3.10 freeze | grep scrapli
nornir-scrapli==2022.7.30
scrapli==2022.7.30
scrapli-cfg==2022.7.30
scrapli-community==2022.7.30
scrapli-netconf==2022.7.30
[root@localhost ~]#
```

截至 2022 年 8 月，Scrapli 支持思科的 IOS-XE、NX-OS、IOS-XR 和 Arista 的 EOS，以及 Juniper 的 JunOS。虽然作者本人没有说明 Scrapli 是否支持思科的 IOS 设备，但是经过测试，笔者发现使用 cisco_iosxe 就能直接支持 IOS 设备。

虽然 Scrapli 官网的介绍里没有提到对国产设备的支持，但是经读者测试和反馈，在安装 scrapli-community 模块后，Scrapli 是支持华为、H3C 等设备的，如下图所示。

Platform/OS	Scrapli Driver	Scrapli Async Driver	Platform Name
Cisco IOS-XE	IOSXEDriver	AsyncIOSXEDriver	cisco_iosxe
Cisco NX-OS	NXOSDriver	AsyncNXOSDriver	cisco_nxos
Cisco IOS-XR	IOSXRDriver	AsyncIOSXRDriver	cisco_iosxr
Arista EOS	EOSDriver	AsyncEOSDriver	arista_eos
Juniper JunOS	JunosDriver	AsyncJunosDriver	juniper junos

2. 使用 Scrapli 登录设备

首先创建一个叫作 scrapli1.py 的脚本，放入下面的代码。

```
from scrapli.driver.core import IOSXEDriver

device = {
   "host": "192.168.2.11",
   "auth_username": "python",
   "auth_password": "123",
   "port": 22,
   "auth_strict_key": False,
}

conn = IOSXEDriver(**device)
conn.open()
responses = conn.send_commands(["show clock", "show ip int brief"])

for response in responses:
  print(response.result)
conn.close()
```

```
[root@localhost Nornir]# cat scrapli1.py
from scrapli.driver.core import IOSXEDriver

device = {
    "host": "192.168.2.11",
    "auth_username": "python",
    "auth_password": "123",
    "port": 22,
    "auth_strict_key": False,
}

conn = IOSXEDriver(**device)
conn.open()
responses = conn.send_commands(["show clock", "show ip int brief"])

for response in responses:
   print(response.result)
conn.close()
[root@localhost Nornir]#
```

代码分段讲解如下。

- Scrapli 的 driver 部分和 NAPALM 很像，因为我们实验用的交换机是思科 IOS 的，这里我们从 scrapli.driver.core 里导入 IOSXEDriver（前面讲到了 IOS-XE driver 同样支持 IOS 设备）。

```
from scrapli.driver.core import IOSXEDriver
```

- 剩余的代码和 Netmiko 非常相似，只是个别参数有些区别，这里的 port 和 auth_strict_key 均为可选参数，其中 port 默认值为 22，auth_strict_key 默认为 True，auth_strict_key 和 Paramiko 的 look_for_keys 参数类似，在生产网络中建议将其设为 True，以提高安全性，不过在实验环境下将其设为 False 可以避免使用 Python 第一次登录某交换机时因为 SSH 主机密匙验证失败的问题。

```
device = {
   "host": "192.168.2.11",
   "auth_username": "python",
   "auth_password": "123",
   "port": 22,
   "auth_strict_key": False,
}

conn = IOSXEDriver(**device)
conn.open()
responses = conn.send_commands(["show clock", "show ip int brief"])
```

```
for response in responses:
  print(response.result)
conn.close()
```

运行脚本，效果如下图所示。

```
[root@localhost Nornir]# python3.10 scrapli1.py
*15:30:27.767 UTC Mon Aug 15 2022
Interface              IP-Address      OK? Method Status                Protocol
GigabitEthernet0/1     unassigned      YES unset  down                  down
GigabitEthernet0/2     unassigned      YES unset  down                  down
GigabitEthernet0/3     unassigned      YES unset  down                  down
GigabitEthernet0/0     192.168.2.11    YES NVRAM  up                    up
[root@localhost Nornir]#
```

需要注意的是，Scrapli 在默认状态下去掉了设备的 hostname 和提示符，也就是打印出的回显内容里看不到“sw1#”这个提示符，关于这个问题后面在讲解 nornir_scrapli 时会给出解决方法。

3. 使用 Scrapli 给设备做配置

创建一个叫作 scrapli2.py 的脚本，放入下面的代码。

```
from scrapli.driver.core import IOSXEDriver


device = {
    "host": "192.168.2.11",
    "auth_username": "python",
    "auth_password": "123",
    “port”: 22,
    “auth_strict_key”: False
}


conn = IOSXEDriver(**device)
conn.open()
output = conn.send_configs(["interface GigabitEthernet0/1", "description Configured by Scrapli"])
print(output.result)
output = conn.send_command("show interface Gi0/1 description")
print(output.result)
```

```
[root@localhost Nornir]# cat scrapli2.py
from scrapli.driver.core import IOSXEDriver

device = {
    "host": "192.168.2.11",
    "auth_username": "python",
    "auth_password": "123",
    "port": 22,
    "auth_strict_key": False
}

conn = IOSXEDriver(**device)
conn.open()
output = conn.send_configs(["interface GigabitEthernet0/1", "description Configured by Scrapli"])
print(output.result)
output = conn.send_command("show interface Gi0/1 description")
print(output.result)

[root@localhost Nornir]#
```

在登录交换机 sw1 后，对其 Gi0/1 端口配置了 description，随后 show interface Gi0/1 description 进行验证。

运行脚本，效果如下图所示。

```
[root@localhost Nornir]# python3.10 scrapli2.py
interface GigabitEthernet0/1
description Configured by Scrapli

Interface                      Status         Protocol Description
Gi0/1                          down           down     Configured by Scrapli
[root@localhost Nornir]#
```

4. nornir_scrapli 的使用

Nornir 的 config、defaults、groups 这几个 YAML 文件的内容和配置与前面一样，这里我们只需要对 hosts.yaml 和 groups.yaml 文件稍作修改即可。

◎　hosts.yaml

```
---
sw1:
  hostname: 192.168.2.11
  username: python
  password: '123'
  groups:
    - cisco
  connection_options:
    scrapli:
      platform: cisco_iosxe
      extras:
        ssh_config_file: True
```

```
            auth_strict_key: False

sw2:
    hostname: 192.168.2.12
    username: python
    password: '123'
    groups:
        - cisco
    connection_options:
        scrapli:
            platform: cisco_iosxe
            extras:
                ssh_config_file: True
                auth_strict_key: False
```

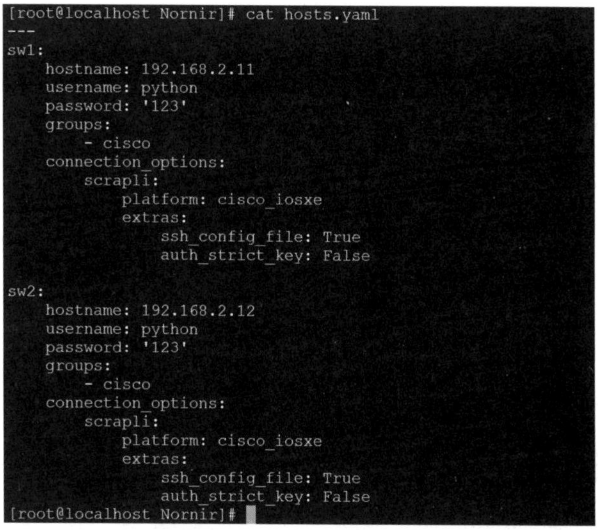

```
[root@localhost Nornir]# cat hosts.yaml
---
sw1:
    hostname: 192.168.2.11
    username: python
    password: '123'
    groups:
        - cisco
    connection_options:
        scrapli:
            platform: cisco_iosxe
            extras:
                ssh_config_file: True
                auth_strict_key: False

sw2:
    hostname: 192.168.2.12
    username: python
    password: '123'
    groups:
        - cisco
    connection_options:
        scrapli:
            platform: cisco_iosxe
            extras:
                ssh_config_file: True
                auth_strict_key: False
[root@localhost Nornir]#
```

可以看到，这里的 hosts.yaml 里，我们启用了 cisco 这个 group，因此还需要将该 group 加入 groups.yaml 里，如下图所示。

```
---
cisco:
    platform: ios

cisco_group1:
    platform: ios
```

```
cisco_group2:
   platform: ios
```

```
[root@localhost Nornir]# cat groups.yaml
---
cisco:
    platform: ios

cisco_group1:
    platform: ios

cisco_group2:
    platform: ios
[root@localhost Nornir]#
```

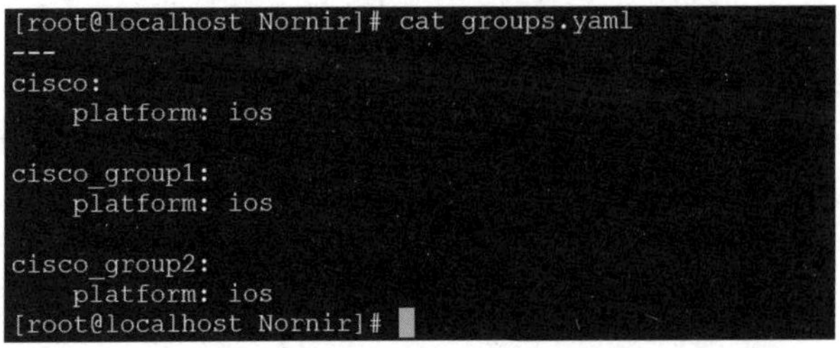

然后创建一个名为 nornir9.py 的脚本，将下列代码写入脚本中。

```
from nornir import InitNornir
from nornir_scrapli.tasks import get_prompt, send_command, send_configs

nr = InitNornir(config_file="config.yaml")

prompt_results = nr.run(task=get_prompt)
config_results = nr.run(task=send_configs, configs=["interface loop99", "description
Nornir loopback"])
command_results = nr.run(task=send_command, command="wr mem")

print(prompt_results["sw1"].result)
print(config_results["sw1"].result)
print(command_results["sw1"].result)
```

```
[root@localhost Nornir]# cat nornir9.py
from nornir import InitNornir
from nornir_scrapli.tasks import get_prompt, send_command, send_configs

nr = InitNornir(config_file="config.yaml")

prompt_results = nr.run(task=get_prompt)
config_results = nr.run(task=send_configs, configs=["interface loop99", "description Nornir loopback"])
command_results = nr.run(task=send_command, command="wr mem")

print(prompt_results["sw1"].result)
print(config_results["sw1"].result)
print(command_results["sw1"].result)
[root@localhost Nornir]#
```

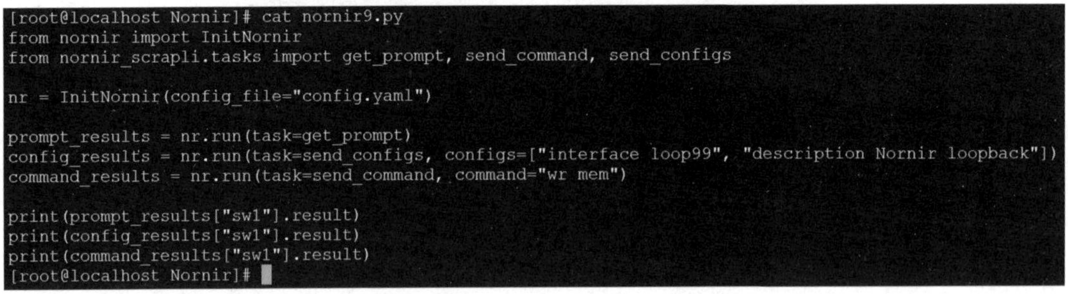

代码分段讲解如下。

- 首先从 nornir_scrapli.tasks 插件中导入 get_prompt、send_command、send_configs 3 个方法，前面提到了 Scrapli 默认在回显内容中省去了主机名和提示符，这就导致了一个问题：同时巡检多台设备时，如果没有主机名和提示符，则我们将不知道回显内容对应的是哪一台设备，因此我们导入 get_prompt 方法来解决这个问题。

```
from nornir import InitNornir
from nornir_scrapli.tasks import get_prompt, send_command, send_configs
```

- 这里我们对 hosts.yaml 中所有的设备做配置，为它们的 loopback 99 端口添加 description，注意 prompt_results、config_results、command_results 这 3 个变量的数据类型为 Nornir 特有的 AggregatedResult，AggregatedResult 类似于字典，我们需要在其后面加上设备名，才能获取具体的回显内容，这里我们只选择打印 sw1 的回显内容。

```
nr = InitNornir(config_file="config.yaml")

prompt_results = nr.run(task=get_prompt)
config_results = nr.run(task=send_configs, configs=["interface loop99", "description
Nornir loopback"])
command_results = nr.run(task=send_command, command="wr mem")

print(prompt_results["sw1"].result)
print(config_results["sw1"].result)
print(command_results["sw1"].result)
```

运行脚本，效果如下图所示。

```
[root@localhost Nornir]# python3.10 nornir9.py
sw1#
interface loop99
description Nornir loopback

Building configuration...
Compressed configuration from 5378 bytes to 2161 bytes[OK]
```

上图中，第一行回显内容中的 sw1#即由 get_prompt 方法提供的。

9.20　实验 9　Nornir 3 + Scrapli（华为设备）

Scrapli 本身没有支持国内厂商设备，不过提供了社区版本 scrapli community，可以设备一些国产设备。简单来说，在不支持国产设备的 Scrapli 中，打上一个叫 scrapli community 的“补丁”，即可开启对国产设备的支持。

考虑到要尽可能覆盖一些主流系统，丰富本书内容，这次实验选取了一台 Ubuntu 20.04.3 LTS 与一台华为真机 S5300 交换机。

Ubuntu（Python 所在，172.25.1.231）同局域网直连（172.25.1.234），两边 ping 通即可。（此例与第 5 章的实验 4 是一致的，只是 Windows 换成了 Ubuntu。）

```
#相对于思科设备实验，需要多安装一个库 scrapli-community
pip install scrapli
pip install scrapli-community
```

Python 脚本代码如下。

```
    from scrapli import Scrapli


    my_device = {
        "hot": "172.25.1.234",
    "auth_username": "python",
    "auth_password": "1234abcd",
    "auth_strict_key": False,
    "platform": "huawei_vrp",          #华为设备使用 huawei_vrp
    "ssh_config_file": "ssh_config"  #补多这一句
}

conn = Scrapli(**my_device)
conn.open()

#操作内容与思科设备实验 Scrapli 保持一致
responses = conn.send_commands(["dis clock", "dis ip int brief"])

for response in responses:
  print(response.result)
conn.close()
```

脚本执行结果如下图所示，实验成功并符合预期。

```
(nornir) zhujiasheng1987@wgjws-desktop:~/ipmanpy/nornir/nornir/scrapli/lab1$ sudo nano lab1-2.py
(nornir) zhujiasheng1987@wgjws-desktop:~/ipmanpy/nornir/nornir/scrapli/lab1$ python3 lab1-2.py
interface GigabitEthernet 0/0/1
description Configured by Scrapli

PHY: Physical
*down: administratively down
(l): loopback
(s): spoofing
(E): E-Trunk down
(b): BFD down
(e): ETHOAM down
(dl): DLDP down
Interface                      PHY     Protocol Description
GE0/0/1                        up      up       Configured by Scrapli
(nornir) zhujiasheng1987@wgjws-desktop:~/ipmanpy/nornir/nornir/scrapli/lab1$
```

Scrapli 测试成功执行后，随着就有很多问题可以继续铺开解决了。在 Ubuntu 系统中，Python 虚拟空间的创建和激活、Scrapli 与 Nornir 的联动、华为部分设备版本需 commit 提交确认等其他情形，请读者自行尝试。

9.21 实验 10 Nornir 3 + TextFSM（思科设备）

第 7 章已经详细讲解了 TextFSM，本节将介绍 TextFSM 在 Nornir 3 中的用法，具体讲解如何使用 Nornir 配合 TextFSM 将思科 IOS 设备 show interface switchport 命令的回显内容用 ntc-templates 给出的模板将其以有序的字典的格式输出。

在 Nornir 中可以直接调用 TextFSM，非常方便。Nornir 中不需要用户自己用 TextFSM 手动写模板，它默认调用 Network To Code 团队打造的 ntc-templates 模板集，开始实验前需要先从 GitHub 上下载 ntc-templates 模板集，具体方法可以参考 6.3.3 节的内容。

ntc-templates 下载完毕后，首先把实验 2 中的脚本 nornir2.py 稍作修改，将其中的 show clock 改成 show interface switchport，然后将脚本保存为 nornir10.py，如下图所示。

```
from nornir import InitNornir
from nornir_netmiko import netmiko_send_command
from nornir_utils.plugins.functions import print_result

nr = InitNornir(config_file="config.yaml")
results = nr.run(netmiko_send_command, command_string='sh interfaces switchport')
print_result(results)
```

```
[root@localhost Nornir]# cat nornir10.py
from nornir import InitNornir
from nornir_netmiko import netmiko_send_command
from nornir_utils.plugins.functions import print_result

nr = InitNornir(config_file="config.yaml")
results = nr.run(netmiko_send_command, command_string='sh interfaces switchport')
print_result(results)

[root@localhost Nornir]#
```

运行脚本，效果如下图所示。

```
[root@localhost Nornir]# python3.10 nornir10.py
netmiko_send_command*************************************************************
* sw1 ** changed : False ********************************************************
vvvv netmiko_send_command ** changed : False vvvvvvvvvvvvvvvvvvvvvvvvvvvvvvvvvvvv INFO
Name: Gi0/1
Switchport: Enabled
Administrative Mode: dynamic desirable
Operational Mode: down
Administrative Trunking Encapsulation: negotiate
Negotiation of Trunking: On
Access Mode VLAN: 1 (default)
Trunking Native Mode VLAN: 1 (default)
Administrative Native VLAN tagging: enabled
Voice VLAN: none
Administrative private-vlan host-association: none
Administrative private-vlan mapping: none
Administrative private-vlan trunk native VLAN: none
Administrative private-vlan trunk Native VLAN tagging: enabled
Administrative private-vlan trunk encapsulation: dot1q
Administrative private-vlan trunk normal VLANs: none
```

```
Administrative private-vlan trunk associations: none
Administrative private-vlan trunk mappings: none
Operational private-vlan: none
Trunking VLANs Enabled: ALL
Pruning VLANs Enabled: 2-1001
Capture Mode Disabled
Capture VLANs Allowed: ALL

Appliance trust: none

Name: Gi0/2
Switchport: Enabled
Administrative Mode: dynamic desirable
Operational Mode: down
Administrative Trunking Encapsulation: negotiate
Negotiation of Trunking: On
Access Mode VLAN: 1 (default)
Trunking Native Mode VLAN: 1 (default)
Administrative Native VLAN tagging: enabled
Voice VLAN: none
Administrative private-vlan host-association: none
Administrative private-vlan mapping: none
Administrative private-vlan trunk native VLAN: none
Administrative private-vlan trunk Native VLAN tagging: enabled
Administrative private-vlan trunk encapsulation: dot1q
Administrative private-vlan trunk normal VLANs: none
Administrative private-vlan trunk associations: none
```

上图为在思科 IOS 交换机上 show interfaces switchport 后的回显内容，可以看到对应的交换机的每个端口的信息量都很大，虽然内容清晰、翔实，但是因为这类回显内容是无序的字符串，我们很难使用 Python 的 re 模块（正则表达式）对其中我们感兴趣的信息做解析，将它们提取出来，所以接下来我们要使用 ntc-templates 模板集下的 cisco_ios_show_interfaces_switchport 这个 TextFSM 模板来将该回显内容转换为有序的字典格式。

在 Nornir 中使用 TextFSM 非常简单，只需在脚本 nornir10.py 的第 6 行代码 results = nr.run(netmiko_send_command, command_string='sh interfaces switchport') 的最后加上 use_textfsm=True 即可，如下图所示。

```
from nornir import InitNornir
from nornir_netmiko import netmiko_send_command
from nornir_utils.plugins.functions import print_result

nr = InitNornir(config_file="config.yaml")
results = nr.run(netmiko_send_command, command_string='sh interfaces switchport',
use_textfsm=True)
print_result(results)
```

```
[root@localhost Nornir]# cat nornir10.py
from nornir import InitNornir
from nornir_netmiko import netmiko_send_command
from nornir_utils.plugins.functions import print_result

nr = InitNornir(config_file="config.yaml")
results = nr.run(netmiko_send_command, command_string='sh interfaces switchport', use_textfsm=True)
print_result(results)

[root@localhost Nornir]#
```

在 nr.run()中调用 use_textfsm 参数后，Nornir 会自动按照 command_string 参数所对应的命令去匹配 ntc-templates 下相应的模板，比如我们在 nr.run()里使用的是 command_string='sh interfaces switchport'，那么 Nornir 会自动在 ntc-templates 下面调用 cisco_ios_show_interfaces_switchport.textfsm 模板将该命令的回显内容从无序的字符串格式转换成有序的字典格式，非常方便。

运行脚本后的效果如下图所示。

```
[root@localhost Nornir]# python3.10 nornir10.py
netmiko_send_command*************************************************************
* sw1 ** changed : False *********************************************************
vvvv netmiko_send_command ** changed : False vvvvvvvvvvvvvvvvvvvvvvvvvvvvvvvvvvvvv INFO
[ { 'access_vlan': '1',
    'admin_mode': 'dynamic desirable',
    'interface': 'Gi0/1',
    'mode': 'down',
    'native_vlan': '1',
    'switchport': 'Enabled',
    'switchport_monitor': '',
    'switchport_negotiation': 'On',
    'trunking_vlans': ['ALL'],
    'voice_vlan': 'none'},
  { 'access_vlan': '1',
    'admin_mode': 'dynamic desirable',
    'interface': 'Gi0/2',
    'mode': 'down',
    'native_vlan': '1',
    'switchport': 'Enabled',
    'switchport_monitor': '',
    'switchport_negotiation': 'On',
    'trunking_vlans': ['ALL'],
    'voice_vlan': 'none'},
  { 'access_vlan': '1',
    'admin_mode': 'dynamic desirable',
    'interface': 'Gi0/3',
    'mode': 'down',
    'native_vlan': '1',
    'switchport': 'Enabled',
    'switchport_monitor': '',
    'switchport_negotiation': 'On',
    'trunking_vlans': ['ALL'],
    'voice_vlan': 'none'}]
^^^^ END netmiko_send_command ^^^^^^^^^^^^^^^^^^^^^^^^^^^^^^^^^^^^^^^^^^^^^^^^^^
* sw2 ** changed : False *********************************************************
vvvv netmiko_send_command ** changed : False vvvvvvvvvvvvvvvvvvvvvvvvvvvvvvvvvvvvv INFO
[ { 'access_vlan': '1',
    'admin_mode': 'dynamic desirable',
    'interface': 'Gi0/1',
```

通过 Nornir 配合 TextFSM 收集到交换机 show interfaces swithport 的回显内容后，下一节实验将介绍如何在 Nornir 中使用 ipdb 模块来对上面得到的这组基于字典、有序的回显内容数据做验证和调用，以便我们在代码中提取需要的内容和参数。

9.22　实验 10　Nornir 3 + TextFSM（华为设备）

在实验 10 中，我们重新回到 Windows 系统上，在 nornir8 文件夹的基础上，复制成 nornir10，继续开展实验，如下图所示。TextFSM 在 Netmiko 详解中已有相关的介绍，这里仅演示在 Nornir 框架中的一些效果。

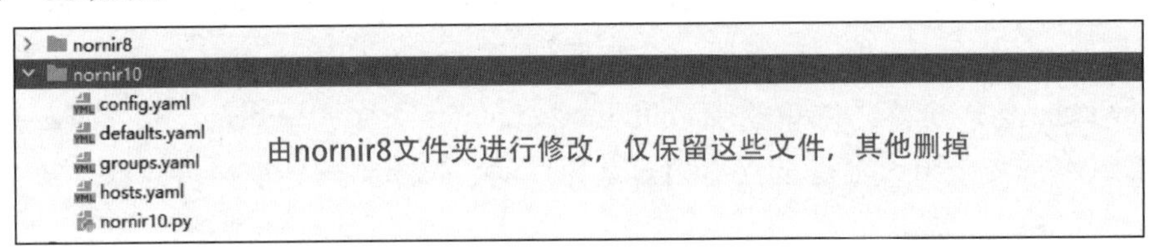

准备 nornir10.py。

```
from nornir import InitNornir
from nornir_netmiko import netmiko_send_command
from nornir_utils.plugins.functions import print_result

nr = InitNornir(config_file="config.yaml")
results = nr.run(netmiko_send_command, command_string='display interface brief')
print_result(results)
```

此时，没有使用 TextFSM 模板，执行结果如下图所示。

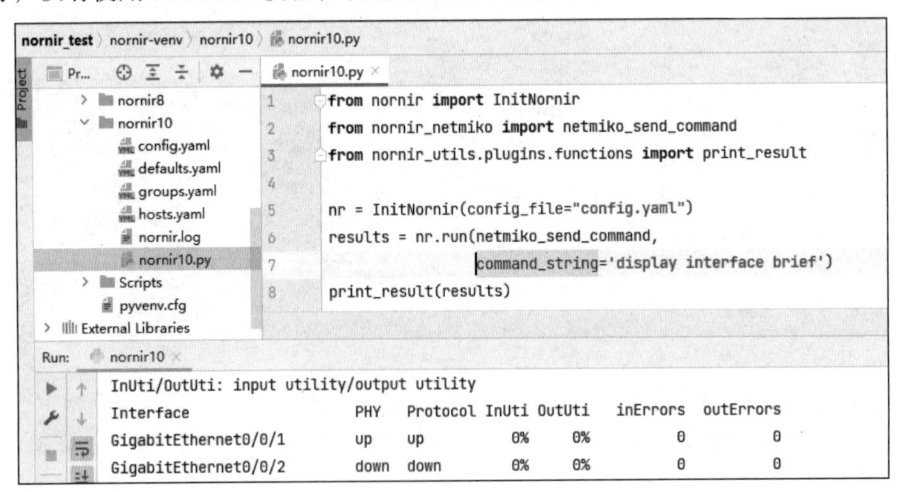

并未对回显数据进行格式化，此时我们在 Python 脚本中多加一个参数，即可让回显结果变成结构化数据，如下图所示。

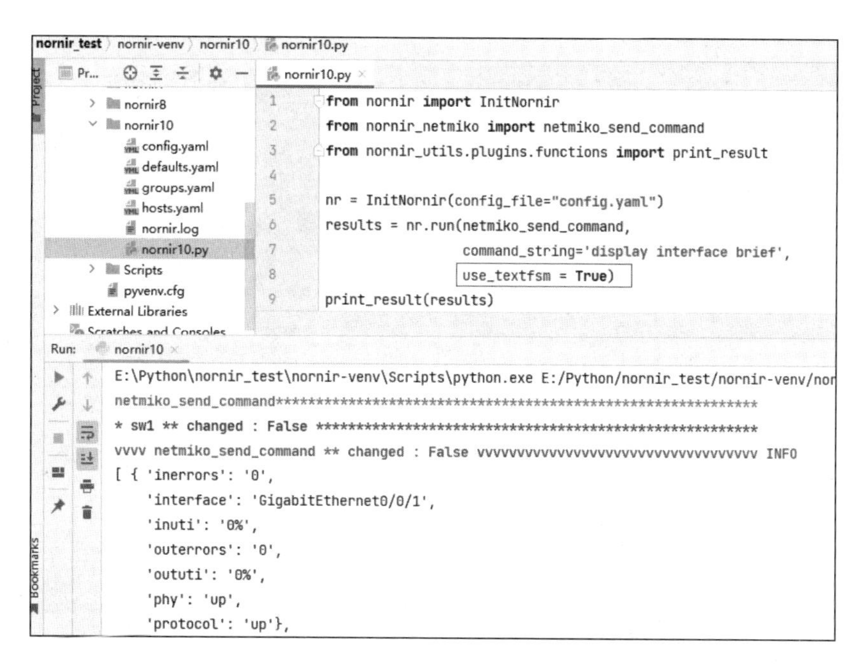

特别强调，这里的实验操作需要配合环境变量设定、ntc-templates 中 TextFSM 模板的拟写等操作，具体参见 Netmiko 详解中联动 TextFSM 的内容。

9.23　实验 11　Nornir 3 + ipdb（思科设备）

ipdb 全称为 IPython pdb，IPython 是一种基于 Python 的交互式解释器，相较于本地的 Python 解释器（比如 IDLE），IPython 提供了更为强大的编辑和交互功能。

知道了 ipdb 中的 i 是什么意思后，我们再来看 pdb。pdb 是 Python 内置的排错器（debug）模块，pdb 提供交互式的界面让用户对代码进行验证和排错，它支持在源码行间设置（有条件的）断点和单步执行、监视堆栈帧、列出源码列表，以及在任何堆栈帧的上下文中运行任意 Python 代码。它还支持事后调试，可以在程序控制下调用。

综上所述，ipdb 是基于 pdb 开发出来的一个针对 IPython 的第三方排错模块，ipdb 之于 IPython 也就是 pdb 之于本地的 Python 解释器，本节将介绍 ipdb 在 Nornir 3 中的用法，讲解如何在 Nornir 中使用 ipdb 模块来对实验 10 中得到的基于字典、有序的回显内容数据做验证和调用，以便我们在代码中提取需要的内容和参数。

开始实验前，首先通过 pip 安装 ipdb 模块。

```
pip install ipdb
```

```
[root@localhost Nornir]# pip3.10 install ipdb
Collecting ipdb
  Downloading ipdb-0.13.9.tar.gz (16 kB)
  Preparing metadata (setup.py) ... done
Requirement already satisfied: setuptools in /usr/local/lib/python3.10/site-packages (from ipdb) (63.2.0)
Collecting ipython>=7.17.0
  Downloading ipython-8.4.0-py3-none-any.whl (750 kB)
     ---------------------------------------- 750.8/750.8 kB  eta 0:00:00
Collecting toml>=0.10.2
  Downloading toml-0.10.2-py2.py3-none-any.whl (16 kB)
Collecting decorator
  Downloading decorator-5.1.1-py3-none-any.whl (9.1 kB)
Collecting jedi>=0.16
  Downloading jedi-0.18.1-py2.py3-none-any.whl (1.6 MB)
     ---------------------------------------- 1.6/1.6 MB  eta 0:00:00
Collecting prompt-toolkit!=3.0.0,!=3.0.1,<3.1.0,>=2.0.0
  Downloading prompt_toolkit-3.0.30-py3-none-any.whl (381 kB)
     ---------------------------------------- 381.7/381.7 kB  eta 0:00:00
Collecting pickleshare
  Downloading pickleshare-0.7.5-py2.py3-none-any.whl (6.9 kB)
Collecting backcall
  Downloading backcall-0.2.0-py2.py3-none-any.whl (11 kB)
Collecting matplotlib-inline
  Downloading matplotlib_inline-0.1.5-py3-none-any.whl (9.4 kB)
Collecting pexpect>4.3
  Downloading pexpect-4.8.0-py2.py3-none-any.whl (59 kB)
     ---------------------------------------- 59.0/59.0 kB  eta 0:00:00
Collecting pygments>=2.4.0
  Downloading Pygments-2.13.0-py3-none-any.whl (1.1 MB)
     ---------------------------------------- 1.1/1.1 MB  eta 0:00:00
Collecting traitlets>=5
  Downloading traitlets-5.3.0-py3-none-any.whl (106 kB)
```

下载 ipdb 后，进入 Python 编辑器 import ipdb，如果没有报错，则说明 ipdb 模块安装成功，如下图所示。

```
[root@localhost Python-3.10.6]# python3.10
Python 3.10.6 (main, Aug 17 2022, 00:29:37) [GCC 8.5.0 20210514 (Red Hat 8.5.0-1
5)] on linux
Type "help", "copyright", "credits" or "license" for more information.
>>> import ipdb
>>> exit()
[root@localhost Python-3.10.6]#
```

注：ipdb 模块依赖于 sqlite3 数据库，如果 import ipdb 后系统提示 module not found error: no module named '_sqlite3'错误，则需要通过下列方法解决。首先回到 Python 3.10 的安装文件夹（参考第 1 章中 1.1.2 节的内容），通过 yum install sqlite-devel 安装 sqlite，然后在 Python 安装文件夹中重新对 Python 进行编译加入该 sqlite 组件，方法如下。

```
cd Python-3.10.6/
yum install sqlite-devel
./configure --enable-loadable-sqlite-extensions && make install
```

```
[root@localhost ~]# cd Python-3.10.6/
[root@localhost Python-3.10.6]# yum install sqlite-devel
Last metadata expiration check: 2:08:11 ago on Tue 16 Aug 2022 10:31:10 PM PDT.
Package sqlite-devel-3.26.0-15.el8.x86_64 is already installed.
Dependencies resolved.
Nothing to do.
Complete!
[root@localhost Python-3.10.6]# ./configure --enable-optimizations --enable-loadable-sqlite-extensions && make install
```

在实验 10 中，我们已经通过 TextFSM 将 show interfaces switchport 的内容以字典的格式输入出来，如下图所示。

```
[root@localhost Nornir]# python3.10 nornir10.py
netmiko_send_command*********************************************************************
* sw1 ** changed : False ****************************************************************
vvvv netmiko_send_command ** changed : False vvvvvvvvvvvvvvvvvvvvvvvvvvvvvvvvvvvvvvvv INFO
[ { 'access_vlan': '1',
    'admin_mode': 'dynamic desirable',
    'interface': 'Gi0/1',
    'mode': 'down',
    'native_vlan': '1',
    'switchport': 'Enabled',
    'switchport_monitor': '',
    'switchport_negotiation': 'On',
    'trunking_vlans': ['ALL'],
    'voice_vlan': 'none'},
  { 'access_vlan': '1',
    'admin_mode': 'dynamic desirable',
    'interface': 'Gi0/2',
    'mode': 'down',
    'native_vlan': '1',
    'switchport': 'Enabled',
    'switchport_monitor': '',
    'switchport_negotiation': 'On',
    'trunking_vlans': ['ALL'],
    'voice_vlan': 'none'},
  { 'access_vlan': '1',
    'admin_mode': 'dynamic desirable',
    'interface': 'Gi0/3',
    'mode': 'down',
    'native_vlan': '1',
    'switchport': 'Enabled',
    'switchport_monitor': '',
    'switchport_negotiation': 'On',
    'trunking_vlans': ['ALL'],
    'voice_vlan': 'none'}]
^^^^ END netmiko_send_command ^^^^^^^^^^^^^^^^^^^^^^^^^^^^^^^^^^^^^^^^^^^^^^^^^^^^^^^^^
* sw2 ** changed : False ****************************************************************
vvvv netmiko_send_command ** changed : False vvvvvvvvvvvvvvvvvvvvvvvvvvvvvvvvvvvvvvvv INFO
[ { 'access_vlan': '1',
    'admin_mode': 'dynamic desirable',
    'interface': 'Gi0/1',
```

在实验环境中，我们用到了 sw1、sw2、sw3、sw4 总共 4 台思科虚拟三层交换机，每台设备都有 Gi0/0、Gi0/1、Gi0/2 总共 3 个物理端口。你肯定会问：在 Nornir 脚本里怎样写才能做过滤，将某个交换机具体的某个参数读取出来呢？比如这里我们想知道且只想知道交换机 sw1 下面的 Gi0/0 的 mode 键是什么值、代码要怎么写呢？

在回答这个问题之前，我们先用 ipdb 来做验证，首先在 norni10.py 的基础上稍作修改，创建 nornir11.py，写入下面的代码。

```
import ipdb
from nornir import InitNornir
from nornir_netmiko import netmiko_send_command
```

```
from nornir_utils.plugins.functions import print_result

nr = InitNornir(config_file="config.yaml")
output = nr.run(netmiko_send_command, command_string='sh interface switchport',
use_textfsm=True)
print_result(output)
print(output)
ipdb.set_trace()
```

```
[root@localhost Nornir]# cat nornir11.py
import ipdb
from nornir import InitNornir
from nornir_netmiko import netmiko_send_command
from nornir_utils.plugins.functions import print_result

nr = InitNornir(config_file="config.yaml")
output = nr.run(netmiko_send_command, command_string='sh interface switchport', use_textfsm=True)
print_result(output)
print(output)
ipdb.set_trace()
[root@localhost Nornir]#
```

代码分段讲解如下。

- 我们引入 ipdb 模块，以便稍后能进入 IPython 解释器，其余代码在实验 2 和实验 10 中已经讲过，不再赘述。

```
import ipdb
from nornir import InitNornir
from nornir_netmiko import netmiko_send_command
from nornir_utils.plugins.functions import print_result

nr = InitNornir(config_file="config.yaml")
output = nr.run(netmiko_send_command, command_string='sh interface switchport',
use_textfsm=True)
print_result(output)
```

- 在 print_result(output)后，我们又于倒数第二行代码用到了 print(output)。之所以加上 print(output)，是为了让大家看到 AggreagtedResult 和 MultiResult 两个 Nornir 中特有的数据类型，配合后文的讲解。最后通过 ipdb.set_trace()来调用 ipdb，在运行脚本后进入 IPython 解释器。

```
print(output)
ipdb.set_trace()
```

运行脚本后的效果如下图所示。

```
{ 'access_vlan': '1',
  'admin_mode': 'dynamic desirable',
  'interface': 'Gi0/3',
  'mode': 'down',
  'native_vlan': '1',
  'switchport': 'Enabled',
  'switchport_monitor': '',
  'switchport_negotiation': 'On',
  'trunking_vlans': ['ALL'],
  'voice_vlan': 'none'}]
^^^^ END netmiko_send_command ^^^^^^^^^^^^^^^^^^^^^^^^^^^^^^^^^^^^^^^^^^^^^^^^^^
AggregatedResult (netmiko_send_command): {'sw1': MultiResult: [Result: "netmiko_send_command"], 'sw2': MultiResult: [Result: "netmiko_send_command"]}
--Return--
None
> /root/Nornir/nornir11.py(10)<module>
      8 print_result output
      9 print output
---> 10 ipdb set_trace

ipdb>
```

因为回显内容过多，这里只截取运行脚本后的末尾部分，其中小线框标注部分（ipdb>命令行）即我们通过 ipdb.set_trace()进入的 IPython 解释器，而大线框标注的部分即 print(output)后返回的内容，可以看到一个 AggregatedResult 和两个 MultiResult（分别接在'sw1'、'sw2'后面）。

```
AggregatedResult (netmiko_send_command): {'sw1': MultiResult: [Result:
"netmiko_send_command"], 'sw2': MultiResult: [Result: "netmiko_send_command"]}
```

重点：在 Nornir 中，AggregatedResult 和 MultiResult 分别对应的是字典和列表两种数据类型，既然是字典和列表，那么针对 AggregatedResult 数据类型，我们可以使用['key_name']来获取键名对应的值，而针对 MultiResult，我们可以用索引号（比如[0]）读取其中的元素。

我们能看到 AggregatedResult 这组字典有两个键，分别是'sw1'和'sw2'（因为此时 hosts.yaml 里只放了这两台交换机），而两个键对应的值又是 MultiResult。接下来我们在 ipdb 中验证，因为我们在运行 nr.run()后将结果保存在了变量 output，首先在 ipdb>后面输入 output，随即得到和 print(output)一样的内容，如下图所示。

```
ipdb> output
AggregatedResult (netmiko_send_command): {'sw1': MultiResult: [Result: "netmiko_send_command"], 'sw2': MultiResult: [Result: "netmiko_send_command"]}
ipdb>
```

继续输入 **output['sw1']**，得到键名'sw1'对应的值（一个 MultiResult）。

```
ipdb> output['sw1']
MultiResult: [Result: "netmiko_send_command"]
ipdb>
```

继续输入 **output['sw1'].result**，随即得到交换机 sw1 下面所有通过 TextFSM 运行 show interfaces switchport 模板后返回的内容（如果要查看交换机 sw2，则用 output['sw2'].result，依此类推），如下图所示。

```
ipdb> output['sw1'].result
[{'interface': 'Gi0/1', 'switchport': 'Enabled', 'switchport_monitor': '', 'switchport_negotiation': 'On', 'mode': 'down', 'admin_mode': 'dynamic desirable', 'access_vlan': '1', 'native_vlan': '1', 'voice_vlan': 'none', 'trunking_vlans': ['ALL']}, {'interface': 'Gi0/2', 'switchport': 'Enabled', 'switchport_monitor': '', 'switchport_negotiation': 'On', 'mode': 'down', 'admin_mode': 'dynamic desirable', 'access_vlan': '1', 'native_vlan': '1', 'voice_vlan': 'none', 'trunking_vlans': ['ALL']}, {'interface': 'Gi0/3', 'switchport': 'Enabled', 'switchport_monitor': '', 'switchport_negotiation': 'On', 'mode': 'down', 'admin_mode': 'dynamic desirable', 'access_vlan': '1', 'native_vlan': '1', 'voice_vlan': 'none', 'trunking_vlans': ['ALL']}]
ipdb>
```

前面提到，在实验环境中，每台交换机都有 Gi0/0、Gi0/1、Gi0/2 3 个物理端口，这里 output['sw1'].result 返回的是一个列表，该列表有 3 个元素，均为字典，每个字典对应一个物理端口，那么 output['sw1'].result[0]返回的是 sw1 的 Gi0/0 下的信息，output['sw1'].result[1]返回的是 sw1 的 Gi0/1 下的信息，output['sw1'].result[2]返回的是 sw1 的 Gi0/2 下的信息，依此类推，在 ipdb 下输入 output['sw1'].result[0]来过滤 sw1 的 Gi0/0 下的信息。

```
ipdb> output['sw1'].result[0]
{'interface': 'Gi0/1', 'switchport': 'Enabled', 'switchport_monitor': '', 'switchport_negotiation': 'On', 'mode': 'down', 'admin_mode': 'dynamic desirable', 'access_vlan': '1', 'native_vlan': '1', 'voice_vlan': 'none', 'trunking_vlans': ['ALL']}
ipdb>
```

因为 output['sw1'].result[0]返回的值是一个字典，继续输入 output['sw1'].result[0]['mode']即得到前面我们要找的“交换机 sw1 下面的 Gi0/0 的 mode 键对应的值”，即 down，如下图所示。

```
ipdb> output['sw1'].result[0]['mode']
'down'
ipdb>
```

使用完后，输入 exit，退出 ipdb。

最后我们把脚本 nornir11.py 稍作修改，去掉 ipdb 及 print_result(output)和 print(output)，只将我们要找的“交换机 sw1 下面的 Gi0/0 的 mode 键对应的值”打印出来，脚本代码如下。

```
import ipdb
from nornir import InitNornir
from nornir_netmiko import netmiko_send_command
from nornir_utils.plugins.functions import print_result
```

```
nr = InitNornir(config_file="config.yaml")
output = nr.run(netmiko_send_command, command_string='sh interface switchport',
use_textfsm=True)

#print_result(output)
#print(output)

#ipdb.set_trace()
print (output['sw1'].result[0]['mode'])
```

```
[root@localhost Nornir]# cat nornir11.py
import ipdb
from nornir import InitNornir
from nornir_netmiko import netmiko_send_command
from nornir_utils.plugins.functions import print_result

nr = InitNornir(config_file="config.yaml")
output = nr.run(netmiko_send_command, command_string='sh interface switchport',
use_textfsm=True)

#print_result(output)
#print(output)

#ipdb.set_trace()
print (output['sw1'].result[0]['mode'])
[root@localhost Nornir]#
```

再次运行代码，可以看到这次脚本只返回了 down 这个数据，实验成功，如下图所示。

```
[root@localhost Nornir]# python3.10 nornir11.py
down
[root@localhost Nornir]#
```

通过本节实验可以看到，Nornir 中使用 TextFSM 后返回的 structured data 由于牵涉到 AggregatedResult 和 MultiResult 两种数据类型，并且数据结构层次分明（但也十分复杂），我们必须借助 ipdb 解释器来“抽丝剥茧”地找出我们想要过滤的数据。下一节实验中，我们将继续借助 ipdb 来了解 Nornir 中最重要的组件之一：inventory。

9.24　实验 11　Nornir 3 + ipdb（华为设备）

在 nornir10 文件夹的基础上，复制成 nornir11，继续开展实验。重新在 CMD 中进入虚拟环境的目录，激活原先的虚拟环境。

```
nornir-venv\Scripts\activate
```

安装 ipdb，等 Sucessfully 的代码如下。

```
pip3 install ipdb
```

```
(nornir-venv) E:\Python\nornir_test>pip3 show ipdb
Name: ipdb
Version: 0.13.9
Summary: IPython-enabled pdb
Home-page: https://github.com/gotcha/ipdb
Author: Godefroid Chapelle
Author-email: gotcha@bubblenet.be
License: BSD
Location: e:\python\nornir_test\nornir-venv\lib\site-packages
Requires: decorator, ipython, setuptools, toml
Required-by:

(nornir-venv) E:\Python\nornir_test>
```

Python 脚本 nornir11.py。

```
from nornir import InitNornir
from nornir_netmiko import netmiko_send_command
from nornir_utils.plugins.functions import print_result
import ipdb

nr = InitNornir(config_file="config.yaml")
results = nr.run(netmiko_send_command,
            command_string='display interface brief',
            use_textfsm = True)
print_result(results)

print(results)
ipdb.set_trace()
```

直接运行一下 python 脚本，结果如下图所示。

```
^^^^ END netmiko_send_command ^^^^^^^^^^^^^^^^^^^^^^^^^^^^^^^^^^^^^^^^^^^^^^^^^^^^^^^
AggregatedResult (netmiko_send_command): {'sw1': MultiResult: [Result: "netmiko_send_command"], 'sw
 'sw3': MultiResult: [Result: "netmiko_send_command"], 'sw4': MultiResult: [Result: "netmiko_send_c
--Return--          print(results)的内容
None
> e:\python\nornir_test\nornir-venv\nornir11\nornir11.py(17)<module>()
     15
     16 print(results)
---> 17 ipdb.set_trace()

ipdb>
```

把 print(results)的结果单独列出来，此时看不到端口 interface 的任何信息，但能看出是一个框架结构。

```
^^^^ END netmiko_send_command ^^^^^^^^^^^^^^^^^^^^^^^^^^^^^^^^^^^^^^^^^^^^^^^^^^^^^
AggregatedResult (netmiko_send_command): {'sw1': MultiResult: [Result:
"netmiko_send_command"], 'sw2': MultiResult: [Result: "netmiko_send_command"], 'sw3':
MultiResult: [Result: "netmiko_send_command"], 'sw4': MultiResult: [Result:
"netmiko_send_command"]}
--Return--
```

AggregatedResult 把所有执行结果聚合在一起；MultiResult 又把单台设备的指令执行结果聚合起来（本例只执行了一条指令）。此时，用 ipdb.set_trace()打开跟 AggregatedResult、MultiResult 等这些特殊类型对话的“话匣子”。

比如，AggregatedResult (netmiko_send_command)后面好像是一个字典，于是考虑用字典键值对索引的方法提取，得到键名'sw1'对应的值，是一个 MultiResult 对象。

```
ipdb> results['sw1']
MultiResult: [Result: "netmiko_send_command"]
```

这类似与 IDLE 的对话模式，华为设备的调测过程与思科设备实验是一致的，读者可做如下尝试。

```
ipdb> results['sw1'].result
[{'interface': 'GigabitEthernet0/0/1', 'phy': 'up', 'protocol': 'up', 'inuti': '0%',
'ututi': '0%', 'inerrors': '0', 'outerrors': '0'}, {'interface':
'GigabitEthernet0/0/2'......
```

逐层拨开，最后检索到自己想要的资源信息。

```
ipdb> results['sw1'].result[0]
{'interface': 'GigabitEthernet0/0/1', 'phy': 'up', 'protocol': 'up', 'inuti': '0%',
'ututi': '0%', 'inerrors': '0', 'outerrors': '0'}
ipdb> results['sw1'].result[0]['interface']
'GigabitEthernet0/0/1'
ipdb>
```

ipdb 是一个工具，在复杂嵌套数据的对象中进行辅助定位，从而捕捉想要的信息。当定位完成后，可将 ipdb 的相关代码去除，回到常规代码中，如下图所示。

```
#print_result(results)
#print(results)
#ipdb.set_trace()
print (results['sw1'].result[0]['interface'],   #只取期望信息
      results['sw1'].result[0]['phy'])
```

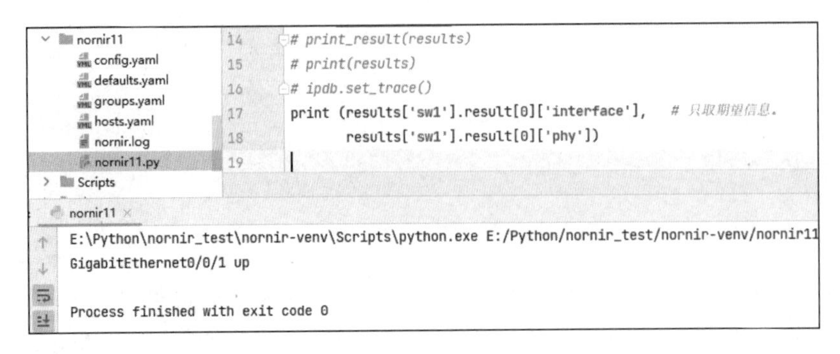

获得这个资源以后，我们可以打印出来，写入文件存档或者进行其他操作，按需执行即可。

9.25　实验 12　Nornir 的 Inventory（思科设备）

Inventory 是 Nornir 中最重要的插件之一，用来管理由 Nornir 操控的网络设备，记录这些设备的 IP 地址、SSH 登录用户名和密码、OS 版本、分组等信息，以及其他一些可选的信息。Nornir 本身自带一个叫作 SimpleInventory 的插件来管理设备，SimpleInventory 插件由 hosts.yaml、groups.yaml 及 defaults.yal 3 个 YAML 文件组成，是的，就是本章开篇提到的那几个重要的 YAML 文件。

在开始实验前，首先将 hosts.yaml 文件的内容恢复成实验 4 中的状态，如下图所示。

```
---
sw1:
   hostname: 192.168.2.11
   username: python
   password: '123'
   platform: ios
   groups:
      - cisco_group1
   data:
      building: '1'
      level: '1'
sw2:
   hostname: 192.168.2.12
   platform: ios
   groups:
      - cisco_group1
   data:
      building: '1'
```

```
        level: '2'
sw3:
    hostname: 192.168.2.13
    platform: ios
    groups:
        - cisco_group2
    data:
        builiding: '2'
        level: '1'
sw4:
    hostname: 192.168.2.14
    platform: ios
    groups:
        - cisco_group2
    data:
        building: '2'
        level: '2'
```

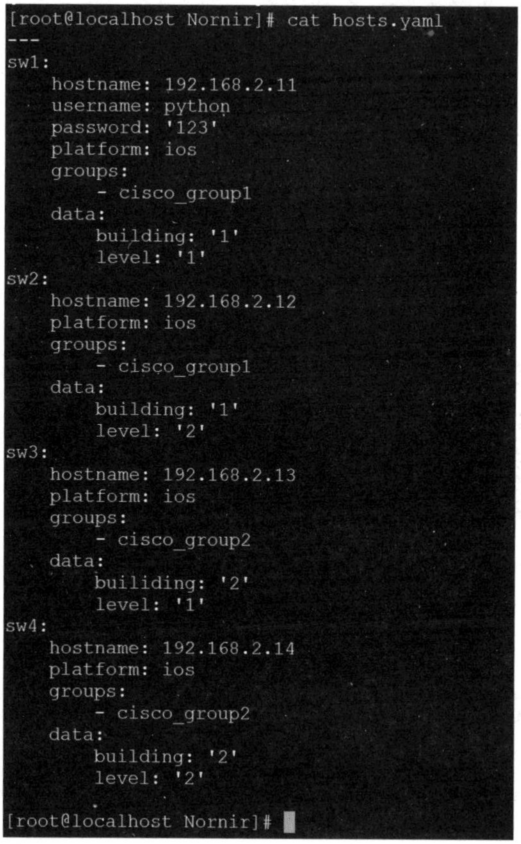

```
[root@localhost Nornir]# cat hosts.yaml
---
sw1:
    hostname: 192.168.2.11
    username: python
    password: '123'
    platform: ios
    groups:
        - cisco_group1
    data:
        building: '1'
        level: '1'
sw2:
    hostname: 192.168.2.12
    platform: ios
    groups:
        - cisco_group1
    data:
        building: '1'
        level: '2'
sw3:
    hostname: 192.168.2.13
    platform: ios
    groups:
        - cisco_group2
    data:
        builiding: '2'
        level: '1'
sw4:
    hostname: 192.168.2.14
    platform: ios
    groups:
        - cisco_group2
    data:
        building: '2'
        level: '2'

[root@localhost Nornir]#
```

在本节实验中，我们将提取出交换机 sw1 对应的 name、hostname、username、password、platform、groups、data 等参数。为了向读者演示如何通过 Inventory 插件获取 hosts.yaml 中设备对应的参数，我们要用到实验 11 中讲到的 ipdb。首先创建一个名为 nornir12.py 的脚本文件，放入下列代码。

```
import ipdb
from nornir import InitNornir
from nornir_netmiko import netmiko_send_command
from nornir_utils.plugins.functions import print_result

nr = InitNornir(config_file='config.yaml')
ipdb.set_trace()
```

```
[root@localhost Nornir]# cat nornir12.py
import ipdb
from nornir import InitNornir
from nornir_netmiko import netmiko_send_command
from nornir_utils.plugins.functions import print_result

nr = InitNornir(config_file='config.yaml')
ipdb.set_trace()
[root@localhost Nornir]#
```

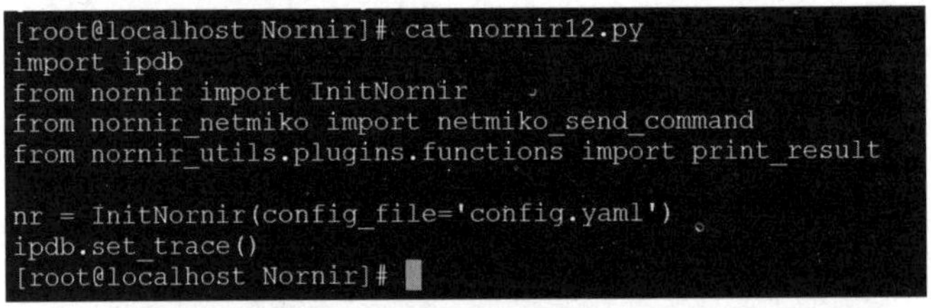

运行脚本后，直接进入 ipdb 解释器，这里我们通过 InitNornir()将 Nornir 初始化，并将它赋值给 nr 变量。要使用 Inventory 插件很简单，只需输入 **nr.inventory.hosts** 即可看到 hosts.yaml 中保存的 4 个交换机的名称（sw1、sw2、sw3、sw4），而从大括号可以判断出 **nr.inventory.hosts 返回的值为一个字典**，如下图所示。

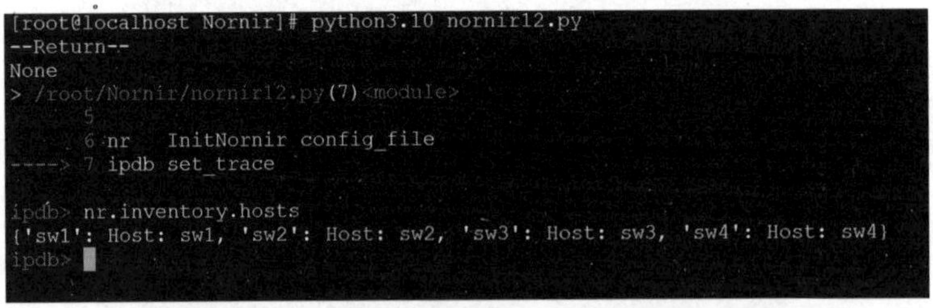

```
[root@localhost Nornir]# python3.10 nornir12.py
--Return--
None
> /root/Nornir/nornir12.py(7)<module>
      5
      6 nr   InitNornir config_file
----> 7 ipdb set_trace

ipdb> nr.inventory.hosts
{'sw1': Host: sw1, 'sw2': Host: sw2, 'sw3': Host: sw3, 'sw4': Host: sw4}
ipdb>
```

既然是字典，那我们就可以通过键名来获取对应的值。进一步输入 nr.inventory.hosts['sw1']，得到交换机 sw1 对应的值，而通过 type()函数，我们可以看到该值的数据类型为 Nornir 特有的 nornir.core.inventory.Host 类型，如下图所示。

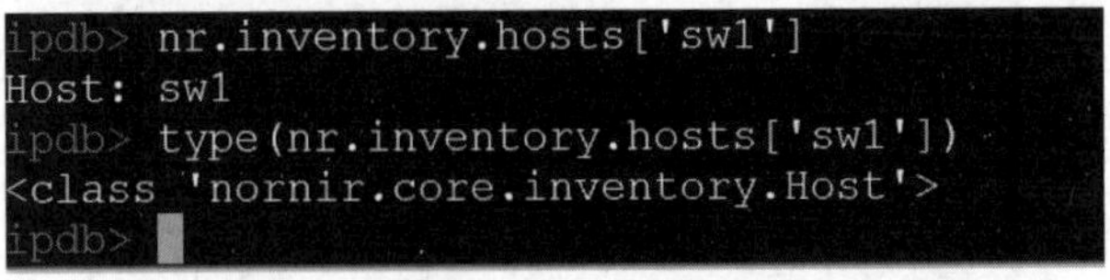

```
ipdb> nr.inventory.hosts['sw1']
Host: sw1
ipdb> type(nr.inventory.hosts['sw1'])
<class 'nornir.core.inventory.Host'>
ipdb>
```

针对该数据类型，我们可以直接调用 hosts.yaml 中对应的参数来获取相应的值，如下图所示。

```
ipdb> nr.inventory.hosts['sw1'].name
'sw1'
ipdb> nr.inventory.hosts['sw1'].hostname
'192.168.2.11'
ipdb> nr.inventory.hosts['sw1'].username
'python'
ipdb> nr.inventory.hosts['sw1'].password
'123'
ipdb> nr.inventory.hosts['sw1'].platform
'ios'
ipdb> nr.inventory.hosts['sw1'].groups
[Group: cisco_group1]
ipdb> nr.inventory.hosts['sw1'].data
{'building': '1', 'level': '1'}
ipdb>
```

我们可以通过查看 hosts.yaml，对上面返回的值一一做对比验证，如下图所示。

```
[root@localhost Nornir]# cat hosts.yaml
---
sw1:
    hostname: 192.168.2.11
    username: python
    password: '123'
    platform: ios
    groups:
        - cisco_group1
    data:
        building: '1'
        level: '1'
```

需要注意的是，nr.inventory.hosts['sw1'].groups 返回值的类型为列表，而 nr.inventory.hosts['sw1'].data 返回值的类型为字典，我们还可以通过索引号和键名继续对它们包含的数据和元素做过滤，这里留给读者自行尝试。

如果要获取其他交换机的数据，则只需将 nr.inventory.hosts['sw1']中的 sw1 替换成 sw2、sw3 等即可，是不是很简单明了？

接下来我们将脚本 nornir12.py 稍作修改，将 sw1 的各项参数全部通过 Inventory 打印出来，代码如下所示。

```
import ipdb
from nornir import InitNornir
from nornir_netmiko import netmiko_send_command
from nornir_utils.plugins.functions import print_result
```

```
nr = InitNornir(config_file='config.yaml')
#ipdb.set_trace()
print (nr.inventory.hosts['sw1'].name)
print (nr.inventory.hosts['sw1'].hostname)
print (nr.inventory.hosts['sw1'].username)
print (nr.inventory.hosts['sw1'].password)
print (nr.inventory.hosts['sw1'].platform)
print (nr.inventory.hosts['sw1'].groups)
print (nr.inventory.hosts['sw1'].data)
```

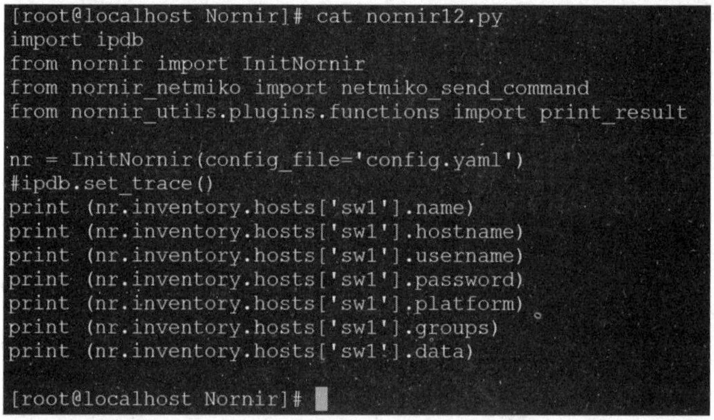

```
[root@localhost Nornir]# cat nornir12.py
import ipdb
from nornir import InitNornir
from nornir_netmiko import netmiko_send_command
from nornir_utils.plugins.functions import print_result

nr = InitNornir(config_file='config.yaml')
#ipdb.set_trace()
print (nr.inventory.hosts['sw1'].name)
print (nr.inventory.hosts['sw1'].hostname)
print (nr.inventory.hosts['sw1'].username)
print (nr.inventory.hosts['sw1'].password)
print (nr.inventory.hosts['sw1'].platform)
print (nr.inventory.hosts['sw1'].groups)
print (nr.inventory.hosts['sw1'].data)

[root@localhost Nornir]#
```

运行脚本，效果如下图所示。

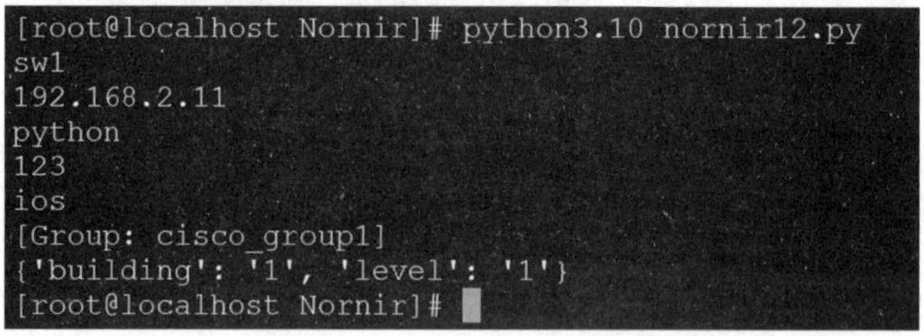

```
[root@localhost Nornir]# python3.10 nornir12.py
sw1
192.168.2.11
python
123
ios
[Group: cisco_group1]
{'building': '1', 'level': '1'}
[root@localhost Nornir]#
```

9.26　实验 12　Nornir 的 Inventory（华为设备）

在 nornir11 文件夹的基础上，复制成 nornir12，继续开展实验。这次实验要通过 Nornir 的 Inventory 配合 ipdb 来探究 sw 的信息，因而再次把 host.yaml 文件重新附上来。

```
---
sw1:
  hostname: 192.168.2.11
```

```
    username: python
    password: '123'
    groups:
        - huawei_group1
    data:
        building: '1'
        level: '1'
sw2:
    hostname: 192.168.2.12
    groups:
        - huawei_group1
    data:
        building: '1'
        level: '2'
sw3:
    hostname: 192.168.2.13
    groups:
        - huawei_group2
    data:
        building: '2'
        level: '1'
sw4:
    hostname: 192.168.2.14
    groups:
        - huawei_group2
    data:
        building: '2'
        level: '2'
```

Python 脚本修改为 nornir12.py。

```
import ipdb
from nornir import InitNornir
from nornir_netmiko import netmiko_send_command
from nornir_utils.plugins.functions import print_result

nr = InitNornir(config_file='config.yaml')
ipdb.set_trace()
```

这次实验的 Python 脚本非常简洁，其中 nr = InitNornir(config_file='config.yaml')是根据

config.yaml 文件，关联了同文件夹中的 hosts.yaml、groups.yaml、defaults.yaml 3 个文件涉及的网络设备信息。由此，InitNornir 函数进行初始化后赋值给变量 nr，完成 Nornir 与网络设备的联动，此后 Nornir 就可以通过其他代码控制网络设备了。ipdb.set_trace()在实验 11 中已经阐述了，主要用于调试使用。

Python 脚本启动执行，如下图所示。

现在有了 nr 这个对象（变量），试一下 dir 能返回一些内容。这里使用了列表推导式，旨在把 nr 对象可供使用的属性或者方法提取出来（过滤掉魔法方法，因为正常使用不建议使用魔法方法）。

```
ipdb> nr
<nornir.core.Nornir object at 0x0000028942880880>
ipdb>
ipdb> [method for method in dir(nr) if not method.startswith('_')]
['close_connections', 'config', 'data', 'dict', 'filter', 'get_validators', 'inventory',
'processors', 'run', 'runner', 'validate', 'with_processors', 'with_runner']
ipdb>
ipdb> nr.inventory
<nornir.core.inventory.Inventory object at 0x000002C07BB53700>
ipdb>
ipdb> [method for method in dir(nr.inventory) if not method.startswith('_')]
['children_of_group', 'defaults', 'dict', 'filter', 'groups', 'hosts', 'schema']
ipdb>
ipdb> nr.inventory.hosts
{'sw1': Host: sw1, 'sw2': Host: sw2, 'sw3': Host: sw3, 'sw4': Host: sw4}
ipdb>
```

```
ipdb> nr.inventory.hosts['sw1']
Host: sw1
ipdb>
ipdb> [method for method in dir(nr.inventory.hosts['sw1']) if not
method.startswith('_')]
['close_connection', 'close_connections', 'connection_options', 'connections', 'data',
'defaults', 'dict', 'extended_data', 'extended_groups', 'get', 'get_connection',
'get_connection_parameters', 'groups', 'has_parent_group', 'hostname', 'items', 'keys',
'name', 'open_connection', 'password', 'platform', 'port', 'schema', 'username',
'values']
ipdb>
```

通过一层一层的调测，就可以找到 Inventory 中 hosts、name 等相关资源。这是一个通法，Nornir 可以这么用，其他框架也可以这么用。

```
ipdb> nr.inventory.hosts['sw1'].name
'sw1'
ipdb> nr.inventory.hosts['sw1'].hostname
'192.168.2.11'
ipdb> nr.inventory.hosts['sw1'].username
'python'
ipdb> nr.inventory.hosts['sw1'].password
'123'
ipdb> nr.inventory.hosts['sw1'].platform
'huawei'
ipdb> nr.inventory.hosts['sw1'].groups
[Group: huawei_group1]
ipdb> nr.inventory.hosts['sw1'].data
{'building': '1', 'level': '1'}
ipdb>
```

对比一下 host.yaml 中关于 sw1 的参数，与我们用 ipdb 和 dir 配合“挖”出来的参数是一致的。当使用 ipdb 配合 dir 摸索一番，定位到自己想要的资源，就可以撤掉它们了，稍微调整下代码。

```
import ipdb
from nornir import InitNornir
from nornir_netmiko import netmiko_send_command
from nornir_utils.plugins.functions import print_result
```

```
nr = InitNornir(config_file='config.yaml')
#ipdb.set_trace()
print (nr.inventory.hosts['sw1'].name)
print (nr.inventory.hosts['sw1'].hostname)
print (nr.inventory.hosts['sw1'].username)
print (nr.inventory.hosts['sw1'].password)
print (nr.inventory.hosts['sw1'].platform)
print (nr.inventory.hosts['sw1'].groups)
print (nr.inventory.hosts['sw1'].data)
```

运行结果如下图所示。

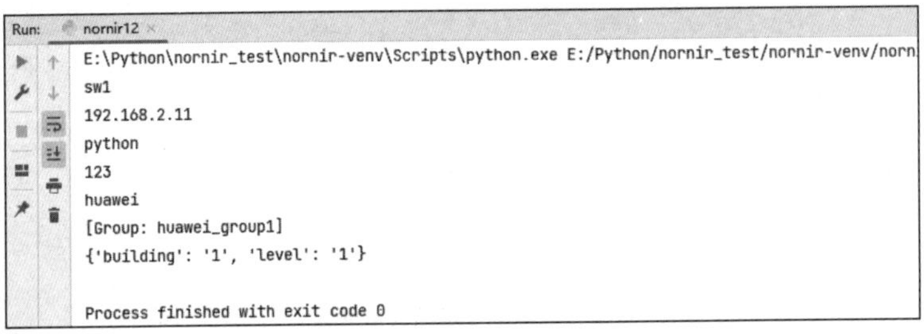

```
Run:   nornir12 ×
E:\Python\nornir_test\nornir-venv\Scripts\python.exe E:/Python/nornir_test/nornir-venv/norn
sw1
192.168.2.11
python
123
huawei
[Group: huawei_group1]
{'building': '1', 'level': '1'}

Process finished with exit code 0
```

9.27　实验 13　Nornir 的 Task（思科设备）

在 Nornir 中，Task（任务）是一组可重复使用的代码，任务通常以一个自定义函数的参数形式出现。它的作用是用来返回写在 hosts.yaml 文件里一个或多个主机的属性（如 IP 地址、主机名），返回值的数据类型为 Nornir 特有的 AggregatedResult、MultiResult 及 Result。

使用 Task 有两种方法：使用或者不使用 from nornir.core.task import Result。

1. 不使用 from nornir.core.task import Result

首先创建脚本 nornir13.py，写入下列代码。

```
from nornir import InitNornir
from nornir_utils.plugins.functions import print_result


nr = InitNornir(config_file="config.yaml")
def get_ip_add(task):
  return (f"The ip address of {task.host.name} is {task.host.hostname}.")


output = nr.run(task=get_ip_add)
```

```
print_result(output)
```

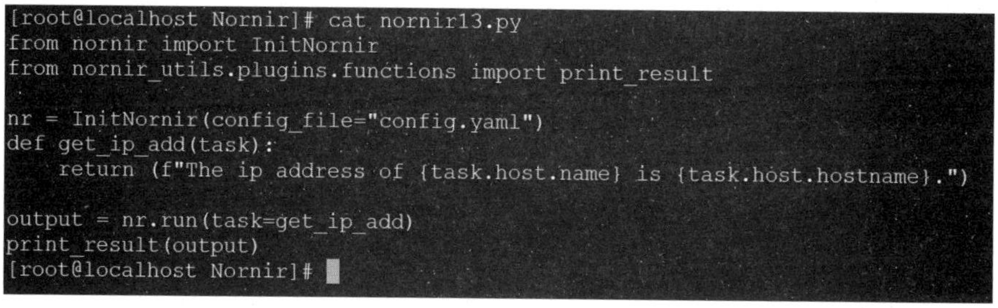

```
[root@localhost Nornir]# cat nornir13.py
from nornir import InitNornir
from nornir_utils.plugins.functions import print_result

nr = InitNornir(config_file="config.yaml")
def get_ip_add(task):
    return (f"The ip address of {task.host.name} is {task.host.hostname}.")

output = nr.run(task=get_ip_add)
print_result(output)
[root@localhost Nornir]#
```

这里我们自定义了一个函数 get_ip_add(task)，该函数的参数 task 就是本节提到的 Task，注意：Task 作为自定义函数的参数时，参数名是用户自定义的，并非必须写作 task，把参数名取为 abc 写成 get_ip_add(abc)也是可以的，只是用 abc 作为参数名没有任何意义罢了。

顾名思义，get_ip_add(task)函数的作用是获取主机的 IP 地址，调用 Task 时，task.host 表示的是 hosts.yaml 文件下所有的主机及其属性，task.host.name 返回主机的 IP 地址，task.host.hostname 返回主机的主机名。

```
def get_ip_add(task):
  return (f"The ip address of {task.host.name} is {task.host.hostname}.")
```

运行脚本，效果如下图所示。

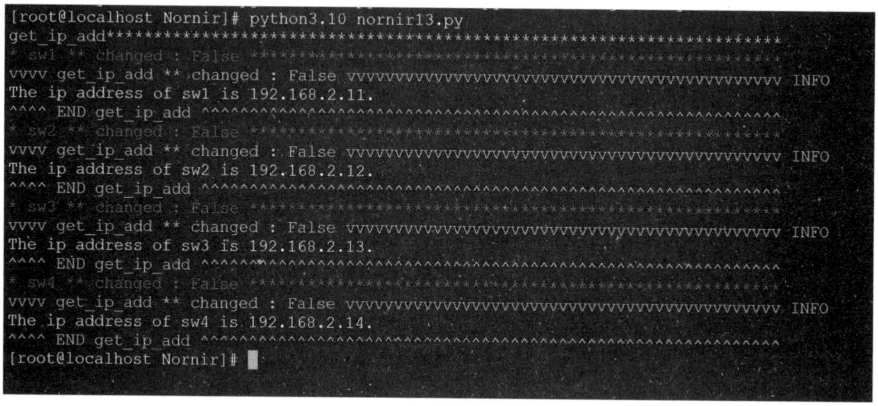

```
[root@localhost Nornir]# python3.10 nornir13.py
get_ip_add**********************************************************************
* sw1 ** changed : False ********************************************************
vvvv get_ip_add ** changed : False vvvvvvvvvvvvvvvvvvvvvvvvvvvvvvvvvvvvvvvvvvvvvv INFO
The ip address of sw1 is 192.168.2.11.
^^^^ END get_ip_add ^^^^^^^^^^^^^^^^^^^^^^^^^^^^^^^^^^^^^^^^^^^^^^^^^^^^^^^^^^^^
* sw2 ** changed : False ********************************************************
vvvv get_ip_add ** changed : False vvvvvvvvvvvvvvvvvvvvvvvvvvvvvvvvvvvvvvvvvvvvvv INFO
The ip address of sw2 is 192.168.2.12.
^^^^ END get_ip_add ^^^^^^^^^^^^^^^^^^^^^^^^^^^^^^^^^^^^^^^^^^^^^^^^^^^^^^^^^^^^
* sw3 ** changed : False ********************************************************
vvvv get_ip_add ** changed : False vvvvvvvvvvvvvvvvvvvvvvvvvvvvvvvvvvvvvvvvvvvvvv INFO
The ip address of sw3 is 192.168.2.13.
^^^^ END get_ip_add ^^^^^^^^^^^^^^^^^^^^^^^^^^^^^^^^^^^^^^^^^^^^^^^^^^^^^^^^^^^^
* sw4 ** changed : False ********************************************************
vvvv get_ip_add ** changed : False vvvvvvvvvvvvvvvvvvvvvvvvvvvvvvvvvvvvvvvvvvvvvv INFO
The ip address of sw4 is 192.168.2.14.
^^^^ END get_ip_add ^^^^^^^^^^^^^^^^^^^^^^^^^^^^^^^^^^^^^^^^^^^^^^^^^^^^^^^^^^^^
[root@localhost Nornir]#
```

2. 使用 from nornir.core.task import Result

首先修改脚本 nornir13.py，写入下列代码。

```
from nornir import InitNornir
from nornir.core.task import Result
from nornir_utils.plugins.functions import print_result
```

```
nr = InitNornir(config_file="config.yaml")
def get_ip_add(task):
  return Result(host=task.host,
          result=f"The ip address of {task.host.name} is {task.host.hostname}",
          changed = True)

output = nr.run(task=get_ip_add)
print_result(output)
```

```
[root@localhost Nornir]# cat nornir13.py
from nornir import InitNornir
from nornir.core.task import Result
from nornir_utils.plugins.functions import print_result

nr = InitNornir(config_file="config.yaml")
def get_ip_add(task):
    return Result(host=task.host,
            result=f"The ip address of {task.host.name} is {task.host.hostname}",
            changed = True)

output = nr.run(task=get_ip_add)
print_result(output)
[root@localhost Nornir]#
[root@localhost Nornir]# 
```

由图中框线部分可知，Result()是 nornir.core.task 的一个子类。顾名思义，它代表的是在 Nornir 里执行每个 Task 后所得到的结果，Result 的相关属性和方法如下图所示。

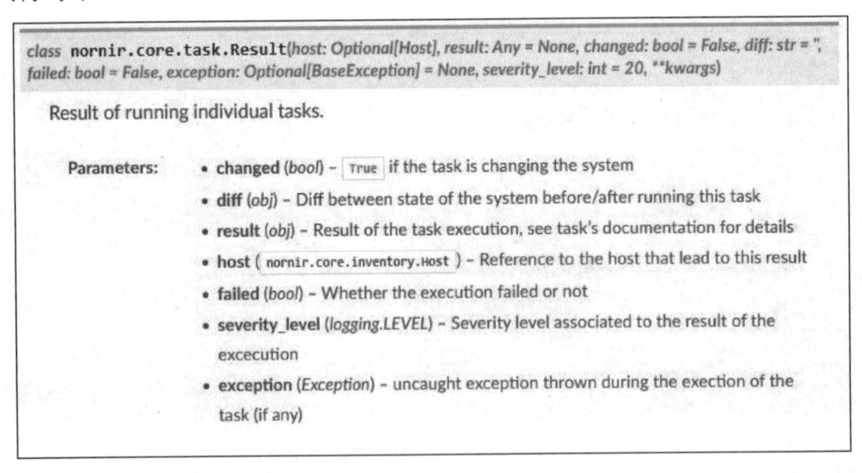

class **nornir.core.task.Result**(*host: Optional[Host], result: Any = None, changed: bool = False, diff: str = '', failed: bool = False, exception: Optional[BaseException] = None, severity_level: int = 20, **kwargs*)

Result of running individual tasks.

Parameters:
- **changed** (*bool*) – `True` if the task is changing the system
- **diff** (*obj*) – Diff between state of the system before/after running this task
- **result** (*obj*) – Result of the task execution, see task's documentation for details
- **host** (`nornir.core.inventory.Host`) – Reference to the host that lead to this result
- **failed** (*bool*) – Whether the execution failed or not
- **severity_level** (*logging.LEVEL*) – Severity level associated to the result of the excecution
- **exception** (*Exception*) – uncaught exception thrown during the exection of the task (if any)

简单来说，只有 host 是 Result()的必选参数，在脚本 nornir13.py 中，我们将 task.host 赋值给 host 参数，即将 hosts.yaml 中所有设备的参数赋值给 host 参数。除了 host，脚本中还额外用到的 result 和 changed 两个参数都是可选的，result 用来打印输出内容，这里我们使用 result=f"The ip address of {task.host.name} is {task.host.hostname}"，用来返回主机的 IP 地址和对应的主机名并将其打印出来，而 changed 参数的数据类型为布尔值，它的作用是提示用户运行某个 Task 后是否对设备的配置做出了修改。这里虽然没有对设备的配置做出任何修改，changed 的值应为 False，但是出于演示的目的，笔者故意将 changed 设为了 True。

```
def get_ip_add(task):
    return Result(host=task.host,
            result=f"The ip address of {task.host.name} is {task.host.hostname}",
            changed = True)
```

运行脚本，效果如下图所示。

```
[root@localhost Nornir]# python3.10 nornir13.py
get_ip_add*********************************************************************
* sw1 ** changed : True *******************************************************
vvvv get_ip_add ** changed : True vvvvvvvvvvvvvvvvvvvvvvvvvvvvvvvvvvvvvvvvvvvvvvv INFO
The ip address of sw1 is 192.168.2.11
^^^^ END get_ip_add ^^^^^^^^^^^^^^^^^^^^^^^^^^^^^^^^^^^^^^^^^^^^^^^^^^^^^^^^^^^^
* sw2 ** changed : True *******************************************************
vvvv get_ip_add ** changed : True vvvvvvvvvvvvvvvvvvvvvvvvvvvvvvvvvvvvvvvvvvvvvvv INFO
The ip address of sw2 is 192.168.2.12
^^^^ END get_ip_add ^^^^^^^^^^^^^^^^^^^^^^^^^^^^^^^^^^^^^^^^^^^^^^^^^^^^^^^^^^^^
* sw3 ** changed : True *******************************************************
vvvv get_ip_add ** changed : True vvvvvvvvvvvvvvvvvvvvvvvvvvvvvvvvvvvvvvvvvvvvvvv INFO
The ip address of sw3 is 192.168.2.13
^^^^ END get_ip_add ^^^^^^^^^^^^^^^^^^^^^^^^^^^^^^^^^^^^^^^^^^^^^^^^^^^^^^^^^^^^
* sw4 ** changed : True *******************************************************
vvvv get_ip_add ** changed : True vvvvvvvvvvvvvvvvvvvvvvvvvvvvvvvvvvvvvvvvvvvvvvv INFO
The ip address of sw4 is 192.168.2.14
^^^^ END get_ip_add ^^^^^^^^^^^^^^^^^^^^^^^^^^^^^^^^^^^^^^^^^^^^^^^^^^^^^^^^^^^^
[root@localhost Nornir]#
```

3. 主任务和子任务

在 Nornir 中使用 Task 的好处是，它允许我们同时在设备里执行多条命令。在前面两个例子中，我们只用了一个 Task，即 get_ip_add(task)，我们将其称为**主任务**（**Main Task**）。

下面举例说明主任务和子任务的区别，首先将 nornir13.py 的代码修改如下。

```
from nornir import InitNornir
from nornir_netmiko import netmiko_send_command
from nornir_utils.plugins.functions import print_result

nr = InitNornir(config_file="config.yaml")
def get_ip_add(task):
    task.run(netmiko_send_command, command_string='show clock')
    task.run(netmiko_send_command, command_string='show vlan summary')
    return (f"The ip address of {task.host.name} is {task.host.hostname}")

output = nr.run(task=get_ip_add)
print_result(output)
```

```
[root@localhost Nornir]# cat nornir13.py
from nornir import InitNornir
from nornir_netmiko import netmiko_send_command
from nornir_utils.plugins.functions import print_result

nr = InitNornir(config_file="config.yaml")
def get_ip_add(task):
    task.run(netmiko_send_command, command_string='show clock')
    task.run(netmiko_send_command, command_string='show vlan summary')
    return (f"The ip address of {task.host.name} is {task.host.hostname}")

output = nr.run(task=get_ip_add)
print_result(output)
[root@localhost Nornir]#
```

我们在主任务 get_ip_add(task)下额外添加了 task.run(netmiko_send_command, command_string='show clock')和 task.run(netmiko_send_command, command_string='show vlan summary')两个任务来分别获取向设备执行 show clock 和 show vlan summary 后的回显内容，我们将这两个任务称为**子任务**（**Subtask**）。

```
def get_ip_add(task):
  task.run(netmiko_send_command, command_string='show clock')
  task.run(netmiko_send_command, command_string='show vlan summary')
  return (f"The ip address of {task.host.name} is {task.host.hostname}")
```

运行脚本，效果如下图所示。

```
[root@localhost Nornir]# python3.10 nornir13.py
get_ip_add**********************************************************************
* sw1 ** changed : False *******************************************************
vvvv get_ip_add ** changed : False vvvvvvvvvvvvvvvvvvvvvvvvvvvvvvvvvvvvvvvvvvvvv INFO
The ip address of sw1 is 192.168.2.11
---- netmiko_send_command ** changed : False ------------------------------------ INFO
*22:20:58.687 UTC Mon Aug 15 2022
---- netmiko_send_command ** changed : False ------------------------------------ INFO
Number of existing VLANs               : 9
 Number of existing VTP VLANs          : 9
 Number of existing extended VLANS     : 0

^^^^ END get_ip_add ^^^^^^^^^^^^^^^^^^^^^^^^^^^^^^^^^^^^^^^^^^^^^^^^^^^^^^^^^^^^^
* sw2 ** changed : False *******************************************************
vvvv get_ip_add ** changed : False vvvvvvvvvvvvvvvvvvvvvvvvvvvvvvvvvvvvvvvvvvvvv INFO
The ip address of sw2 is 192.168.2.12
---- netmiko_send_command ** changed : False ------------------------------------ INFO
*23:31:34.668 UTC Mon Aug 15 2022
---- netmiko_send_command ** changed : False ------------------------------------ INFO
Number of existing VLANs               : 9
 Number of existing VTP VLANs          : 9
 Number of existing extended VLANS     : 0

^^^^ END get_ip_add ^^^^^^^^^^^^^^^^^^^^^^^^^^^^^^^^^^^^^^^^^^^^^^^^^^^^^^^^^^^^^
* sw3 ** changed : False *******************************************************
vvvv get_ip_add ** changed : False vvvvvvvvvvvvvvvvvvvvvvvvvvvvvvvvvvvvvvvvvvvvv INFO
The ip address of sw3 is 192.168.2.13
---- netmiko_send_command ** changed : False ------------------------------------ INFO
*23:34:54.933 UTC Mon Aug 15 2022
---- netmiko_send_command ** changed : False ------------------------------------ INFO
Number of existing VLANs               : 8
 Number of existing VTP VLANs          : 8
 Number of existing extended VLANS     : 0

^^^^ END get_ip_add ^^^^^^^^^^^^^^^^^^^^^^^^^^^^^^^^^^^^^^^^^^^^^^^^^^^^^^^^^^^^^
* sw4 ** changed : False *******************************************************
vvvv get_ip_add ** changed : False vvvvvvvvvvvvvvvvvvvvvvvvvvvvvvvvvvvvvvvvvvvvv INFO
The ip address of sw4 is 192.168.2.14
---- netmiko_send_command ** changed : False ------------------------------------ INFO
*22:18:07.090 UTC Mon Aug 15 2022
---- netmiko_send_command ** changed : False ------------------------------------ INFO
Number of existing VLANs               : 9
 Number of existing VTP VLANs          : 9
 Number of existing extended VLANS     : 0

^^^^ END get_ip_add ^^^^^^^^^^^^^^^^^^^^^^^^^^^^^^^^^^^^^^^^^^^^^^^^^^^^^^^^^^^^^
[root@localhost Nornir]#
```

4. 理解 AggregatedResult 和 MultiResult

前面提到子任务的概念，也许你会问：当对不同子任务的回显内容做解析时，要怎么区分它们呢？这就涉及 AggregatedResult 和 MultiResult 的知识点。首先将 nornir13.py 的代码修改如下。

```
from nornir import InitNornir
from nornir_netmiko import netmiko_send_command
from nornir_utils.plugins.functions import print_result

nr = InitNornir(config_file="config.yaml")
def get_ip_add(task):
    task.run(netmiko_send_command, command_string='show clock')
    task.run(netmiko_send_command, command_string='show vlan summary')
    return (f"The ip address of {task.host.name} is {task.host.hostname}")

output = nr.run(task=get_ip_add)
#print_result(output)
print (type(output))
print (output)
```

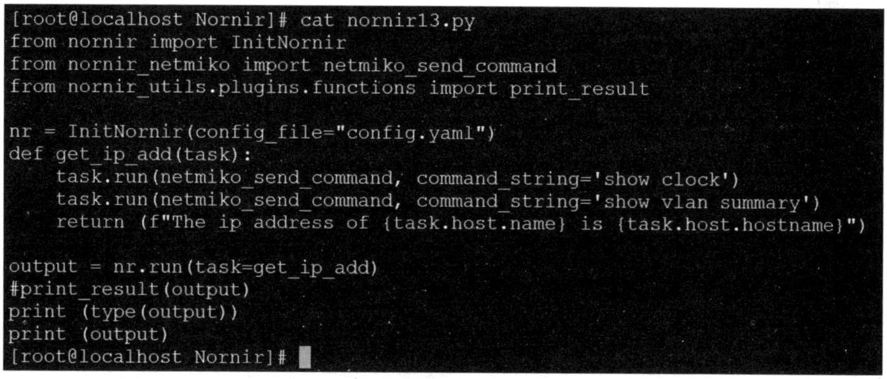

```
[root@localhost Nornir]# cat nornir13.py
from nornir import InitNornir
from nornir_netmiko import netmiko_send_command
from nornir_utils.plugins.functions import print_result

nr = InitNornir(config_file="config.yaml")
def get_ip_add(task):
    task.run(netmiko_send_command, command_string='show clock')
    task.run(netmiko_send_command, command_string='show vlan summary')
    return (f"The ip address of {task.host.name} is {task.host.hostname}")

output = nr.run(task=get_ip_add)
#print_result(output)
print (type(output))
print (output)
[root@localhost Nornir]#
```

我们将 print_result(output)去掉，将其替代为 print(type(output))和 print (output)，目的是让读者看到 AggregatedResult、MultiResult 及 Result 几种特殊的数据类型。前面已经介绍过 Result，那么什么是 AggregatedResult 和 MultiResult 呢？

- AggregatedResult：在实验 9 中已经讲到，AggregatedResult 类似于字典，它包含了执行主任务和子任务后在所有涉及的设备上所获取的数据，每个设备的设备名均为 AggregatedResult 的键，而改键对应的值则为一组 MultiResult。

- MultiResult：在实验 9 中已经讲到，MultiResult 类似于列表，它包含了执行主任务和子任务后在所有涉及的设备上所获取的数据。每一个主任务和子任务返回的 Result 作为元素存在于 MultiResult，可以通过对 MultiResult 使用.result 来获取主任务和子任务返回的结果，如果 MultiResult 里有多于一个任务，则还需在.result 前面加上索引号。

运行脚本，效果如下图所示。

```
[root@localhost Nornir]# python3.10 nornir13.py
<class 'nornir.core.task.AggregatedResult'>
AggregatedResult (get_ip_add): {'sw1': MultiResult: [Result: "get_ip_add", Resu
t: "netmiko_send_command", Result: "netmiko_send_command"], 'sw2': MultiResult:
[Result: "get_ip_add", Result: "netmiko_send_command", Result: "netmiko_send_co
mand"], 'sw3': MultiResult: [Result: "get_ip_add", Result: "netmiko_send_comma
", Result: "netmiko_send_command"], 'sw4': MultiResult: [Result: "get_ip_add",
esult: "netmiko_send_command", Result: "netmiko_send_command"]}
[root@localhost Nornir]#
```

从 print(type(output))返回的<class 'nornir.core.task.AggregatedResult'>可以看到，执行 get_ip_add(task)任务后返回值为类似于字典的 AggregatedResult。

从 print(output)返回的内容可以看到，AggregatedResult 的键名为设备在 hosts.yaml 里的主机名（sw1、sw2、sw3、sw4），每个键对应的值均为 MultiResult。MultiResult 中的元素即每个主任务和子任务运行后的 Result。每个交换机都对应了 3 个 Result。

- Result: "get_ip_add" →主任务。
- Result: "netmiko_send_command" →子任务(show clock)。
- Result: "netmiko_send_command" →子任务(show vlan summary)。

9.28　实验 13　Nornir 的 Task（华为设备）

在 nornir12 文件夹的基础上，复制成 nornir13，继续开展实验。这个实验其实思科设备也行，华为设备也可，没有差异，更多的是对 Nornir 的 Task 的逻辑的理解和补充。

在下面的 Python 脚本中，我们把“自定义”函数中的 task 改成 task_2022，这样可以避免跟 result 那一行的 task 造成混淆。

```
from nornir import InitNornir
from nornir_netmiko import netmiko_send_command
from nornir_utils.plugins.functions import print_result
from nornir.core.task import Task, Result
```

```
nr = InitNornir(config_file='config.yaml')

print(type(nr))

def host_parm(task_2022:Task) -> Result:
    return (task_2022.host.name,
        task_2022.host.hostname,
        task_2022.host.username,
        task_2022.host.password,
        task_2022.host.platform,
        task_2022.host.groups,
        task_2022.host.data)

result = nr.run(task=host_parm)
print_result(result)
```

在代码中，Task 和 Result 只是限定标注类型（即告诉你函数是什么对象，其结果是什么对象，说白了就是“注释”）而已。

类型标注应该还好理解一点。Task 自定义函数配合'nornir.core.Nornir'对象的 run()函数就有点抽象。其过程相当于 run 函数调用了 Nornir 框架里的一个自定义 Task 函数 host_parm()，其实这个函数还可以继续调用其他 Task 函数（此处不涉及），像递归吗？

Task 函数中的第一个参数，叫作“上下文”（调用时当前的环境快照），就更抽象了。能不能这么理解？比如看电影，买票后进入电影放映厅时，此时你自己是否买了爆米花买了饮料、是否戴了 3D 眼镜、电影院壁灯是否亮、广告开始没有、多少人就座了，凡此种种，就是你看电影这个自定义函数的上下文，这些就是此时的环境快照值。你进放映厅的瞬间就是这么一个快照。随着电影的放映，爆米花可能被你吃了、饮料被你喝了、广告可能播完了、3D 眼镜可能被你摘掉了，因为这个电影是 2D 的……

而你进商场，又像不像是个 run()函数？你想看电影，就商场.run(task=电影)；你想夹娃娃，就商场.run(task=夹娃娃)；你想吃肯德基，就商场.run(task=肯德基)……

对于编程功底不深的网络工程师来说，如果实在难以理解，那就先按套路来码放参数，先满足网络运维任务，再慢慢体会。

9.29　实验 14　使用 Nornir 按需批量修改交换机配置（思科设备）

在本章最后一节中，我们将使用 Nornir 来做一个综合性实验。实验的目的是使用 Nornir 配合 TextFSM 找出 hosts.yaml 里所有交换机各自的 trunk 和 access 端口，将它们的 trunk 端口的 description 统一设为 Trunk Port (Nornir)，将 access 端口的 description 根据它所在的 VLAN ID 统一设为 Access Port to VLAN xxx (Nornir)，比如一个 access 端口所在 VLAN 为 999，那么就将它的 description 设为 Access Port to VLAN 999 (Nornir)。

在笔者的实验环境中，每台交换机都有 Gi0/0、Gi0/1、Gi0/2 总共 3 个物理端口，如下图所示。

```
S1#show ip int brief
Interface              IP-Address      OK? Method Status                Protocol
GigabitEthernet0/0     unassigned      YES unset  up                    up
GigabitEthernet0/1     unassigned      YES unset  up                    up
GigabitEthernet0/2     unassigned      YES unset  up                    up
Loopback0              1.1.1.1         YES NVRAM  up                    up
Loopback1              11.11.11.11     YES NVRAM  up                    up
Vlan1                  192.168.2.11    YES NVRAM  up                    up
S1#
```

其中 Gi0/0 为 access port（VLAN 1），Gi0/1 为 trunk port，Gi0/2 为 access port（VLAN 999），3 个端口目前都还没有配置 description（hosts.yaml 中其余的 sw2、sw3、sw4 的配置和 sw1 一样），如下图所示。

```
S1#show run int gi0/0
Building configuration...

Current configuration : 71 bytes
!
interface GigabitEthernet0/0
 media-type rj45
 negotiation auto
end

S1#show run int gi0/1
Building configuration...

Current configuration : 135 bytes
!
interface GigabitEthernet0/1
 switchport trunk encapsulation dot1q
 switchport mode trunk
 media-type rj45
 no negotiation auto
end

S1#show run int gi0/2
Building configuration...

Current configuration : 102 bytes
!
interface GigabitEthernet0/2
 switchport access vlan 999
 media-type rj45
 no negotiation auto
end

S1#
```

在开始实验前，我们先将实验 10 中的脚本 nornir10.py 运行一次，该脚本会配合 TextFSM 模板，将 show interface switchport 命令的回显内容以字典的形式输出，如下图所示。

```
[root@localhost Nornir]# cat nornir10.py
from nornir import InitNornir
from nornir_netmiko import netmiko_send_command
from nornir_utils.plugins.functions import print_result

nr = InitNornir(config_file="config.yaml")
results = nr.run(netmiko_send_command, command_string='sh interface switchport', use_textfsm=True)
print_result(results)
[root@localhost Nornir]#
[root@localhost Nornir]# python3 nornir10.py
netmiko_send_command*************************************************************
* sw1 ** changed : False ******************************************************
vvvv netmiko_send_command ** changed : False vvvvvvvvvvvvvvvvvvvvvvvvvvvvvvvvvvvvvv INFO
[ { 'access_vlan': '1',
    'admin_mode': 'dynamic desirable',
    'interface': 'Gi0/0',
    'mode': 'static access',
    'native_vlan': '1',
    'switchport': 'Enabled',
    'switchport_monitor': '',
    'switchport_negotiation': 'On',
    'trunking_vlans': ['ALL'],
    'voice_vlan': 'none'},
  { 'access_vlan': '1',
    'admin_mode': 'trunk',
    'interface': 'Gi0/1',
    'mode': 'trunk',
    'native_vlan': '1',
    'switchport': 'Enabled',
    'switchport_monitor': '',
    'switchport_negotiation': 'On',
    'trunking_vlans': ['ALL'],
    'voice_vlan': 'none'},
  { 'access_vlan': '999',
    'admin_mode': 'dynamic desirable',
    'interface': 'Gi0/2',
    'mode': 'static access',
    'native_vlan': '1',
    'switchport': 'Enabled',
    'switchport_monitor': '',
    'switchport_negotiation': 'On',
    'trunking_vlans': ['ALL'],
    'voice_vlan': 'none'}]
^^^^ END netmiko_send_command ^^^^^^^^^^^^^^^^^^^^^^^^^^^^^^^^^^^^^^^^^^^^^^^^^^^^^^
```

运行 nornir10.py 后，要重点注意其回显内容中 access_vlan、mode 两个键。可以看到，Gi0/0 的 access_vlan 键对应的值为 1，mode 键对应的值为 static access，说明 Gi0/0 为 access port（VLAN 1）；同理，Gi0/1 的 access_vlan 键对应的值为 1，mode 键对应的值为 trunk，说明 Gi0/1 为 trunk port；Gi0/2 的 access_vlan 键对应的值为 999，mode 键对应的值为 static access，说明 Gi0/2 为 access port（VLAN 999）。

首先创建脚本 nornir14.py，写入下列代码。

```
from nornir import InitNornir
from nornir_netmiko import netmiko_send_command, netmiko_send_config
from nornir_utils.plugins.functions import print_result

nr = InitNornir(config_file="config.yaml")
output = nr.run(netmiko_send_command, command_string='sh interface switchport',
use_textfsm=True)
for switch in output.keys():
  for i in range(3):
     trunk_cmd = ['interface ' + output[switch].result[i]['interface'], 'description
Trunk Port (Nornir)']
     access_cmd = ['interface ' + output[switch].result[i]['interface'], 'description
Access Port to VLAN ' + output[switch].result[i]['access_vlan'] + ' (Nornir)']
     if 'trunk' in output[switch].result[i]['mode']:
         nr.run(netmiko_send_config, config_commands = trunk_cmd)
     elif 'access' in output[switch].result[i]['mode']:
         nr.run(netmiko_send_config, config_commands = access_cmd)
results = nr.run(netmiko_send_command, command_string='sh interface description')
print_result(results)
```

```
[root@localhost Nornir]# cat nornir14.py
from nornir import InitNornir
from nornir_netmiko import netmiko_send_command, netmiko_send_config
from nornir_utils.plugins.functions import print_result

nr = InitNornir(config_file="config.yaml")
output = nr.run(netmiko_send_command, command_string='sh interface switchport', use_textfsm=True)
for switch in output.keys():
    for i in range(3):
        trunk_cmd = ['interface ' + output[switch].result[i]['interface'], 'description Trunk Port (Nornir)']
        access_cmd = ['interface ' + output[switch].result[i]['interface'], 'description Access Port to VLAN ' +
            output[switch].result[i]['access_vlan'] + ' (Nornir)']
        if 'trunk' in output[switch].result[i]['mode']:
            nr.run(netmiko_send_config, config_commands = trunk_cmd)
        elif 'access' in output[switch].result[i]['mode']:
            nr.run(netmiko_send_config, config_commands = access_cmd)
results = nr.run(netmiko_send_command, command_string='sh interface description')
print_result(results)
[root@localhost Nornir]#
```

代码分段讲解如下。

- 下面这部分代码和实验 10 的代码类似，都是用 TextFSM 将 show interface switchport 命令的回显内容以 JSON 格式输出并赋值给变量 output。

```
from nornir import InitNornir
```

```
from nornir_netmiko import netmiko_send_command, netmiko_send_config
from nornir_utils.plugins.functions import print_result

nr = InitNornir(config_file="config.yaml")
output = nr.run(netmiko_send_command, command_string='sh interface switchport',
use_textfsm=True)
```

- 变量 output 的数据类型为类似字典的 AggregateResult，因此我们可以调用字典自带的 keys()方法，该方法返回的是一个列表，该列表内的元素为字典里所有的键名，也就是 sw1、sw2、sw3、sw4。然后我们用 for 循环遍历该列表，再用另外一个嵌套 for 循环遍历每台交换机下的所有物理端口，因为每台交换机总共有 3 个物理端口（Gi0/0、Gi0/1、Gi0/2），因此用 range(3)来生成 0、1、2 三个整数，对应 Gi0/0、Gi0/1、Gi0/2 中的 0、1、2。随后我们为所有 trunk 和 access 端口准备好它们各自需要的配置命令，分别赋值给 trunk_cmd 和 access_cmd 两个变量。最后通过 if 语句来判断每个物理端口其 mode 键名对应的值，如果值里包含 trunk，则对该端口执行 trunk_cmd 变量里的命令，如果值里包含 access，则对该端口执行 access_cmd 变量里的命令。

```
for switch in output.keys():
  for i in range(3):
     trunk_cmd = ['interface ' + output[switch].result[i]['interface'], 'description Trunk
Port (Nornir)']
     access_cmd = ['interface ' + output[switch].result[i]['interface'], 'description
Access Port to VLAN ' + output[switch].result[i]['access_vlan'] + ' (Nornir)']
     if 'trunk' in output[switch].result[i]['mode']:
        nr.run(netmiko_send_config, config_commands = trunk_cmd)
     elif 'access' in output[switch].result[i]['mode']:
        nr.run(netmiko_send_config, config_commands = access_cmd)
```

- 最后对所有交换机执行 show interface description 命令，来验证每个物理端口是否已经配置好了相应的 description。

```
results = nr.run(netmiko_send_command, command_string='sh interface description')
print_result(results)
```

运行脚本，效果如下图所示。

```
[root@localhost Nornir]# python3 nornir14.py
netmiko_send_command*****************************************************************
* sw1 ** changed : False ************************************************************
vvvv netmiko_send_command ** changed : False vvvvvvvvvvvvvvvvvvvvvvvvvvvvvvvvvvvvvvv INFO

Interface                      Status         Protocol Description
Gi0/0                          up             up       Access Port to VLAN 1 (Nornir)
Gi0/1                          up             up       Trunk Port (Nornir)
Gi0/2                          up             up       Access Port to VLAN 999 (Nornir)
Lo0                            up             up
Lo1                            up             up
Vl1                            up             up
^^^^ END netmiko_send_command ^^^^^^^^^^^^^^^^^^^^^^^^^^^^^^^^^^^^^^^^^^^^^^^^^^^^^^^^
* sw2 ** changed : False ************************************************************
vvvv netmiko_send_command ** changed : False vvvvvvvvvvvvvvvvvvvvvvvvvvvvvvvvvvvvvvv INFO

Interface                      Status         Protocol Description
Gi0/0                          up             up       Access Port to VLAN 1 (Nornir)
Gi0/1                          up             up       Trunk Port (Nornir)
Gi0/2                          up             up       Access Port to VLAN 999 (Nornir)
Lo0                            up             up
Lo1                            up             up
Lo99                           up             up
Vl1                            up             up
^^^^ END netmiko_send_command ^^^^^^^^^^^^^^^^^^^^^^^^^^^^^^^^^^^^^^^^^^^^^^^^^^^^^^^^
* sw3 ** changed : False ************************************************************
vvvv netmiko_send_command ** changed : False vvvvvvvvvvvvvvvvvvvvvvvvvvvvvvvvvvvvvvv INFO

Interface                      Status         Protocol Description
Gi0/0                          up             up       Access Port to VLAN 1 (Nornir)
Gi0/1                          up             up       Trunk Port (Nornir)
Gi0/2                          up             up       Access Port to VLAN 999 (Nornir)
Lo0                            up             up
Lo1                            up             up
Lo99                           up             up
Vl1                            up             up
^^^^ END netmiko_send_command ^^^^^^^^^^^^^^^^^^^^^^^^^^^^^^^^^^^^^^^^^^^^^^^^^^^^^^^^
* sw4 ** changed : False ************************************************************
vvvv netmiko_send_command ** changed : False vvvvvvvvvvvvvvvvvvvvvvvvvvvvvvvvvvvvvvv INFO

Interface                      Status         Protocol Description
Gi0/0                          up             up       Access Port to VLAN 1 (Nornir)
Gi0/1                          up             up       Trunk Port (Nornir)
Gi0/2                          up             up       Access Port to VLAN 999 (Nornir)
Lo0                            up             up
Lo1                            up             up
Lo99                           up             up
Vl1                            up             up
^^^^ END netmiko_send_command ^^^^^^^^^^^^^^^^^^^^^^^^^^^^^^^^^^^^^^^^^^^^^^^^^^^^^^^^
[root@localhost Nornir]#
```

从上图可以看到，我们已经成功通过 Nornir 按需批量地修改了 sw1 ~ sw4 四台交换机每个物理端口的 description。

9.30　实验 14　使用 Nornir 按需批量修改交换机配置（华为设备）

综合实验结合华为设备做了一点调整。在 nornir13 文件夹的基础上，复制成 nornir14，继续开展实验。

调整实验目的如下。

（1）Nornir 结合 TextFSM，找出 hosts.yaml 设备中的 trunk 和 access 端口。（复习实验 10

知识）

（2）Nornir 用 description 命令将 trunk 端口描述配置为 Trunk Port (Nornir)；

（3）Nornir 用 description 命令将 access 端口描述配置为 Access Port to VLAN xxx (Nornir)。xxx 为 VLAN 号。

（4）Nornir 检查结果。

搭建好实验环境后，sw1 ~ sw4 都手工设置如下配置。

```
sys
vlan batch 999 1111 2222

interface GigabitEthernet0/0/1
 port link-type access
#
interface GigabitEthernet0/0/2
 port link-type trunk
port trunk allow-pass vlan 1111 2222
#
interface GigabitEthernet0/0/3
 port link-type access
 port default vlan 999
```

在 ntc-tempales 中 index 文件 huawei 条目附近增加如下条目。

```
huawei_display_current-configuration_interface_gi.textfsm, .*, huawei, display
current-configuration | be interface GigabitEthernet
```

```
index - 记事本
文件(F)  编辑(E)  格式(O)  查看(V)  帮助(H)
hp_procurve_show_trunks.textfsm, .*, hp_procurve, sh[[ow]] tr[[unks]]
hp_procurve_show_vlans.textfsm, .*, hp_procurve, sh[[ow]] vl[[ans]]
hp_procurve_show_arp.textfsm, .*, hp_procurve, sh[[ow]] ar[[p]]

huawei_display_current-configuration_interface_gi.textfsm, .*, huawei, display current-configuration | be interface GigabitEthernet
huawei_vrp_display_interface_description.textfsm, .*, huawei_vrp, disp[[lay]] inter[[face]] des[[cription]]
huawei_vrp_display_interface_brief.textfsm, .*, huawei_vrp, disp[[lay]] inter[[face]] br[[ief]]
huawei_vrp_display_lldp_neighbor.textfsm, .*, huawei_vrp, disp[[lay]] lldp nei[[ghbor]]
huawei_vrp_display_temperature.textfsm, .*, huawei_vrp, disp[[lay]] tem[[perature]]
huawei_vrp_display_port_vlan.textfsm, .*, huawei_vrp, disp[[lay]] port vl[[an]]
huawei_vrp_display_version.textfsm, .*, huawei_vrp, disp[[lay]] ver[[sion]]
huawei_display_interface_brief.textfsm, .*, huawei, disp[[lay]] int[[erface]] br[[ief]]

juniper_junos_show_chassis_cluster_interfaces.textfsm, .*, juniper_junos, sh[[ow]] ch[[assis]] c[[luster]] i[[nterface]]
```

huawei_display_current-configuration_interface_gi.textfsm 文件的 TextFSM 模板内容如下。

```
Value INTERFACE (GigabitEthernet\d+/\d+/\d+)
Value Required MODE (access|trunk)
Value VLAN (\d+)

Start
  ^interface GigabitEthernet -> Continue.Record
  ^interface ${INTERFACE}
  ^ port link-type ${MODE}
  ^ port default vlan ${VLAN}
```

这里强调一下：

（1）要提取的端口及配置信息，其最后一行是不确定的，但第一行 interface Gi……是确定的，所以我们需要使用 Continue.Record。

（2）在本次实验中，sw 有 24 个 GigabitEthernet 端口，但实验只配了 3 个端口，此时 Value Required MODE (access|trunk)中的 Required 关键字可过滤掉其他端口。

Python 实验脚本 nornir14.py 代码如下。

```
import os
from nornir import InitNornir
from nornir_netmiko import netmiko_send_command
from nornir_utils.plugins.functions import print_result

os.environ['NET_TEXTFSM']=r'C:\Program
Files\Python310\Lib\site-packages\ntc_templates\templates'   #在脚本中临时指定系统变量的
方法，路径视情况调整

nr = InitNornir(config_file="config.yaml")
output = nr.run(netmiko_send_command, command_string='display current-configuration |
be interface GigabitEthernet',use_textfsm = True)
print_result(output)
```

代码运行结果部分如下图所示。

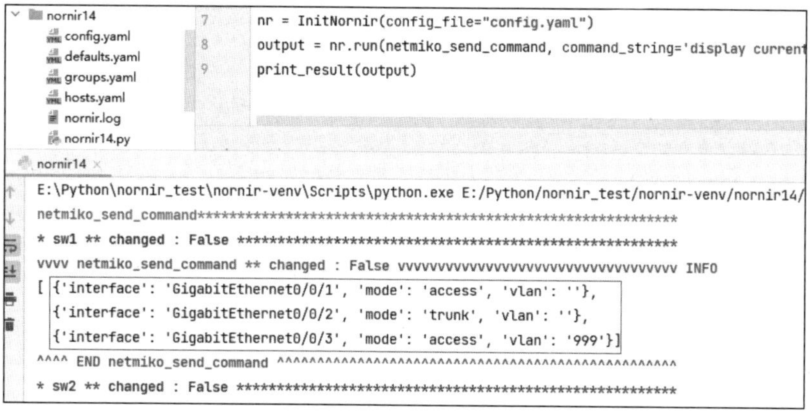

```
nornir14
  config.yaml
  defaults.yaml
  groups.yaml
  hosts.yaml
  nornir.log
  nornir14.py

7    nr = InitNornir(config_file="config.yaml")
8    output = nr.run(netmiko_send_command, command_string='display current
9    print_result(output)

nornir14 ×
E:\Python\nornir_test\nornir-venv\Scripts\python.exe E:/Python/nornir_test/nornir-venv/nornir14/
netmiko_send_command*****************************************************************
* sw1 ** changed : False ***********************************************************
vvvv netmiko_send_command ** changed : False vvvvvvvvvvvvvvvvvvvvvvvvvvvvvvvvvvv INFO
[ {'interface': 'GigabitEthernet0/0/1', 'mode': 'access', 'vlan': ''},
  {'interface': 'GigabitEthernet0/0/2', 'mode': 'trunk', 'vlan': ''},
  {'interface': 'GigabitEthernet0/0/3', 'mode': 'access', 'vlan': '999'}]
^^^^ END netmiko_send_command ^^^^^^^^^^^^^^^^^^^^^^^^^^^^^^^^^^^^^^^^^^^^^^^^^^^^^^
* sw2 ** changed : False ***********************************************************
```

代码调测很难一步到位，需要花时间、花精力一点一点研究透彻。接下来，我们需要让 Nornir 来执行配置下发工作。

现在，根据需求，增加原有的代码内容，新建一个脚本文件 nornir14-2.py。

```
import os
from nornir import InitNornir
from nornir_netmiko import netmiko_send_command,netmiko_send_config
from nornir_utils.plugins.functions import print_result

os.environ['NET_TEXTFSM']=r'C:\Program
Files\Python310\Lib\site-packages\ntc_templates\templates'

nr = InitNornir(config_file="config.yaml")
output = nr.run(netmiko_send_command, command_string='display current-configuration |
be interface GigabitEthernet',use_textfsm = True)

for switch in output.keys():
    #print(output[switch].result)  #这种可以用来调测，探寻返回内容，具体操作思路在实验 11、12
中已经介绍过
    for i in range(len(output[switch].result)):
        trunk_cmd = ['interface ' + output[switch].result[i]['interface'], 'description
Trunk Port (Nornir)']
        access_cmd = ['interface ' + output[switch].result[i]['interface'], 'description
Access Port to VLAN ' +
                      (output[switch].result[i]['vlan'] if
output[switch].result[i]['vlan'] else '1') + ' (Nornir)']
        #print(trunk_cmd,',', access_cmd)
```

```
        if 'trunk' in output[switch].result[i]['mode']:
            nr.run(netmiko_send_config, config_commands = trunk_cmd)
            print(switch,output[switch].result[i]['interface'],'已配置完成')
        elif 'access' in output[switch].result[i]['mode']:
            nr.run(netmiko_send_config, config_commands = access_cmd)
            print(switch, output[switch].result[i]['interface'], '已配置完成')

results = nr.run(netmiko_send_command, command_string='display interface description
| inc Nornir')
print_result(results)
```

代码调整成 for i in range(len(output[switch].result)):，这样会比较具有通用性。For 部分是循环，就没有 Nornir 本身的多线程了，所以脚本运行过程会比较耗时，因此加了一句提示，这样脚本执行过程可打点显示。

```
E:\Python\nornir_test\nornir-venv\Scripts\python.exe
E:/Python/nornir_test/nornir-venv/nornir14/nornir14-2.py
sw1 GigabitEthernet0/0/1 已配置完成
sw1 GigabitEthernet0/0/2 已配置完成
sw1 GigabitEthernet0/0/3 已配置完成
sw2 GigabitEthernet0/0/1 已配置完成
sw2 GigabitEthernet0/0/2 已配置完成
sw2 GigabitEthernet0/0/3 已配置完成
sw3 GigabitEthernet0/0/1 已配置完成
sw3 GigabitEthernet0/0/2 已配置完成
sw3 GigabitEthernet0/0/3 已配置完成
sw4 GigabitEthernet0/0/1 已配置完成
sw4 GigabitEthernet0/0/2 已配置完成
sw4 GigabitEthernet0/0/3 已配置完成
netmiko_send_command*************************************************************
* sw1 ** changed : False ********************************************************
vvvv netmiko_send_command ** changed : False vvvvvvvvvvvvvvvvvvvvvvvvvvvvvvvvv INFO
PHY: Physical
*down: administratively down
(l): loopback
(s): spoofing
(b): BFD down
(e): ETHOAM down
(dl): DLDP down
```

```
(d): Dampening Suppressed
Interface                     PHY     Protocol Description
GE0/0/1                       up      up       Access Port to VLAN 1 (Nornir)
GE0/0/2                       down    down     Trunk Port (Nornir)
GE0/0/3                       down    down     Access Port to VLAN 999 (Nornir)
^^^^ END netmiko_send_command ^^^^^^^^^^^^^^^^^^^^^^^^^^^^^^^^^^^^^^^^^^^^^^^^^^^^
* sw2 ** changed : False ********************************************************
vvvv netmiko_send_command ** changed : False vvvvvvvvvvvvvvvvvvvvvvvvvvvvvvvvv INFO
PHY: Physical
*down: administratively down
(l): loopback
(s): spoofing
(b): BFD down
(e): ETHOAM down
(dl): DLDP down
(d): Dampening Suppressed
Interface                     PHY     Protocol Description
GE0/0/1                       up      up       Access Port to VLAN 1 (Nornir)
GE0/0/2                       down    down     Trunk Port (Nornir)
GE0/0/3                       down    down     Access Port to VLAN 999 (Nornir)
^^^^ END netmiko_send_command ^^^^^^^^^^^^^^^^^^^^^^^^^^^^^^^^^^^^^^^^^^^^^^^^^^^^
* sw3 ** changed : False ********************************************************
vvvv netmiko_send_command ** changed : False vvvvvvvvvvvvvvvvvvvvvvvvvvvvvvvvv INFO
PHY: Physical
*down: administratively down
(l): loopback
(s): spoofing
(b): BFD down
(e): ETHOAM down
(dl): DLDP down
(d): Dampening Suppressed
Interface                     PHY     Protocol Description
GE0/0/1                       up      up       Access Port to VLAN 1 (Nornir)
GE0/0/2                       down    down     Trunk Port (Nornir)
GE0/0/3                       down    down     Access Port to VLAN 999 (Nornir)
^^^^ END netmiko_send_command ^^^^^^^^^^^^^^^^^^^^^^^^^^^^^^^^^^^^^^^^^^^^^^^^^^^^
* sw4 ** changed : False ********************************************************
vvvv netmiko_send_command ** changed : False vvvvvvvvvvvvvvvvvvvvvvvvvvvvvvvvv INFO
PHY: Physical
```

```
*down: administratively down
(l): loopback
(s): spoofing
(b): BFD down
(e): ETHOAM down
(dl): DLDP down
(d): Dampening Suppressed
Interface                   PHY     Protocol Description
GE0/0/1                     up      up       Access Port to VLAN 1 (Nornir)
GE0/0/2                     down    down     Trunk Port (Nornir)
GE0/0/3                     down    down     Access Port to VLAN 999 (Nornir)
^^^^ END netmiko_send_command ^^^^^^^^^^^^^^^^^^^^^^^^^^^^^^^^^^^^^^^^^^^^^^^^^^^^^

Process finished with exit code 0
```

实验部分如下图所示。

```
E:\Python\nornir_test\nornir-venv\Scripts\python.exe E:/Python/nornir_test/nornir-venv/nornir14/nor
sw1 GigabitEthernet0/0/1 已配置完成
sw1 GigabitEthernet0/0/2 已配置完成
```

```
(d): Dampening Suppressed
Interface                   PHY     Protocol Description
GE0/0/1                     up      up       Access Port to VLAN 1 (Nornir)
GE0/0/2                     down    down     Trunk Port (Nornir)
GE0/0/3                     down    down     Access Port to VLAN 999 (Nornir)
^^^^ END netmiko_send_command ^^^^^^^^^^^^^^^^^^^^^^^^^^^^^^^^^^^^^^^^^^^^^^^^^^^^^
* sw2 ** changed : False ****************************************************************
vvvv netmiko_send_command ** changed : False vvvvvvvvvvvvvvvvvvvvvvvvvvvvvvvvvvvvvv INFO
PHY: Physical
*down: administratively down
```

至此，本章 Nornir 的 14 个实验介绍完毕，通过一些生产场景逻辑抽象，层层引出知识点。除了学习知识点，希望读者能把一些学习思路和实践方法慢慢掌握起来。

第 10 章

NETCONF 详解

2002 年 6 月，互联网架构委员会（Internet Architeture Board，IAB）举办了一场主题为 Network Management 的研讨会，邀请了一大帮研发界业内名气响当当的人物（大多来自 IETF），共同商议讨论彼时普遍使用的网络管理协议，找出它们各自的优势和缺点，并列出一个优秀的网络管理协议必须具备的特征，准备为开发下一代网络管理协议（NextGen Network Management Protocol）做准备。2003 年 5 月，IETF 发布了 RFC 3535（Overview of the 2002 IAB Network Management Workshop），正式提出了下一代网络管理协议应该具备的七大特征，如下图所示。

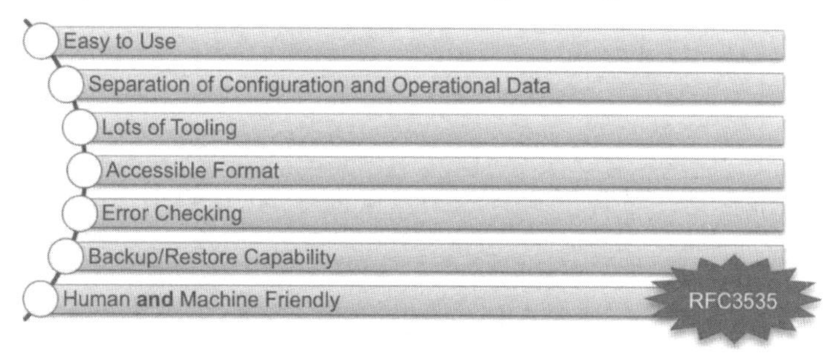

在这个背景下，IETF 于 2006 年 12 月率先发布了 RFC 4741，也就是 NETCONF 这个基于 XML，用来替代 CLI、SNMP 的网络配置和管理协议。在随后几年加加改改进行一番修订后，IETF 又于 2011 年 6 月以 RFC 6241 作为终稿将其再次发布。

虽然 NETCONF 年龄不算小，但是倒回 10 年前，你要是问一个网络工程师什么是 NETCONF（包括 YANG），相信没有几位能回答上来，因为当时主流的思科 IOS 设备根本就不支持 NETCONF，当时的思科对 NETCONF 并不“感冒”，而是自己闭门造车，分别在 2007 年和 2012 年开发了 WSMA（Web Service Management Agent）和 onePK 这两个 API。那些年跟随思科的步伐，一路从 CCNA、CCNP、CCIE 学习起来的网络工程师压根儿就没听说过还有这么一个书上不会教、考试不会考、工作中也基本用不到的非常小众的技术。十年河东十年河西，这些年随着 SDN 和网络运维自动化技术的强势崛起，NETCONF 终于在诞生 10 多年后在传统计算机网络领域重新进入人们的视野，本章就来讲解一下这门“老树发新芽“的技术。

10.1　NETCONF 的理论部分

关于 NETCONF 的理论部分，最权威的肯定还是 RFC 6241，这里做一下总结。

（1）NETCONF 的协议框架分为四层：由低到高，分别为安全传输（Secure Transport）、消息（Messages）、操作（Operations）和内容（Content），如下图所示。

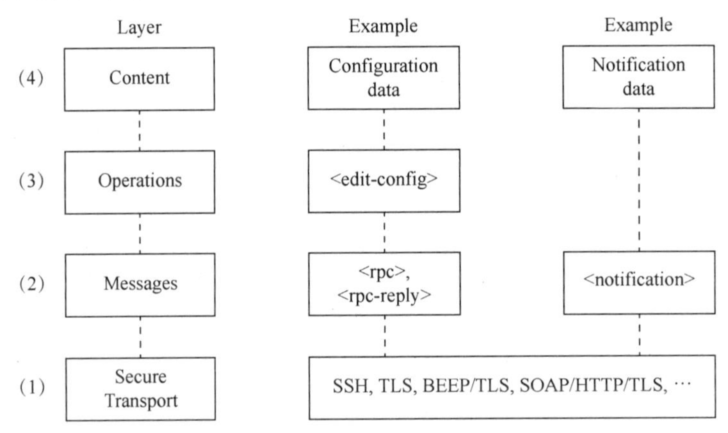

（2）安全传输层最常见、最常用的是 SSH，这也是 NETCONF 和同为最近几年很流行的 REST（基于 Web）之间最大的区别。

（3）消息层基于 RPC（Remote Procedure Call，远程调用）协议，其作用是提供一个简易的不依赖于传输层，生成 RPC 和通知消息框架的通信协议。在后面的 NETCONF 实验中，我们会多次看到<rpc>、<rpc-reply>这两个代表 RPC 的标签。

（4）操作层定义了一组用来配置、复制、删除设备命令及获取设备信息的基本操作，基本操作包括 get、get-config、edit-config、copy-config、delete-config、lock、unlock、close-session 和 kill-session，这些基本操作都是在 XML 语言中被调用的，比如<get>、<edit-config>等，后面实验部分会讲到。

（5）内容层由管理数据内容的数据模型定义，该数据模型也就是后面会讲到的 YANG。

（6）NETCONF 的编码格式基于 XML，基于 XML 来做网络管理主要是看中了 XML 强大的数据表示能力（对初学者来说，XML 其实并不怎么友好，不如 JSON、YAML 那么易读和容易上手，这是笔者个人的观点）。

（7）NETCONF 网络架构由客户端和服务器组成，客户端就是运行 NETCONF 的主机，服务器就是被操控的交换机或者路由器，NETCONF 默认使用端口 830，可以更改。

（8）NETCONF 协议中还有一个重要的概念叫作数据集（Datastores），数据集的作用是用来存储一份配置数据的备份，确保设备能从开机时的初始状态进入它能正常运行时的工作状态（即改了配置后需要 write memroy，不然重启设备后配置会丢失）。数据集分为 Running（类似于思科的 running config）、Start-up（类似于思科的 startup config）和 Candidate（思科 IOS-XR 会用到）3 种，其中只有 Running 是强制使用的。

10.2　YANG 的理论部分

了解了 NETCONF 的理论部分，下面再来看 YANG。在讲 YANG 之前，我们首先需要知道什么是数据模型。

10.2.1　什么是数据模型

数据模型的作用是描述一组具有统一标准的数据，并用明确的参数和模型来将数据的呈现标准化、规范化。举个例子，不同厂商之间对一些行业术语的规范及具体参数存在较大差异，比如用来区分路由协议优先级，在思科这边被称作管理距离（Administrative Distance）的东西，到了 Juniper 那边被改叫作 Route Preference，并且它们分配给各个路由协议的 AD 和 RP 值也不一样，比如 RIP 在思科的 AD 为 120，在 Juniper 的 RP 为 100；又比如，OSPF 在思科的 AD 为 110，而在 Juniper 这边，OSPF 还被分为 OSPF Interal（RP 值为 10）和 OSPF External（RP 值为 150）。虽然不同厂商之间对某些技术标准存在这样或那样的差异，但是对于绝大部分技术，它们还是遵守统一标准的，比如 VLAN。

用来描述 VLAN 的数据无外乎如下几点。

（1）VLAN ID（1 ~ 4096 的整数，rw 类型）。

（2）VLAN Name（用字符串代表的 VLAN 名称，rw 类型）。

（3）VLAN State（用枚举表示的 down/up 或 shutdown/no shutdown 表示的 VLAN 状态，ro 类型）。

由此可以看到，类似于这种明确定义了数据内容及其数据类型和范围，以及 rw、ro 类型的一套模型，就被叫作数据模型。

10.2.2　YANG 模型

YANG（英文全称 Yet Another Next Generation），是一种以网络为中心的数据模型语言（Network-centric data modeling language），由 IETF 于 2010 年 10 月（也就是 NETCONF 终稿发布之前的一年）在 RFC 6020 中被提出，其诞生之初的目的很明确，是专门为 NETCONF 量身打造的建模语言，不过现在也被 REST 和其他协议所采用。

YANG 的模型分为标准（Open 或者 Standard）和私有（Native）两种类型，其中标准类型中比较著名的有 IETF、IEEE、OpenConfig 等由国际知名组织制定的模型（IETF 制定的 YANG 模型最常见）。除标准 YANG 模型外，各厂商又根据自家产品的不同设计了私有的 YANG 模型，比如思科、Juniper、华为、Fujitsu、诺基亚等都有自家的 YANG 模型。

其中思科的 YANG 模型又分为 Cisco Common 和 Cisco Platform Specific，其中前者为思科所有 OS（IOS-XE、IOS-XR、NX-OS）通用的 YANG 模型，后者为一些 OS 独占的 YANG 模型，如下图所示。

YANG

Open and Native YANG Models

- Standard definition(IETF,ITU, OpenConfig,etc.)
- Compliant with standard,i.e. “Policy”
  ietf-diffserv-policy.yang
  ietf-diffserv-classifer.yang
  ietf-diffserv-target.yang

- Cisco definition
- Common across Cisco platforms,

  i.e. “OTV” on IOS-XE and NX-OS

- Cisco definition
- Unique to specific Cisco platform,

  i.e. “BGP” extensions on IOS-XE

10.2.3　YANG 模块

模块（Module）是 YANG 定义的基本单位，一个模块可以用来定义一个单一的数据类型，也可以增加一个现有的数据模型和其他节点。下图是一个叫作 ietf-interfaces，由 IETF 制定，用来描述设备端口参数的 YANG 模块，可以看到 YANG 模块是以树形结构展开的。

```
module: ietf-interfaces
  +--rw interfaces
  |  +--rw interface* [name]
  |     +--rw name                        string
  |     +--rw description?                string
  |     +--rw type                        identityref
  |     +--rw enabled?                    boolean
  |     +--rw link-up-down-trap-enable?   enumeration {if-mib}?
  |     +--ro admin-status                enumeration {if-mib}?
  |     +--ro oper-status                 enumeration
  |     +--ro last-change?                yang:date-and-time
  |     +--ro if-index                    int32 {if-mib}?
  |     +--ro phys-address?               yang:phys-address
  |     +--ro higher-layer-if*            interface-ref
  |     +--ro lower-layer-if*             interface-ref
  |     +--ro speed?                      yang:gauge64
  |     +--ro statistics
  |        +--ro discontinuity-time    yang:date-and-time
  |        +--ro in-octets?            yang:counter64
  |        +--ro in-unicast-pkts?      yang:counter64
  |        +--ro in-broadcast-pkts?    yang:counter64
  |        +--ro in-multicast-pkts?    yang:counter64
  |        +--ro in-discards?          yang:counter32
  |        +--ro in-errors?            yang:counter32
  |        +--ro in-unknown-protos?    yang:counter32
  |        +--ro out-octets?           yang:counter64
  |        +--ro out-unicast-pkts?     yang:counter64
  |        +--ro out-broadcast-pkts?   yang:counter64
  |        +--ro out-multicast-pkts?   yang:counter64
  |        +--ro out-discards?         yang:counter32
  |        +--ro out-errors?           yang:counter32
```

10.2.4　从 GitHub 下载 YANG 模块

GitHub 上存放着大部分标准和私有的 YANG 模型，截至 2022 年 8 月，标准模型中包含 IETF、IEEE、ETSI 等国际知名组织和协会制定的 YANG 模块，私有模型中包含思科、Juniper、华为、诺基亚及富士通在内的厂商自己制定的 YANG 模块，如下图所示。

我们可以通过 git clone 来下载上述所有标准和私有的 YANG 模型及其对应的模块。

```
git clone https://github.com/YangModels/yang.git
```

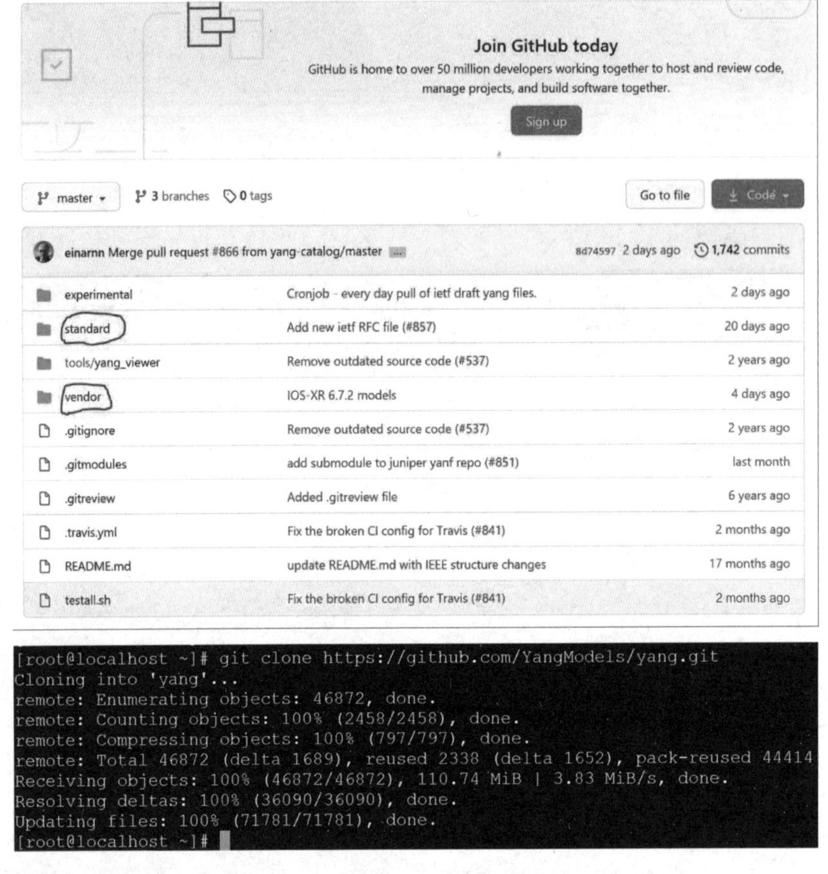

下载完毕后，当前目录下多出来一个名叫 yang 的子目录，其中 yang/standard/ietf/RFC 下包含了全部由 IETF 制定的 YANG 模块，其中也包含我们上面举例的 ietf-interfaces 模块。

```
[root@localhost RFC]# ls
iana-bfd-types@2021-10-21.yang
iana-bfd-types.yang
iana-crypt-hash@2014-08-06.yang
iana-crypt-hash.yang
iana-dots-signal-channel@2020-05-28.yang
iana-dots-signal-channel@2021-09-02.yang
iana-dots-signal-channel.yang
iana-hardware@2018-03-13.yang
iana-hardware.yang
iana-if-type@2014-05-08.yang
iana-if-type@2017-01-19.yang
iana-if-type.yang
iana-routing-types@2017-12-04.yang
iana-routing-types@2018-10-29.yang
iana-routing-types.yang
```

```
iana-tunnel-type@2019-11-16.yang
iana-tunnel-type.yang
ietf-access-control-list@2019-03-04.yang
ietf-access-control-list.yang
ietf-acldns@2019-01-28.yang
ietf-acldns.yang
ietf-alarms@2019-09-11.yang
ietf-alarms-x733@2019-09-11.yang
ietf-alarms-x733.yang
ietf-alarms.yang
ietf-bfd@2021-10-21.yang
[root@localhost RFC]# ls | grep interfaces
ietf-interfaces@2014-05-08.yang
ietf-interfaces@2018-02-20.yang
ietf-interfaces.yang
[root@localhost RFC]#
```

用 cat ietf-interfaces.yang 查看该模块的内容，如下图所示。

```
[root@localhost RFC]# cat ietf-interfaces.yang
module ietf-interfaces {
  yang-version 1.1;
  namespace "urn:ietf:params:xml:ns:yang:ietf-interfaces";
  prefix if;

  import ietf-yang-types {
    prefix yang;
  }

  organization
    "IETF NETMOD (Network Modeling) Working Group";

  contact
    "WG Web:   <https://datatracker.ietf.org/wg/netmod/>
     WG List:  <mailto:netmod@ietf.org>

     Editor:   Martin Bjorklund
               <mailto:mbj@tail-f.com>";

  description
    "This module contains a collection of YANG definitions for
     managing network interfaces.

     Copyright (c) 2018 IETF Trust and the persons identified as
     authors of the code.  All rights reserved.

     Redistribution and use in source and binary forms, with or
     without modification, is permitted pursuant to, and subject
     to the license terms contained in, the Simplified BSD License
     set forth in Section 4.c of the IETF Trust's Legal Provisions
     Relating to IETF Documents
     (https://trustee.ietf.org/license-info).

     This version of this YANG module is part of RFC 8343; see
     the RFC itself for full legal notices.";

  revision 2018-02-20 {
    description
```

```
    "Updated to support NMDA.";
  reference
    "RFC 8343: A YANG Data Model for Interface Management";
}

revision 2014-05-08 {
  description
    "Initial revision.";
  reference
    "RFC 7223: A YANG Data Model for Interface Management";
}
```

你也许会问，怎么和 10.2.3 节中举例的 ietf-interfaces 模块内容不一样？这是因为上面的例子中我们是以树形格式（Tree View Format）展开 ietf-interfaces 模块的，而这里用 cat 查看的是该模块的源码。我们可以用 pyang 这个 Python 模块来以树形格式的形式查看一个 YANG 模块的具体结构。

10.2.5　pyang 模块

pyang 是 Python 专为 YANG 开发的一个第三方模块，主要有三个功能：（1）用来验证 YANG 模块代码的准确性；（2）将 YANG 模块转换成其他格式（比如我们前面讲到的树形格式）；（3）从 YANG 模块中生成代码。这里笔者主要介绍如何使用 pyang 来将刚才的 ietf-interfaces.yang 模块转化为树形格式。

首先通过 pip 下载安装 pyang，如下图所示。

```
pip3.10 install pyang
```

```
[root@localhost RFC]# pip3.10 install pyang
Collecting pyang
  Downloading pyang-2.5.3-py2.py3-none-any.whl (592 kB)
     ---------------------------------------- 592.9/592.9 kB        eta 0:00:00
Requirement already satisfied: lxml in /usr/local/lib/python3.10/site-packages (from
Installing collected packages: pyang
Successfully installed pyang-2.5.3
WARNING: Running pip as the 'root' user can result in broken permissions and conflict
ttps://pip.pypa.io/warnings/venv

[      ] A new release of pip available:        -> 22.2.2
[      ] To update, run: pip install --upgrade pip
[root@localhost RFC]#
```

然后回到 yang/standard/ietf/RFC 下，使用命令 pyang -f tree ietf-interfaces.yang 就能将 ietf-interfaces 这个 YANG 模块转换成树形模式查看，如下图所示。

```
[root@localhost RFC]# pyang -f tree ietf-interfaces.yang
module: ietf-interfaces
  +--rw interfaces
  |  +--rw interface* [name]
  |     +--rw name                        string
  |     +--rw description?                string
  |     +--rw type                        identityref
  |     +--rw enabled?                    boolean
  |     +--rw link-up-down-trap-enable?   enumeration {if-mib}?
  |     +--ro admin-status                enumeration {if-mib}?
  |     +--ro oper-status                 enumeration
  |     +--ro last-change?                yang:date-and-time
  |     +--ro if-index                    int32 {if-mib}?
  |     +--ro phys-address?               yang:phys-address
  |     +--ro higher-layer-if*            interface-ref
  |     +--ro lower-layer-if*             interface-ref
  |     +--ro speed?                      yang:gauge64
  |     +--ro statistics
  |        +--ro discontinuity-time    yang:date-and-time
  |        +--ro in-octets?            yang:counter64
  |        +--ro in-unicast-pkts?      yang:counter64
  |        +--ro in-broadcast-pkts?    yang:counter64
  |        +--ro in-multicast-pkts?    yang:counter64
  |        +--ro in-discards?          yang:counter32
  |        +--ro in-errors?            yang:counter32
  |        +--ro in-unknown-protos?    yang:counter32
  |        +--ro out-octets?           yang:counter64
  |        +--ro out-unicast-pkts?     yang:counter64
  |        +--ro out-broadcast-pkts?   yang:counter64
  |        +--ro out-multicast-pkts?   yang:counter64
  |        +--ro out-discards?         yang:counter32
  |        +--ro out-errors?           yang:counter32
  x--ro interfaces-state
     x--ro interface* [name]
```

10.3　NETCONF 的实验部分

介绍完 NETCONF 的理论部分，接下来介绍 NETCONF 实验。

10.3.1　实验拓扑和实验环境

NETCONF 的实验环境很简单，就是一台 CentOS 8 主机和一台思科 3850 的交换机直连，如下图所示。

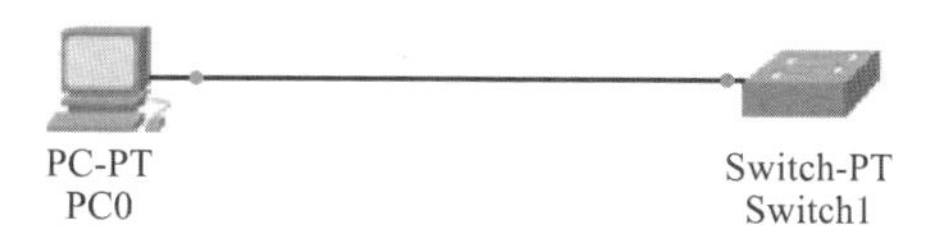

主机操作系统：CentOS 8

交换机型号：WS-C3850-48P-S

交换机 OS 版本：IOS-XE 17.03.01

主机 IP: 172.16.224.27

交换机 IP: 172.16.224.28

实验涉及的端口为交换机的 GigabitEthernet1/0/33，其初始配置如下图所示。

```
Current configuration : 79 bytes
!
interface GigabitEthernet1/0/33
 no switchport
 no ip address
 shutdown
end
```

在开始实验前，先提几点使用 NETCONF 需要注意的地方。

- IOS-XE 版本必须为 16.3 以上。
- NETCONF 默认使用 TCP 端口 830，可以更改。
- NETCONF 的传输协议基于 SSH，在交换机上必须开启 SSH，并且用户特权级别必须为 15。
- 如果交换机配置了 AAA+TACACS（比如说思科的 ISE），必须保证配置了“aaa authorization exec default local”，然后使用交换机本地创建的用户名（非 ISE 上的用户名），否则在主机端会出现“Permission denied, please try again.”，在交换机端会出现“%DMI-5-AUTHENTICATION_FAILED:”的错误，如下图所示。

```
[root@localhost ~]# ssh parry@172.16.224.28 -p 830 netconf
parry@172.16.224.28's password:
Permission denied, please try again.
parry@172.16.224.28's password:
Permission denied, please try again.
```

```
Sep 22 06:15:09.595: %DMI-5-AUTHENTICATION_FAILED: Switch 1 R0/0: dmiauthd: Authentication failure from 10.126.76.15:36757 for netconf over ssh.
Sep 22 06:15:17.964: %DMI-5-AUTHENTICATION_FAILED: Switch 1 R0/0: dmiauthd: Authentication failure from 10.126.76.15:36757 for netconf over ssh.
Sep 22 06:15:39.066: %DMI-5-AUTHENTICATION_FAILED: Switch 1 R0/0: dmiauthd: Authentication failure from 10.126.76.15:36758 for netconf over ssh.
Sep 22 06:17:08.456: %DMI-5-AUTH_PASSED: Switch 1 R0/0: dmiauthd: User 'test' authenticated successfully from 10.126.76.15:36780 and was authorized for netconf over ssh. External groups: PRIV15
```

10.3.2　实验步骤

（1）首先在交换机上开启 NETCONF，方法很简单，在全局模式下输入 netconf-yang 和 netconfg ssh 即可（可选，前面理论部分已经讲到了，NETCONF 默认使用的是 SSH 作为安全传输层协议，所以 netconf ssh 这条命令默认就是开启的，不需要特意配置）。其余的交换机初始配置，比如开启 SSH 和创建特权级别为 15 的用户请读者自行完成。

```
SW(config)#netconf-yang
SW(config)#netconf ssh （默认是开启的）
```

（2）回到主机，通过 SSH 访问交换机的 830 端口。

```
[root@localhost ~]# ssh -p 830 test@172.16.224.28
test@172.16.224.28's password:
```

（3）连接成功后，交换机会回复一个 hello 包，以及一长串能力集（capabilities），所谓能力集就是该交换机支持的 yang model：既有思科的，也有 IETF 的，由于回显内容过多，下图只截取部分以作演示。

```
[root@localhost ~]# ssh -p 830 test@172.16.224.28
test@172.16.224.28's password:
<?xml version="1.0" encoding="UTF-8"?>
<hello xmlns="urn:ietf:params:xml:ns:netconf:base:1.0">
<capabilities>
<capability>urn:ietf:params:netconf:base:1.0</capability>
<capability>urn:ietf:params:netconf:base:1.1</capability>
<capability>urn:ietf:params:netconf:capability:writable-running:1.0</capability>
<capability>urn:ietf:params:netconf:capability:xpath:1.0</capability>
<capability>urn:ietf:params:netconf:capability:validate:1.0</capability>
<capability>urn:ietf:params:netconf:capability:validate:1.1</capability>
<capability>urn:ietf:params:netconf:capability:rollback-on-error:1.0</capability>
<capability>urn:ietf:params:netconf:capability:notification:1.0</capability>
<capability>urn:ietf:params:netconf:capability:interleave:1.0</capability>
<capability>urn:ietf:params:netconf:capability:with-defaults:1.0?basic-mode=explicit&amp;also-supported=report-all-tagged</capability>
<capability>urn:ietf:params:netconf:capability:yang-library:1.0?revision=2016-06-21&amp;module-set-id=04a4985ba3a91dd024b765809aef8439</capability>
<capability>http://tail-f.com/ns/netconf/actions/1.0</capability>
<capability>http://tail-f.com/ns/netconf/extensions</capability>
<capability>http://cisco.com/ns/cisco-xe-ietf-ip-deviation?module=cisco-xe-ietf-ip-deviation&amp;revision=2016-08-10</capability>
<capability>http://cisco.com/ns/cisco-xe-ietf-ipv4-unicast-routing-deviation?module=cisco-xe-ietf-ipv4-unicast-routing-deviation&amp;revision=2015-09-11</capability>
<capability>http://cisco.com/ns/cisco-xe-ietf-ipv6-unicast-routing-deviation?module=cisco-xe-ietf-ipv6-unicast-routing-deviation&amp;revision=2015-09-11</capability>
<capability>http://cisco.com/ns/cisco-xe-ietf-ospf-deviation?module=cisco-xe-ietf-ospf-deviation&amp;revision=2018-02-09</capability>
<capability>http://cisco.com/ns/cisco-xe-ietf-routing-deviation?module=cisco-xe-ietf-routing-deviation&amp;revision=2016-07-09</capability>
<capability>http://cisco.com/ns/cisco-xe-openconfig-acl-deviation?module=cisco-xe-openconfig-acl-deviation&amp;revision=2017-08-25</capability>
<capability>http://cisco.com/ns/mpls-static/devs?module=common-mpls-static-devs&amp;revision=2015-09-11</capability>
<capability>http://cisco.com/ns/nvo/devs?module=nvo-devs&amp;revision=2015-09-11</capability>
```

（4）随后我们使用下列 XML 代码向交换机回复一个 hello 包（注：交换机收到 hello 包后不会做出任何回应），之后主机和交换机之间的 netconf session 即建立成功。

```
<?xml version="1.0" encoding="UTF-8"?>
<hello xmlns="urn:ietf:params:xml:ns:netconf:base:1.0">
   <capabilities>
      <capability>urn:ietf:params:netconf:base:1.0</capability>
   </capabilities>
</hello>]]>]]>
```

```
<capability>
        urn:ietf:params:netconf:capability:notification:1.1
      </capability>
</capabilities>
<session-id>30</session-id></hello>]]>]]>

<?xml version="1.0" encoding="UTF-8"?>
<hello xmlns="urn:ietf:params:xml:ns:netconf:base:1.0">
    <capabilities>
        <capability>urn:ietf:params:netconf:base:1.0</capability>
    </capabilities>
</hello>]]>]]>
```

这里简单讲解下 XML 语言的格式和内容，XML 指令以<?xml version="1.0" encoding="UTF-8"?>开头，其中：

- <?：表示一条 XML 指令的开始。
- xml：表示此文件是 XML 文件。
- version：NETCONF 协议版本号。
- encoding：字符集编码格式，当前仅支持 UTF-8 编码。
- ?>：表示一条 XML 指令的结束。

在<hello xmlns="urn:ietf:params:xml:ns:netconf:base:1.0">里：

- hello 表示主机（NETCONF 客户端）向交换机（NETCONF 服务器）回复的 hello 包。
- xmlns 是 XML Namespace（XML 命名空间）的缩写，在 XML 里，元素名称（可以把它理解为 Python 中的变量名）是由开发者自己定义的，当将两个不同的文档合并成一个文档时，如果两个文档里使用相同的元素名，就会发生命名冲突。比如说 A 和 B 两人分别开发了一个 XML 文档，A 用 name 作为元素名称来描述一个人名，而 B 也用 name 作为元素名称来描述一个地名，虽然 A、B 两人描述的东西不一样，但是他俩定义的元素名称（name）是一样的，当 A、B 两人的文档被合并成一个文档时，就会造成冲突，而 XML 命名空间的作用就是避免 XML 内的元素冲突。
- urn:ietf:params:xml:ns:netconf:base:1.0， 在 RFC 4741 中规定了所有涉及 NETCONF 协议的元素必须以 urn:ietf:params:xml:ns:netconf:base:1.0 作为 XML 命名空间。
- <capabilities>代表能力集，<capability>代表具体的能力，是 NETCONF 客户端和服务器互相向对方说明自己具备什么能力的作用，这里的<capability>urn:ietf:params:netconf:base:1.0</capability>表示客户端告诉服务器：我具备使用 IETF 制定的 NETCONF 协议的能力。
- 和 Python 不一样，XML 并没有严格要求缩进，但是为了美观和可读性，一般我们用两个空格或者四个空格做缩进。

- 注意代码最尾部的]]>]]>，它是 NETCONF 客户端和 NETCONF 服务器用来表示消息请求或回复终止的部分（用来告知对方“我话说完了，轮到你了”），并不是 XML 语言本身的一部分，后面会说明为什么它很重要。

（5）发送下列 XML 代码向交换机查询 Gi1/0/33 端口的配置。

```
<?xml version="1.0" encoding="UTF-8"??>
<nc:rpc message-id="101" xmlns:nc="urn:ietf:params:xml:ns:netconf:base:1.0">
  <nc:get>
    <nc:filter type="subtree">
      <native xmlns="http://cisco.com/ns/yang/Cisco-IOS-XE-native">
      <interface>
        <GigabitEthernet>
        <name>1/0/33</name>
        <description></description>
          <ip></ip>
        </GigabitEthernet>
       </interface>
       </native>
    </nc:filter>
  </nc:get>
</nc:rpc>
]]>]]>
```

随即交换机会回复一个 rpc-reply 包（下图框线中的部分），后面的内容即我们请求的 Gi1/0/33 端口的配置信息，包括该端口下的 description 和 IP 地址（为空）。

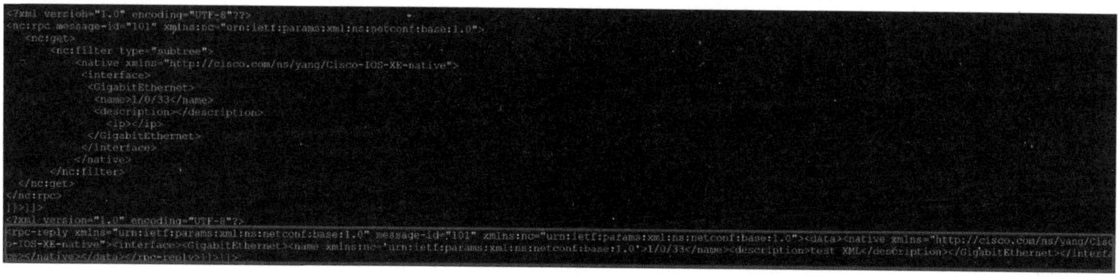

```
<?xml version="1.0" encoding="UTF-8"??>
<nc:rpc message-id="101" xmlns:nc="urn:ietf:params:xml:ns:netconf:base:1.0">
  <nc:get>
    <nc:filter type="subtree">
      <native xmlns="http://cisco.com/ns/yang/Cisco-IOS-XE-native">
      <interface>
        <GigabitEthernet>
        <name>1/0/33</name>
        <description></description>
          <ip></ip>
        </GigabitEthernet>
       </interface>
       </native>
    </nc:filter>
  </nc:get>
</nc:rpc>
]]>]]>
<?xml version="1.0" encoding="UTF-8"?>
<rpc-reply xmlns="urn:ietf:params:xml:ns:netconf:base:1.0" message-id="101" xmlns:nc="urn:ietf:params:xml:ns:netconf:base:1.0"><data><native xmlns="http://cisco.com/ns/yang/Cisc
o-IOS-XE-native"><interface><GigabitEthernet><name xmlns:nc='urn:ietf:params:xml:ns:netconf:base:1.0'>1/0/33</name><description>test XML</description></GigabitEthernet></interf
ce></native></data></rpc-reply>]]>]]>
```

（6）如果你觉得交换机回复的 rpc-reply 不易读，可以将它复制粘贴到 Code Beautify 网站，用 Code Beautify 中 XML Viewer 提供的 Tree View 或者 Beautify / Format 让交换机的 rpc-reply 内容更易读，如下图所示。

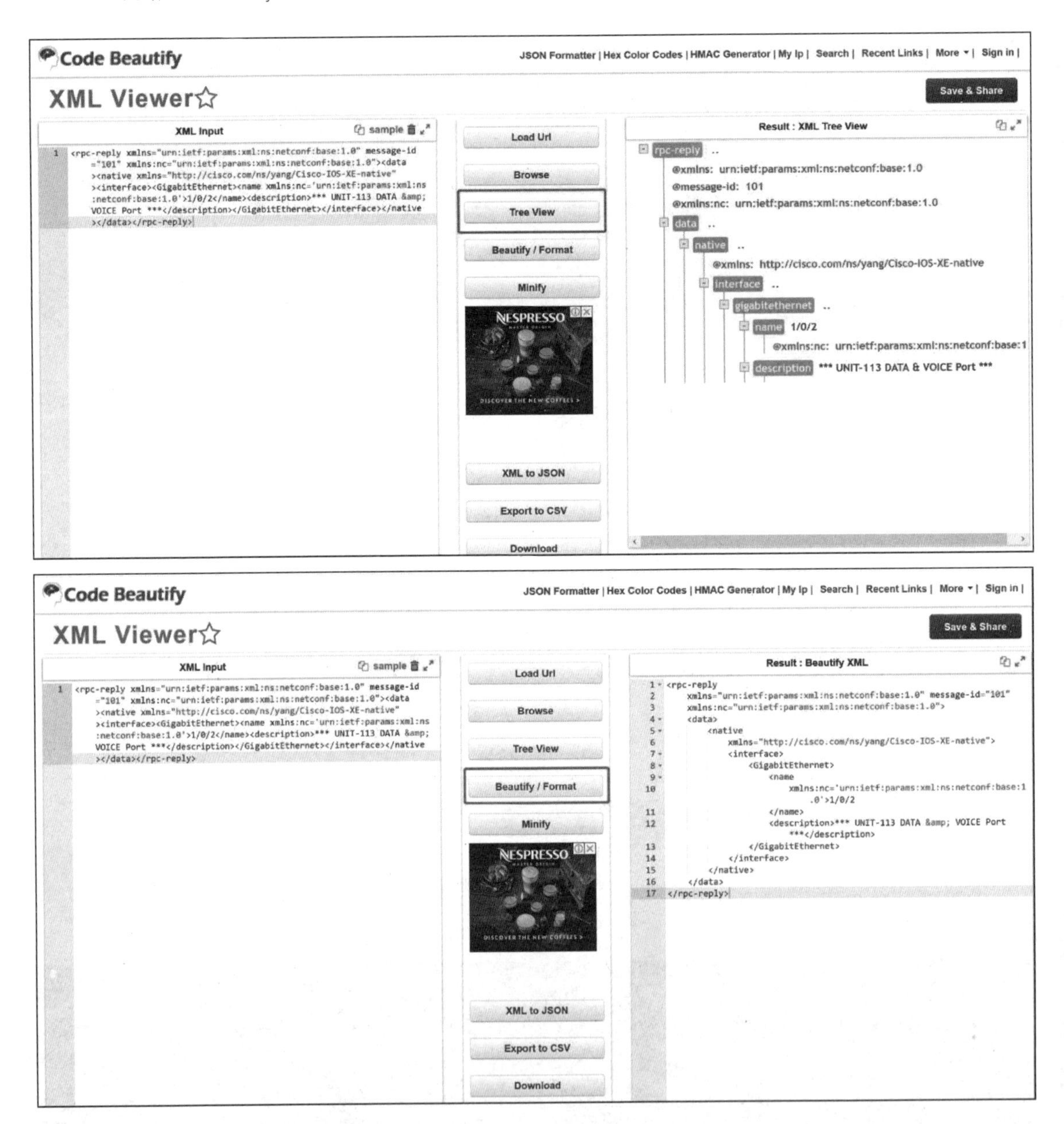

注意，复制 RPC-REPLY 内容时不要把末尾的]]>]]>一并复制了（它们不是 XML 语言的一部分），不然 Code Beautify 会报错，如下图所示。

```
<rpc-reply xmlns="urn:ietf:params:xml:ns:netconf:base:1.0" message-id="101"
o-IOS-XE-native"><interface><GigabitEthernet><name xmlns:nc='urn:ietf:params
n></GigabitEthernet></interface></native></data></rpc-reply>]]>]]>
```

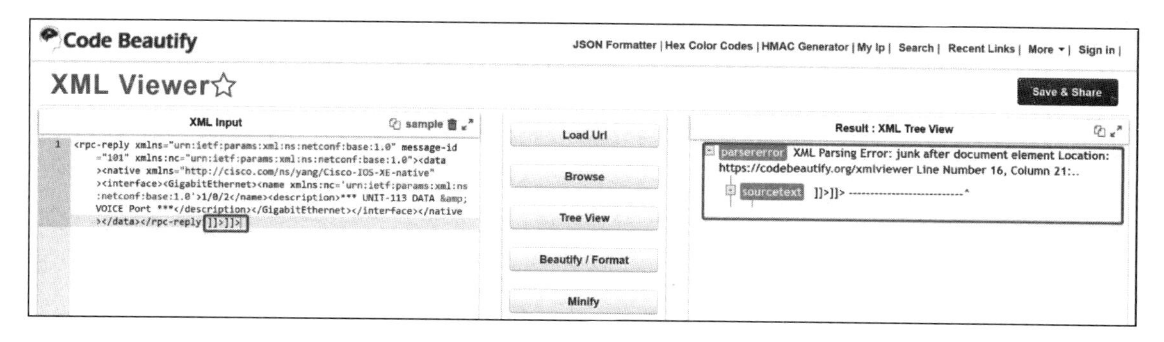

（7）接下来我们用下面这段 XML 代码来查询交换机的 running configuration。这段 XML 代码里笔者故意没做缩进，就是为了演示 XML 没有严格要求代码缩进。

```
<?xml version="1.0"?>
<rpc xmlns="urn:ietf:params:xml:ns:netconf:base:1.0" message-id="102">
<get-config>
<source>
<running/>
</source>
</get-config>
</rpc>]]>]]>
```

（8）用下面这段 XML 代码来为端口 Gi1/0/33 添加一个 description，内容为 test XML，表示该 description 是我们通过 NETCONF 配置的。

```
<?xml version="1.0"? encoding="UTF-8"??>
<rpc message-id="102" xmlns="urn:ietf:params:xml:ns:netconf:base:1.0">
   <edit-config>
      <target>
         <startup/>
      </target>
      <config>
         <native xmlns="http://cisco.com/ns/yang/Cisco-IOS-XE-native">
            <interface>
               <GigabitEthernet>
                  <name>1/0/33</name>
                  <description>test XML</description>
               </GigabitEthernet>
            </interface>
         </native>
      </config>
```

```
    </edit-config>
</rpc>
]]>]]>
```

这里，我们使用了<edit-config>这个 NETCONF 的理论部分讲到的基本操作，通常<edit-config>下面会接一个<target>，表示需要修改配置的目标，<startup/>表示数据集，这里我们要更改的是 startup 这个数据集（也可以使用 running/），后面的<native xmlns="http://cisco.com/ns/yang/Cisco-IOS-XE-native">表示使用了思科的私有 YANG 模型。

回到交换机，验证 Gi1/0/33 的配置，如下图所示。

```
Current configuration : 101 bytes
!
interface GigabitEthernet1/0/33
 description test XML
 no switchport
 no ip address
 shutdown
end
```

（9）最后我们用下面这段 XML 代码为 Gi1/0/33 配置 10.1.1.2 /24 这个 IP 地址。

```
<?xml version="1.0"? encoding="UTF-8"??>
<rpc message-id="103" xmlns="urn:ietf:params:xml:ns:netconf:base:1.0">
    <edit-config>
        <target>
            <running/>
        </target>
        <config>
            <native xmlns="http://cisco.com/ns/yang/Cisco-IOS-XE-native">
               <interface>
                    <GigabitEthernet>
                         <name>1/0/33</name>
                         <ip>
                             <address>
                                 <primary>
                                  <address>10.1.1.2</address>
                                  <mask>255.255.255.0</mask>
                                </primary>
                             </address>
                         </ip>
                    </GigabitEthernet>
               </interface>
```

```
            </native>
        </config>
    </edit-config>
</rpc>
]]>]]>
```

最后回到交换机上验证 Gi1/0/33 的配置，可以看到该端口已经被配上了 10.1.1.2 25 这个 IP 地址，并且子网掩码为 255.255.255.0，如下图所示。

```
Current configuration : 121 bytes
!
interface GigabitEthernet1/0/33
 description test XML
 no switchport
 ip address 10.1.1.2 255.255.255.0
 shutdown
end
```

关于 NETCONF 和 YANG 的理论和实战就先介绍到这里，可以看到 NETCONF 这种通过复制粘贴 XML 代码的操作方式对传统网络工程师来说很不适应。虽然不太好用，但 NETCONF 确实是 NetDevOps 时代网络工程师们必须掌握的技能之一。下一节中将继续讲解 ncclient 这个 Python 专为 NETCONF 客户端开发的第三方模块，看看 NETCONF 结合 ncclient 会不会让 NETCONF 的操作变得简单一些。

10.4　ncclient

前面讲到了 NETCONF 和 YANG 的历史和理论，以及它们之间的关系。通过实验介绍了怎么在思科的 IOS-XE 设备上启用 NETCONF，怎么在 NETCONF 客户端（即主机）上通过 SSH 协议以默认的 830 端口登录思科的 IOS-XE 交换机，怎样通过发送 XML 代码的方式来对设备的配置数据和设备信息做查询和更改，以及如何使用 Code Beautify 网站提供的 XML Viewer 来美化 XML 代码格式，以及以树形结构展开代码。

很多第一次接触 NETCONF 和 XML 的人（尤其是那些过去没有编程基础，而且也没接触过 HTML 这个与 XML 非常类似的标记语言的传统网络工程师）对这种通过手动复制粘贴 XML 命名空间、标签的代码，向 NETCONF 服务器设备（路由器、交换机）发送 RPC 请求消息，等待设备回复 rpc-reply 响应消息的操作方式非常不适应，这是完全正常的。如果仅靠这种方式来和设备互动，其效率甚至还不如 CLI，好在这个痛点可以通过 Python 的 ncclient 模块来解决。

10.4.1 ncclient 简介

ncclient 是一个开源的 Python 模块，用来在 NETCONF 客户端开发各种和 NETCONF 相关的网络运维脚本和应用程序。最早的作者是曾在彭博社和 Facebook 工作过的软件工程师 Shikar Bhushan，目前维持这个项目的是 Leonidas Poulopoulos 和 Einar Nilsen-Nygaard，截至 2022 年 8 月最新的版本为 0.6.11，ncclient 对系统环境有如下要求。

- Python 2.7 或 Python 3.4+
- setuptools 0.6+
- Paramiko 1.7+
- lxml 3.3.0+
- libxml2
- libxslt

ncclient 可以通过 pip 下载安装，如下图所示。

```
[root@localhost ~]# pip3.10 install ncclient
Requirement already satisfied: ncclient in /usr/local/lib/python3.10/site-packag
es (0.6.13)
Requirement already satisfied: setuptools>0.6 in /usr/local/lib/python3.10/site-
packages (from ncclient) (63.2.0)
Requirement already satisfied: paramiko>=1.15.0 in /usr/local/lib/python3.10/sit
e-packages (from ncclient) (2.11.0)
Requirement already satisfied: lxml>=3.3.0 in /usr/local/lib/python3.10/site-pac
kages (from ncclient) (4.9.1)
Requirement already satisfied: six in /usr/local/lib/python3.10/site-packages (f
rom ncclient) (1.16.0)
Requirement already satisfied: bcrypt>=3.1.3 in /usr/local/lib/python3.10/site-p
ackages (from paramiko>=1.15.0->ncclient) (3.2.2)
Requirement already satisfied: cryptography>=2.5 in /usr/local/lib/python3.10/si
te-packages (from paramiko>=1.15.0->ncclient) (37.0.4)
Requirement already satisfied: pynacl>=1.0.1 in /usr/local/lib/python3.10/site-p
ackages (from paramiko>=1.15.0->ncclient) (1.5.0)
Requirement already satisfied: cffi>=1.1 in /usr/local/lib/python3.10/site-packa
ges (from bcrypt>=3.1.3->paramiko>=1.15.0->ncclient) (1.15.1)
Requirement already satisfied: pycparser in /usr/local/lib/python3.10/site-packa
ges (from cffi>=1.1->bcrypt>=3.1.3->paramiko>=1.15.0->ncclient) (2.21)
WARNING: Running pip as the 'root' user can result in broken permissions and con
flicting behaviour with the system package manager. It is recommended to use a v
irtual environment instead: https://pip.pypa.io/warnings/venv

[      ] A new release of pip available:        -> 22.2.2
[      ] To update, run: pip install --upgrade pip
[root@localhost ~]#
```

10.4.2　ncclient 实战应用（get_config）

为了方便读者更直观地了解 ncclient 的应用和操作，ncclient 的实战应用将以 Windows 下的 Python 解释器 IDLE 运行演示，并且将使用同样为 IOS-XE（版本为 17.03.01）的思科 CSR1000v 路由器做演示。实战应用部分将分别介绍如何使用 ncclient 的 get_config、edit_config 来获取设备配置、修改设备配置及删除设备配置，本节介绍如何使用 get_config 来获取设备配置。

（1）首先完成模块导入部分，manager 是 ncclient 中最重要的对象，用来帮助我们通过 NETCONF 协议连接设备，而 lxml 模块的作用后面会讲到。

```
Python 3.10.6 (tags/v3.10.6:9c7b4bd, Aug  1 2022, 21:53:49) [MSC v.1932 64 bit (AMD64)]
on win32
Type "help", "copyright", "credits" or "license()" for more information.
from ncclient import manager
from lxml import etree
```

```
IDLE Shell 3.10.6
File Edit Shell Debug Options Window Help
Python 3.10.6 (tags/v3.10.6:9c7b4bd, Aug  1 2022, 21:53:49) [MSC v.1932 64 bit (
AMD64)] on win32
Type "help", "copyright", "credits" or "license()" for more information.
>>> from ncclient import manager
>>> from lxml import etree
>>>
```

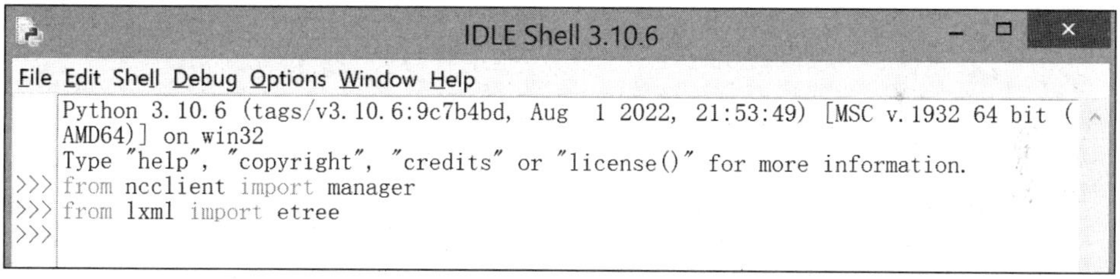

（2）调用 manager 对象下的 connect()函数，连接我们要通过 NETCONF 操控的设备，并赋值给变量 m。顾名思义，connect()函数中的各种参数都能理解。其中 device_params 用来规定设备的类型和操作系统，比如思科的设备就分了 iosxe、csr、iosxr、nxos 等，这里我们要登录的是一台使用 IOS-XE 的 CSR1000v 交换机，所以 device_params 这里放'name'='csr'。另外，有些设备的 running 配置过大，有时通过 NETCONF 的 get_config 操作来获取设备的全部配置时会引起超时的现象，因为默认的 timeout 只有 60s 很容易超时，这里一般都会通过 timeout 改成 300s 比较保险（当然实验环境下没什么配置，或者生产环境下配置很少的设备不用担心这点）。

```
>>> m = manager.connect(host='x.x.x.x',
...                     port=830,
...                     username='***',
...                     password='***',
```

```
...                     hostkey_verify=False,
...                     device_params={'name':'csr'}
...                     timeout=300):
```

连接之后，在设备上输入命令 show user，如果可以看到 NETCONF（ONEP），即代表设备和 Python 主机的 NETCONF 连接成功，如下图所示。

```
    Line       User       Host(s)              Idle       Location
*  2 vty 0     -          idle                 00:00:00

  Interface    User               Mode         Idle     Peer Address
  unknown      NETCONF(ONEP)      com.cisco.ne 00:00:44 0x73a4e004:0x00000000
  unknown      a(ONEP)            com.cisco.sy 00:00:45 0x73a4e008:0x00000000
```

（3）在 10.3.2 节中提到过，通过 NETCONF 手动连接目标设备后，目标设备首先会发一个 hello 包给我们，该 hello 包包含了该设备所具备的所有能力（capability），如下图所示。

```
[root@localhost ~]# ssh -p 830 netconf@172.16.224.28
netconf@172.16.224.28's password:
<?xml version="1.0" encoding="UTF-8"?>
<hello xmlns="urn:ietf:params:xml:ns:netconf:base:1.0">
<capabilities>
<capability>urn:ietf:params:netconf:base:1.0</capability>
<capability>urn:ietf:params:netconf:base:1.1</capability>
<capability>urn:ietf:params:netconf:capability:writable-running:1.0</capability>
<capability>urn:ietf:params:netconf:capability:xpath:1.0</capability>
<capability>urn:ietf:params:netconf:capability:validate:1.0</capability>
<capability>urn:ietf:params:netconf:capability:validate:1.1</capability>
<capability>urn:ietf:params:netconf:capability:rollback-on-error:1.0</capability
>
<capability>urn:ietf:params:netconf:capability:notification:1.0</capability>
<capability>urn:ietf:params:netconf:capability:interleave:1.0</capability>
<capability>urn:ietf:params:netconf:capability:with-defaults:1.0?basic-mode=expl
icit&amp;also-supported=report-all-tagged</capability>
<capability>urn:ietf:params:netconf:capability:yang-library:1.0?revision=2016-06
-21&amp;module-set-id=46bdd15a7205161479645d1423ae664b</capability>
<capability>http://tail-f.com/ns/netconf/actions/1.0</capability>
<capability>http://tail-f.com/ns/netconf/extensions</capability>
```

由于这里是通过 ncclient 连接的设备，因此我们“肉眼”看不到该 hello 包，以及设备的能力集，可以使用 manager 对象的 **server_capabilities** 属性来获取 NETCONF 服务器端的能力集（capabilities）。

```
>>> for capability in m.server_capabilities:
        print (capability)
```

这里调用 manager 的 server_capabilities 属性，该属性返回的是一个可迭代的对象，包含了 NETCONF 服务器的能力集，配合 for 循环即可遍历，并打印出该设备所具备的所有能力，如下图所示。

根据设备型号和操作系统版本的不同，设备返回的能力的数量是不一样的，通常每个设备都会返回数百上千个能力，这里只截取部分以做演示使用，如下图所示。

```
>>> for capability in m.server_capabilities:
...         print (capability)
...
urn:ietf:params:netconf:base:1.0
urn:ietf:params:netconf:base:1.1
urn:ietf:params:netconf:capability:writable-running:1.0
urn:ietf:params:netconf:capability:rollback-on-error:1.0
urn:ietf:params:netconf:capability:validate:1.0
urn:ietf:params:netconf:capability:validate:1.1
urn:ietf:params:netconf:capability:xpath:1.0
urn:ietf:params:netconf:capability:notification:1.0
urn:ietf:params:netconf:capability:interleave:1.0
urn:ietf:params:netconf:capability:with-defaults:1.0?basic-mode=explicit&also-supported=report-all-tagged,report-all
urn:ietf:params:netconf:capability:yang-library:1.0?revision=2016-06-21&module-set-id=83d9b8e0bc0de8d1b825b64384570040
http://tail-f.com/ns/netconf/actions/1.0
http://cisco.com/ns/cisco-xe-ietf-ip-deviation?module=cisco-xe-ietf-ip-deviation&revision=2016-08-10
http://cisco.com/ns/cisco-xe-ietf-ipv4-unicast-routing-deviation?module=cisco-xe-ietf-ipv4-unicast-routing-deviation&r
http://cisco.com/ns/cisco-xe-ietf-ipv6-unicast-routing-deviation?module=cisco-xe-ietf-ipv6-unicast-routing-deviation&r
http://cisco.com/ns/cisco-xe-ietf-ospf-deviation?module=cisco-xe-ietf-ospf-deviation&revision=2018-02-09
http://cisco.com/ns/cisco-xe-ietf-routing-deviation?module=cisco-xe-ietf-routing-deviation&revision=2016-07-09
http://cisco.com/ns/cisco-xe-openconfig-acl-deviation?module=cisco-xe-openconfig-acl-deviation&revision=2017-08-25
http://cisco.com/ns/cisco-xe-openconfig-aft-deviation?module=cisco-xe-openconfig-aft-deviation&revision=2018-12-05
http://cisco.com/ns/cisco-xe-openconfig-isis-deviation?module=cisco-xe-openconfig-isis-deviation&revision=2018-12-05
http://cisco.com/ns/cisco-xe-openconfig-lldp-deviation?module=cisco-xe-openconfig-lldp-deviation&revision=2018-07-25
http://cisco.com/ns/cisco-xe-openconfig-mpls-deviation?module=cisco-xe-openconfig-mpls-deviation&revision=2019-06-27
http://cisco.com/ns/cisco-xe-openconfig-segment-routing-deviation?module=cisco-xe-openconfig-segment-routing-deviation
http://cisco.com/ns/cisco-xe-routing-openconfig-system-management-deviation?module=cisco-xe-routing-openconfig-system-
http://cisco.com/ns/mpls-static/devs?module=common-mpls-static-devs&revision=2015-09-11
```

（4）在 10.2.4 节中，我们将 GitHub 上所有的 YANG 模块都下载到了本地，然后通过记事本或者 cat 等命令直接查看其内容，同样的操作我们可以调用 manager 的 get_schema()函数完成，这里以 Cisco-IOS-XE-interface-common 这个 YANG 模块为例。

```
>>>schema = m.get_schema('Cisco-IOS-XE-interface-common')
>>>print (schema)
```

```
>>> schema = m.get_schema('Cisco-IOS-XE-interface-common')
>>> print (schema)
<?xml version="1.0" encoding="UTF-8"?>
<rpc-reply xmlns="urn:ietf:params:xml:ns:netconf:base:1.0" message-id="urn:uuid:
3b077120-c36f-440c-80a1-16cfcd60f563" xmlns:nc="urn:ietf:params:xml:ns:netconf:b
ase:1.0"><data xmlns='urn:ietf:params:xml:ns:yang:ietf-netconf-monitoring'><![CD
ATA[module Cisco-IOS-XE-interface-common {
  namespace "http://cisco.com/ns/yang/Cisco-IOS-XE-interface-common";
  prefix ios-ifc;

  import cisco-semver {
    prefix cisco-semver;
  }

  organization
    "Cisco Systems, Inc.";
  contact
    "Cisco Systems, Inc.
     Customer Service

     Postal: 170 W Tasman Drive
     San Jose, CA 95134

     Tel: +1 1800 553-NETS

     E-mail: cs-yang@cisco.com";
  description
    "Cisco XE Native Interfaces Common Yang model.
     Copyright (c) 2016-2020 by Cisco Systems, Inc.
     All rights reserved.";

  revision 2020-07-01 {
    description
      "- Added interface type MFR and serial-subinterface";
    cisco-semver:module-version "1.3.0";
  }
  revision 2020-03-01 {
    description
      "- Added Obsolete interface grouping for obsolete interfaces";
```

（5）NETCONF 协议的第三层（操作层）定义了一组用来配置、复制、删除设备命令及获取设备信息的基本操作，基本操作包括 get、get-config、edit-config、copy-config、delete-config、lock、unlock、close-session 和 kill-session。

ncclient 的 manager 对象支持所有上述基本操作，首先来看如何通过 get-config 来获取设备的 running config，方法很简单，两行代码即可搞定。

```
>>> running_config = m.get_config('running')
>>> print (running_config)
```

```
>>> running_config = m.get_config('running')
>>>
>>>
>>> print (running_config)
```


```
<?xml version="1.0" encoding="UTF-8"?>
<rpc-reply xmlns="urn:ietf:params:xml:ns:netconf:base:1.0" message-id="urn:uuid:
81408c29-2f74-4df6-98bf-a06391ca5c26" xmlns:nc="urn:ietf:params:xml:ns:netconf:b
ase:1.0"><data><native xmlns="http://cisco.com/ns/yang/Cisco-IOS-XE-native"><ver
sion>17.3</version><boot-start-marker/><boot-end-marker/><banner><motd><banner>
Welcome to the DevNet Sandbox for CSR1000v and IOS XE

The following programmability features are already enabled:
  - NETCONF
  - RESTCONF

Thanks for stopping by.
</banner></motd></banner><memory><free><low-watermark><processor>71556</processo
r></low-watermark></free></memory><call-home><contact-email-addr xmlns="http://c
isco.com/ns/yang/Cisco-IOS-XE-call-home">sch-smart-licensing@cisco.com</contact-
email-addr><tac-profile xmlns="http://cisco.com/ns/yang/Cisco-IOS-XE-call-home">
<profile><CiscoTAC-1><active>true</active><destination><transport-method>http</t
ransport-method></destination></CiscoTAC-1></profile></tac-profile></call-home><
service><timestamps><debug><datetime><msec></msec></datetime></debug><log><datet
ime><msec/></datetime></log></timestamps><call-home/></service><platform><consol
e xmlns="http://cisco.com/ns/yang/Cisco-IOS-XE-platform"><output>virtual</output
></console><qfp xmlns="http://cisco.com/ns/yang/Cisco-IOS-XE-platform"><utilizat
ion><monitor><load>80</load></monitor></utilization></qfp><punt-keepalive xmlns=
"http://cisco.com/ns/yang/Cisco-IOS-XE-platform"><disable-kernel-core>true</disa
ble-kernel-core></punt-keepalive></platform><hostname>csr1000v-1</hostname><enab
le><secret><type>9</type><secret>$9$GNcSrWbM1PyCqU$9.BCza34ClqbgyABGzRV1v5hjCWha
oN9K.gqxCtcCvE</secret></secret></enable><username><name>developer</name><privil
ege>15</privilege><secret><encryption>9</encryption><secret>$9$oNguEA9um9vRx.$Ms
Dk0DOy1rzBjKAcySWdNjoKcA7GetG9YNnKOs8S67A</secret></secret></username><username>
<name>root</name><privilege>15</privilege><secret><encryption>9</encryption><sec
ret>$9$IRHgr7MCAEFNLk$Zf0rXoRLFhh6gHIxhLPfOC9KzjIWISv4KqkNAR51kmI</secret></secr
et></username><ip><vrf><name>cut-a</name><rd>1:111</rd><route-target><direction>
export</direction><target>65535:1</target></route-target><route-target><directio
n>import</direction><target>65535:1</target></route-target></vrf><domain><name>l
ab.devnetsandbox.local</name></domain><forward-protocol><protocol>nd</protocol><
/forward-protocol><ftp><passive/></ftp><multicast><route-limit xmlns="http://cis
```


很显然，设备返回的 rpc-reply 的内容基本不是一个正常人类可读的，如 10.3.2 节中所提到的，我们可以将该内容复制粘贴到 Code Beautify 的 XML Viewer 里进行美化或者以树形结构展开，如下图所示。

（6）如果你不喜欢这种一大串以 XML 代码展示的 running config，只想获取设备的部分配置和信息，比如 hostname，或者某个端口下的 IP 地址、掩码、description 等内容，这时我们就需要设定一个过滤器来从 running config 中过滤出我们想要的配置及其对应的 XML 命名空间和标签。下面演示如何过滤出设备的 hostname，如下图所示。

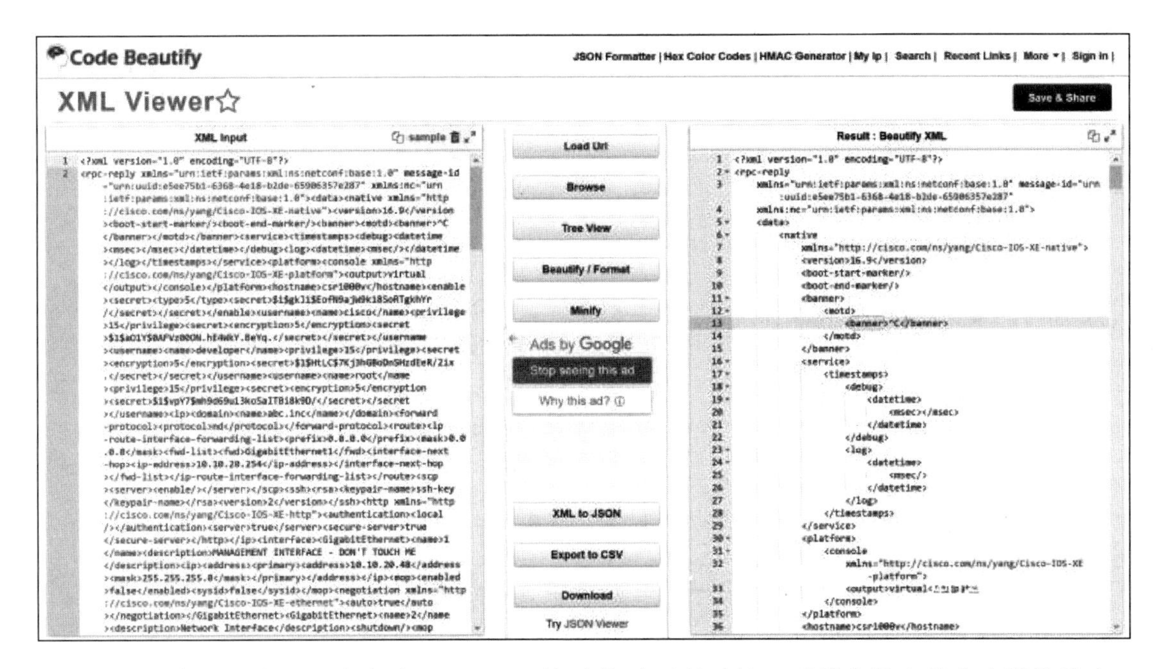

首先创建一个变量，取名为 FILTER，然后将下面所示的三引号字符串的内容赋值给它。

```
>>> FILTER = """
<filter xmlns="urn:ietf:params:xml:ns:netconf:base:1.0">
   <native xmlns="http://cisco.com/ns/yang/Cisco-IOS-XE-native">
        <hostname></hostname>
   </native>
</filter>"""

>>>filtered = m.get_config('running', FILTER)
>>>print (filtered)
```

这里我们使用<filter>这个标签来表示过滤，下面的<native xmlns="http://cisco.com/ns/yang/Cisco-IOS-XE-native">表示思科私有的 YANG 模型，该 XML 命名空间的内容"http://cisco.com/ns/yang/Cisco-IOS-XE-native"在思科所有的 IOS-XE 设备中都通用，因为设备的 hostname 对应的标签为<hostname>，这里我们将它写出来表示我们只想过滤出 hostname 段内容。

然后我们继续调用第 5 步里讲到的 get_config()方法，将 FILTER 作为第二个参数放入，然后赋值给 filtered 变量，将其打印出即可得到过滤后的 rpc-reply 消息，刚才数百行的 running config 现在只显示了两排，得到了我们想要的设备的 hostname，即 csr1000v-1，如下图所示。

```
>>> FILTER = """
... <filter xmlns="urn:ietf:params:xml:ns:netconf:base:1.0">
...     <native xmlns="http://cisco.com/ns/yang/Cisco-IOS-XE-native">
...             <hostname></hostname>
...     </native>
... </filter>"""
...
>>> filtered = m.get_config('running', FILTER)
>>> print (filtered)
<?xml version="1.0" encoding="UTF-8"?>
<rpc-reply xmlns="urn:ietf:params:xml:ns:netconf:base:1.0" message-id="urn:uuid:
2640a089-e533-45d8-a18d-5aefd0fa8ead" xmlns:nc="urn:ietf:params:xml:ns:netconf:b
ase:1.0"><data><native xmlns="http://cisco.com/ns/yang/Cisco-IOS-XE-native"><hos
tname>csr1000v-1</hostname></native></data></rpc-reply>
>>>
```

注：笔者演示所用的 IOS-XE 版本为 17.x.x，如果读者的 IOS-XE 版本为 16.x.x，则<filter xmlns="urn:ietf:params:xml:ns:netconf:base:1.0">写为<filter>即可（下同），如下。

```
>>> FILTER = """
<filter>
    <native xmlns="http://cisco.com/ns/yang/Cisco-IOS-XE-native">
        <hostname></hostname>
    </native>
</filter>"""
```

这时如果我们继续 print type(filtered)，会发现它的数据类型并不是字符串，而是 ncclient 下的一个类，如下图所示。

```
>>> print type(filtered)
<class 'ncclient.operations.retrieve.GetReply'>
>>>
```

如果读者觉得该 rpc-reply 的内容的格式不易读，并且想把该内容转换成字符串，可以调用实验第 1 步讲到的 lxml 函数下的 etree 对象，然后调用 etree 的 tostring()函数，即可将该内容转换成字符串格式，并以更人性化的树形结构展示。

```
>>>filtered_string = etree.tostring(filtered.data, pretty_print=True)
>>>print (filtered_string.decode( 'utf-8'))
```

```
>>> filtered_string = etree.tostring(filtered.data, pretty_print=True)
>>> print (filtered_string.decode('utf-8'))
<data xmlns="urn:ietf:params:xml:ns:netconf:base:1.0" xmlns:nc="urn:ietf:params:
xml:ns:netconf:base:1.0">
  <native xmlns="http://cisco.com/ns/yang/Cisco-IOS-XE-native">
    <hostname>csr1000v-1</hostname>
  </native>
</data>

>>>
```

注：因为 Python 3 中 etree.tostring()函数返回的值为 byte，这里需要对其使用 decode('utf-8')将其转换成字符串，如果是 Python 2 的话，则可以免去 decode('utf-8')，因为 Python 2 中

etree.tostring()函数返回的值为字符串。

（7）如果不希望看到那么多多余的 XML 标签和 XML 命名空间，只希望 NETCONF 把我们真正想要的 hostname（也就是 csr1000v-1）以字符串的形式返回，则加入下面两行代码即可。

```
>>>hostname =
filtered.data.find('.//{http://cisco.com/ns/yang/Cisco-IOS-XE-native}hostname')
>>>print (hostname.text)
```

继续对过滤后的内容调用.data.find()函数，该函数的参数始终以.//开头，后面配上{命名空间}标签的形式来匹配该标签里的实际内容，这里的{http://cisco.com/ns/yang/Cisco-IOS-XE-native}即命名空间，hostname 即标签，我们将结果赋值给变量 hostname，然后调用其属性 text 并将其打印，即得到<hostname></hostname>中的内容，即 csr1000v-1。

```
>>> hostname = filtered.data.find('.//{http://cisco.com/ns/yang/Cisco-IOS-XE-native}
    hostname')
>>> print (hostname.text)
    csr1000v-1
>>> |
```

（8）我们来看如何从 running config 中过滤出 GigabitEthernet 1 端口下的 description、IP 地址及子网掩码 3 个信息。开始之前，首先在 CLI 上确认 GigabitEthernet 1 当前的配置，如下图所示。

```
csr1000v-1#show run int gi1
Building configuration...

Current configuration : 171 bytes
!
interface GigabitEthernet1
 description MANAGEMENT INTERFACE - DON'T TOUCH ME
 ip address 10.10.20.48 255.255.255.0
 negotiation auto
 no mop enabled
 no mop sysid
end
```

同第 6 步一样，首先创建一个 XML 过滤器并将其赋值给 FILTER 变量，调用 get_config()获取目标设备的 running config，将 FILTER 作为第二个参数放入，然后使用 etree.tostring()将其转化为字符串格式后打印出来。

```
>>>FILTER = """
<filter xmlns="urn:ietf:params:xml:ns:netconf:base:1.0">
   <native xmlns="http://cisco.com/ns/yang/Cisco-IOS-XE-native">
```

```
    <interface>
        <GigabitEthernet>
            <name>1</name>
        </GigabitEthernet>
    </interface>
  </native>
</filter>"""
>>> filtered = m.get_config('running', FILTER)
>>> filtered_string = etree.tostring(filtered.data, pretty_print=True)
>>> print (filtered_string.decode('utf-8'))
```

```
>>> filtered = m.get_config('running', FILTER)
>>> filtered_string = etree.tostring(filtered.data, pretty_print=True)
>>> print (filtered_string.decode('utf-8'))
<data xmlns="urn:ietf:params:xml:ns:netconf:base:1.0" xmlns:nc="urn:ietf:params:xml:ns:netconf:base:1.0">
  <native xmlns="http://cisco.com/ns/yang/Cisco-IOS-XE-native">
    <interface>
      <GigabitEthernet>
        <name>1</name>
        <description>MANAGEMENT INTERFACE - DON'T TOUCH ME</description>
        <ip>
          <address>
            <primary>
              <address>10.10.20.48</address>
              <mask>255.255.255.0</mask>
            </primary>
          </address>
        </ip>
        <logging>
          <event>
            <link-status/>
          </event>
        </logging>
        <mop>
          <enabled>false</enabled>
          <sysid>false</sysid>
        </mop>
        <negotiation xmlns="http://cisco.com/ns/yang/Cisco-IOS-XE-ethernet">
          <auto>true</auto>
        </negotiation>
      </GigabitEthernet>
    </interface>
  </native>
</data>

>>>
```

从上面的截图中可以看到，<address>这个标签出现了两次，后面过滤的时候不能再简单地通过 filtered.data.find('.//{http://cisco.com/ns/yang/Cisco-IOS-XE-native}address') 来获取 10.10.20.48 这个 IP 地址了，因为会引起歧义，处理的方法如下。

```
>>> description =
filtered.data.find('.//{http://cisco.com/ns/yang/Cisco-IOS-XE-native}description')
>>> primary =
filtered.data.find('.//{http://cisco.com/ns/yang/Cisco-IOS-XE-native}primary')
>>> ip_addr =
```

```
primary.find('.//{http://cisco.com/ns/yang/Cisco-IOS-XE-native}address')
>>> mask = primary.find('.//{http://cisco.com/ns/yang/Cisco-IOS-XE-native}mask')
>>> print (description.text)
>>> print (ip_addr.text)
>>> print (mask.text)
```

description 部分很好理解，因为<description>标签只出现了一次，所以我们直接用 filtered.data.find('.//{http://cisco.com/ns/yang/Cisco-IOS-XE-native}description')找出即可。

而关于 address 这部分，前面讲到，因为<address>标签出现了两次，直接过滤会引起歧义，解决的方法是调用两次 find()函数，第一次用 filtered.data.find('.//{http://cisco.com/ns/yang/Cisco-IOS-XE-native}primary')找到<primary>标签，将其赋值给 primary 变量，然后再次通过 primary.find('.//{http://cisco.com/ns/yang/Cisco-IOS-XE-native}address')和 primary.find('.//{http://cisco.com/ns/yang/Cisco-IOS-XE-native}mask')分别找出位于<primary>下面的<address>和<mask>即可，如下图所示。

```
>>> description = filtered.data.find('.//{http://cisco.com/ns/yang/Cisco-IOS-XE-native}description')
>>> primary = filtered.data.find('.//{http://cisco.com/ns/yang/Cisco-IOS-XE-native}primary')
>>> ip_addr = primary.find('.//{http://cisco.com/ns/yang/Cisco-IOS-XE-native}address')
>>> mask = primary.find('.//{http://cisco.com/ns/yang/Cisco-IOS-XE-native}mask')
>>> print (description.text)
MANAGEMENT INTERFACE - DON'T TOUCH ME
>>> print (ip_addr.text)
10.10.20.48
>>> print (mask.text)
255.255.255.0
>>> |
```

（9）接下来看如何用 ncclient 一次找出目标设备所有端口的端口号、description、IP 地址及子网掩码。首先来看当前目标设备上所有的端口及其配置，如下图所示。

```
csr1000v-1#show ip int b
Interface              IP-Address      OK? Method Status                Protocol
GigabitEthernet1       10.10.20.48     YES NVRAM  up                    up
GigabitEthernet2       100.1.100.200   YES other  up                    up
GigabitEthernet3       unassigned      YES NVRAM  administratively down down
csr1000v-1#show run int gi1
Building configuration...

Current configuration : 171 bytes
!
interface GigabitEthernet1
 description MANAGEMENT INTERFACE - DON'T TOUCH ME
 ip address 10.10.20.48 255.255.255.0
 negotiation auto
 no mop enabled
 no mop sysid
end
```

```
csr1000v-1#show run int gi2
Building configuration...

Current configuration : 215 bytes
!
interface GigabitEthernet2
 description test
 ip address 100.1.100.200 255.255.255.0
 negotiation auto
 vlan-range dot1q 102 104
 !
 vlan-id dot1q 100
 !
 no mop enabled
 no mop sysid
 spanning-tree portfast
end

csr1000v-1#show run int gi3
Building configuration...

Current configuration : 138 bytes
!
interface GigabitEthernet3
 description Network Interface
 no ip address
 shutdown
 negotiation auto
 no mop enabled
 no mop sysid
end

csr1000v-1#
```

在解释器里输入如下 ncclient 代码。

```
FILTER = """
<filter xmlns="urn:ietf:params:xml:ns:netconf:base:1.0">
    <native
        xmlns="http://cisco.com/ns/yang/Cisco-IOS-XE-native">
        <interface></interface>
    </native>
</filter>"""

filtered = m.get_config('running', FILTER)
filtered_string = etree.tostring(filtered.data, pretty_print=True)
print (filtered_string.decode('utf-8'))
```

因为是要过滤出所有的端口，因此过滤器下直接放<interface></interface>来过滤出所有的端口。这里调用 etree.tostring()将它打印出来，<interface></interface>中间的部分即目标设备现在所拥有的 3 个端口 GigabitEthernet 1、GigabitEthernet 2 和 GigabitEthernet 3 的配置信息，如下图所示。

```
>>> FILTER = """
... <filter xmlns="urn:ietf:params:xml:ns:netconf:base:1.0">
...         <native
...                 xmlns="http://cisco.com/ns/yang/Cisco-IOS-XE-native">
...                 <interface></interface>
...         </native>
... </filter>"""
...
>>> filtered = m.get_config('running', FILTER)
>>> filtered_string = etree.tostring(filtered.data, pretty_print=True)
>>> print (filtered_string.decode('utf-8'))
<data xmlns="urn:ietf:params:xml:ns:netconf:base:1.0" xmlns:nc="urn:ietf:params:xml:ns:netconf:base:1.0">
  <native xmlns="http://cisco.com/ns/yang/Cisco-IOS-XE-native">
    <interface>
      <GigabitEthernet>
        <name>1</name>
        <description>MANAGEMENT INTERFACE - DON'T TOUCH ME</description>
        <ip>
          <address>
            <primary>
              <address>10.10.20.48</address>
              <mask>255.255.255.0</mask>
            </primary>
          </address>
        </ip>
        <logging>
          <event>
            <link-status/>
          </event>
        </logging>
        <mop>
          <enabled>false</enabled>
          <sysid>false</sysid>
        </mop>
        <negotiation xmlns="http://cisco.com/ns/yang/Cisco-IOS-XE-ethernet">
          <auto>true</auto>
        </negotiation>
      </GigabitEthernet>
      <GigabitEthernet>
        <name>2</name>
        <description>test</description>
        <ip>
          <address>
            <primary>
              <address>100.1.100.200</address>
              <mask>255.255.255.0</mask>
            </primary>
          </address>
        </ip>
        <logging>
          <event>
            <link-status/>
          </event>
        </logging>
        <mop>
          <enabled>false</enabled>
          <sysid>false</sysid>
        </mop>
        <negotiation xmlns="http://cisco.com/ns/yang/Cisco-IOS-XE-ethernet">
          <auto>true</auto>
        </negotiation>
        <spanning-tree xmlns="http://cisco.com/ns/yang/Cisco-IOS-XE-spanning-tree">
          <portfast/>
        </spanning-tree>
      </GigabitEthernet>
      <GigabitEthernet>
        <name>3</name>
        <description>Network Interface</description>
        <shutdown/>
        <logging>
          <event>
            <link-status/>
          </event>
        </logging>
        <mop>
          <enabled>false</enabled>
          <sysid>false</sysid>
        </mop>
        <negotiation xmlns="http://cisco.com/ns/yang/Cisco-IOS-XE-ethernet">
          <auto>true</auto>
        </negotiation>
      </GigabitEthernet>
    </interface>
  </native>
</data>
```

接下来继续针对<interface>调用 filtered.data.find('.//{http://cisco.com/ns/yang/Cisco-IOS-XE-native}interface')，并将其赋值给 interface_container 变量，然后使用 iter()函数，该函数返回的值是一个可迭代的对象，方便我们使用 for 循环其中的内容，代码如下。

```
>>> interface_container = filtered.data.find('.//{http://cisco.com/ns/yang/Cisco-IOS-XE-native}interface')
>>> for name in interface_container.iter('{http://cisco.com/ns/yang/Cisco-IOS-XE-native}name'):
      print (name.text)
>>> for address in interface_container.iter('{http://cisco.com/ns/yang/ Cisco-IOS-XE-native}address'):
      print (address.text)
>>> for description in interface_container.iter('{http://cisco.com/ns/yang/Cisco-IOS-XE-native}description'):
   print (description.text)
>>> for mask in
interface_container.iter('{http://cisco.com/ns/yang/Cisco-IOS-XE-native}mask'):
   print (mask.text)
```

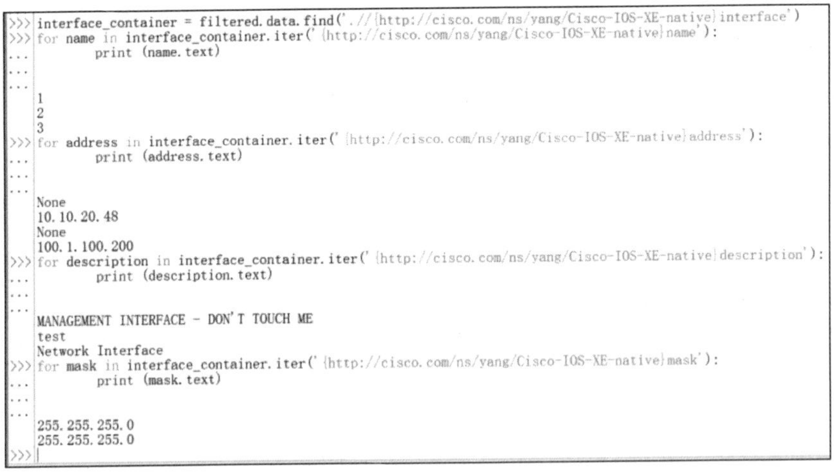

```
>>> interface_container = filtered.data.find('.//{http://cisco.com/ns/yang/Cisco-IOS-XE-native}interface')
>>> for name in interface_container.iter('{http://cisco.com/ns/yang/Cisco-IOS-XE-native}name'):
...         print (name.text)
...
...

1
2
3
>>> for address in interface_container.iter('{http://cisco.com/ns/yang/Cisco-IOS-XE-native}address'):
...         print (address.text)
...
...

None
10.10.20.48
None
100.1.100.200
>>> for description in interface_container.iter('{http://cisco.com/ns/yang/Cisco-IOS-XE-native}description'):
...         print (description.text)
...
...

MANAGEMENT INTERFACE - DON'T TOUCH ME
test
Network Interface
>>> for mask in interface_container.iter('{http://cisco.com/ns/yang/Cisco-IOS-XE-native}mask'):
...         print (mask.text)
...
...

255.255.255.0
255.255.255.0
>>>
```

至于为什么这里打印的每个端口的 IP 地址上都有一个 None，这个问题留给读者自己去思考（提示：答案在第 8 步中可以找到）。

10.4.3　ncclient 实战应用（edit_config）

知道如何使用 get_config 获取设备的配置后，接下来看如何使用 edit_config 来修改和删除设备配置。

（1）首先我们将通过 edit_config 将交换机的主机名从 csr1000v-1 修改为 csr1000v-2，代码如下。

```
CONFIGURATION = """
<config xmlns="urn:ietf:params:xml:ns:netconf:base:1.0">
    <native
        xmlns="http://cisco.com/ns/yang/Cisco-IOS-XE-native">
        <hostname>csr1000v-2</hostname>
    </native>
</config>"""

DATA = m.edit_config(CONFIGURATION, target='running')
print (DATA)
```

我们创建一个名为 CONFIGURATION 的变量，将修改交换机主机名的 XML 语言赋值给它，随后调用 ncclient.manager.connect 下的 edit_config()函数将配置发送给交换机，目标参数为 running（target='running'），也就是思科交换机的 running-configuration，并将回显内容赋值给变量 DATA，最后打印出 DATA，可以看到交换机回复了一个 rpc-reply 包，内容为<ok/>，表示配置顺利完成，如下图所示。

```
>>> CONFIGURATION = """
... <config xmlns="urn:ietf:params:xml:ns:netconf:base:1.0">
...         <native
...                 xmlns="http://cisco.com/ns/yang/Cisco-IOS-XE-native">
...                 <hostname>csr1000v-2</hostname>
...         </native>
... </config>"""
...
>>> DATA = m.edit_config(CONFIGURATION, target='running')
>>> print (DATA)
    <?xml version="1.0" encoding="UTF-8"?>
    <rpc-reply xmlns="urn:ietf:params:xml:ns:netconf:base:1.0" message-id="urn:uuid:
    5fb017d9-20d4-4143-ad2a-34e9b6c1e99f" xmlns:nc="urn:ietf:params:xml:ns:netconf:b
    ase:1.0"><ok/></rpc-reply>
>>>
```

登录交换机验证，可以看到，此时设备的主机名已经从 csr1000v-1 变成了 csr1000v-2，如下图所示。

（2）我们再次通过 edit_config 来修改交换机 GigabitEthernet2 端口的配置，将该端口的 description 改为 Configured by Netconf edit_config，将该端口的 IP 地址配置为 10.10.10.10/24。

在开始实验前，首先看一下当前 GigabitEthernet2 端口的配置，如下图所示。

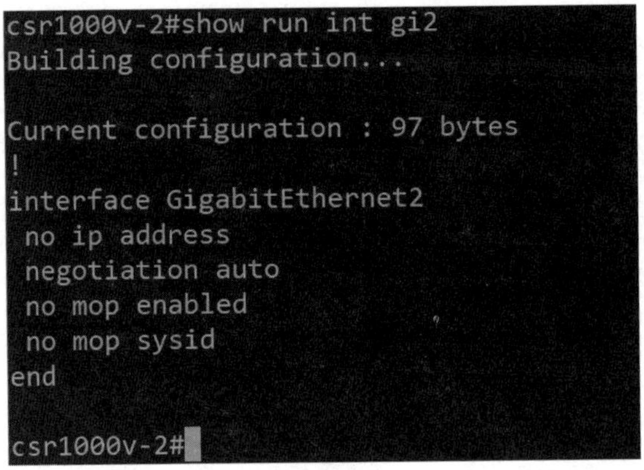

```
csr1000v-2#show run int gi2
Building configuration...

Current configuration : 97 bytes
!
interface GigabitEthernet2
 no ip address
 negotiation auto
 no mop enabled
 no mop sysid
end

csr1000v-2#
```

然后回到 IDLE 解释器，写入下列代码。

```
CONFIGURATION = """
<config xmlns="urn:ietf:params:xml:ns:netconf:base:1.0">
<interfaces
      xmlns="urn:ietf:params:xml:ns:yang:ietf-interfaces">
      <interface>
         <name>GigabitEthernet2</name>
         <description>Configured by Netconf edit_config</description>
         <type
            xmlns:ianaift="urn:ietf:params:xml:ns:yang:iana-if-type">ianaift:
ethernetCsmacd
         </type>
         <enabled>true</enabled>
         <ipv4
            xmlns="urn:ietf:params:xml:ns:yang:ietf-ip">
            <address>
               <ip>10.10.10.10</ip>
               <netmask>255.255.255.0</netmask>
            </address>
         </ipv4>
         <ipv6
            xmlns="urn:ietf:params:xml:ns:yang:ietf-ip">
         </ipv6>
      </interface>
</interfaces>
```

```
</config>"""

DATA = m.edit_config(CONFIGURATION, target='running')
print (DATA)
```

```
>>> DATA = m.edit_config(CONFIGURATION, target='running')
>>> CONFIGURATION = """
... <config xmlns="urn:ietf:params:xml:ns:netconf:base:1.0">
... <interfaces
...         xmlns="urn:ietf:params:xml:ns:yang:ietf-interfaces">
...         <interface>
...             <name>GigabitEthernet2</name>
...             <description>Configured by Netconf edit_config</description>
...             <type
...                 xmlns:ianaift="urn:ietf:params:xml:ns:yang:iana-if-type">ianaift
:ethernetCsmacd
...             </type>
...             <enabled>true</enabled>
...             <ipv4
...                 xmlns="urn:ietf:params:xml:ns:yang:ietf-ip">
...                 <address>
...                     <ip>10.10.10.10</ip>
...                     <netmask>255.255.255.0</netmask>
...                 </address>
...             </ipv4>
...             <ipv6
...                 xmlns="urn:ietf:params:xml:ns:yang:ietf-ip">
...             </ipv6>
...         </interface>
... </interfaces>
... </config>"""
>>> DATA = m.edit_config(CONFIGURATION, target='running')
>>> print (DATA)
<?xml version="1.0" encoding="UTF-8"?>
<rpc-reply xmlns="urn:ietf:params:xml:ns:netconf:base:1.0" message-id="urn:uuid:
9347354e-2d3f-4405-8560-ad98efe6d67d" xmlns:nc="urn:ietf:params:xml:ns:netconf:b
ase:1.0"><ok/></rpc-reply>
>>>
                                                                    Ln: 148 Col
```

这里笔者将 YANG 模型从思科特有的<native xmlns="http://cisco.com/ns/yang/Cisco-IOS-XE-native">改为了标准的 IETF 的 YANG 模型<interfaces xmlns="urn:ietf:params:xml:ns:yang:ietf-interfaces">，目的是向读者演示在 NETCONF 中，我们有多种 YANG 模型可用来获取、修改、删除目标设备的配置。如果要使用思科的 YANG 模型来完成上述操作，代码如下。

```
CONFIGURATION = """
<config xmlns="urn:ietf:params:xml:ns:netconf:base:1.0">
    <native
        xmlns="http://cisco.com/ns/yang/Cisco-IOS-XE-native">
            <interface>
                <GigabitEthernet>
                    <name>2</name>
                    <description> Configured by Netconf edit_config </description>
                    <ip>
                        <address>
                            <primary>
```

```
                            <address>10.10.10.10</address>
                            <mask>255.255.255.0</mask>
                        </primary>
                    </address>
                </ip>
                <logging>
                    <event>
                        <link-status/>
                    </event>
                </logging>
                <mop>
                    <enabled>false</enabled>
                    <sysid>false</sysid>
                </mop>
                <negotiation
                    xmlns="http://cisco.com/ns/yang/Cisco-IOS-XE-ethernet">
                    <auto>true</auto>
                </negotiation>
            </GigabitEthernet>
        </interface>
    </native>
</config>"""

DATA = m.edit_config(CONFIGURATION, target='running')
print (DATA)
```

最后登录交换机做验证，发现此时 GigabitEthernet2 的 description 已经变为 Configured by Netconf edit_config，端口的 IP 地址也被设为了 10.10.10.10 /24，如下图所示。

```
csr1000v-2#show run int gi2
Building configuration...

Current configuration : 167 bytes
!
interface GigabitEthernet2
 description Configured by Netconf edit_config
 ip address 10.10.10.10 255.255.255.0
 negotiation auto
 no mop enabled
 no mop sysid
end

csr1000v-2#
```

（3）我们继续使用 edit_config，通过思科私有的 YANG 模型对交换机配置一个 username，将其取名为 netconf，特权等级设为 15，secret 密码设为 netconf，代码如下。

```
CONFIGURATION = """
<config xmlns="urn:ietf:params:xml:ns:netconf:base:1.0">
    <native
        xmlns="http://cisco.com/ns/yang/Cisco-IOS-XE-native">
    <username>
     <name>netconf</name>
     <privilege>15</privilege>
     <secret>
          <encryption>0</encryption>
          <secret>netconf</secret>
     </secret>
    </username>
    </native>
</config>"""

DATA = m.edit_config(CONFIGURATION, target='running')
print (DATA)
```

```
>>> CONFIGURATION = """
... <config xmlns="urn:ietf:params:xml:ns:netconf:base:1.0">
...     <native
...         xmlns="http://cisco.com/ns/yang/Cisco-IOS-XE-native">
...     <username>
...         <name>netconf</name>
...         <privilege>15</privilege>
...         <secret>
...                 <encryption>0</encryption>
...                 <secret>netconf</secret>
...         </secret>
...     </username>
...     </native>
... </config>"""
...
>>>
>>> DATA = m.edit_config(CONFIGURATION, target='running')
>>> print (DATA)
    <?xml version="1.0" encoding="UTF-8"?>
    <rpc-reply xmlns="urn:ietf:params:xml:ns:netconf:base:1.0" message-id="urn:uuid:
    0bef145d-5167-499f-8e58-762cf01cfb36" xmlns:nc="urn:ietf:params:xml:ns:netconf:b
    ase:1.0"><ok/></rpc-reply>
>>>
```

配置完成后，登录交换机验证，如下图所示。

```
csr1000v-1#show run | i username
username developer privilege 15 secret 9 $9$oNguEA9um9vRx.$MsDk0DOy1rzBjKAcySWdNjoKcA7GetG9YNnKOs8S67A
username root privilege 15 secret 9 $9$IRHgr7MCAEFNLk$ZfOrXoRLFhh6gHIxhLPfOC9KzjIWISv4KqkNAR51kmI
username netconf privilege 15 secret 9 $14$KTP6$TSI4Qer3W4iJi.$mqedqyc8/ZjbebTktBoghYnBJBSMtz25RUlpp8a6kQM
csr1000v-1#
```

（4）最后我们使用 edit_config，继续使用思科的私有 YANG 模型将刚才配置的 netconfg 这个 username 从交换机中删除，代码如下。

```
CONFIGURATION = """
<config xmlns="urn:ietf:params:xml:ns:netconf:base:1.0">
    <native
        xmlns="http://cisco.com/ns/yang/Cisco-IOS-XE-native">
    <username operation='delete'>
     <name>netconf</name>
     <privilege>15</privilege>
     <secret>
          <encryption>0</encryption>
          <secret>netconf</secret>
     </secret>
    </username>
    </native>
</config>"""

DATA = m.edit_config(CONFIGURATION, target='running')
print (DATA)
```

```
>>> CONFIGURATION = """
... <config xmlns="urn:ietf:params:xml:ns:netconf:base:1.0">
...     <native
...         xmlns="http://cisco.com/ns/yang/Cisco-IOS-XE-native">
...     <username operation='delete'>
...         <name>netconf</name>
...         <privilege>15</privilege>
...         <secret>
...                 <encryption>0</encryption>
...                 <secret>netconf</secret>
...         </secret>
...     </username>
...     </native>
... </config>"""
>>> DATA = m.edit_config(CONFIGURATION, target='running')
>>> print (DATA)
<?xml version="1.0" encoding="UTF-8"?>
<rpc-reply xmlns="urn:ietf:params:xml:ns:netconf:base:1.0" message-id="urn:uuid:
7ca5b60b-f5f4-4c25-85ad-babca12b74f0" xmlns:nc="urn:ietf:params:xml:ns:netconf:b
ase:1.0"><ok/></rpc-reply>
>>>
```

可以看到，通过 edit_config 删除交换机 username 的方法很简单，只需要在<username>后加上一个 operatoin='delete'参数即可。最后登录设备做验证，如下图所示。

```
csr1000v-1#show run | i username
username developer privilege 15 secret 9 $9$oNguEA9um9vRx.$MsDk0DOy1rzBjKAcySWdNjoKcA7GetG9YNnKOs8S67A
username root privilege 15 secret 9 $9$IRHgr7MCAEFNLk$ZfOrXoRLFhh6gHIxhLPfOC9KzjIWISv4KqkNAR51kmI
csr1000v-1#
```

10.5 NETCONF 实验（华为设备）

前面已对 NETCONF 基础知识及思科设备进行了讲解，下面通过对一台华为设备进行讲解，提供一种国产设备切入 NETCONF 的思路。这种思路不限于某个厂商的某个具体型号。

就国产设备而言，NETCONF 与网工的传统运维逻辑有点相左。本节试图从一名初学者的视角，切入这块内容的学习。虽然按惯例依然以华为某设备拟写文章，但更多的是分享方法，其他厂商的其他型号设备，也大体如此！相对于传统 CLI 屏幕抓取的自动化，在 NETCONF 自动化方面，如何能有效地检索对应的产品手册尤为重要。

10.5.1 实验拓扑

本节选取的华为 CE12800 交换机在现网还算主流，在 ENSP 上也有其身影（尽管部分新功能可能开发还不到位，不稳定）。如果把其并入本书惯用的实验拓扑，则如下图所示。

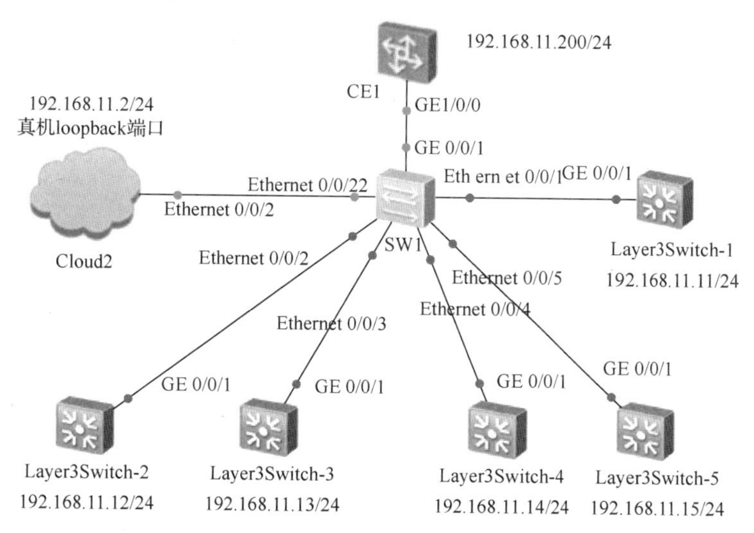

NETCONF 服务端：CE1（192.168.11.200/24），即 NETCONF Agent。

NETCONF 客户端：PC（192.168.11.2/24），即 NETCONF Client。

NETCONF 服务端和客户端在局域网内点对点直连。

10.5.2　实验目的

（1）在 CE12800 华为交换机探索 NETCONF 功能，搭建实验环境。

（2）使用 Python 的 ncclient 模块进行设备的简单联动。

10.5.3　启动 NETCONF

1. 资料准备

打开华为 HedEx Lite（传说中的“海带丝”），进入“文档管理”，然后开始用“12800”的关键字搜索，结合 disp ver 查到的版本信息 V200R005C10 下载一个最新的（如果版本点差异，就找一个最接近的。如果没有这些工具，就用产品文档手册，即设备的说明书。其他国产品牌设备也如此）。

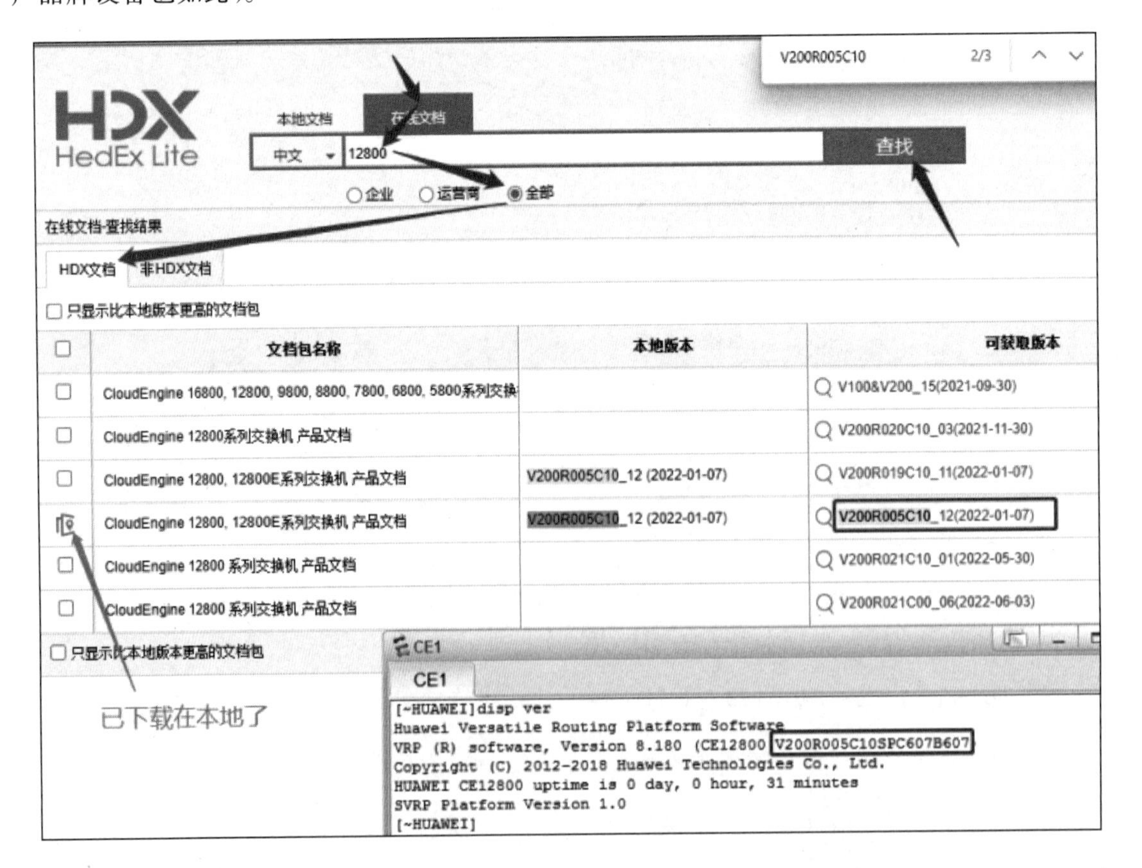

打开以后通常可以搜索一下“NETCONF”，看到“NETCONF 简介”，如下图所示。

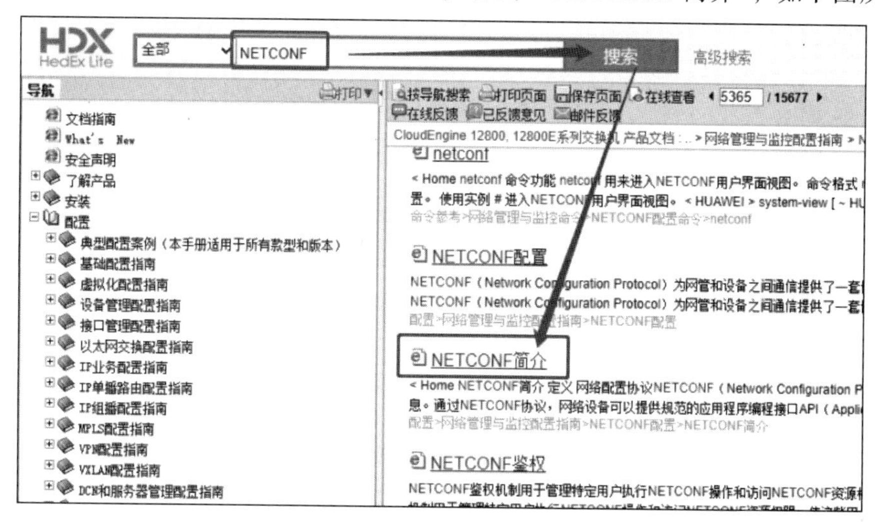

点进去，即可以在导航栏中看到 NETCONF 相关的目录结构，如下图所示。

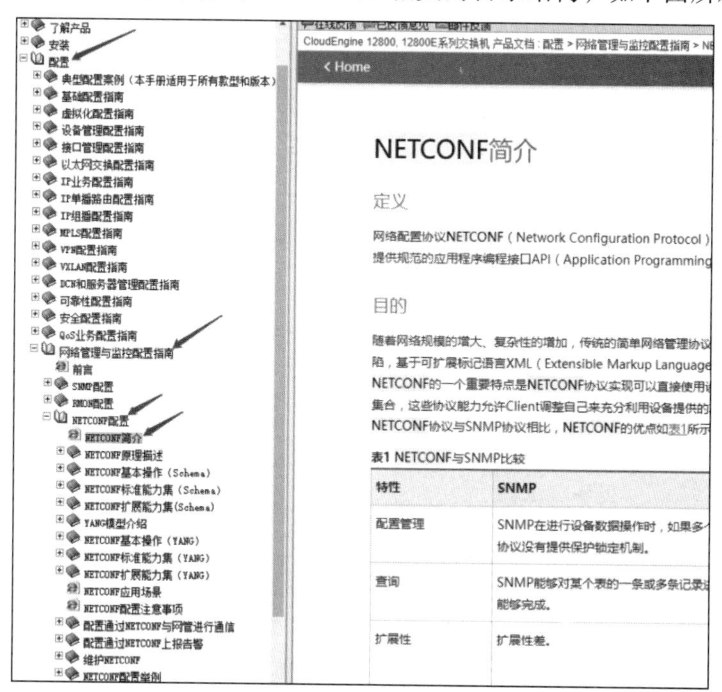

换一台华为设备如此，换成华三、中兴、锐捷等设备也如此，找到产品手册结合相关 RFC 的通用 NETCONF 内容，检索阅读产品文档，进行“知识扫盲”。此后，看到一些“配置举例”等相关字眼，即可实现最小启动实践。

2. CE1 连通性

```
sys
sysname CE1
interface GE1/0/0
 undo portswitch
 undo shutdown
 ip address 192.168.11.200 255.255.255.0
commit
```

从 PC（Windows，已安装 Python）上可 ping 通 CE1 时，网络连通性问题即解决。

3. CE1 SSH 配置

NETCONF 实际依赖 SSH，我们需要开启 SSH 服务器功能，并配置一个 SSH 账号。

```
aaa
 local-user python password irreversible-cipher Admin@123
 local-user python service-type ssh
 local-user python level 3

user-interface vty 0 4
 authentication-mode aaa
 protocol inbound all
commit

#SSH 配置
stelnet server enable
ssh authentication-type default password
rsa local-key-pair create
ssh server publickey  rsa
ssh user python
ssh user python authentication-type password
ssh user python service-type all
ssh authorization-type default aaa
commit
```

SSH 终端如 CRT 手工可登录即可。如果现网使用，则可以配置专门的 NETCONF 账号。

4. 开启 NETCONF 功能

```
[~CE1]snetconf server enable
Info: Succeeded in starting the SNETCONF server on SSH port 22.
```

```
[*CE1]com
[~CE1]
```

查看手册，还可以使用 830 专用端口，读者可以自行尝试，这里我们继续使用 22 端口。

5. 检查 NETCONF 功能

以下检查命令均摘自产品手册。

```
display ssh user-information #可查看到 SSH 用户信息
display ssh server status     #可查看到 SSH 服务器全局配置信息（限于篇幅，请自行尝试）
display netconf capability   #可查看到 CE1 支持的能力（限于篇幅，请自行尝试）
display netconf session      #查看 CE1 与 PC 之间活动的 NETCONF 会话信息（现在还没有）
```

6. 联动 API 文档

开启对应功能后，需要检索到对应的 API 信息，依然是产品手册，如下图所示。

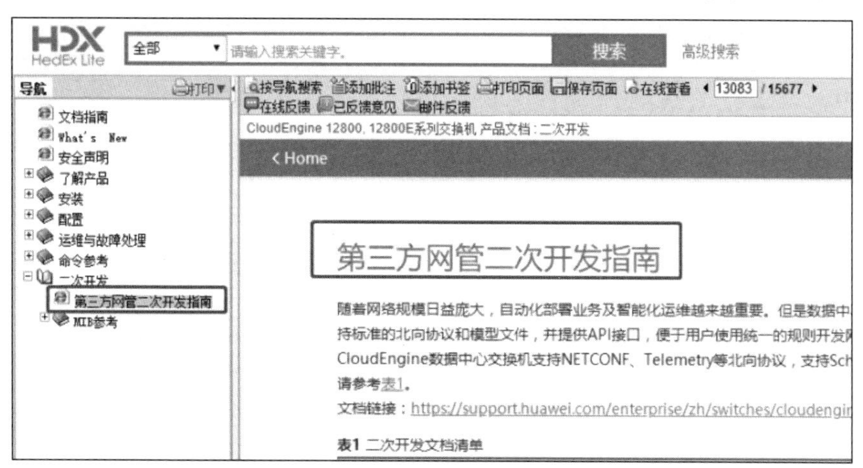

如下图所示，可以下载对应的文档。

文档链接：https://support.huawei.com/enterprise/zh/switches/cloudengine-12800-16800-pid-252837173?category=developer-documents

表1 二次开发文档清单

文档类型	文档清单
二次开发指南	NETCONF Schema API二次开发指南
	Telemetry二次开发指南
API参考	NETCONF Schema API参考
	NETCONF YANG API参考
	RESTful API参考

把相关资料下载到本地后，里面会有一些示例，依然跟着示例做。比如，《CloudEngine 12800, 12800E V200R005C10 NETCONF Schema API 二次开发指南》（以下简称“二次开发指南”）也介绍 ncclient 模块，即可转入 ncclient，如下图所示。

10.5.4　联动 ncclient

ncclient 联机的操作，依然还是“增删改查”系列动作，直接看二次开发指南中的例子，如下图所示。

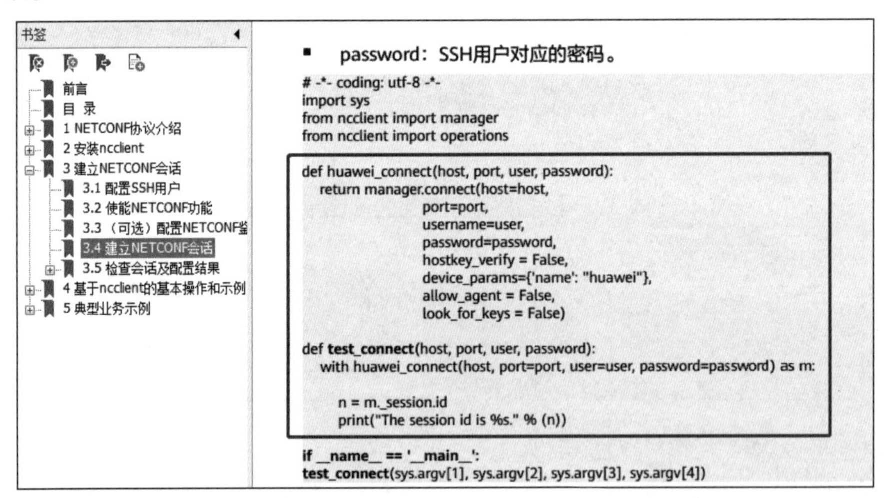

二次开发指南给的案例还是相对复杂些的，因为它要考虑逻辑完整性和规范性。从初学角度出发，我们可以简化一下！Python 脚本的 netconf_connect.py 文件内容如下。

```
#netconf_connect.py
from ncclient import manager


host = '192.168.2.200'
port = '22'
```

```
user = 'python'
password = 'Admin@1234'

def huawei_connect(host, port, user, password):
    return manager.connect(host=host,
                        port=port,
                        username=user,
                        password=password,
                        hostkey_verify = False,
                        device_params={'name': "huawei"},
                        allow_agent = False,
                        look_for_keys = False)

m = huawei_connect(host, port, user, password)
```

运行代码，结果如下图所示。

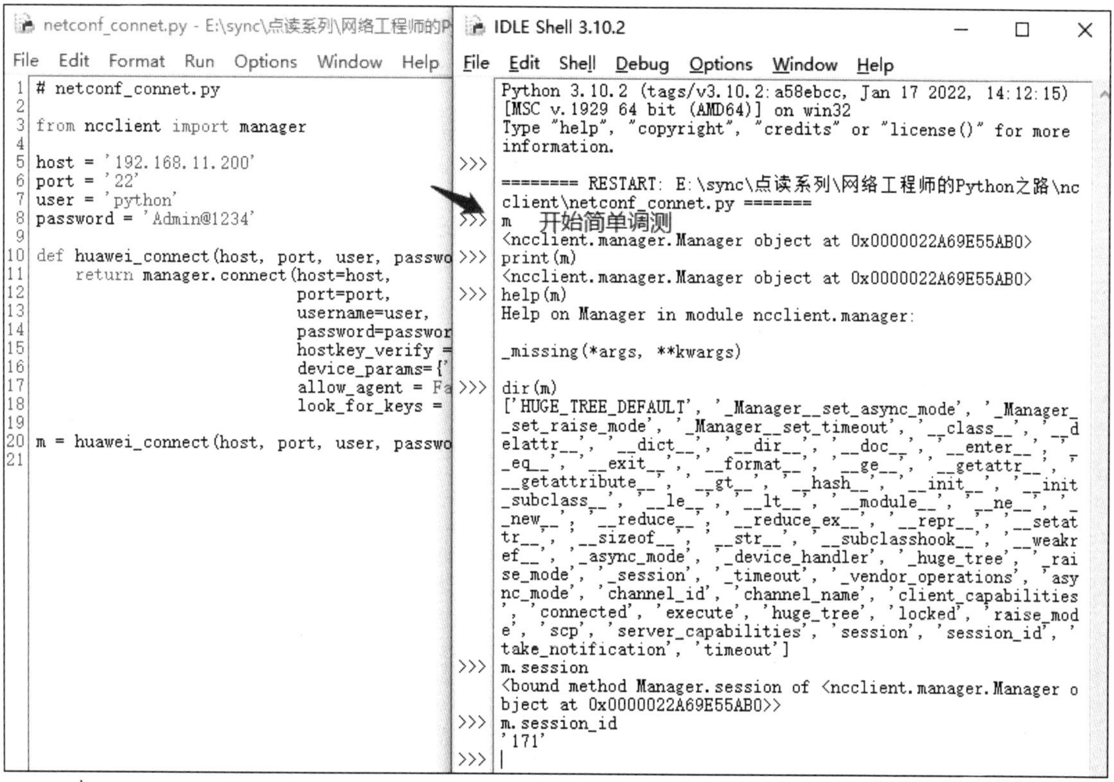

```
netconf_connet.py - E:\sync\点读系列\网络工程师的P
File  Edit  Format  Run  Options  Window  Help
1  # netconf_connet.py
2
3  from ncclient import manager
4
5  host = '192.168.11.200'
6  port = '22'
7  user = 'python'
8  password = 'Admin@1234'
9
10 def huawei_connect(host, port, user, passwo
11     return manager.connect(host=host,
12                            port=port,
13                            username=user,
14                            password=passwor
15                            hostkey_verify =
16                            device_params={'
17                            allow_agent = Fa
18                            look_for_keys =
19
20 m = huawei_connect(host, port, user, passwo
21
```

```
IDLE Shell 3.10.2
File  Edit  Shell  Debug  Options  Window  Help
Python 3.10.2 (tags/v3.10.2:a58ebcc, Jan 17 2022, 14:12:15)
[MSC v.1929 64 bit (AMD64)] on win32
Type "help", "copyright", "credits" or "license()" for more
information.
>>>
======== RESTART: E:\sync\点读系列\网络工程师的Python之路\nc
client\netconf_connet.py =======
>>> m    开始简单调测
<ncclient.manager.Manager object at 0x0000022A69E55AB0>
>>> print(m)
<ncclient.manager.Manager object at 0x0000022A69E55AB0>
>>> help(m)
Help on Manager in module ncclient.manager:

_missing(*args, **kwargs)

>>> dir(m)
['HUGE_TREE_DEFAULT', '_Manager__set_async_mode', '_Manager_
_set_raise_mode', '_Manager__set_timeout', '__class__', '__d
elattr__', '__dict__', '__dir__', '__doc__', '__enter__', '_
_eq__', '__exit__', '__format__', '__ge__', '__getattr__', '
__getattribute__', '__gt__', '__hash__', '__init__', '__init
_subclass__', '__le__', '__lt__', '__module__', '__ne__', '_
_new__', '__reduce__', '__reduce_ex__', '__repr__', '__setat
tr__', '__sizeof__', '__str__', '__subclasshook__', '__weakr
ef__', '_async_mode', '_device_handler', '_huge_tree', '_rai
se_mode', '_session', '_timeout', '_vendor_operations', 'asy
nc_mode', 'channel_id', 'channel_name', 'client_capabilities
', 'connected', 'execute', 'huge_tree', 'locked', 'raise_mod
e', 'scp', 'server_capabilities', 'session', 'session_id', '
take_notification', 'timeout']
>>> m.session
<bound method Manager.session of <ncclient.manager.Manager o
bject at 0x0000022A69E55AB0>>
>>> m.session_id
'171'
>>> |
```

对初学者来说，最安全的就是“查”，获取设备信息，因此我们从这里切入。我们再查产品手册，这里已经很清楚地说明了步骤 1 做什么、步骤 2 做什么，如下图所示。虽然，我们此时可能还不知道这些到底是做什么的，不过可以依葫芦画瓢。

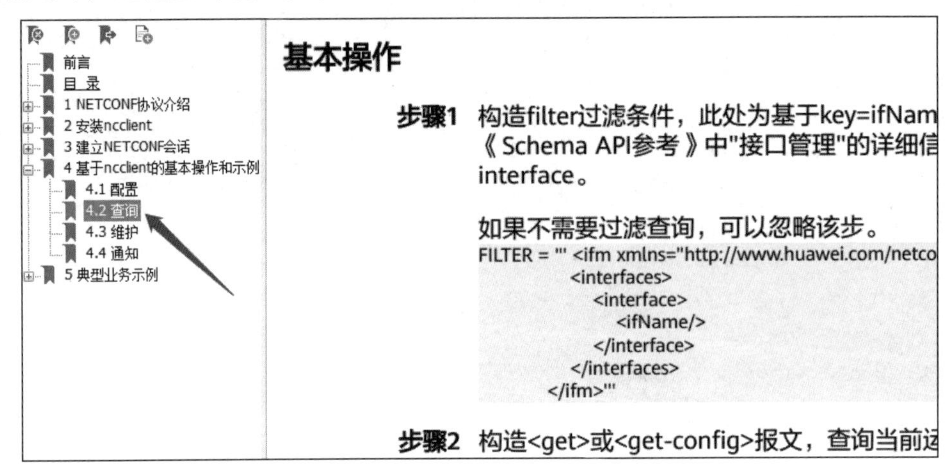

增加一部分 Python 代码，把查询的内容加进去。手册中使用到查端口，再次按手册上的示例，将代码进行精简。

```
#netconf_connect_get.py

from ncclient import manager

host = '192.168.2.200'
port = '22'
user = 'python'
password = 'Admin@1234'

def huawei_connect(host, port, user, password):
    return manager.connect(host=host,
                           port=port,
                           username=user,
                           password=password,
                           hostkey_verify = False,
                           device_params={'name': "huawei"},
                           allow_agent = False,
                           look_for_keys = False)
```

```
m = huawei_connect(host, port, user, password)


FILTER = ''' <ifm xmlns="http://www.huawei.com/netconf/vrp" content-version="1.0"
format-version="1.0">
<interfaces>
<interface>
<ifName/>
</interface>
</interfaces>
</ifm>'''

#get_reply1 或 get_reply2 选一个就好
get_reply1 = m.get(("subtree", FILTER))
#get_reply2 = m.get_config(source='running',filter=("subtree",FILTER))

print(get_reply1)
#m.close  #不用上下文 with 管理器的话，我们最后 close
```

运行一下看看结果。没错，成功了！这就是最小启动，如下图所示。

```
def huawei_connect(host, port, user, password):
    return manager.connect(host=host,
                           port=port,
                           username=user,
                           password=password,
                           hostkey_verify = False,
                           device_params={'name': "huawei"},
                           allow_agent = False,
                           look_for_keys = False)

m = huawei_connect(host, port, user, password)


FILTER = ''' <ifm xmlns="http://www.huawei.com/netconf/vrp" c
<interfaces>
<interface>
<ifName/>
</interface>
</interfaces>
</ifm>'''

# get_reply1 或 get_reply2 选一个就好
get_reply1 = m.get(("subtree", FILTER))
#get_reply2 = m.get_config(source='running',filter=("subtree"

print(get_reply1)
#m.close  # 不用上下文 with 管理器的话，我们最后close一下
```

```
</interface>
<interface>
  <ifName>GE1/0/0</ifName>
</interface>
<interface>
  <ifName>GE1/0/1</ifName>
</interface>
<interface>
  <ifName>GE1/0/2</ifName>
</interface>
<interface>
  <ifName>GE1/0/3</ifName>
</interface>
<interface>
  <ifName>GE1/0/4</ifName>
</interface>
<interface>
  <ifName>GE1/0/5</ifName>
</interface>
<interface>
  <ifName>GE1/0/6</ifName>
</interface>
<interface>
  <ifName>GE1/0/7</ifName>
</interface>
<interface>
  <ifName>GE1/0/8</ifName>
</interface>
<interface>
  <ifName>GE1/0/9</ifName>
```

当最小启动完成后，改成 830 端口，检查端口 IP 地址、板卡信息等就可以配合产品手册和 API 手册逐步开展起来，包括其他“增删改查”，请读者自行尝试。

从 NETCONF 概念出发到设备 ncclient 成功联动，需要参考的资料很多，如何有效地检索、提炼资料是一种自学能力，需要平时逐步内化。有时候会走弯路，但不要怕，要做的就是静下来“啃”资料。初学阶段，重要的不是深挖过多的细节，而是提纲挈领，实验最小启动。实验成功启动后，就会有成就感，随之可在某个点位上“抛锚作业”，进行深挖。慢慢地，这个过程就形成了“正反馈”。有了正反馈后，学习，实践，提升，再学习，再实践，再提升，这便是所谓的“螺旋上升”。如果这期间再配合梳理分享，则效果会更佳！

第 11 章

RESTCONF 详解

作为很常见的 API，RESTCONF 和 NETCONF 在 NetDevOps 领域的地位可以说是并驾齐驱的，本章将分为两部分来介绍 RESTCONF 的应用，第一部分用 Postman 这个 API 工具实战操作 RESTCONF，第二部分用 Python 的 requests 模块来讲解，以实验的方式带读者快速入手 RESTCONF 的实际使用。

11.1　RESTCONF 简介

在介绍 RESTCONF 之前，先来回顾一下 NETCONF 的知识点，因为两者关系密切。

之前在介绍 NETCONF 的时候，我们知道在 NETCONF 中定义了数据集（datastore）的概念，数据集的作用是存储一份配置数据的备份，确保设备能从开机时的初始状态进入它能正常运行时的工作状态（类似于改了配置后需要 write memory，不然重启设备后配置会丢失）。而针对数据集的操作又引入了 CRUD 这一概念，所谓 CRUD 是 Create、Read、Update、Delete 4 种操作的首字母缩写，用来表示数据集支持创建、读取、更新、删除 4 种操作。除此之外，在 NETCONF 的基础上，IETF 又在 RFC 6020 中提出了 YANG 这个专为 NETCONF 打造的建模语言，YANG 语言定义了数据存储内容、配置、状态数据、RPC 操作，以及事件通知的标准，为 CRUD 访问的数据集里的数据制定了统一的格式。

而根据 RFC 8040 的定义，所谓 RESTCONF 就是通过 HTTP 协议对采用 YANG 模型的“概念上的数据集”（conceptual datastore）进行 CRUD 操作，理论上 RESTCONF 和 NETCONF 做的是同样的事（消息层都基于 RPC 协议），只是实现的方法不同，NETCONF 基于 SSH 协议，而 RESTCONF 则基于 HTTP 协议。

另外，RESTCONF 相较于 NETCONF 有一个最大的优势是它的输出值的数据类型为 JSON，和 NETCONF 输出的 XML 数据类型比起来更易读。

11.1.1　HTTP 方法和 CRUD 的对应关系

HTTP 中总共有 GET、POST、PUT、PATCH 及 DELETE 5 种请求方式，它们与 CRUD 的对应关系及在 NetDevOps 中的应用如下表所示。

HTTP方法	CRUD操作	NetDevOps中的应用举例
GET	READ	读取交换机配置
POST	CREATE	添加交换机配置
PUT	UPDATE	替换交换机配置
PATCH	UPDATE	更改交换机配置
DELETE	DELETE	删除交换机配置

11.1.2　Postman

Postman 是目前最知名的 RESTCONF 工具之一，其图标如右图所示。最早的作者是印度人 Abhinav Asthana，其创建 Postman 的初衷是打造一个简化 API 测试功能的工具。后来 Postman 一炮而红，Abhinav 的前同事 Ankit Sobti 和 Abhijit Kane 也加入他的团队。现在 Postman 已经拿到了 INSIGHT PARTNERS 2.25 亿美元的 D 轮融资，总部设在美国旧金山，两个分部设在印度班加罗尔，是当前最火热的 API 工具之一，思科的 DevNet 认证教程中关于 RESTCONF 的内容都是基于 Postman 进行演示和讲解的。

Postman 的使用是免费的（也有付费版，只是作为网络工程师的我们暂时用不上），支持 Windows、Linux 和 MacOS 等主流操作系统，读者可以根据自身情况在其官网下载。

除 Postman 外，也可以使用 curl、HTTPie 及 Python 的 requests 模块对网络设备进行基于 RESTCONF 的 CRUD 操作，本节将介绍 RESTCONF 在 Postman 中的使用方式，下节将介绍如何通过 requests 模块在 Python 中使用 RESTCONF。

11.2　RESTCONF 实验（Postman）

11.2.1　实验环境

实验设备：思科 Catalyst 9300 交换机（真机）1 台

交换机型号：C9300-48U

交换机操作系统版本：IOS-XE 16.12.4 （思科从 IOS-XE 16.3 开始支持 RESTCONF，更早的版本不支持）

主机操作系统：Windows 8.1

Postman 版本：9.0.9

11.2.2　交换机初始配置

思科 IOS-XE 设备开启 RESTCONF 非常简单，但是在有 AAA（TACACS 或者 RADIUS）的环境里有一些额外的配置需要完成，这一步非常卡人，思科从 IOS-XE 16.3 才开始支持 RESTCONF，目前就连思科 TAC 都不提供技术支持，网上包括思科官方给的资料都很少、很模糊，如下图所示。

RESTCONF

RESTCONF provides a means to programmatically interact with a device - in a model-based, machine-consumable, easy to understand and standards-based way. RESTCONF is defined by an RFC as of the time of the IOS-XE 16.3 release. RESTCONF is described by the IETF Draft located here

As of IOS-XE 16.3, RESTCONF is available in the code, but as of yet not supported by Cisco-TAC. The CLI is also hidden from the parser.

The structure of data exchanged using the RESTCONF interface is defined (in advance) using YANG models. Management systems using YANG can directly access managed resources in a single operation. Familiar REST verbs are included with RESTCONF, like GET, POST, PUT, PATCH, etc.

For background on model based management, please see Model-Based Management Introduction.

这里笔者将给出 RESTCONF 在使用本地用户的非 AAA 环境和使用 AAA 的 TACACS 或 RADIUS 的环境下的初始配置。

◎　本地用户（非 AAA 环境）

```
config terminal
RESTCONF
ip http server
ip http secure-server
username test privilege 15 secret test
```

在思科 IOS-XE 上启用 RESTCONF 很简单，只需要输入命令 RESTCONF 启用 RESTCONF，开启 HTTP 和 HTTPS 服务，然后配置一个特权级别为 15 的用户名即可。

◎　AAA 环境（假设使用思科的 ISE）

```
config terminal
```

```
RESTCONF
ip http server
ip http secure-server

aaa group server tacacs+ ISE
aaa authentication login default group ISE local
aaa authentication login NOAUTH none
aaa authorization exec default group ISE local
aaa session-id common
```

这里 aaa authentication login default 和 aaa authorization exec default 这两个思科默认自带的 authentcation list 及 authorization list 是非常容易忽略的配置，请务必确保你的 AAA 配置中有它们，不然使用 RESTCONF 的时候会一直卡在因为权限不够而禁止访问的地方。

11.2.3　Postman 初始配置

鉴于不少读者是初次使用 Postman，下面给出详细的使用步骤。

（1）第一次进入 Postman 后，在 Start with something new 下点击 Create New，如果提示注册用户名，请按需完成（注册用户名是免费的），如下图所示。

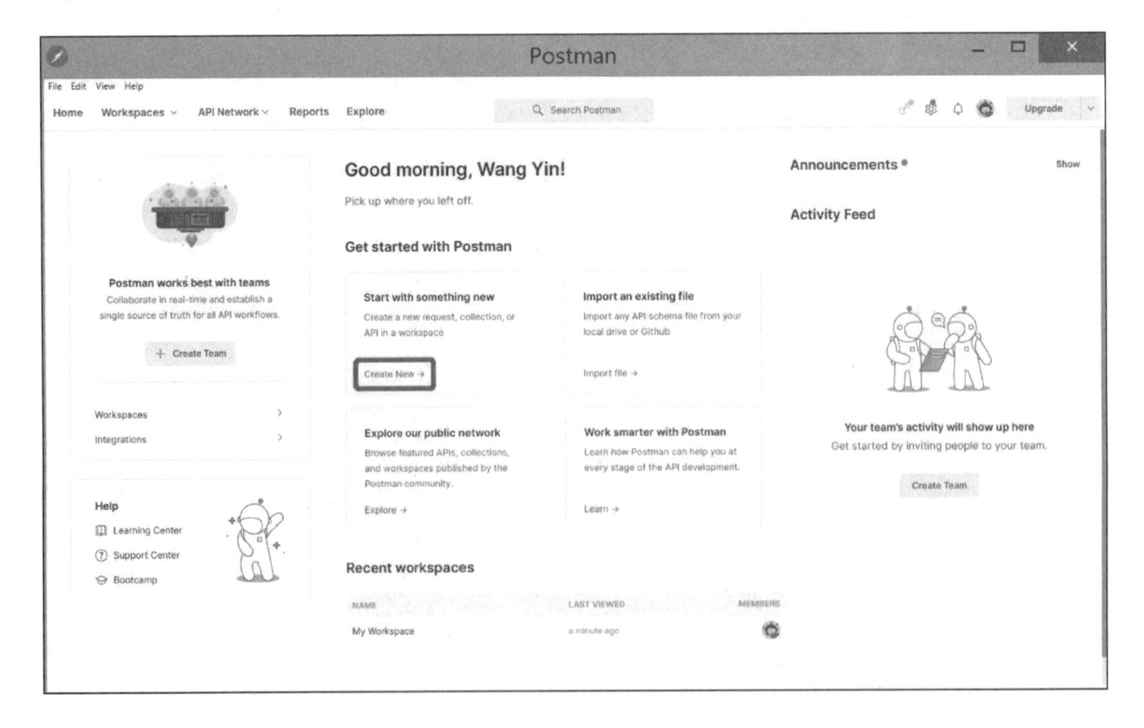

（2）选择 HTTP Request，如下图所示。

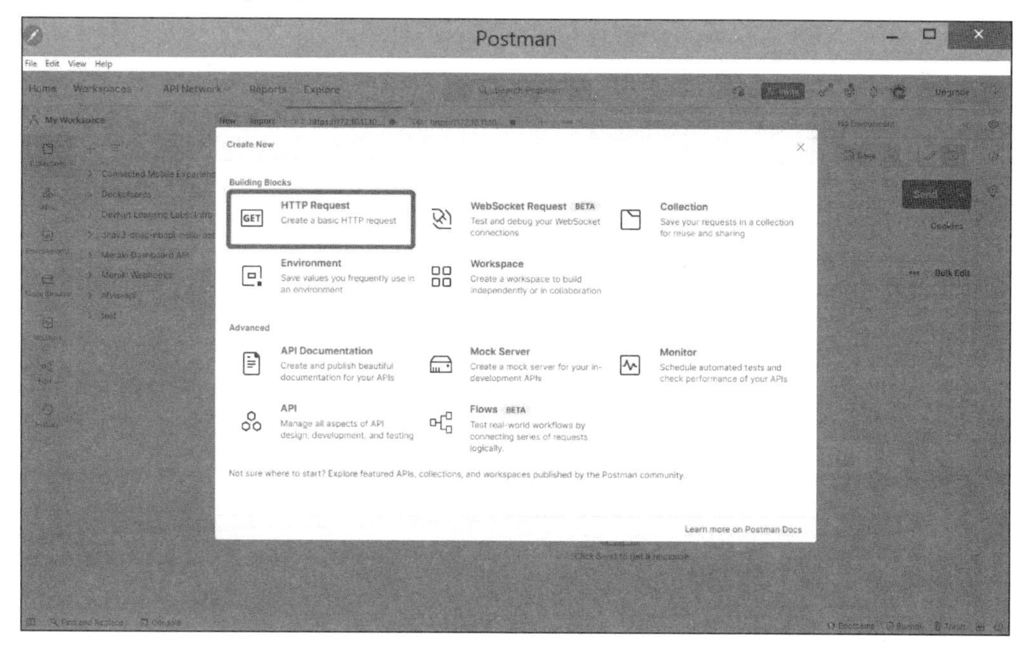

（3）进入后，你会看见一个叫作 GET Untitled Request 的任务标签，点击该标签下面的 Headers，添加 Content-Type: application/yang-data+json 和 Accept: application/yang-data+json 两组键值对，如下图所示。

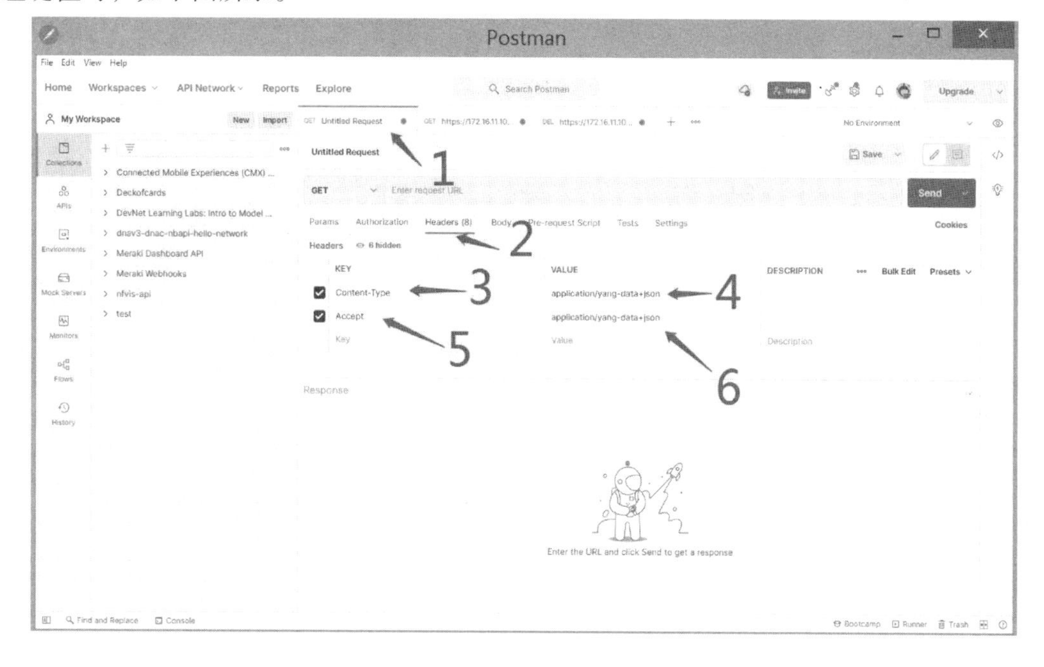

（4）点击 Authorization，在 Type 处选择 Basic Auth，如下图所示。

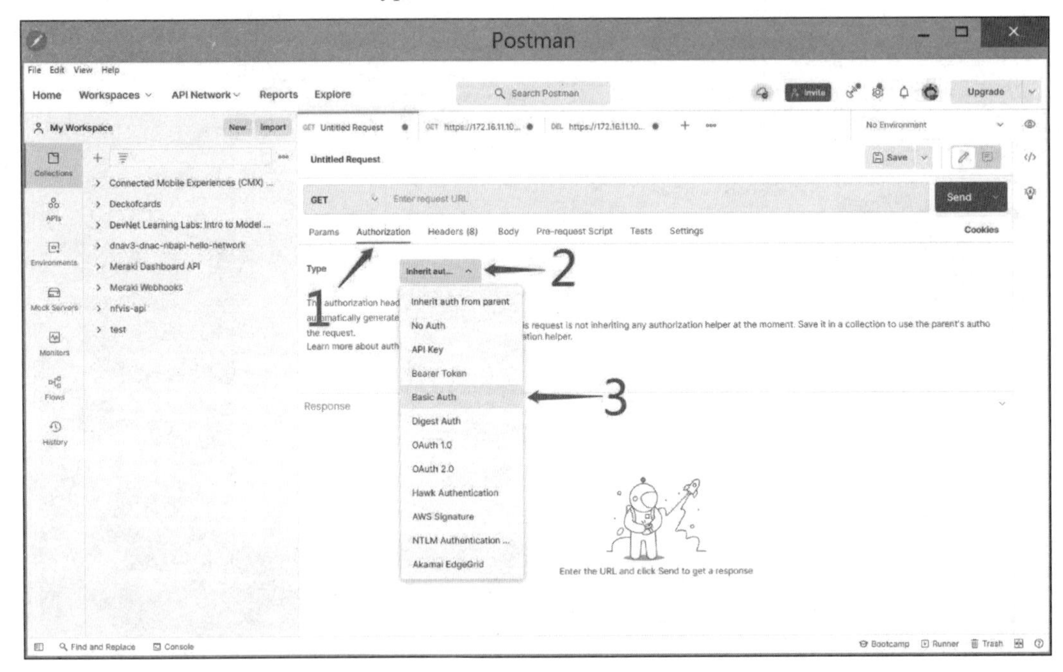

（5）之后根据自己的环境按需输入交换机本地用户名和密码或者 TACACS/RADIUS 用户名和密码，如下图所示。

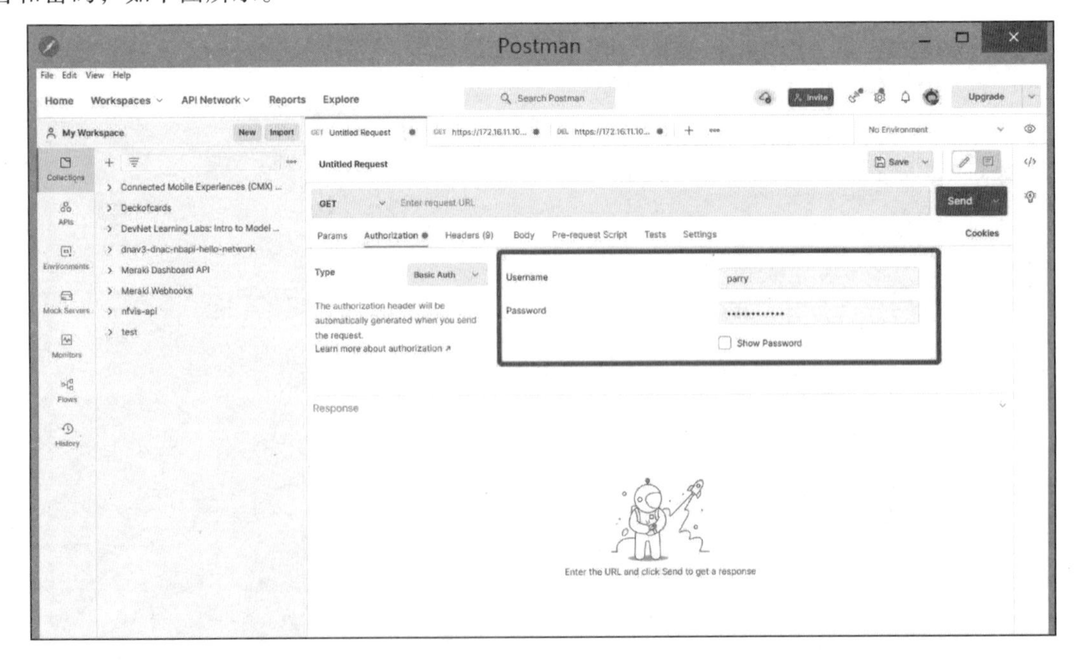

11.2.4 通过 GET 方法获取交换机配置

Postman 初始配置完毕后，首先我们来看下如何在 Postman 上通过 RESTCONF 来获取交换机的 running 配置（也就是命令行里的 show run）。

首先确保你在 Postman 中选择的是 GET 方法，然后在地址栏里输入下面的地址。

```
https://xx.xx.xx.xx/RESTCONF/data/Cisco-IOS-XE-native:native
```

这里用到了思科的 Cisco-IOS-XE-native:native 的 YANG 模型。单击 Send 按钮后，在下面的 Body 部分可以看到以 JSON 格式返回的交换机的 running 配置，如下图所示。

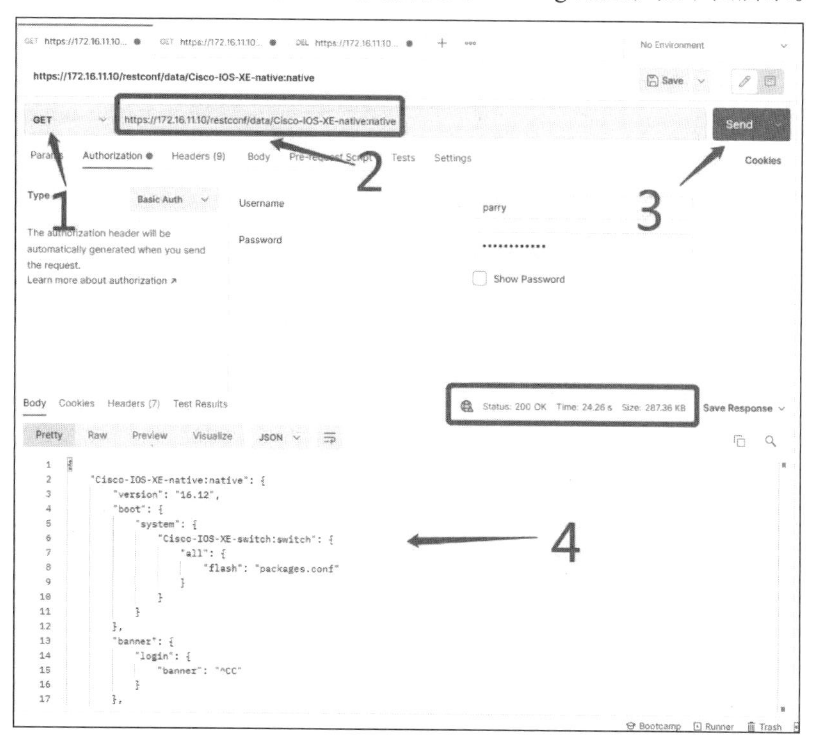

注意，通过 RESTCONF 的 GET 方法获取交换机 running 配置的耗时会根据交换机 running 配置文件的大小而异，笔者的 9300 真机因为配置较多，所以花了 24.26s，如果你们的 running 配置文件也较大，那么请耐心等待，不要误认为是 Postman “卡住了”。

接着我们还可以修改 URL，继续通过 GET 方法配合 Cisco-IOS-XE-native:native 这个 YANG 模型来提取交换机的主机名、操作系统版本信息、所有端口、VLAN 等信息，方法如下。

1. 获取交换机主机名

```
https://xx.xx.xx.xx/RESTCONF/data/Cisco-IOS-XE-native:native/hostname
```

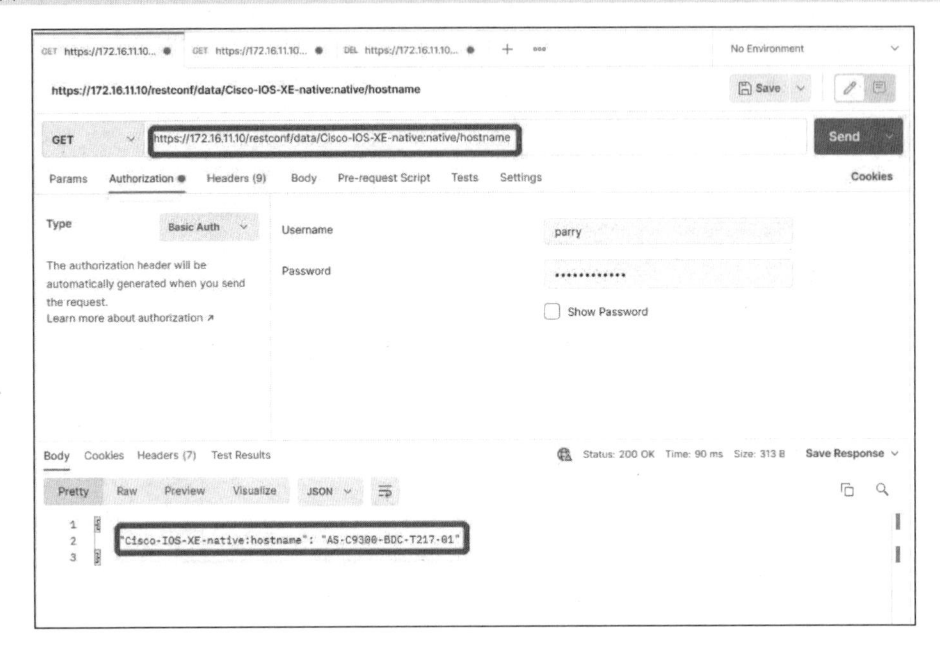

2. 获取交换机操作系统版本信息

```
https://xx.xx.xx.xx/RESTCONF/data/Cisco-IOS-XE-native:native/version
```

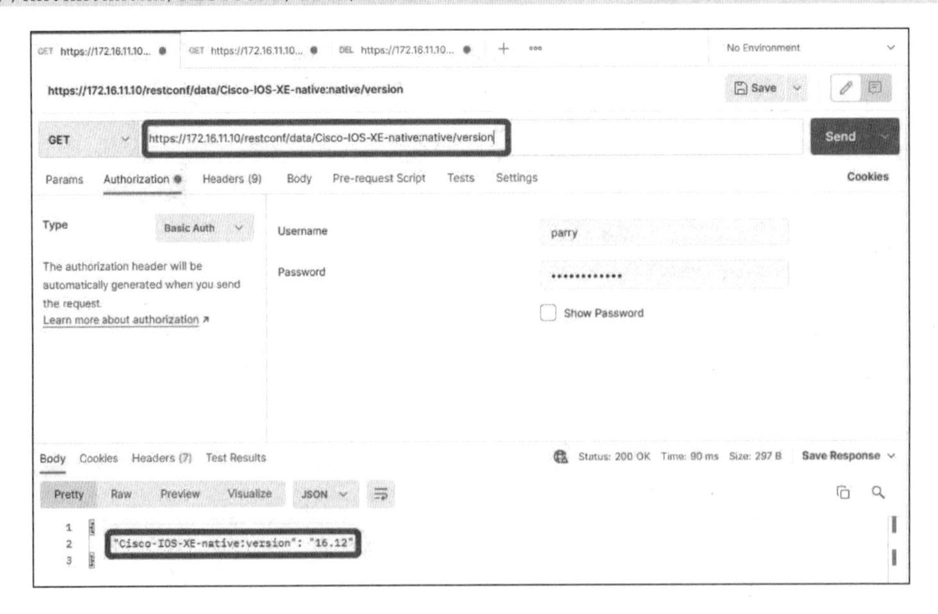

3. 获取交换机端口信息

```
https://xx.xx.xx.xx/RESTCONF/data/Cisco-IOS-XE-native:native/interface
```

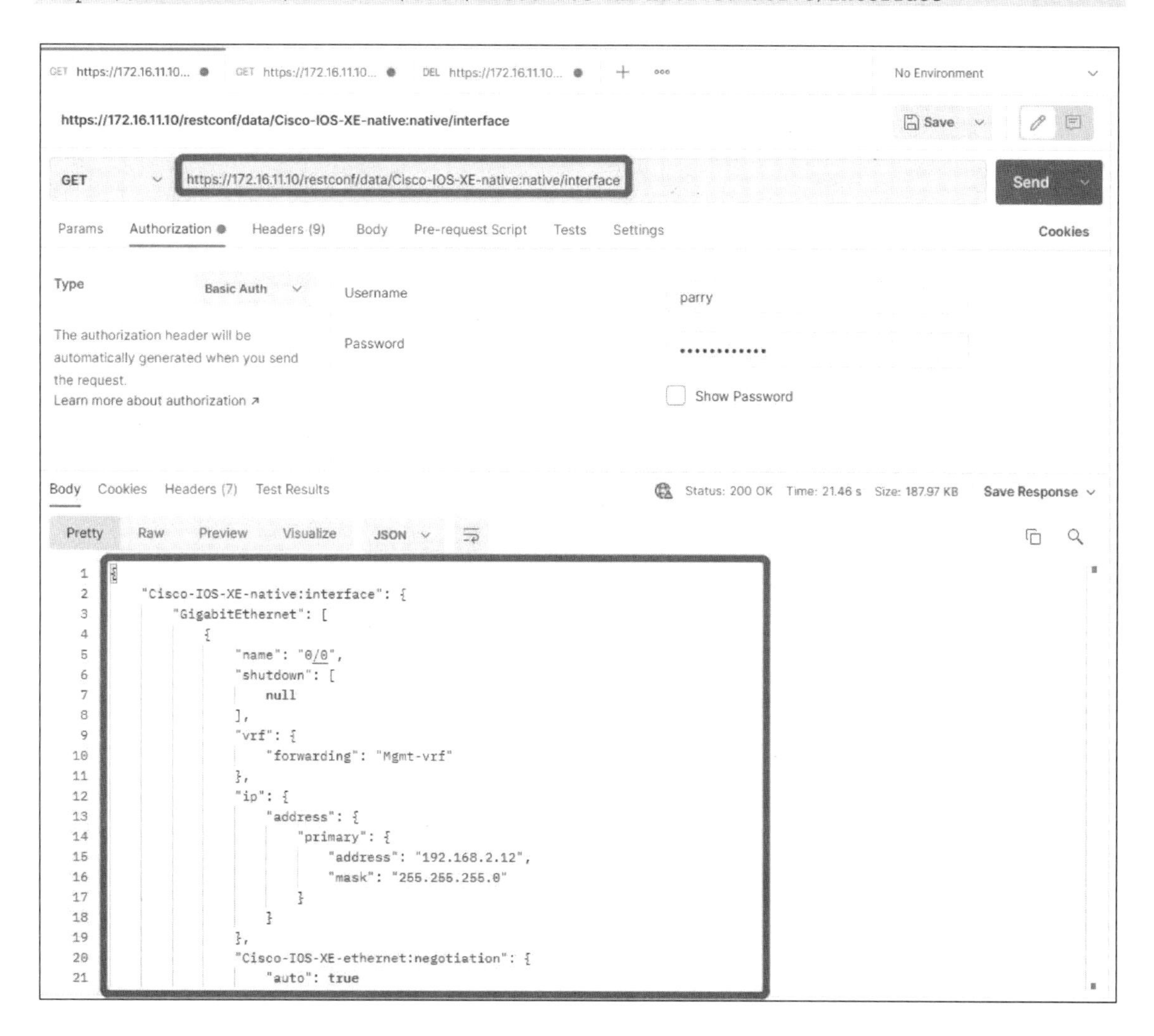

4. 获取交换机 VLAN

```
https://xx.xx.xx.xx/RESTCONF/data/Cisco-IOS-XE-native:native/vlan
```

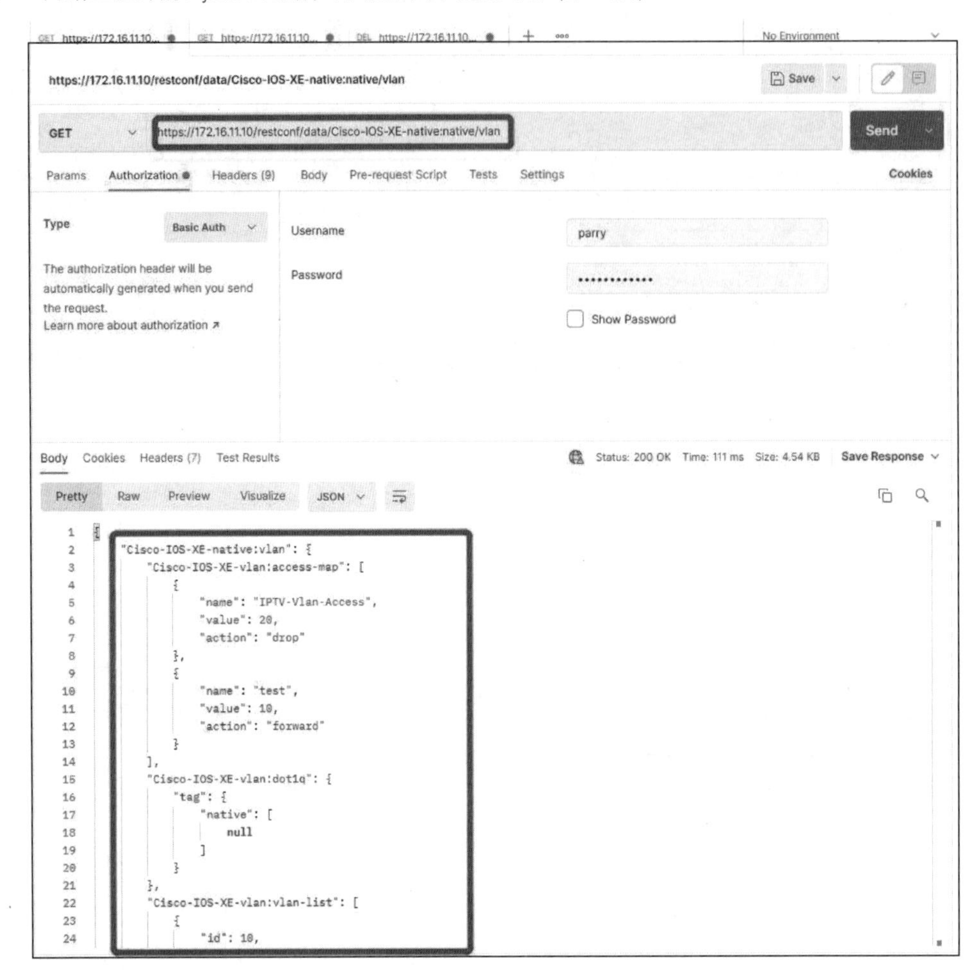

5. 使用?depth=修改深度

我们知道 YANG 模型是树形结构的，Cisco-IOS-XE-native:native 这个思科私有的 YANG 模型自然也不例外。上面我们获取的 running 配置内容过多，返回的以 JSON 格式展现的配置一层套了一层又一层，比如 Cisco-IOS-XE-native:native 下面有 version、boot、banner 等，而 boot 下面又有 system，system 下面又有 Cisco-IOS-XE-switch:switch，Cisco-IOS-XE-switch:switch 下面又有 all 等，有些时候我们其实并不需要如此详尽的信息。

RESTCONF 支持在 URL 里使用?depth=来修改深度，?depth=1 表示只返回一层的内容，?depth=2 表示返回两层的内容，依此类推。我们尝试对交换机的 running 配置使用?depth=1，效果如下图所示。

```
https://xx.xx.xx.xx/RESTCONF/data/Cisco-IOS-XE-native:native?depth=1
```

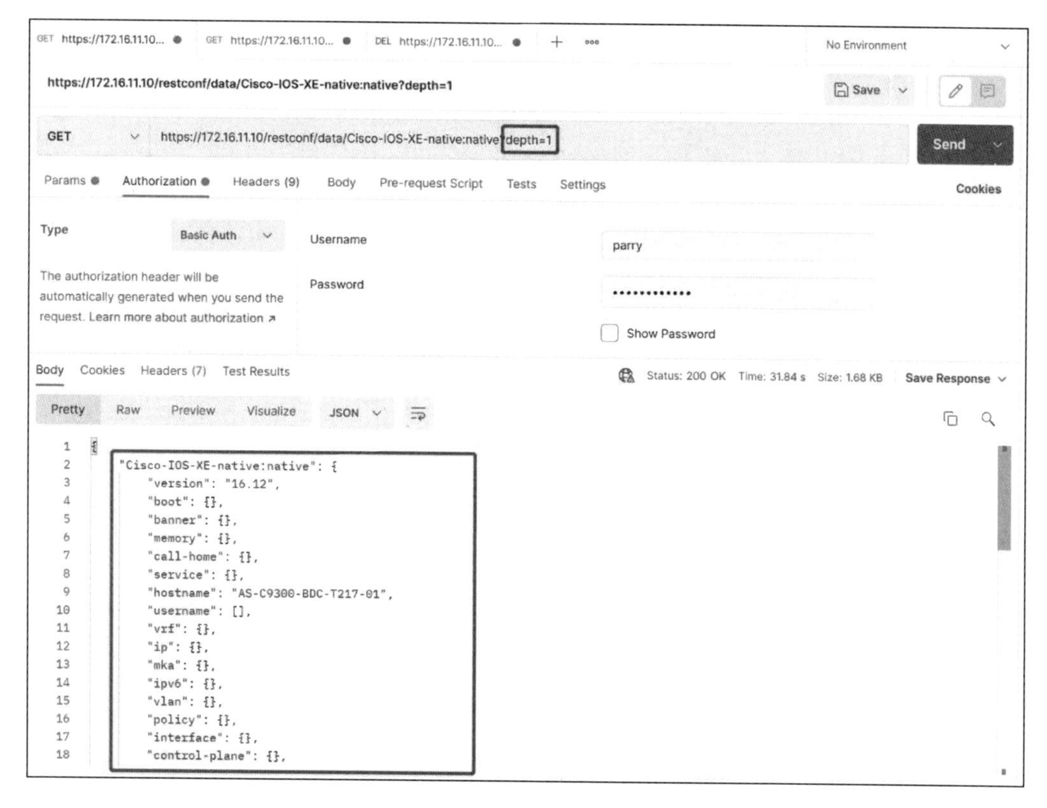

至于?depth=2、?depth=3 等就留给读者自行去尝试。

11.2.5　通过 PATCH 方法更改交换机配置

前面在介绍 CRUD 和 HTTP 的 5 种方法对比时已经提到，PATCH 和 PUT 都可以用来为交换机做配置，它们的区别是前者是用来“更改”配置的，而后者是用来“替换”配置。

这里以 9300 交换机中的 Gi0/0 端口为对象，来具体演示 PATCH 和 PUT 两者的区别，看下到底要怎么理解“更改”和“替换”。

首先来看 Gi0/0 端口当前的配置，可以看到该端口目前是没有 IP 地址的，如下图所示。

```
AS-C9300-BDC-T217-01#show run int gi0/0
Building configuration...

Current configuration : 104 bytes
!
interface GigabitEthernet0/0
 vrf forwarding Mgmt-vrf
 no ip address
 shutdown
 negotiation auto
end
```

手动为它添加一个 IP 地址：192.168.2.11，如下图所示。

```
AS-C9300-BDC-T217-01#conf t
Enter configuration commands, one per line.  End with CNTL/Z.
AS-C9300-BDC-T217-01(config)#int gi0/0
AS-C9300-BDC-T217-01(config-if)#ip add 192.168.2.11 255.255.255.0
AS-C9300-BDC-T217-01(config-if)#end
AS-C9300-BDC-T217-01#show run int gi0/0
Building configuration...

Current configuration : 128 bytes
!
interface GigabitEthernet0/0
 vrf forwarding Mgmt-vrf
 ip address 192.168.2.11 255.255.255.0
 shutdown
 negotiation auto
end
```

回到 Postman，通过以下 GET 方法得到 Gi0/0 配置 IP 地址后的状态。

```
https://x.x.x.x/RESTCONF/data/Cisco-IOS-XE-native:native/interface/GigabitEthernet=
0%2F0
```

注意，这里我们通过在 https://x.x.x.x/RESTCONF/data/Cisco-IOS-XE-native:native/interface 后加上“/GigabitEthernet=0%2F0”过滤出 Gi0/0 端口。**在 RESTCONF 中，分隔符/需要用%2F 来表示，因此 GigabitEthernet=0/0 需要写成 Gigabitethernet=0%2F0**，切记，如下图所示。

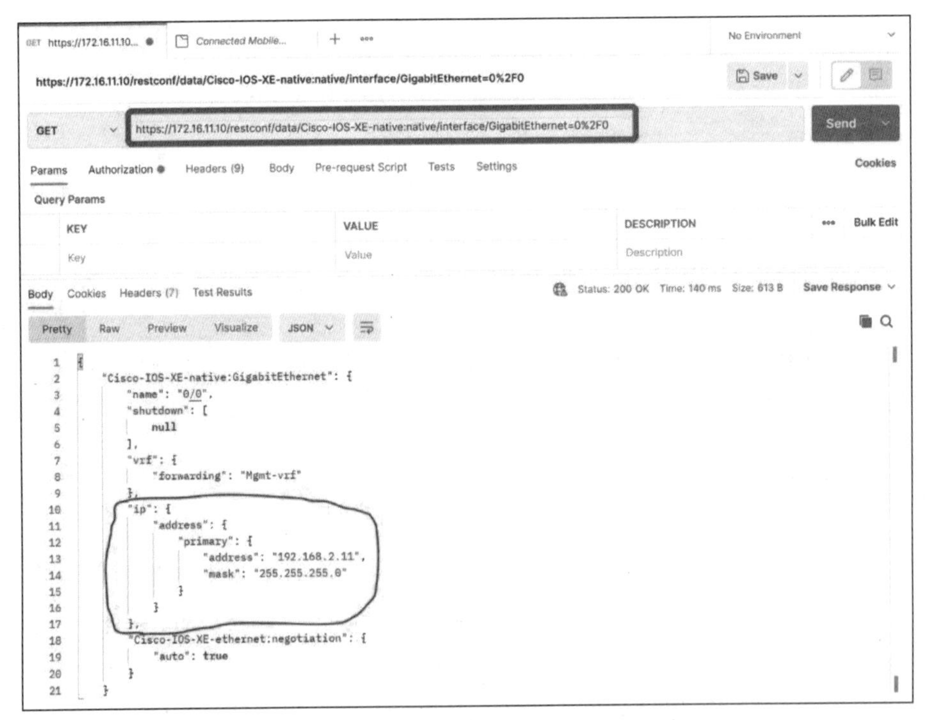

然后我们将上面执行 GET 方法得到的这段 Gi0/0 端口的 JSON 代码全选并复制，如下图所示。

Body　Cookies　Headers (7)　Test Results

Pretty　Raw　Preview　Visualize　JSON

```json
{
    "Cisco-IOS-XE-native:GigabitEthernet": {
        "name": "0/0",
        "shutdown": [
            null
        ],
        "vrf": {
            "forwarding": "Mgmt-vrf"
        },
        "ip": {
            "address": {
                "primary": {
                    "address": "192.168.2.11",
                    "mask": "255.255.255.0"
                }
            }
        },
        "Cisco-IOS-XE-ethernet:negotiation": {
            "auto": true
        }
    }
}
```

接下来在 Postman 里依次将 GET 方法改为 PATCH，然后单击 Body，选择 raw，最后将刚才复制的 JSON 代码粘贴到 raw 下的代码框里，如下图所示。

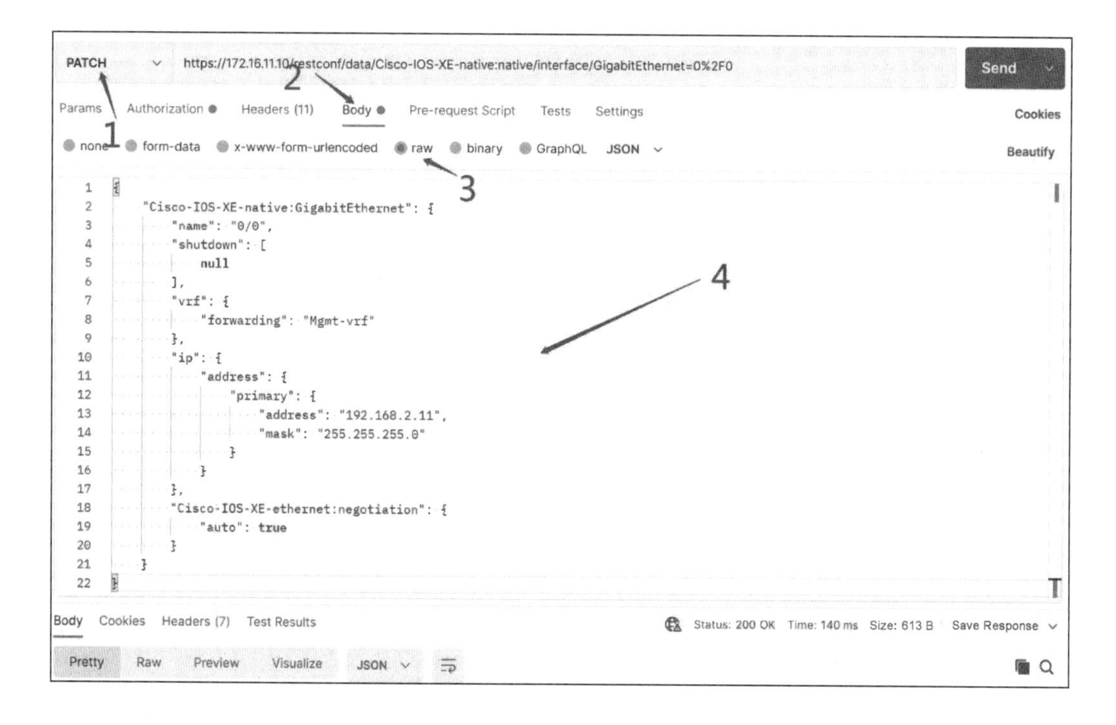

接着我们在 raw 代码框里面将 192.168.2.11 改为 192.168.2.12，然后单击 Send 按钮，如下图所示。

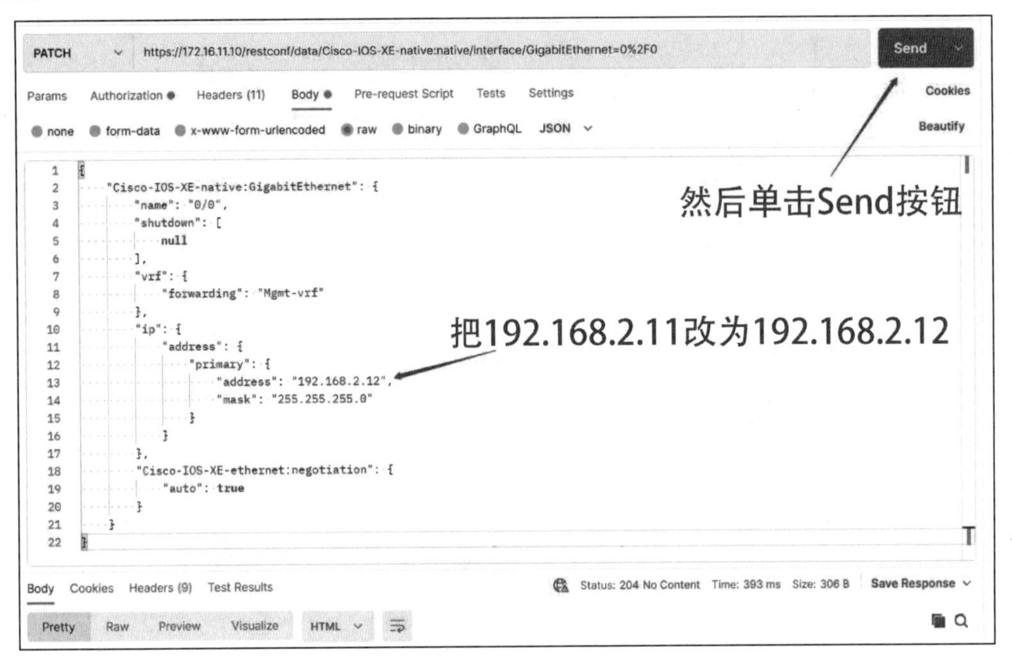

最后回到交换机上，可以看到 Gi0/0 端口的 IP 地址已经从 192.168.2.11 变为 192.168.2.12，如下图所示。

```
AS-C9300-BDC-T217-01#show run int gi0/0
Building configuration...

Current configuration : 128 bytes
!
interface GigabitEthernet0/0
 vrf forwarding Mgmt-vrf
 ip address 192.168.2.11 255.255.255.0
 shutdown
 negotiation auto
end

AS-C9300-BDC-T217-01#show run int gi0/0
Building configuration...

Current configuration : 128 bytes
!
interface GigabitEthernet0/0
 vrf forwarding Mgmt-vrf
 ip address 192.168.2.12 255.255.255.0
 shutdown
 negotiation auto
end

AS-C9300-BDC-T217-01#
```

11.2.6　通过 PUT 方法替换交换机配置

下面我们来看如何通过 PUT 方法替换交换机配置，看它和 PATCH 方法的差异在哪。首先我们手动为 Gi0/0 添加一个 secondary IP - 192.168.3.12，如下图所示。

```
AS-C9300-BDC-T217-01#show run int gi0/0
Building configuration...

Current configuration : 128 bytes
!
interface GigabitEthernet0/0
 vrf forwarding Mgmt-vrf
 ip address 192.168.2.12 255.255.255.0
 shutdown
 negotiation auto
end

AS-C9300-BDC-T217-01#
AS-C9300-BDC-T217-01#
AS-C9300-BDC-T217-01#conf t
Enter configuration commands, one per line.  End with CNTL/Z.
AS-C9300-BDC-T217-01(config)#int gi0/0
AS-C9300-BDC-T217-01(config-if)#ip add 192.168.3.12 255.255.255.0 se
AS-C9300-BDC-T217-01(config-if)#ip add 192.168.3.12 255.255.255.0 secondary
AS-C9300-BDC-T217-01(config-if)#end
AS-C9300-BDC-T217-01#show run int gi0/0
Building configuration...

Current configuration : 177 bytes
!
interface GigabitEthernet0/0
 vrf forwarding Mgmt-vrf
 ip address 192.168.3.12 255.255.255.0 secondary
 ip address 192.168.2.12 255.255.255.0
 shutdown
 negotiation auto
end

AS-C9300-BDC-T217-01#
```

回到 POSTMAN，用 GET 方法查看 Gi0/0 状态，可以看到新添加的 secondary IP，如下图所示。

接下来打开 Postman 左侧的 History，找到之前我们为 Gi0/0 配置的 JSON 代码，然后单击 Send 按钮以 PATCH 方法重新配置一遍，如下图所示。

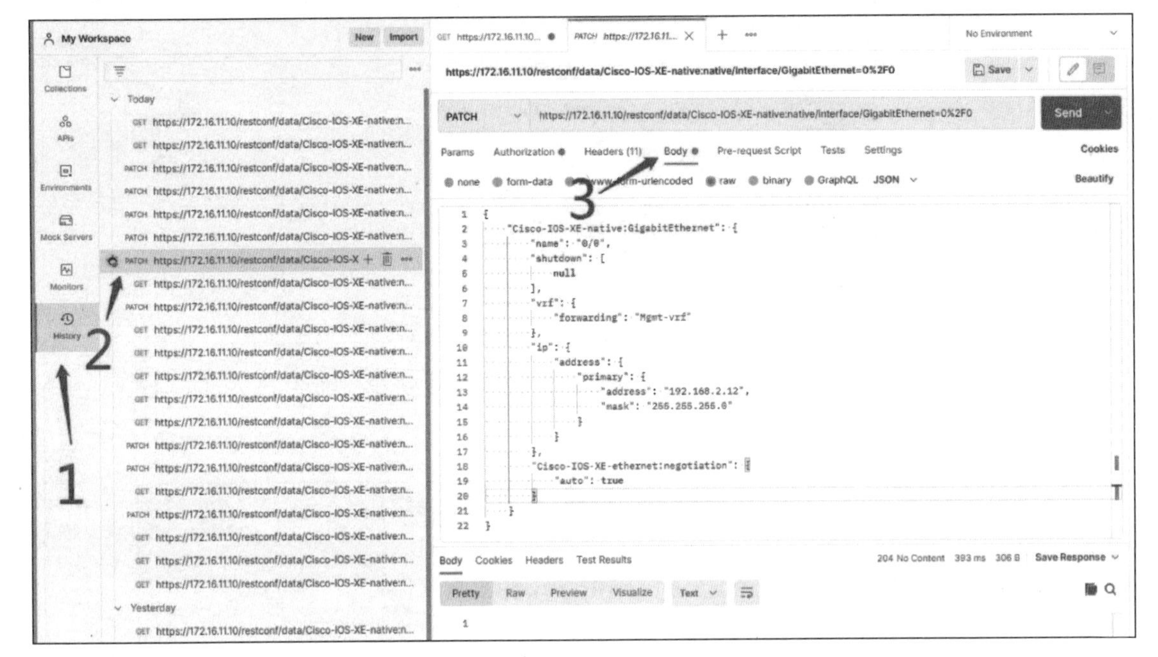

这时回到交换机上，发现 **192.168.3.12 这个 secondary IP 依然存在，原因就在于 PATCH 是不具备覆盖配置的功能的，只能在现有配置的基础上添加配置或修改已有的配置**，如下图所示。

```
AS-C9300-BDC-T217-01#show run int gi0/0
Building configuration...

Current configuration : 177 bytes
!
interface GigabitEthernet0/0
 vrf forwarding Mgmt-vrf
 ip address 192.168.3.12 255.255.255.0 secondary
 ip address 192.168.2.12 255.255.255.0
 shutdown
 negotiation auto
end

AS-C9300-BDC-T217-01#
```

这时我们把 PATCH 改为 PUT，单击 Send 按钮再配置一次，如下图所示。

回到交换机下，发现之前 Gi0/0 的配置已经被 PUT 覆盖，192.168.3.12 这个 secondary IP 已经不在了，如下图所示。

```
AS-C9300-BDC-T217-01#show run int gi0/0
Building configuration...

Current configuration : 177 bytes
!
interface GigabitEthernet0/0
 vrf forwarding Mgmt-vrf
 ip address 192.168.3.12 255.255.255.0 secondary
 ip address 192.168.2.12 255.255.255.0
 shutdown
 negotiation auto
end

AS-C9300-BDC-T217-01#show run int gi0/0
Building configuration...

Current configuration : 128 bytes
!
interface GigabitEthernet0/0
 vrf forwarding Mgmt-vrf
 ip address 192.168.2.12 255.255.255.0
 shutdown
 negotiation auto
end

AS-C9300-BDC-T217-01#
```

可以看出相较于 PATCH，PUT 提供的覆盖配置的功能更贴合思科“基于意图的网络”（Intent-Based Networking）的概念，即“不管设备当前是什么样的配置，用户只想看到设备上出现且仅出现用户所需要的配置”。

11.2.7　通过 DELETE 方法来删除设备配置

最后我们来看怎么通过 DELETE 方法删除交换机的配置。这里以 VLAN 配置举例，首先确认当前 9300 交换机下没有配置 VLAN 2，如下图所示。

```
AS-C9300-BDC-T217-01#show vlan b

VLAN Name                             Status    Ports
---- -------------------------------- --------- ---------
1    default                          active    Gi1/0/12,
                                                Gi6/0/33,
10   L2VPNTest                        active
49   VLAN0049                         active
421  VLAN0421                         active
477  DATA_Level_0                     active    Gi2/0/12
```

在 Postman 上通过 GET 方法验证，在 Cisco-IOS-XE-native:native 这个 YANG 模型下要过滤出具体的单个 VLAN，比如 VLAN 2，方式如下图所示。

```
https://x.x.x.x/RESTCONF/data/Cisco-IOS-XE-native:native/vlan/vlan-list=2
```

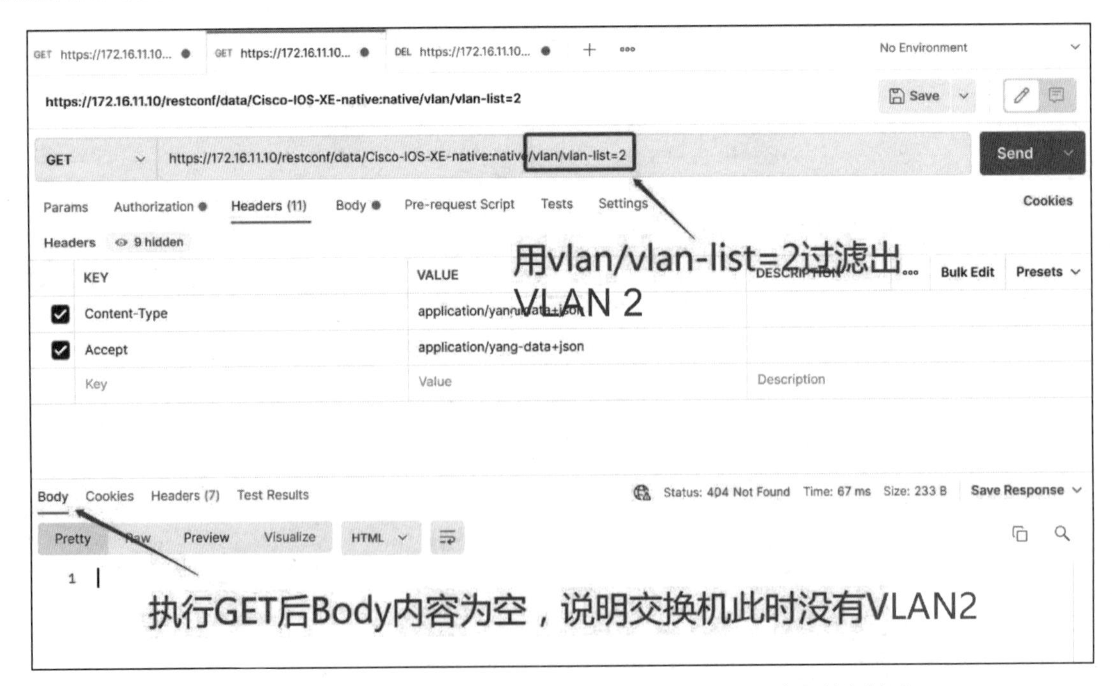

回到交换机，手动添加 VLAN 2，将其取名为 RESTCONF，如下图所示。

```
AS-C9300-BDC-T217-01#conf t
Enter configuration commands, one per line.  End with CNTL/Z.
AS-C9300-BDC-T217-01(config)#vlan 2
AS-C9300-BDC-T217-01(config-vlan)#name RestConf
AS-C9300-BDC-T217-01(config-vlan)#end
AS-C9300-BDC-T217-01#
```

再次回到 Postman 上通过 GET 方法验证，可以看到出现了新创建的 VLAN 2，如下图所示。

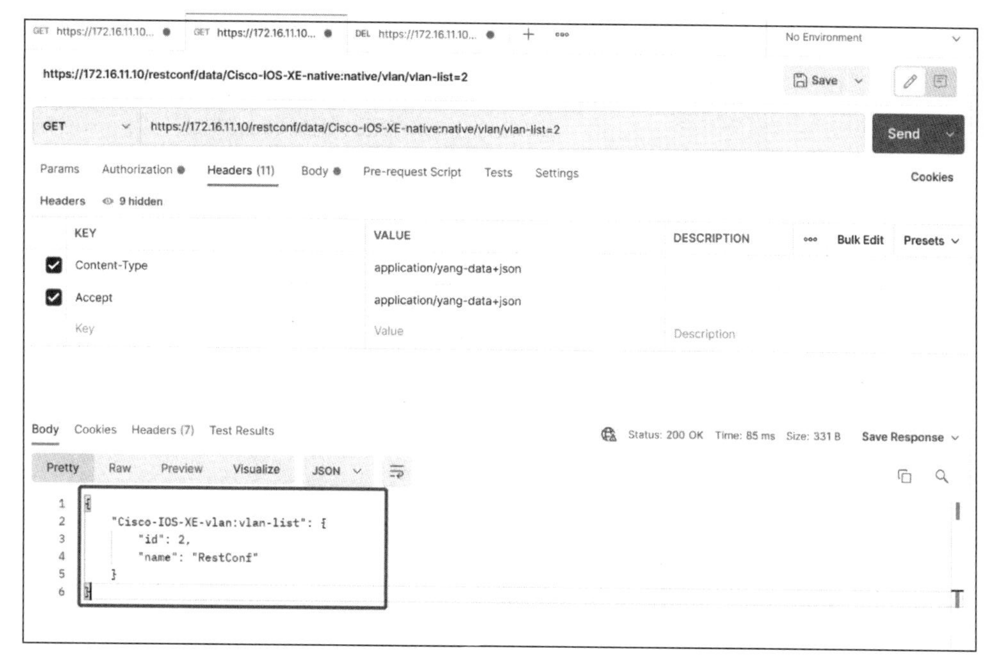

下面我们通过 DELETE 方法删除刚刚创建的 VLAN 2，方法很简单，只需要在 Postman 中将 GET 方法改为 DELETE 即可，如下图所示。

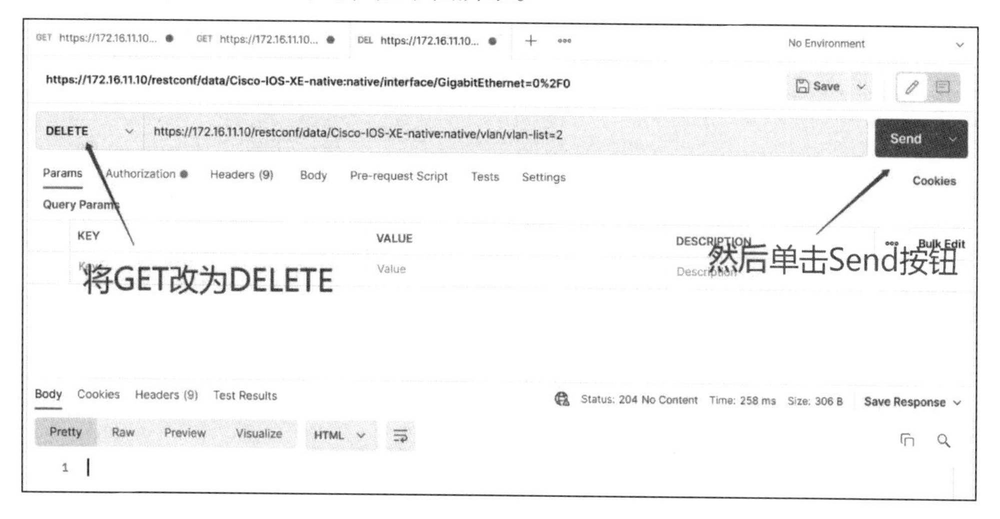

回到交换机上进行 show vlan brief 验证，可以看到刚刚配置好的 VLAN 2 已经通过 RESTCONF 被删除，如下图所示。

```
AS-C9300-BDC-T217-01(config)#vlan 2
AS-C9300-BDC-T217-01(config-vlan)#name RestConf
AS-C9300-BDC-T217-01(config-vlan)#end
AS-C9300-BDC-T217-01#
AS-C9300-BDC-T217-01#show vlan b

VLAN Name                             Status    Ports
---- -------------------------------- --------- ----------
1    default                          active    Gi1/0/12,
                                                Gi6/0/33,
10   L2VPNTest                        active
49   VLAN0049                         active
421  VLAN0421                         active
477  DATA_Level_0                     active    Gi2/0/12
```

11.3　RESTCONF 实验（Requests）

知道了如何使用 Postman 来通过 RESTCONF 对网络设备执行 CRUD 的操作后，本节将继续讲解如何使用 Python 中的 Requests 模块来实现同样的功能。

11.3.1　Requests 模块简介

Requests 模块是 Python 的第三方模块，在它诞生之前，Python 已经有了 urllib、urllib2 这些用来发送 HTTP 请求、获取 HTTP 响应数据，可以满足绝大部分基于 HTTP 协议操作的内置模块。但是 urllib 和 urllib2 都有一个缺点，就是它们的 API 不够简洁，使用起来很麻烦。而号称“为人类服务的 HTTP”的 Requests 就是为解决 urllib 和 urllib2 这一痛点而诞生的。

除继承了 urllib 和 urllib2 的所有特性外，Requests 的底层实际上是 urllib3。Requests 支持 HTTP 连接保持和连接池，支持使用 cookie 保持会话，支持文件上传，支持自动确定响应内容的编码，支持国际化的 URL 和 POST 数据自动编码等。

目前 Requests 是 GitHub 上下载数量最多的模块之一，每周平均的下载量达到了 1400 万次，另外截至 2022 年，GitHub 上还有超过 50 万个其他第三方模块是依赖 Requests 开发的，这绝对是目前 Python 上最靠谱、最受欢迎的 HTTP 模块，而基于 HTTP 协议的 RESTCONF 正好用得上它。

11.3.2　HTTP 基础知识回顾

首先来简单回顾一下 HTTP 的相关基础知识。在 HTTP 中，要让客户端成功地向 HTTP

服务器发送请求，需要具备 URI（Uniform Resource Identifier）、HTTP Method（方法）、HTTP Header（头部）及 HTTP Body（主体）4 个必要条件。上一节在讲 Postman 时，我们已经提到了 HTTP 方法（GET、POST、PUT、PATCH、DELETE），这里再来简单讲下 URI、HTTP 头部和 HTTP 主体。

1. RESTCONF 中的 URI

URL（Uniform Resource Locator，统一资源定位符），也就是俗称的网址，没有人会感到陌生，URL 特指互联网上的网站地址。但是在 RESTCONF 中用到的网址叫作 URI，URL 是 URI 的子集，两者具体的区别不在本节讨论的范围内，读者可以自己扩展阅读。

在 RESTCONF 中，URI 的标准格式如下。

```
https://<ADDRESS>/<ROOT>/data/<[YANG_MODULE:]CONTAINER>/ <LEAF>[?<OPTIONS>]
```

在对网络设备进行 CRUD 操作的应用中，上面 RESTCONF URI 中各个组成部分的解释如下。

- ADDRESS 为设备（路由器、交换机）的 IP 地址。
- ROOT 是所有 RESTCONF 请求的入口，在接入任何 RESTCONF 服务器之前必须指定 ROOT。根据 RFC 8010 的定义，所有 RESTCONF 服务器必须启用一个叫作 /.well-known/host-meta 的资源，用以开启 ROOT。通常在 NetDevOps 的应用中（比如思科的设备），这里的 ROOT 我们会写成 RESTCONF。
- data，顾名思义是 RESTCONF API 中用来指定数据（data）的资源类型（resource type），用以 RPC 协议的运行。
- [YANG_MODULE:]CONTAINER，其中 YANG_MODULE（可选）为 YANG 模型，CONTAINER 为我们使用的基础模式容器。
- LEAF 则为 YANG 模型里的分支元素，比如上一节讲 Postman 时我们用到的 hostname、version、interface、vlan 等。
- 最后[?<OPTIONS>]则是可选的参数，对 HTTP 响应的内容做各种额外的过滤处理（不是所有厂商的设备都支持），常见的 option 如下。

  a. depth = unbounded：在上一节讲 Postman 的时候，我们已经使用了?depth=x 来改变 HTTP 响应中返回数据的深度（x 为整数），如果没有指定 depth 数，则其默认值为 unbounded，表示对 HTTP 响应中返回的数据不做任何处理。

b. content = [all, config, nonconfig]:：content 参数用来控制返回数据的类型，如果没有指定 content 参数，则使用默认值，即 all。

c. fields = expr：fields 用来限制 HTTP 相应内容中 LEAF 的范围。

这里以 RFC 8343 里举例提到的 ietf-interfaces 这个 YANG 模型为例来展示 RESTCONF URI 和 YANG 模型的对应关系，如下图所示。

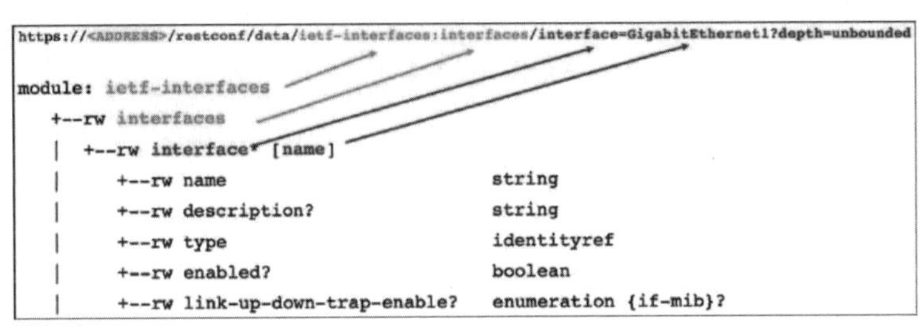

2. HTTP 头部

HTTP 头部是 RESTCONF 请求和响应最重要的部分之一，其提供的信息可以帮助我们排错。HTTP 头部中又包含了请求头部（request header）和响应头部（response header）。请求头部的内容遵循 name:value 的格式（类似 key:value 键值对），如果一个 name 对应多个 value，则 value 之间以逗号隔开，举例如下。

```
Host: www.github.com
Connection: Keep-Alive
Accept: image/gif, image/jpeg, */*
Accept-Language: us-en, fr, cn
```

响应头部的内容同样遵循 name:value 的格式，举例如下。

```
Content-Type: text/html
Content-Length: 35
Connection: Keep-Alive
Keep-Alive: timeout=15, max=100
The response message body contains the resource data
requested.
```

部分常见的请求头部和响应头部内容讲解如下。

- WWW-Authenticate：在响应头部中出现，由 HTTP 服务器发出，用来要求客户端提供验证其身份的凭证（用户名和密码以冒号隔开，写作用户名：冒号的格式）。

WWW-Authenticate 通常接在响应代码 401 后面，响应代码 401 的含义为“unauthorized”（未授权），举例如下。

```
HTTP/1.1 401 Authorization Required
www-Authenticate: Basic realm= "test"
```

- Authorization：在请求头部中出现，客户端在收到服务器发来的要求其提供用户名和密码的信息后，客户端向服务器发送验证其身份的凭证，如果验证方式是基础验证（Basic Authentication）（HTTP 的验证部分后面会提到），则客户端提供的用户名:密码，比如 admin:123，会以 base64 的形式编码为 YWRtaW46MTIz，然后再发送给服务器，举例如下。

```
Authorization: Basic YWRtaW46MTIz
```

- Accept：在请求头部中出现，表示客户端希望从服务器接收的数据类型，常见的如 image/gif、image/jpeg 等。在上一节讲 Postman 的时候，我们已经用到了 Accept : application/yang-data+json，用来表示客户端希望从服务器接收到以 JSON 格式展现的 YANG 模型数据。注意 RESTCONF 和 NETCONF 最大的区别之一在于：NETCONF 只支持 XML 形式的数据格式，而 RESTCONF 可以同时支持 XML 和 JSON 的数据格式。如果 RESTCONF 客户端希望从服务器接收以 XML 格式展现的 YANG 模型数据，那么将 Accept: application/yang-data+json 改写为 Accept : application/yang-data+xml 即可，举例如下。

```
Accept : application/yang-data+json
```

- Content-type：用来表示客户端向服务器发送的请求的媒体类型（media type），比较常见的 Content-type 为 text/HTML，而我们在上一节讲 Postman 的时候，用到了 Content-type : application/yang-data+json，举例如下。

```
Content-type : application/yang-data+json
```

- HTTP 响应码：在响应头部中出现，也叫 HTTP 状态码（HTTP Status Codes），常见的如 200（表示服务器成功处理了客户端的请求）、201（请求已被服务器处理并创建了新的资源）、202（服务器已接收请求，但尚未处理）、204（服务器成功处理了请求，但没有返回任何内容）、401（未授权，服务器要求身份验证）、404（服务器找不到请求的网页）等，更多的 HTTP 响应码读者可自行扩展阅读。

3. HTTP 主体

HTTP 主体存在于 HTTP 包中的尾部，HTTP 请求和响应都有 HTTP 主体，但并不是所有请求和响应都有主体。在 HTTP 请求中，GET、HEAD、OPTIONS 等方法都没有主体，而像 PUT、PATCH、POST、DELETE 这些需要将数据发送到服务器以便更新的则会用到主体。

下面是一个 POST 请求方式中使用主体的例子。

```
POST /test HTTP/1.1
Host: foo.example
Content-Type: application/x-www-form-urlencoded
Content-Length: 27

field1=value1&field2=value2
```

最后一行的 field1=value1&field2=value2 即为 POST 请求的主体部分。

而某些 HTTP 响应中也是不含主体部分的，比如含状态码 201、204 的响应。下面是一个 HTTP 响应中使用主体的例子。

```
HTTP/1.1 200 OK
Date: Mon, 21 March 2022 23:26:07 GMT
Server: Apache/2.2.8 (Ubuntu) mod_ssl/2.2.8 OpenSSL/0.9.8g
Last-Modified: Mon, 21 March 2022 22:04:35 GMT
ETag: "45b6-834-49130cc1182c0"
Accept-Ranges: bytes
Content-Length: 13
Connection: close
Content-Type: text/html

Hello world!
```

最后一行的 Hello world！即响应的主体部分。

11.3.3 Requests 实验环境

实验设备：思科 Catalyst 9300 交换机（真机）1 台

交换机型号：C9300-48U

交换机操作系统版本：IOS-XE 16.12.4 （思科从 IOS-XE 16.3 开始支持 RESTCONF，更

早的版本不支持）

主机操作系统：Windows 10

IDE：Sublime Text 3

Requests 版本：2.28.1 （通过 pip 下载安装）

交换机的初始配置和 11.2 节（RESTCONF 实验讲解（Postman））完全一样。

11.3.4　通过 GET 方法获取交换机配置

首先创建一个 py 脚本，写入下列代码。

```
import requests
from requests.auth import HTTPBasicAuth
import json

AUTH = HTTPBasicAuth('xyz', '123')
MEDIA_TYPE = 'application/yang-data+json'
HEADERS = { 'Accept': MEDIA_TYPE, 'Content-Type': MEDIA_TYPE }

def get_request(url):
    response = requests.get(url, auth=AUTH, headers=HEADERS, verify=False)
    print("API: ", url)
    print(response.status_code)
    if response.status_code in [200, 202, 204]:
        print("Successful")
    else:
        print("Error in API Request")
    output = json.loads(response.text)  #response.text 类型是字符串，用 json.loads 将其转化为 JSON Object,也就是字典
    print (output)
url = "https://x.x.x.x/RESTCONF/data/Cisco-IOS-XE-native:native/vlan"
get_request(url)
```

代码分段讲解如下。

- HTTP 认证分为 Basic、Session、Token3 种方式，其中最常用的是 Basic Authentication。思科的 IOS-XE 设备使用的是 Basic Authentication，这里通过 from requests.auth import

HTTPBasicAuth 将其导入，以便简化调用 HTTPBasicAuth()函数的方式。而这里导入 json 的原因会在后面讲到。

```
import requests
from requests.auth import HTTPBasicAuth
import json
```

- 这里调用 HTTPBasicAuth()，写入登录 9300 交换机的用户名和密码后，将其赋值给变量 AUTH，读者可以按需将'xyz'替换为自己的用户名，将'123'替换为自己的密码（下同）。另外，两个变量 MEDIA_TYPE 和 HEADERS 的作用也顾名思义，前面讲到了 HTTP 头部中 Accept 和 Content-Type 的作用，这里我们希望从 HTTP 服务器（也就是 9300 交换机）接收以 JSON 格式展现的 YANG 模型数据，因此使用 application/yang-data+json，并作为值将其赋给 Accept 和 Content-Type。

```
AUTH = HTTPBasicAuth('xyz', '123')
MEDIA_TYPE = 'application/yang-data+json'
HEADERS = { 'Accept': MEDIA_TYPE, 'Content-Type': MEDIA_TYPE }
```

- 之后我们自定义一个叫作 get_request()的函数，该函数的参数即我们要通过 GET 请求方式来访问的 URL。在 requests 模块中实现 GET 请求方式很简单，只需调用 requests.get()函数并放入 url、auth、headers 及 verify 4 个参数即可，其中 url 参数会在调用函数时传入，auth 和 headers 两个参数已经在前面定义好了，而 verify=False 表示忽略对 SSL 证书的验证。然后我们将调用 requests.get()后得到响应的数据赋值给变量 response，通过 response.status_code 得到交换机返回的 HTTP 状态码，如果状态码为 200、202、204，则打印“Successful”，表明试验成功；如果返回其他状态码，则打印“Error in API Request”。最后我们调用 response.text 得到具体的回显内容并将它打印出来。这里注意：response.text 的数据类型为字符串，不像字典那样方便我们对内容做过滤，因此我们调用 json.loads 将其转换为 JSON Object，也就是字典，方便后续对回显内容做解析。

```
def get_request(url):
    response = requests.get(url, auth=AUTH, headers=HEADERS, verify=False)
    print("API: ", url)
    print(response.status_code)
    if response.status_code in [200, 202, 204]:
        print("Successful")
```

```
    else:
        print("Error in API Request")
    output = json.loads(response.text)  #response.text 类型是字符串，用 json.loads 将其转化为 JSON Object,也就是字典
    print (output)
```

- 最后定义我们要访问的 URL，也就是 RESTCONF 中的 URI（读者可将 x.x.x.x 按需替换为自己交换机的 IP 地址），这里我们希望得到交换机的 VLAN 相关的信息，并最终调用之前定义的 get_request()函数来向交换机发送 GET 请求。

```
url = "https://x.x.x.x/RESTCONF/data/Cisco-IOS-XE-native:native/vlan"
get_request(url)
```

运行脚本，效果如下图所示。

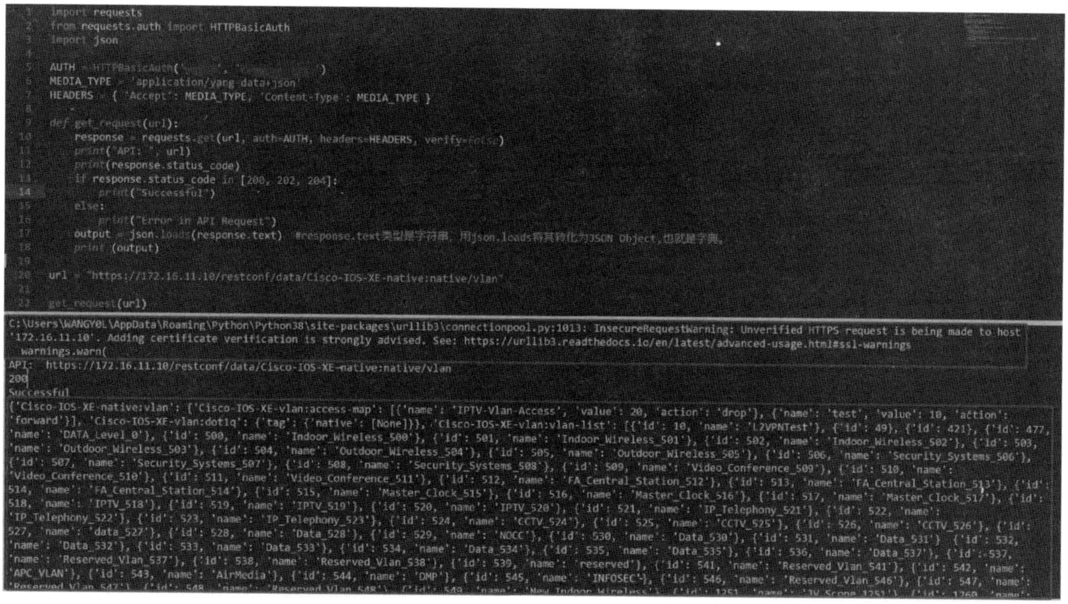

脚本运行成功，我们通过 RESTCONF 以 GET 请求方式得到了交换机所有的 VLAN 相关信息。但是你会发现两个问题。

（1）运行脚本后，Python 会返回"C:\Users\WANGY0L\AppData\Roaming\Python\Python310\site-packages\urllib3\connectionpool.py:1013: InsecureRequestWarning: Unverified HTTPS request is being made to host '172.16.11.10'. Adding certificate verification is strongly advised. See: Advanced Usage - urllib3 2.0.0.dev0 documentation warnings.warn"的告警，原因是我们跳过了验证 SSL 证书的环节。

（2）输出的字典内容格式并不宜阅读。

解决办法如下。

（1）引入 urllib3 模块（requests 的底层就是 urllib3），在脚本中调用 urllib3.disable_warnings (urllib3.exceptions.InsecureRequestWarning)来关闭告警。

（2）代表回显内容的 output 变量是字典类型的，所以输出的格式不好看。有两种方法可以让回显内容变得更加美观易读：可以调用 pprint 模块，使用 pprint(output)来打印；也可以将 print(output)替换为 print(response.text)，直接将原本的响应内容以字符串形式输出（但是在做解析时依然使用字典类型的 output 变量）。

修改后的代码如下，这里笔者在第 22 行使用 print(response.text)，读者也可自行尝试使用 pprint(output)。

```
import requests
from requests.auth import HTTPBasicAuth
import json
from pprint import pprint
import urllib3

urllib3.disable_warnings(urllib3.exceptions.InsecureRequestWarning) #关闭跳过 SSL 验证后
的告警

AUTH = HTTPBasicAuth('xyz', '123')
MEDIA_TYPE = 'application/yang-data+json'
HEADERS = { 'Accept': MEDIA_TYPE, 'Content-Type': MEDIA_TYPE }

def get_request(url):
    response = requests.get(url, auth=AUTH, headers=HEADERS, verify=False)
    print("API: ", url)
    print(response.status_code)
    if response.status_code in [200, 202, 204]:
        print("Successful")
    else:
        print("Error in API Request")
    output = json.loads(response.text)  #response.text 类型是字符串，用 json.loads 将其转化
为 JSON Object,也就是字典
    print (output)

url = "https://x.x.x.x/RESTCONF/data/Cisco-IOS-XE-native:native/vlan"
get_request(url)
```

再次运行脚本，效果如下图所示。

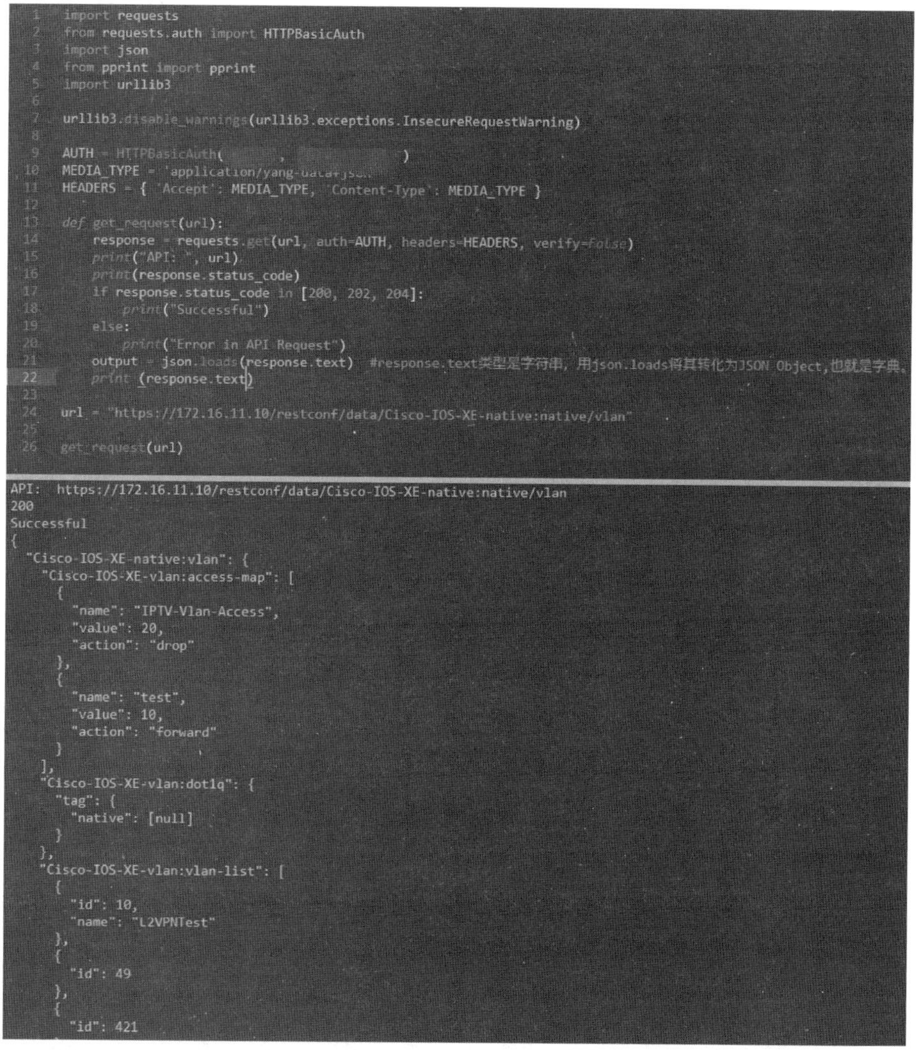

```
import requests
from requests.auth import HTTPBasicAuth
import json
from pprint import pprint
import urllib3

urllib3.disable_warnings(urllib3.exceptions.InsecureRequestWarning)

AUTH = HTTPBasicAuth(          ,           )
MEDIA_TYPE = 'application/yang-data+json'
HEADERS = { 'Accept': MEDIA_TYPE, 'Content-Type': MEDIA_TYPE }

def get_request(url):
    response = requests.get(url, auth=AUTH, headers=HEADERS, verify=False)
    print("API: ", url)
    print(response.status_code)
    if response.status_code in [200, 202, 204]:
        print("Successful")
    else:
        print("Error in API Request")
    output = json.loads(response.text)  #response.text类型是字符串，用json.loads将其转化为JSON Object，也就是字典。
    print (response.text)

url = "https://172.16.11.10/restconf/data/Cisco-IOS-XE-native:native/vlan"

get_request(url)
```

```
API:  https://172.16.11.10/restconf/data/Cisco-IOS-XE-native:native/vlan
200
Successful
{
  "Cisco-IOS-XE-native:vlan": {
    "Cisco-IOS-XE-vlan:access-map": [
      {
        "name": "IPTV-Vlan-Access",
        "value": 20,
        "action": "drop"
      },
      {
        "name": "test",
        "value": 10,
        "action": "forward"
      }
    ],
    "Cisco-IOS-XE-vlan:dot1q": {
      "tag": {
        "native": [null]
      }
    },
    "Cisco-IOS-XE-vlan:vlan-list": [
      {
        "id": 10,
        "name": "L2VPNTest"
      },
      {
        "id": 49
      },
      {
        "id": 421
```

11.3.5　使用?depth=修改深度

?depth=的使用方法和 Postman 里讲到的一样，不再赘述。只需要在 URL 最后面加上?depth=x，写成 https://x.x.x.x/RESTCONF/data/Cisco-IOS-XE-native:native/vlan?depth=x 即可，代码如下。

```
import requests
from requests.auth import HTTPBasicAuth
```

```
import json
from pprint import pprint
import urllib3

urllib3.disable_warnings(urllib3.exceptions.InsecureRequestWarning)

AUTH = HTTPBasicAuth('xyz', '123')
MEDIA_TYPE = 'application/yang-data+json'
HEADERS = { 'Accept': MEDIA_TYPE, 'Content-Type': MEDIA_TYPE }

def get_request(url):
    response = requests.get(url, auth=AUTH, headers=HEADERS, verify=False)
    print("API: ", url)
    print(response.status_code)
    if response.status_code in [200, 202, 204]:
        print("Successful")
    else:
        print("Error in API Request")
    output = json.loads(response.text)  #response.text 类型是字符串，用 json.loads 将其转化为 JSON Object,也就是字典
    print (output)
url = "https://x.x.x.x/RESTCONF/data/Cisco-IOS-XE-native:native/vlan?depth=1"
get_request(url)
```

运行脚本，效果如下图所示。

```
1  import requests
2  from requests.auth import HTTPBasicAuth
3  import json
4  from pprint import pprint
5  import urllib3
6
7  urllib3.disable_warnings(urllib3.exceptions.InsecureRequestWarning)
8
9  AUTH = HTTPBasicAuth
10 MEDIA_TYPE = 'application/yang
11 HEADERS = { 'Accept': MEDIA_TYPE, 'Content-Type': MEDIA_TYPE }
12
13 def get_request(url):
14     response = requests.get(url, auth=AUTH, headers=HEADERS, verify=False)
15     print("API: ", url)
16     print(response.status_code)
17     if response.status_code in [200, 202, 204]:
18         print("Successful")
19     else:
20         print("Error in API Request")
21     output = json.loads(response.text)  #response.text类型是字符串，用json.loads将其转化为JSON Object,也就是字典。
22     print (response.text)
23
24 url = "https://172.16.11.10/restconf/data/Cisco-IOS-XE-native:native/vlan?depth=1"
25
26 get_request(url)
```

```
API:  https://172.16.11.10/restconf/data/Cisco-IOS-XE-native:native/vlan?depth=1
200
Successful
{
  "Cisco-IOS-XE-native:vlan": {
    "Cisco-IOS-XE-vlan:access-map": [
    ],
    "Cisco-IOS-XE-vlan:dot1q": {
    },
    "Cisco-IOS-XE-vlan:vlan-list": [
    ]
  }
}
```

11.3.6　通过 PATCH 方法更改交换机配置

同讲 Postman 的时候一样，我们继续以 9300 交换机的 Gi0/0 为实验对象，向交换机发送 PATCH 和 PUT 请求来改变 Gi0/0 的配置。

当前 Gi0/0 下没有配置 IP 地址，如下图所示。

```
AS-C9300-BDC-T217-01#show run int gi0/0
Building configuration...

Current configuration : 104 bytes
!
interface GigabitEthernet0/0
 vrf forwarding Mgmt-vrf
 no ip address
 shutdown
 negotiation auto
end
```

下面我们尝试通过 PATCH 方式为 Gi0/0 配置 IP 地址：192.168.2.11 /24。首先创建一个 py 文件，写入下面的代码。

```
import requests
from requests.auth import HTTPBasicAuth
import urllib3

urllib3.disable_warnings(urllib3.exceptions.InsecureRequestWarning)
AUTH = HTTPBasicAuth('xyz', '123')
MEDIA_TYPE = 'application/yang-data+json'
HEADERS = { 'Accept': MEDIA_TYPE, 'Content-Type': MEDIA_TYPE }
url =
"https://x.x.x.x/RESTCONF/data/Cisco-IOS-XE-native:native/interface/GigabitEthernet
=0%2f0"
payload = """
{
```

```
  "Cisco-IOS-XE-native:GigabitEthernet": {
    "ip": {
      "address": {
        "primary": {
          "address": "192.168.2.11",
          "mask": "255.255.255.0"
        }
      }
    }
  }
}
"""
response = requests.patch(url, headers=HEADERS, auth=AUTH, data=payload, verify=False)
print(response.status_code)
```

代码分段讲解如下。

- 前面和之前基本一样，只需注意在 RESTCONF 中/符号是以%2f 来表达的，所以 GigabiEthernet0/0 在 URL 中要写成 GigabiEthernet=0%2f0，而不是 GigabiEthernet=0/0。

```
import requests
from requests.auth import HTTPBasicAuth
import urllib3

urllib3.disable_warnings(urllib3.exceptions.InsecureRequestWarning)
AUTH = HTTPBasicAuth('xyz', '123')
MEDIA_TYPE = 'application/yang-data+json'
HEADERS = { 'Accept': MEDIA_TYPE, 'Content-Type': MEDIA_TYPE }
url =
"https://x.x.x.x/RESTCONF/data/Cisco-IOS-XE-native:native/interface/GigabitEthernet
=0%2f0"
```

- 因为是向交换机发送 PATCH 请求，所以要加上 payload（也就是前面讲到的 body），将要为 gi0/0 做的配置写入进去。

```
payload = """
{
  "Cisco-IOS-XE-native:GigabitEthernet": {
    "ip": {
      "address": {
```

```
        "primary": {
          "address": "192.168.2.11",
          "mask": "255.255.255.0"
        }
      }
    }
  }
}
"""
```

- 最后调用 requests.patch()向设备发送 PATCH 请求，更改 Gi0/0 的配置，并打印出交换机返回的响应码（状态码），如果配置成功，则我们会收到状态码 204，表示 HTTP 服务器成功处理了请求，但没有返回任何内容。

```
response = requests.patch(url, headers=HEADERS, auth=AUTH, data=payload, verify=False)
print(response.status_code)
```

运行脚本，效果如下图所示。

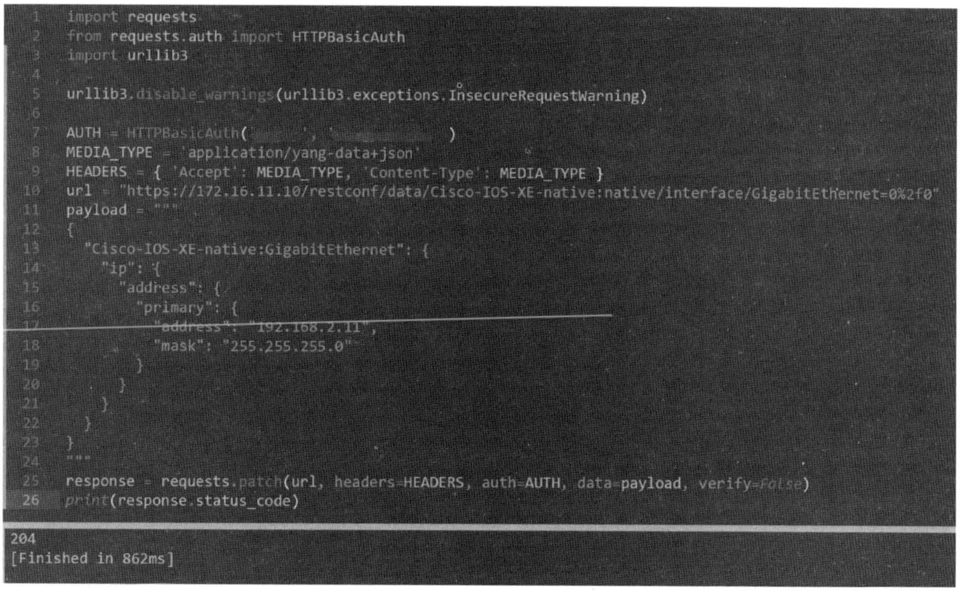

```
import requests
from requests.auth import HTTPBasicAuth
import urllib3

urllib3.disable_warnings(urllib3.exceptions.InsecureRequestWarning)

AUTH = HTTPBasicAuth(        ,         )
MEDIA_TYPE = 'application/yang-data+json'
HEADERS = { 'Accept': MEDIA_TYPE, 'Content-Type': MEDIA_TYPE }
url = "https://172.16.11.10/restconf/data/Cisco-IOS-XE-native:native/interface/GigabitEthernet=0%2f0"
payload = """
{
  "Cisco-IOS-XE-native:GigabitEthernet": {
    "ip": {
      "address": {
        "primary": {
          "address": "192.168.2.11",
          "mask": "255.255.255.0"
        }
      }
    }
  }
}
"""
response = requests.patch(url, headers=HEADERS, auth=AUTH, data=payload, verify=False)
print(response.status_code)

204
[Finished in 862ms]
```

注意，这里返回了状态码 204，说明配置修改成功，回到交换机上进行验证（读者也可以自己另外写一段代码，通过 GET 请求方式来验证），如下图所示。

```
AS-C9300-BDC-T217-01#show run int gi0/0
Building configuration...

Current configuration : 104 bytes
!
interface GigabitEthernet0/0
 vrf forwarding Mgmt-vrf
 no ip address
 shutdown
 negotiation auto
end

AS-C9300-BDC-T217-01#
AS-C9300-BDC-T217-01#show run int gi0/0
Building configuration...

Current configuration : 128 bytes
!
interface GigabitEthernet0/0
 vrf forwarding Mgmt-vrf
 ip address 192.168.2.11 255.255.255.0
 shutdown
 negotiation auto
end

AS-C9300-BDC-T217-01#
```

11.3.7　通过 PUT 方法替换交换机配置

PUT 和 PATCH 的区别在前面已经提到过，PUT 是用来替换网络设备的配置（具有覆盖的功能），PATCH 是用来修改网络设备的配置（不具有覆盖的功能）。

在开始 PUT 的实验之前，我们先用 PATCH 方法为 Gi0/0 添加一个 secondary IP: 192.168.3.12 255.255.255.0（在讲 Postman 的时候，这一步我们是手动在交换机上操作的），代码如下。

```
import requests
from requests.auth import HTTPBasicAuth
import urllib3

urllib3.disable_warnings(urllib3.exceptions.InsecureRequestWarning)

AUTH = HTTPBasicAuth('xyz', '123')
MEDIA_TYPE = 'application/yang-data+json'
HEADERS = { 'Accept': MEDIA_TYPE, 'Content-Type': MEDIA_TYPE }
url = "https://x.x.x.x/restconf/data/Cisco-IOS-XE-native:native/interface/
GigabitEthernet=0%2f0"
payload = """
{
    "Cisco-IOS-XE-native:GigabitEthernet": {
```

```
        "ip": {
            "address": {
                "secondary": [
                    {
                        "address": "192.168.3.12",
                        "mask": "255.255.255.0",
                        "secondary": [
                            null
                        ]
                    }
                ]
            }
        }
    }
}
"""
response = requests.patch(url, headers=HEADERS, auth=AUTH, data=payload, verify=False)
print(response.status_code)
```

运行脚本，效果如下图所示。

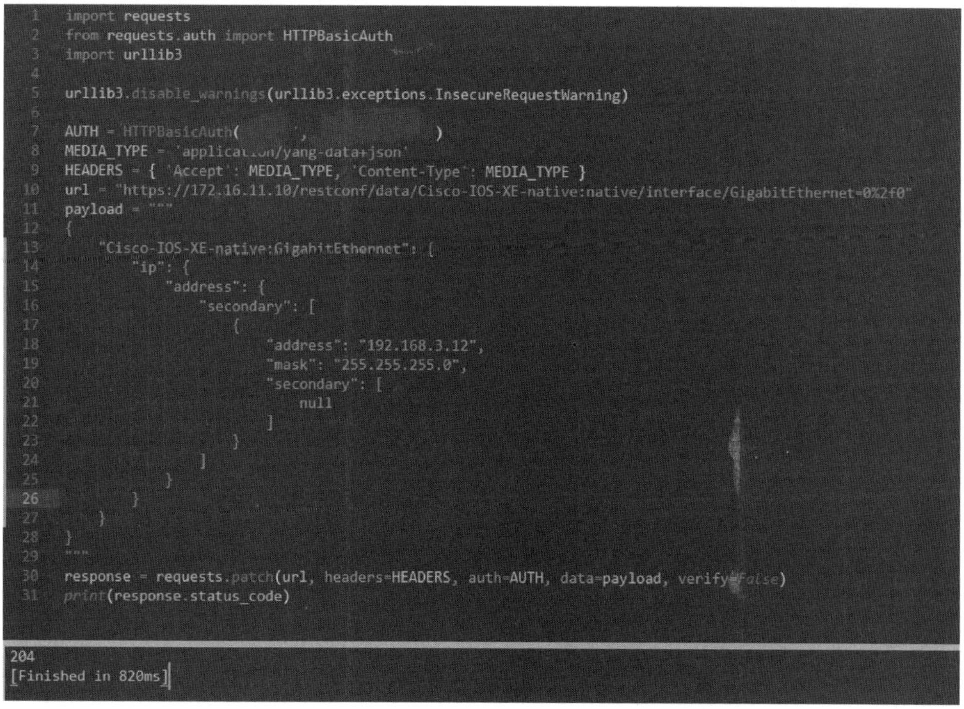

```
1  import requests
2  from requests.auth import HTTPBasicAuth
3  import urllib3
4
5  urllib3.disable_warnings(urllib3.exceptions.InsecureRequestWarning)
6
7  AUTH = HTTPBasicAuth(          ,          )
8  MEDIA_TYPE = 'application/yang-data+json'
9  HEADERS = { 'Accept': MEDIA_TYPE, 'Content-Type': MEDIA_TYPE }
10 url = "https://172.16.11.10/restconf/data/Cisco-IOS-XE-native:native/interface/GigabitEthernet=0%2f0"
11 payload = """
12 {
13     "Cisco-IOS-XE-native:GigabitEthernet": {
14         "ip": {
15             "address": {
16                 "secondary": [
17                     {
18                         "address": "192.168.3.12",
19                         "mask": "255.255.255.0",
20                         "secondary": [
21                             null
22                         ]
23                     }
24                 ]
25             }
26         }
27     }
28 }
29 """
30 response = requests.patch(url, headers=HEADERS, auth=AUTH, data=payload, verify=False)
31 print(response.status_code)

204
[Finished in 820ms]
```

返回交换机上进行验证，如下图所示。

```
AS-C9300-BDC-T217-01#show run int gi0/0
Building configuration...

Current configuration : 177 bytes
!
interface GigabitEthernet0/0
 vrf forwarding Mgmt-vrf
 ip address 192.168.3.12 255.255.255.0 secondary
 ip address 192.168.2.11 255.255.255.0
 shutdown
 negotiation auto
end

AS-C9300-BDC-T217-01#
```

然后创建另外一个脚本来实验 PUT 方法，代码如下。

```
import requests
from requests.auth import HTTPBasicAuth
import urllib3

urllib3.disable_warnings(urllib3.exceptions.InsecureRequestWarning)

AUTH = HTTPBasicAuth('xyz', '123')
MEDIA_TYPE = 'application/yang-data+json'
HEADERS = { 'Accept': MEDIA_TYPE, 'Content-Type': MEDIA_TYPE }
url = "https://x.x.x.x/restconf/data/Cisco-IOS-XE-native:native/interface/
GigabitEthernet=0%2f0"
payload = """
{
    "Cisco-IOS-XE-native:GigabitEthernet": {
        "name": "0/0",
        "shutdown": [
            null
        ],
        "vrf": {
            "forwarding": "Mgmt-vrf"
        },
        "ip": {
            "address": {
                "primary": {
                    "address": "192.168.2.11",
                    "mask": "255.255.255.0"
                }
```

```
            }
        },
        "Cisco-IOS-XE-ethernet:negotiation": {
            "auto": true
        }
    }
}
"""
response = requests.put(url, headers=HEADERS, auth=AUTH, data=payload, verify=False)
print(response.status_code)
```

这里我们用 PUT 方法“复原”了添加 secondary IP 之前的配置。

运行脚本，效果如下图所示。

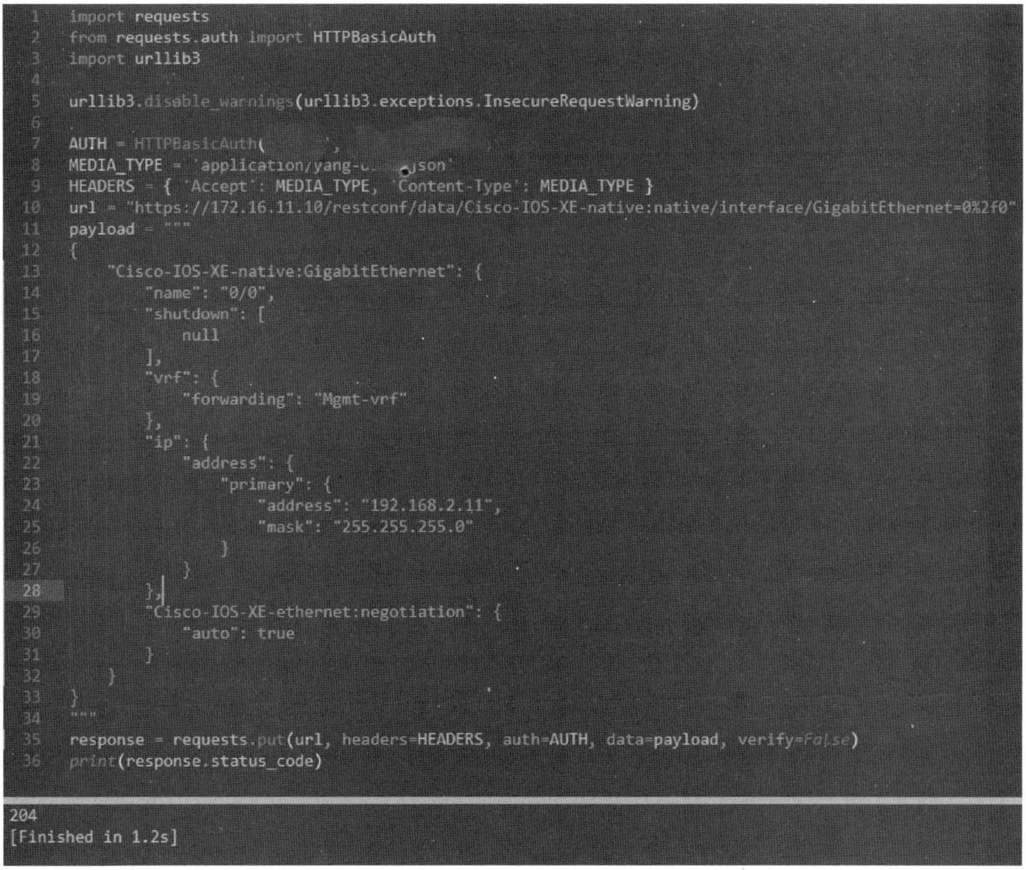

```
1   import requests
2   from requests.auth import HTTPBasicAuth
3   import urllib3
4
5   urllib3.disable_warnings(urllib3.exceptions.InsecureRequestWarning)
6
7   AUTH = HTTPBasicAuth(        ',        )
8   MEDIA_TYPE = 'application/yang-d    json'
9   HEADERS = { 'Accept': MEDIA_TYPE, 'Content-Type': MEDIA_TYPE }
10  url = "https://172.16.11.10/restconf/data/Cisco-IOS-XE-native:native/interface/GigabitEthernet=0%2f0"
11  payload = """
12  {
13      "Cisco-IOS-XE-native:GigabitEthernet": {
14          "name": "0/0",
15          "shutdown": [
16              null
17          ],
18          "vrf": {
19              "forwarding": "Mgmt-vrf"
20          },
21          "ip": {
22              "address": {
23                  "primary": {
24                      "address": "192.168.2.11",
25                      "mask": "255.255.255.0"
26                  }
27              }
28          },
29          "Cisco-IOS-XE-ethernet:negotiation": {
30              "auto": true
31          }
32      }
33  }
34  """
35  response = requests.put(url, headers=HEADERS, auth=AUTH, data=payload, verify=False)
36  print(response.status_code)

204
[Finished in 1.2s]
```

回到交换机上进行验证，可以看到 Gi0/0 的配置回到了配置 secondary IP 之前的状态，如下图所示。

```
AS-C9300-BDC-T217-01#show run int gi0/0
Building configuration...

Current configuration : 177 bytes
!
interface GigabitEthernet0/0
 vrf forwarding Mgmt-vrf
 ip address 192.168.3.12 255.255.255.0 secondary
 ip address 192.168.2.11 255.255.255.0
 shutdown
 negotiation auto
end

AS-C9300-BDC-T217-01#show run int gi0/0
Building configuration...

Current configuration : 128 bytes
!
interface GigabitEthernet0/0
 vrf forwarding Mgmt-vrf
 ip address 192.168.2.11 255.255.255.0
 shutdown
 negotiation auto
end

AS-C9300-BDC-T217-01#
```

11.3.8　通过 POST 方法添加交换机配置

之前我们用 PATCH 方法为 Gi0/0 添加了 IP 和 secondary IP，然后用 PUT 方法“复原”了配置 secondary IP 之前的配置，由此可以看出，不管是 PATCH 方法，还是 PUT 方法，它们的作用都是修改设备已经存在的配置（Gi0/0 下面的配置是交换机出厂时就已经有了的），而要给交换机添加当前不存在的配置时，比如说给交换机添加一个环回端口，例如 interface loopback 100，就需要用到 POST 方法。

下面我们通过 POST 方法为交换机配置 interface loopback 100，并为其添加 IP 地址：1.1.1.1 255.255.255.255，首先在交换机上确认当前并没有 100 这个环回端口，如下图所示。

```
AS-C9300-BDC-T217-01#show run int loopback 100
                                           ^
% Invalid input detected at '^' marker.

AS-C9300-BDC-T217-01#
```

然后创建一个 py 脚本，写入下列代码。

```
import requests
from requests.auth import HTTPBasicAuth
import urllib3

urllib3.disable_warnings(urllib3.exceptions.InsecureRequestWarning)

AUTH = HTTPBasicAuth('xyz', '123')
```

```
MEDIA_TYPE = 'application/yang-data+json'
HEADERS = { 'Accept': MEDIA_TYPE, 'Content-Type': MEDIA_TYPE }
url = "https://x.x.x.x/restconf/data/Cisco-IOS-XE-native:native/interface"
payload = """
{
    "Cisco-IOS-XE-native:Loopback": {
      "name": "100",
      "ip": {
          "address": {
              "primary": {
                  "address": "1.1.1.1",
                  "mask": "255.255.255.255"
              }
          }
      }
    }
}
"""
response = requests.post(url, headers=HEADERS, auth=AUTH, data=payload, verify=False)
print(response.status_code)

if response.status_code >= 400:
    print("同样的配置，POST 方法只能使用一次！")

url =
"https://x.x.x.x/restconf/data/Cisco-IOS-XE-native:native/interface/Loopback=100"
response = requests.get(url, headers=HEADERS, auth=AUTH, verify=False)
print(response.text)
```

需要注意的是：POST 方法和 PUT、PATCH 方法在使用上区别不大，都需要用到 payload，但区别是同样的配置，PUT 和 PATCH 不管你使用多少次都是允许的，而使用 POST 的话只能为设备做一次新配置，如果你在创建新配置后再运行一次同样的脚本，交换机那边会返回 4XX 系列的错误，所以笔者在代码里添加了下面一段代码，用来提醒读者。

```
if response.status_code >= 400:
    print("同样的配置，POST 方法只能使用一次！")
```

在代码的最后，我们用 GET 方法做了验证，证实 interface loopback 100 创建成功。

运行脚本，效果如下图所示。

```
1  import requests
2  from requests.auth import HTTPBasicAuth
3  import urllib3
4
5  urllib3.disable_warnings(urllib3.exceptions.InsecureRequestWarning)
6
7  AUTH = HTTPBasicAuth(          ,              )
8  MEDIA_TYPE = 'application/yang-data+json'
9  HEADERS = { 'Accept': MEDIA_TYPE, 'Content-Type': MEDIA_TYPE }
10 url = "https://172.16.11.10/restconf/data/Cisco-IOS-XE-native:native/interface"
11 payload = """
12 {
13     "Cisco-IOS-XE-native:Loopback": {
14       "name": "100",
15       "ip": {
16           "address": {
17               "primary": {
18                   "address": "1.1.1.1",
19                   "mask": "255.255.255.255"
20               }
21           }
22       }
23     }
24 }
25 """
26 response = requests.post(url, headers=HEADERS, auth=AUTH, data=payload, verify=False)
27 print(response.status_code)
28
29 if response.status_code >= 400:
30     print("同样的配置，POST方法只能使用一次！")
31
32 url = "https://172.16.11.10/restconf/data/Cisco-IOS-XE-native:native/interface/Loopback=100"
33 response = requests.get(url, headers=HEADERS, auth=AUTH, verify=False)
34 print(response.text)
```

```
201
{
  "Cisco-IOS-XE-native:Loopback": {
    "name": 100,
    "ip": {
      "address": {
        "primary": {
          "address": "1.1.1.1",
          "mask": "255.255.255.255"
        }
      }
    }
  }
}
[Finished in 1.3s]
```

回到交换机上做验证，如下图所示。

```
AS-C9300-BDC-T217-01#show run int loopback 100
                                          ^
% Invalid input detected at '^' marker.

AS-C9300-BDC-T217-01#show run int loopback 100
Building configuration...

Current configuration : 65 bytes
!
interface Loopback100
 ip address 1.1.1.1 255.255.255.255
end

AS-C9300-BDC-T217-01#
```

11.3.9　通过 DELETE 方法删除交换机配置

最后来看如何使用 Requests 模块通过 DELETE 方法删除交换机配置。我们把刚才创建的环回端口 100 删除，代码如下。

```
import requests
from requests.auth import HTTPBasicAuth
import urllib3

urllib3.disable_warnings(urllib3.exceptions.InsecureRequestWarning)

AUTH = HTTPBasicAuth('xyz', '123')
MEDIA_TYPE = 'application/yang-data+json'
HEADERS = { 'Accept': MEDIA_TYPE, 'Content-Type': MEDIA_TYPE }
url = "https://x.x.x.x/restconf/data/Cisco-IOS-XE-native:native/interface/
Loopback=100"
payload = """
{
    "Cisco-IOS-XE-native:Loopback": {
      "name": "100",
      "ip": {
          "address": {
              "primary": {
                  "address": "1.1.1.1",
                  "mask": "255.255.255.255"
              }
          }
      }
    }
}
"""
response = requests.delete(url, headers=HEADERS, auth=AUTH, data=payload, verify=False)
print(response.status_code)

if response.status_code >= 400:
    print("同样的配置，DELETE 方法只能使用一次！")

url = "https://x.x.x.x/restconf/data/Cisco-IOS-XE-native:native/interface/
```

```
Loopback=100"
response = requests.get(url, headers=HEADERS, auth=AUTH, verify=False)
print(response.text)
```

这段代码和 POST 方法基本一样，只是将 requests.post()替换成了 requests.delete()，并且同样用下面这段代码提醒读者同样的配置只能用 DELETE 方法删除一次，如果第二次运行同样的脚本，则交换机会返回 4XX 系列的状态码。

```
if response.status_code >= 400:
    print("同样的配置，DELETE 方法只能使用一次！")
```

运行脚本后，可以看到最后用 GET 方法做验证的时候，print(response.text)返回的内容为空，说明环回端口 100 已经被删除，如下图所示。

```
1   import requests
2   from requests.auth import HTTPBasicAuth
3   import urllib3
4
5   urllib3.disable_warnings(urllib3.exceptions.InsecureRequestWarning)
6
7   AUTH = HTTPBasicAuth(                    )
8   MEDIA_TYPE = 'application/yang-data+json'
9   HEADERS = { 'Accept': MEDIA_TYPE, 'Content-Type': MEDIA_TYPE }
10  url = "https://172.16.11.10/restconf/data/Cisco-IOS-XE-native:native/interface/Loopback=100"
11  payload = """
12  {
13      "Cisco-IOS-XE-native:Loopback": {
14        "name": "100",
15        "ip": {
16            "address": {
17                "primary": {
18                    "address": "1.1.1.1",
19                    "mask": "255.255.255.255"
20                }
21            }
22        }
23      }
24  }
25  """
26  response = requests.delete(url, headers=HEADERS, auth=AUTH, data=payload, verify=False)
27  print(response.status_code)
28
29  if response.status_code >= 400:
30      print("同样的配置，DELETE方法只能使用一次！")
31
32  url = "https://172.16.11.10/restconf/data/Cisco-IOS-XE-native:native/interface/Loopback=100"
33  response = requests.get(url, headers=HEADERS, auth=AUTH, verify=False)
34  print(response.text)

204

[Finished in 2.3s]
```

回到交换机上进行验证，如下图所示。

```
AS-C9300-BDC-T217-01#show run int loopback 100
                                            ^
% Invalid input detected at '^' marker.

AS-C9300-BDC-T217-01#show run int loopback 100
Building configuration...

Current configuration : 65 bytes
!
interface Loopback100
 ip address 1.1.1.1 255.255.255.255
end

AS-C9300-BDC-T217-01#show run int loopback 100
                                            ^
% Invalid input detected at '^' marker.

AS-C9300-BDC-T217-01#
```

11.4　RESTCONF 实验（华为设备）

前面已对 RESTCONF 的基础知识及思科设备进行了介绍，下面通过对一台华为设备进行探讨，提供一种国产设备切入 RESTCONF 的思路。这种思路不限于某个厂商的某个具体型号。

11.4.1　实验拓扑

与 NETCONF 内容类似，我们依然选用 CE12800 作为实验设备，如下图所示。

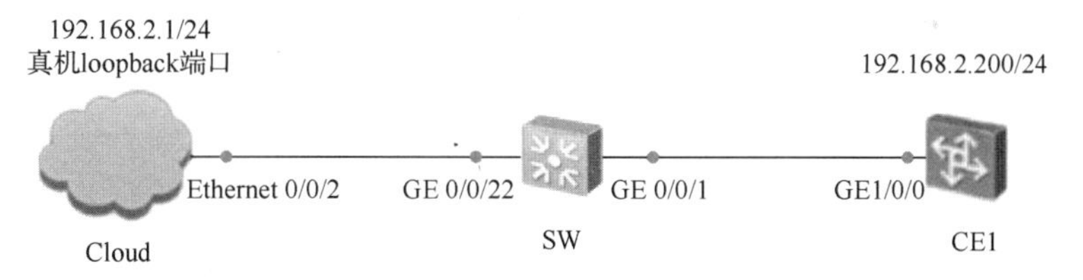

RESTCONF 服务端：CE1（192.168.2.200/24）

RESTCONF 客户端：PC（192.168.2.1/24）

拓扑结构与 NETCONF 的实验是一模一样的，实际上也是 PC-CE1 直连拓扑。

11.4.2　实验目的

- 在 CE12800 华为交换机探索 RESTCONF 功能，实验环境搭建。
- 在 Windows 上使用 Postman 软件进行设备简单联动。

11.4.3　启动 RESTCONF

1. 资料准备

与 NETCONF 查找资料方式相同，检索产品手册找到相关的信息，如下图所示。

2. CE1 基础配置

我们需要在 CE1 上做些基础配置，处理连通性，如下图所示。

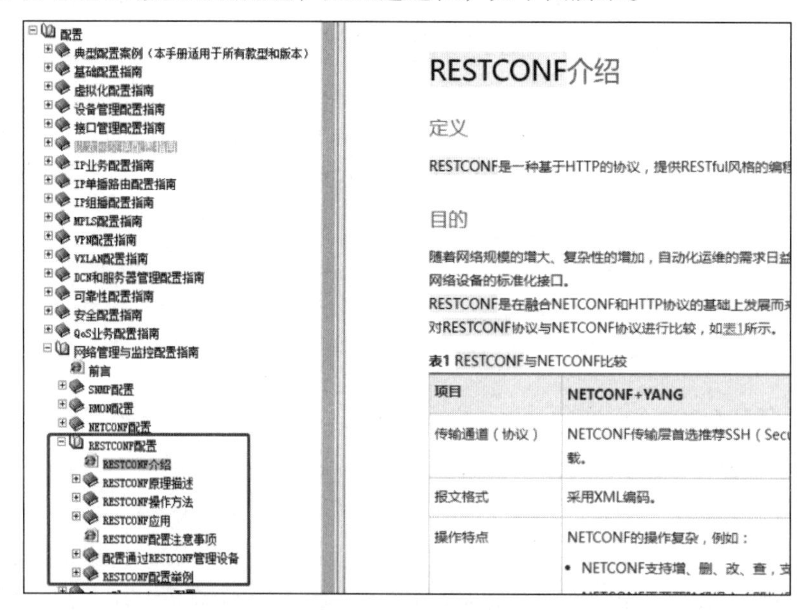


```
sys
sysname CE1
interface GE1/0/0
 undo portswitch
 undo shutdown
 ip address 192.168.2.200 255.255.255.0
commit
```

3. 开启 RESTCONF 功能

如果要启用 HTTPS，则需要加载数字证书，现网生产可以找到设备厂商做支撑。这里为简单演示起见，采用 HTTP 代替。这些命令同样来自设备对应的产品手册中的示例。

```
1. 创建 AAA 用户
<HUAWEI> system-view
[~HUAWEI] aaa
[~HUAWEI-aaa] local-user huawei123 password cipher Huawei@123_h_w
[*HUAWEI-aaa] local-user huawei123 service-type http
[*HUAWEI-aaa] quit
[*HUAWEI] commit

3. 在 Server 端使能 HTTP 服务器功能
[~HUAWEI] http
[*HUAWEI-http] service restconf
[*HUAWEI-http-service-restconf] server enable
[*HUAWEI-http-service-restconf] server port 8080  #这条可选，先不要配，默认是 80 端口，可以改成 8080 端口
[*HUAWEI-http-service-restconf] commit
[~HUAWEI-http-service-restconf] quit
[~HUAWEI-http] quit
```

PC（192.168.2.1）可打开一个浏览器输入地址测试，如下图所示。

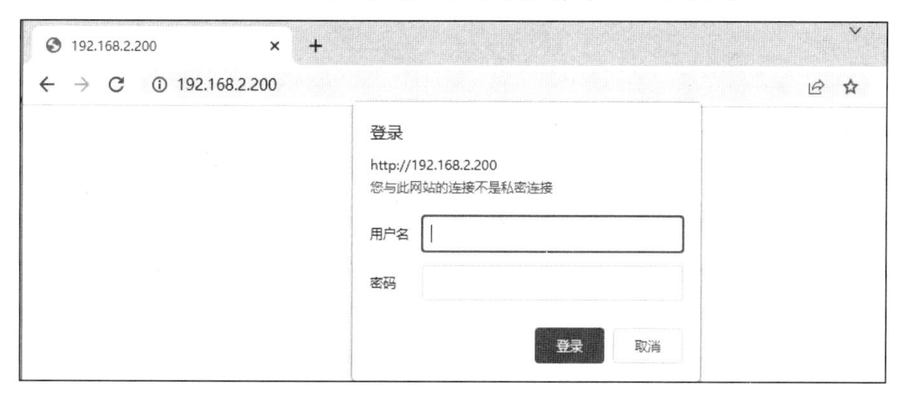

输入用户名和密码，虽然看到很多 error 字眼，但很明确已经有响应了，如下图所示。

192.168.2.200
不安全 | 192.168.2.200


```
{
 "ietf-restconf:errors" : {
   "error" : [
    {
      "error-type" : "application",
      "error-tag" : "invalid-value"
    }
  ]
 }
}
```

这个时候已经成功拉通了 RESTCONF 服务器和 RESTCONF 客户端。为了确认是否真的已经开启 HTTP，可以尝试几个动作。

（1）再次关闭 HTTP 中的 server enable，看浏览器是否无法打开。

（2）在 CE1 联机口上抓包，看是否有 HTTP 报文交互。

（3）修改 server port 其他端口，比如 8080。

（4）流量器打开“开发者选项”（按 F12 键），看交互情况。

限于篇幅，这些操作请读者自行尝试。

11.4.4　联动 Postman

Postman 的相关操作与 11.2 中思科设备实验保持一致，认证栏填写用户名和密码后，单击发送（Send）按钮，如下图所示。

HTTP 返回的错误码 400，其含义是“请求消息体错误”。于是，我们使用 Postman 工具，就是为了构建这个消息体，让消息体正确。

json 或 xml 两种都可以用来设置报文格式，这里我们先用 xml。

```
Content-Type: application/yang-data+xml
Accept: application/yang-data+xml
```

```
-----或者-----（后面再调整尝试）
Content-Type: application/yang-data+json
Accept: application/yang-data+json
```

HTTP 头部构建两个键值对，如下图所示。

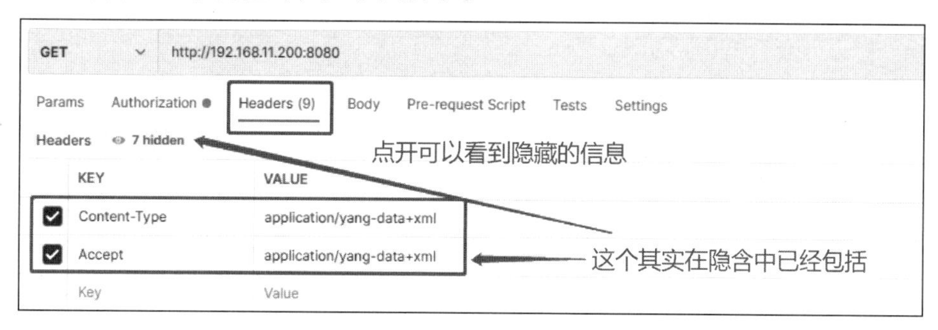

有了消息体，我们在手册中查对应的 URI，才能锚定到对应的资源上。继续查手册，如下图所示。

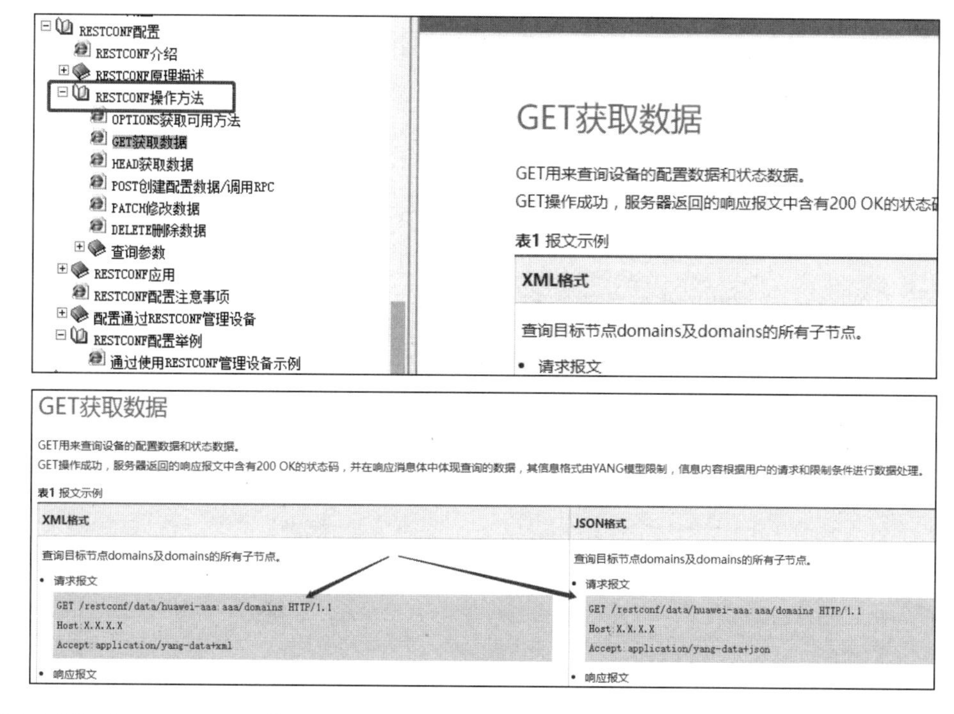

现在可以锁定 URI 了，即/restconf/data/huawei-aaa:aaa/domains，这是手册上的一个示例，组合一下，变成 URL，如下图所示。

```
http://192.168.11.200:8080/restconf/data/huawei-aaa:aaa/domains
```

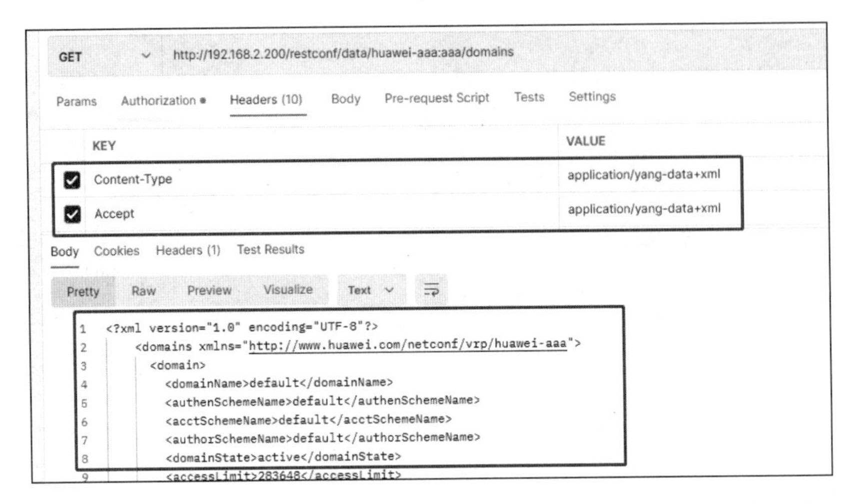

手册示例是查询 AAA 中 domian 的相关信息。我们在 AAA 中的 abc 域上多配置一条指令（access-limit10），然后在 Postman 工具中重新查一下，在响应报文中就可以看到这些我们手工配置的信息了，如下图所示。

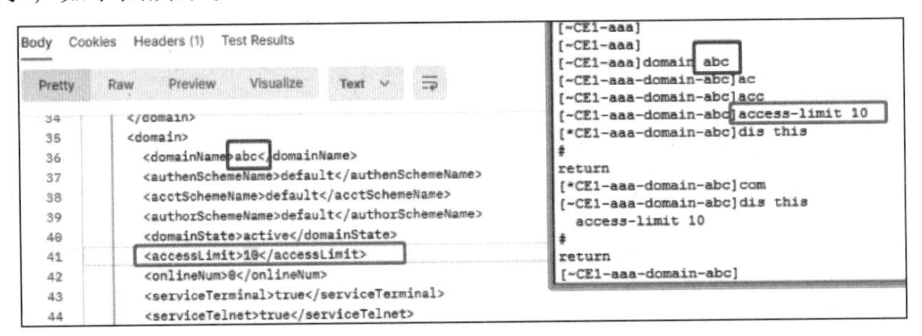

我们把 HTTP 头部的 Content-Type 和 Accept 由 xml 格式改为 json 格式，接着再进行测试，结果同样符合预期，如下图所示。

简单尝试成功以后，我们可以按手册上介绍的其他内容进行尝试，主要以查询为主，比如调整返回深度。

```
http://192.168.2.200/restconf/data/huawei-aaa:aaa/domains?depth=2
```

限于篇幅，不再截图，请读者自行尝试。

实践 RESTCONF 功能，我们除了关心消息体构建，关键的还有 URI，即增删改查（CRUD）需要定位到具体哪个资源节点。这个时候必须找到 API 手册，如下图所示。

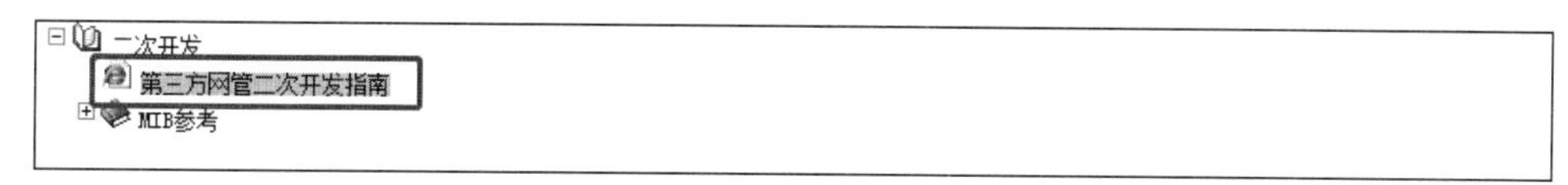

找到 API 手册，下载到本地备用，如下图所示。

CloudEngine 12800, 8800, 7800, 6800, 5800 V200R005C10 RESTful API参考 2021-08-12
CloudEngine 12800, 12800E V200R005C10 NETCONF YANG API参考 2021-09-17
CloudEngine 12800, 12800E V200R005C10 NETCONF Schema API参考 2021-09-07

浏览 API 手册，检索到对应的资源，如下图所示。

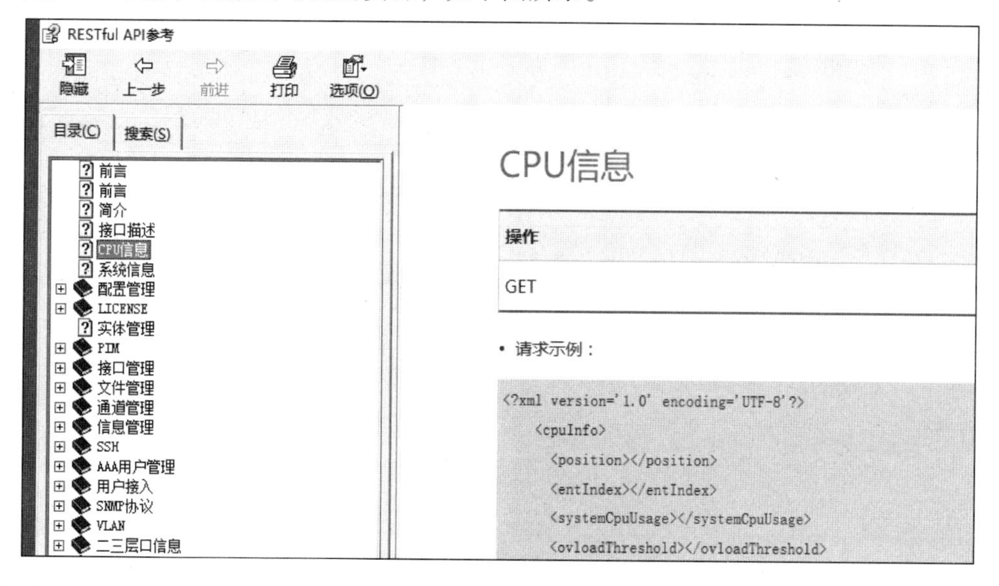

如果是真机，就可以这么组合使用了。但是，模拟器网元，经反复测试，发现可能只有部分实现了 RESTCONF 功能。

这样，我们使用 Postman 工具已经实现了设备 RESTCONF 功能测试，那么，Python 其他第三方库（如 Requests 模块）也如上述思科设备章节中类似，读者可自行尝试。

最后，RESTCONF 与 NETCONF 的学习和实践路径很类似，虽然相关组织都试图定义出统一的标准，但各厂商在实现上或者说各自理解有偏差，或者说都想各具特色，多少有点差异。但是，无论怎么变化，其出发点都是相关的 RFC，而归宿又是各厂商硬件型号软件版本对应的产品手册。当有一定理论概念后，赶紧把产品手册拿到手，对照看，跟着做，以最小实现的方式驱动起来，再展开编织，拓宽边界。另一个思路就是要把这些学到的知识和方式，投入生产实践中，待生产开花时，又再次进行文章梳理，终得果实。